T0266545

by John McPhee

Irons in the Fire

The Ransom of Russian Art

Assembling California

Looking for a Ship

The Control of Nature

Rising from the Plains

Table of Contents

La Place de la Concorde Suisse

In Suspect Terrain

Basin and Range

Giving Good Weight

Coming into the Country

The Survival of the Bark Canoe

Pieces of the Frame

The Curve of Binding Energy

The Deltoid Pumpkin Seed

Encounters with the Archdruid

The Crofter and the Laird

Levels of the Game

A Roomful of Hovings

The Pine Barrens

Oranges

The Headmaster

A Sense of Where You Are

The John McPhee Reader

The Second John McPhee Reader

Annals of the Former World

JOHN McPHEE

Farrar, Straus and Giroux

New York

ANNALS OF THE FORMER WORLD

Farrar, Straus and Giroux
18 West 18th Street, New York 10011

Printed in the United States of America
Published in 1998 by Farrar, Straus and Giroux
First paperback edition, 2000

Large parts of the text of this book originally appeared, in different form, in The New Yorker

Maps by Allan Cartography, Medford, Oregon
Geologic Time Scale by Tom Funk
Coordinated by Carmen Gomezplata and Natasha Wimmer
Index by Julie Kawabata

Note: In 1993, the University of New Mexico Press published Lady's Choice: Ethel Waxham's Journals and Letters, 1905–1910, *compiled and edited by her granddaughters Barbara Love and Frances Love Froidevaux.*

Library of Congress Cataloging-in-Publication Data
McPhee, John A.
Annals of the former world / John McPhee. —1st ed.
p. cm.
Includes index.
ISBN-13: 978-0-374-51873-8
ISBN-10: 0-374-51873-4 (pbk.)
1. Geology—United States. I. Title.
QE77.M38 1998
557.3—dc21 97-39660

Designed by Cynthia Krupat

www.fsgbooks.com

31

The author thanks the Albert P. Sloan Foundation for support in the completion and preparation of this book, and the Geological Society of America for administering the grant.

Yolanda Whitman

Annals of

the Former World

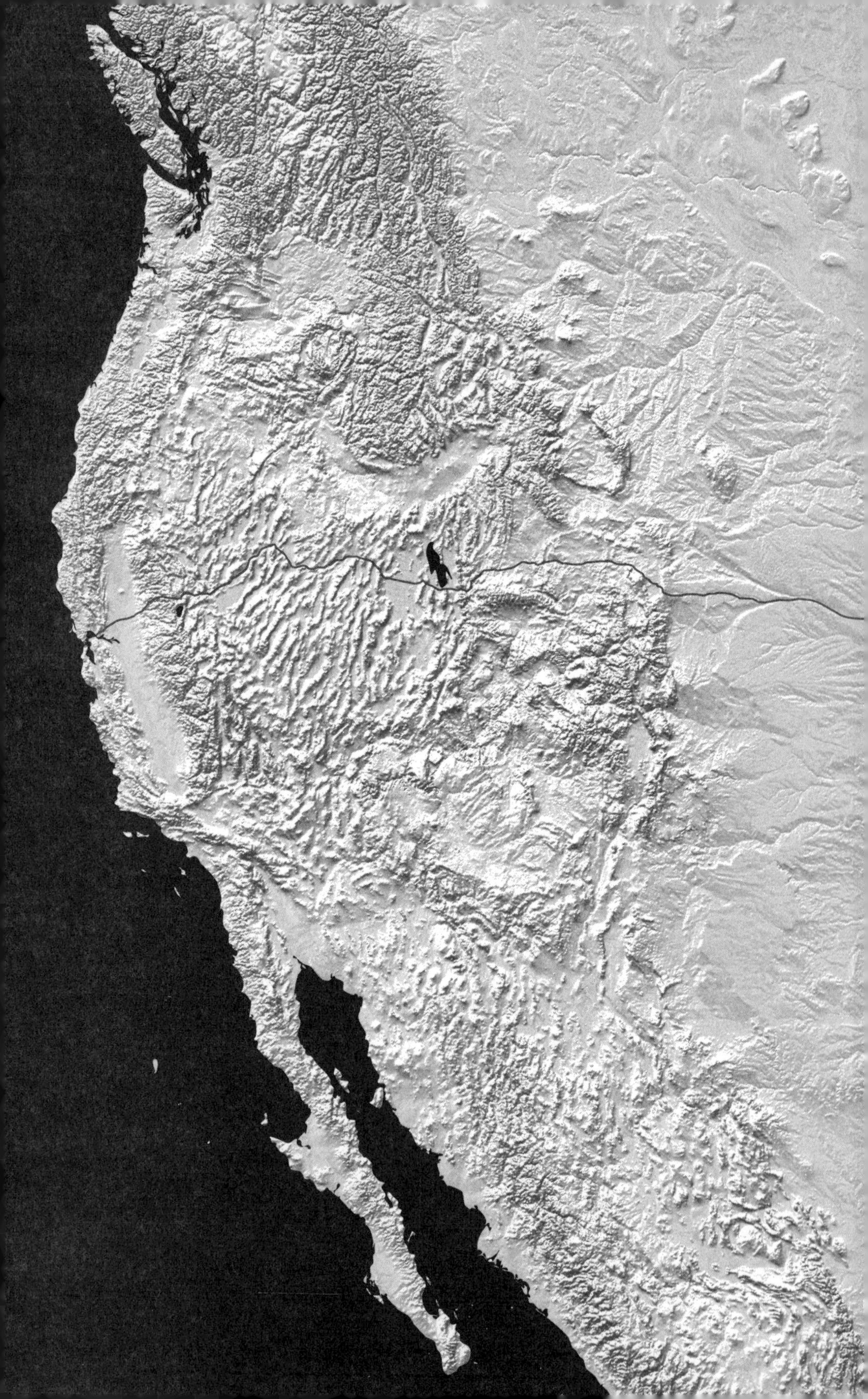

A Narrative Table of Contents

In 1978, I began a series of journeys across the United States in the company of geologists, with the purpose of doing a piece of writing that would describe not only the rock exposed in roadcuts but the geologists with whom I travelled. The result was meant to be a sort of cross section of North America at about the fortieth parallel, and a picture of the science. On (and away from) Interstate 80, I travelled for a year, sometimes traversing the country all the way, but generally covering a segment with this or that geologist—with Professor Kenneth Deffeyes, of Princeton University; with the sedimentologist Karen Kleinspehn, then a Princeton graduate student and now a professor at the University of Minnesota; with Anita Harris, of the United States Geological Survey; with David Love, of the Survey; and with the tectonicist Eldridge Moores, of the University of California, Davis. At the end of that first series of trips, after I transcribed my notes and developed a structure for the over-all composition, I discovered that I had outlined something that would keep me writing for more years than I wanted to spend consecutively on the subject. The structure had four main sections, which stood apart as well as together, so I decided to write them at intervals, always turning to other fields before returning to geology and to more wanderings with the respective geologists. The additional travel led as far away from Interstate 80 as mainland Greece, the island of Cyprus, a mining camp in Arizona, and the San Andreas Fault from end to end. The topic reached beyond the American cross section into world ophiolites and global tectonics. It somehow involved Robert Louis Stevenson, Chief Washakie, and Theodore Roosevelt (twice). It somehow involved Winona of the Lenape, William Tecumseh Sherman, William Penn, and Johann Augustus Sutter.

The controlling element—the theme that guided the basic structure—was plate tectonics. The plate-tectonics revolution had occurred in the nineteen-sixties, and when I set out on my travels the concept was not without detractors. I wanted to see who was opposed, and why, and in what ways the new theory was being tested and applied. There would be plenty of other matters discussed, but plate theory was paramount. The structure is not linear—not a straightforward trip from New York to San Francisco on the interstate. It jumps about the country. For example, it begins in New Jersey and leaps to Nevada, because the tectonics in New Jersey two hundred million years ago are being recapitulated by the tectonics in Nevada today.

Now, with its completion, the project represents the stratum of time 1978–1998. Its first four components were published in 1981, 1983, 1986, and 1993. A fifth and final essay appears in this volume to fill a significant gap. Nowhere in the structure I have referred to was there much of anything about the midcontinent. It was there in a broad and general way. It was there in a number of time lines shot across the country in various periods, epochs, and ages. But it was not there in any kind of rock-to-rock progression. In mid-America, there are few surface rocks. That initial over-all composition, purporting to traverse the nation, deliberately overlooked a large piece of the nation—Chicago to Cheyenne. If the rocks were scarce, the tectonics were scarcer. For more than a billion years, little to nothing had happened there. Even so, I felt a measure of guilt about the omission, and contemplated what to do to close that epic caesura. The answer came with the geophysical insights of recent years, the combined advances in many fields, from radiometric dating to computer science, that have enabled geologists to see the midcontinent itself forming and developing in the Precambrian eons. This was the basement of the world, under construction. I thought it a good idea to travel between Cheyenne and Chicago down there. I did so, in a way, with W. R. Van Schmus, of the University of Kansas.

Because the entire composition in all its parts was written in the form of journeys, set pieces, flashbacks, biographical sketches, and histories of the human and lithic kind—intended as an unfolding piece of writing and not as a catalogue of geologic topics—the text firmly refrained from offering a way in which a reader could easily turn to something like the basic set piece on plate theory or the basic set piece on geologic time. In the interest of the composition, such topics were not discretely labelled. That is why I am writing a narrative table of contents. In this inclusive volume, I am trying to have things both ways. While leaving the text unparcelled and continuous, I want to explain up front not only how the project came to be, and how it evolved across the twenty years, but also what's what and where.

Basin and Range, as the opening story, is the primer. It contains the long set piece on the nature and history of plate tectonics—what it is, who figured it out, and how. Pages 115–31.

Basin and Range also includes the long set piece on time. The time scale we more or less take for granted did not exist in the early nineteenth century. In fifty years or so, it was gradually assembled by amateurs (often medical doctors) who pieced this to that, saw which came earlier, and gave names to distinctive zones of time. As you try to follow the changing face of the earth, the role of time is of course all-important, and time in its quantity is very hard to sense. Pages 69–99.

In college, I majored in English. In college and in high school, I took various introductory courses in physics, chemistry, biology, and geology, but only out of idle interest or to discharge distributional requirements. Like all writing, writing about geology is masochistic, mind-fracturing self-enslaved labor—a description that intensifies when the medium is rock. What then could explain such behavior? Why would someone out of one culture try to make prose out of the other? Why would someone who majored in English choose to write about rocks? Why would a person who works for something called a Humanities Council and teaches a university course called Humanistic Studies 440 undertake to write about geology? I believe those questions are answered in one paragraph from *Basin and Range*. Pages 31–32.

With brief exceptions, I have lived all my life in Princeton, New Jersey, where I was educated in the public schools and at the university. When I was seventeen, I went off to Deerfield Academy, in Massachusetts, where a geologist named Frank Conklin presented his subject in a first-rate full-year course. Even then, I was an English-major designate, but in the decades of writing that followed—highly varied non-fiction writing, often involving natural scenes—the geology lay there to be tapped. Sooner or later in many of my projects, geology would be touched upon in one way or another, and I would ask the geologists of the Princeton faculty to help me get it right. There were some geological passages in books like *The Pine Barrens* and *Encounters with the Archdruid*, for example, and there were more in *Coming into the Country*, arising from a question I had long meant to ask. Obviously, the placer gold in the drainages of the Yukon was there because weather had broken up mountains and bestrewn the gold in the gravels of streams. That I thought I understood. But I wondered what had put the gold in the mountains in the first place. I called the Geology Department and talked with a professor who said he could not begin to answer the question. He had a preoccupying interest in Jurassic leaves. "Call Ken Deffeyes," he said. "Deffeyes knows, or thinks he knows." For me, Deffeyes put the gold in the mountains.

A year or so later, in a random conversation with this same eclectic petrologue, I asked if he thought we might find a Talk of the Town piece for *The New Yorker* in a roadcut near the city. We could look at the blast-exposed face of the rock, read its history, and tell it in the first-person plural. While we were still planning this short trip, I asked him if there would not be an even better story in a journey north from roadcut to roadcut—for example, up the Northway's stunning route through the Adirondacks.

"Not on this continent," said Deffeyes. "If you want to do that sort of thing on this continent, go west—go across the structure."

In one moment, bounding and rash, my thoughts raced to San Francisco with roadcuts lining the route like billboards, each with its own message. "Why not go all the way?" I said to him. Two weeks later, we were looking for silver in Nevada.

Deffeyes has stood beside this project for twenty flattering years, always seeming to assume that my comprehension and capabilities are twice their actual size, never showing the slightest sign of stress, or even awareness, when he is talking six to eight metres above my head. As widely read as if he were a professor of comparative literature, he intuitively understood the goal I had set up: to present his science and its practitioners in a form and manner that was meant to arrest the attention of other people while achieving acceptability in the geologic community. I was naïve even to think of such a thing, and a nervous wreck for months on end, but I learned a lot in twenty years. Deffeyes searched his mind and the geologic literature, and suggested—for the Appalachians, the Rocky Mountains, and California—geologists I might travel with. He called them, interested them in what I was trying to do, and asked if they would help. Since I travelled with all of them in that first year, they in turn became twenty-year counsellors as well as companions.

Several transcontinental time lines are drawn at selected moments in the text—glimpses of paleogeography, sweeping pictures of the United States as it appeared at some far gone date in the former world. There is a late Triassic journey in *Basin and Range* (28–31), and time lines from the Mississippian and Pennsylvanian periods (92–95) are a part of the deep-time set piece. As a way of introducing the idea of time lines, *Basin and Range* first presents a rapid transcontinental traverse through the physiographic provinces of the here-and-now (25–28). In *In Suspect Terrain*, there's a Cambrian and Ordovician pair (189–91), and a pair from the Silurian (199–201). The earliest plants to appear on land came after the first and before the second of those Silurian time lines. In *Rising from the Plains*, an Eocene time line (409–10) starts from both east and west and meets in a huge lake in what is now Wyoming.

After *Basin and Range* deals with plate theory in presentational fashion, Anita Harris, of *In Suspect Terrain*, in several ways attacks it.

In Suspect Terrain was constructed in four panels: 1. the biography of Anita Harris; 2. the Delaware Water Gap as a fragment of the Appalachians (understand a fragment and you'll have gone a long way toward understanding the whole); 3. the Appalachians and plate tectonics; 4. the theory of continental glaciation (used here to contrast its lack of acceptance in the nineteenth century with the experience of plate theory in the twentieth). The four panels are not presented as such, but they are distinct for anyone who cares to notice: 1. pages 147–82; 2. pages 182–209; 3. pages 209–44; 4. pages

254–75. Note the gap between 244 and 254. It contains short set pieces on coal and petroleum. The narrative, at that point, is in western Pennsylvania, and western Pennsylvania is a prime place to go into both of those subjects. The Delaware Water Gap panel is a freestanding experiment, a composition within a composition. Human history there (a few thousand years) is set in duet with the geologic history, to help make some sort of point. The ambition of the text relates closely to the George Inness painting on page 146. Tell me what made that scene and you will tell me what made the eastern United States. Look over the shoulder of the painter and see how it was done.

Geologists write "terrain" when they mean topography and "terrane" when they are referring to a piece of country many miles deep. When I first published *In Suspect Terrain*, I wrote, "I am not a geologist and I refuse to cooperate." Terrane, actually, has been a word in the English language at least since the mid-nineteenth century. Webster knows what it means. But I had not so much as looked it up, and I was bullheaded, savoring the ambiguities that danced around the single spelling. I now have a changed mind, an improved attitude. Terrain is topography. Terrane is a large chunk of the earth, in three dimensions. I have changed the text wherever necessary. My last-ditch holdout position, though, is the title *In Suspect Terrain*. It stays, retaining its meanings.

Anita Harris grew up in the Williamsburg section of Brooklyn and frankly went into geology in order to get out of the city. Within the profile of her is a profile of New York City geology (157–67). Her international reputation is mainly the result of paleontological discoveries that have enhanced the search for oil. I accompanied her as she collected carbonate rocks from New Jersey to Indiana. In the context of Appalachian history—among mountains that are thought by many to represent the suturing of two continents—her cautionary remarks about plate theory are given unrestrained expression, notably on pages 147–49, 217–32, and 274–75.

Rising from the Plains is primarily about Wyoming, which includes within its borders an exceptional range of geology. It's about the roadcuts of the interstate but also about Jackson Hole and the Tetons and the Powder River Basin and the Wind River Basin and the Laramie Range and David Love and his father and especially his mother, who educated her children at Love Ranch, a very long ride from neighbors, in the geographical center of Wyoming. She was born in 1882 and died long before I would have had a chance to meet her, but she is probably the most arresting personality I have encountered in the course of my professional work. You will find the story of the Laramide Orogeny—the rising of the Rocky Mountains—on pages 310–12, the burial and exhumation of the Rockies on pages 313–16, a set piece on the geologic history of Jackson Hole and the Tetons (understand a fragment . . .) on pages 366–78, and a set piece on the theory of geophysical hot spots (such as Yellowstone, Hawaii, Bermuda, Iceland, Tristan da Cunha, Mt. Cameroon) on pages 388–403, preceded by a passage on the tension between field geology and "black-box geology" (380–86).

Sometimes it is said of geologists that they reflect in their professional styles the sort of country in which they grew up. Nowhere could that be better exampled than in the life of a geologist born in the center of Wyoming. The passages on Love Ranch and the years of David Love's upbringing are on pages 281–82, 287–94, 299–308, and 332–56.

In the unspectacular setting of Rawlins, Wyoming, a person can see in one sweeping glance a spread of time far greater than the time represented in the walls of the Grand Canyon. On pages 294–97, a verbal rock column reaches down at Rawlins through those 2.6 billion years.

The final sixth of *Rising from the Plains* is an environmental montage of tensions between geological discovery and environmental preservation. David Love, exploration geologist and passionate defender of wild Wyoming, contains within himself the essence of the struggle, as it is exemplified by coal (404–8), oil shale (412–13), trona (415–16), oil and gas in the Overthrust Belt (417–19), oil in Yellowstone Park (419–21), and—one of his signal discoveries, close to home—sedimentary uranium (421–25).

Assembling California came thirteen years after *Basin and Range*, and the wait was prudent, for, as every grandchild knows, California is challenged only by Alaska as a national showcase of active tectonics. I had scarcely begun writing *Assembling California* when, in 1989, the Loma Prieta earthquake occurred, and inserted itself with prominence in the text. In 1992, other temblors took place at Big Bear, Landers, and Joshua Tree, the latter two evidently initiating a new fault line (587–88) and confirming a prediction made in *Basin and Range* by Ken Deffeyes (137–43).

Assembling California begins and ends at the same point on the Pacific Coast, and in what amounts to a long and digressive flashback traverses the state east to west in the company of Eldridge Moores, whose tectonic hypotheses are on the applied outer boundary of the theory of plate tectonics, where he reconstructs former worlds. When Anita Harris, in *In Suspect Terrain*, alludes to "the plate-tectonics boys," the group would include Eldridge Moores. He has suggested, for example, that Arizona and Antarctica were once conjoined. This did not prevent the Geological Society of America from electing him its president in 1995. The simple itinerant structure of *Assembling California* includes two long set pieces—one near the start and the other near the finish—illustrating the extensive effects of two very different geological events: the gold rush of the eighteen-forties and fifties (454–72) and the Loma Prieta earthquake, of 1989 (603–20).

Moores is an ophiolitologist, an expert on ocean-crustal rock, which asks or answers large questions when it is found detached and lying on continents. An introduction to the nature and complexity of ophiolites (476–511) is followed by subflashbacks to Cyprus (511–19) and Greece (519–26), where Moores has done research for decades and where transported rock of the ocean floor stands as mountains. With the exception of some veneer, this is not sedimentary rock derived from continents and laid down in the sea; this is igneous rock from magma chilled at ocean spreading centers, and rock of the mantle below. A large piece of it, an exotic terrane, is a part of California known in geology as the Smartville Block (479–80, 484–91, 502, 504, 506).

Son of a gold miner, Moores grew up in the almost alpine setting of Crown King, Arizona (526–35), and now lives in the Great Central Valley of California (535–44), whose geologic story has few (if any) parallels among valleys of the world. The Coast Ranges, with their own odd story (544–54), are only a few miles west of Moores' home in Davis. A long set piece on world ophiolites and global tectonics—a narrative of maps in motion, of evolving and dissolving lands, including every plate and continent (554–70)—is the result of a heady conversation in the Louis Martini winery in the Napa Valley.

San Francisco geology is introduced in the roadcuts of the approaching Interstate 80 and pursued on foot among the hills of the city (570–81). A set piece that traverses California the long way, north-south, is about the San Andreas system, which is actually a family of faults (581–603). Among them is the Hayward Fault, which could be a source of considerable trauma for San Francisco, Oakland, Berkeley, and all other Bay Area cities, not to mention Hayward (600–2).

Crossing the Craton describes Nebraska by visiting Colorado, because in Colorado you see the basement of Nebraska bent up into the air. The fact that the journey takes place in the company of a geochronologist from the University of Kansas can only enhance the description. Between Chicago and Cheyenne, the most arresting geophysical feature is the Midcontinent Rift (624, 628–29, 636, 651, 654–55, 658–59), which opened about a fourth of the way back through the history of the earth—1.1 billion years—and serves as an abyssal edge for a look down through deeper and deeper time. The oldest rock yet found on earth (630, 632, 648) has an age close to four billion years, some six hundred million years younger than the earth itself. After reaching back to the earliest beginnings (630–31), the story turns around and comes forward through the Archean Eon, while island arcs accrete and small cratons form (631–33). At the end of the Archean, in the very general neighborhood of 2.5 billion years before the present, great and unrepeatable changes occur in the behavior of the earth, including the precipitation of banded iron and the beginning of modern plate tectonics (634).

In the early Proterozoic Eon, seven small cratons collide, conjoin as the Canadian Shield (633, 636–38). Younger island arcs eventually drift in and dock against the shield, forming large parts of Nebraska and Colorado (638–45). These novel views into Precambrian eons are the result of advances in radiometric dating (645–50) and, among other things, the measurement and interpretation of magnetic and gravity anomalies (651–54), all of it anchored and restrained by well cores (654–56).

More arcs accrete. A coastal plate boundary like the Andean margin of South America develops along what is now a northeasterly trend through New Mexico and Kansas (650–51). About halfway through the Proterozoic, a baffling series of great plutons (each analogous to the Sierra Nevada batholith of relatively modern times) perforates North America from one side to the other and is mysteriously unaccompanied by the building of mountains, as plutons generally are, almost by definition (652–53).

A time line at 1.1 billion years comes in from the eastern and western margins of the continent and converges in the active and growing Midcontinent Rift (657–58). When the spreading stops under Iowa, the granites of Pikes Peak almost unaccountably appear in Colorado, the last tectonic event in Precambrian North America (660).

An editor from whose counsel I have benefitted since the early phases of this project is Sara Lippincott, of Pasadena, California, who left *The New Yorker* in 1993 in order to become a free-lance editor of books. When Sara lived in New York, her idea of a perfect vacation was to get in an airplane and visit Caltech. She did that often. Now that she lives in Pasadena, she teaches at Caltech (a writing course). And from the beginning of *Basin and Range* to the Pikes Peak conclusion of *Crossing the Craton*, Sara has been the editor of this book. In the twenty years following the first journeys that are described here, professional attitudes toward plate theory evolved in different ways. The text, as it evolved, reflects that, and in preparing it for comprehensive publication we have attempted to preserve the sense of growing acceptance. Elsewhere, we have freely added material, adjusted the time scale, and tried to keep pace with the constant refining of radiometric dates. The text has been meshed, melded, revised, in some places cut, and everywhere studied for repetition. For the most part, I have eliminated the repetitions, but I also chose to modify some of them and simply leave others standing. Reminders and repetitions can be as useful in this subject as they are in ballads. Rock carries its own epithets, its own refrains. In it, you see things happening again, and now again. *Annals of the Former World* in selected places echoes what it has said before. Diapirs are redefined; plate theory is repeatedly and in different ways explained. "Geology repeats itself!" Anita Harris likes to say. Anita likes to say it so much that after a minute or two she says, "Geology repeats itself!"

For convenience, this should repeat itself, too. To wit:

Table of Contents

BOOK 1: BASIN AND RANGE

BOOK 2: IN SUSPECT TERRAIN

BOOK 3: RISING FROM THE PLAINS

BOOK 4: ASSEMBLING CALIFORNIA

BOOK 5: CROSSING THE CRATON

Book 1

Basin and Range

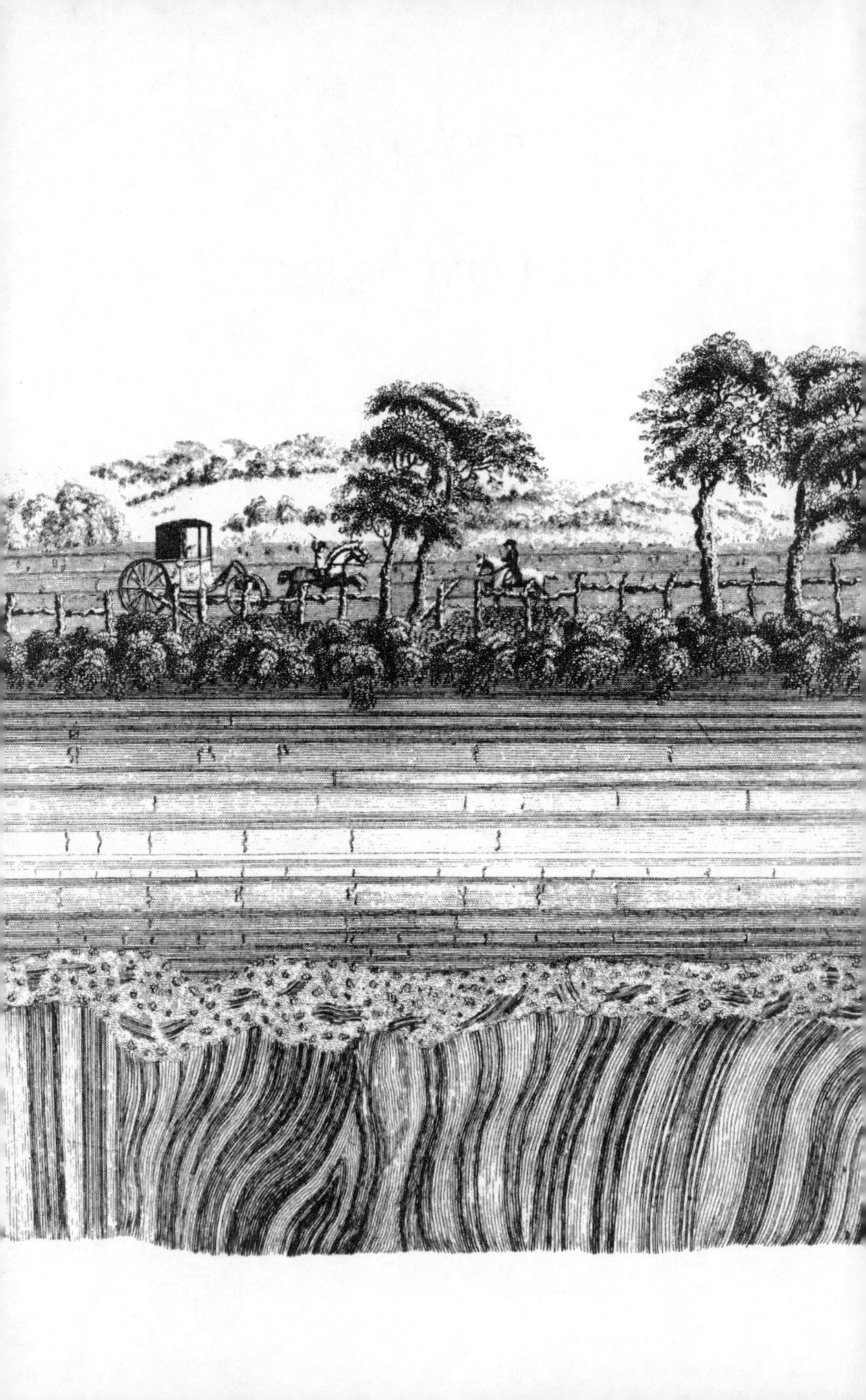

The poles of the earth have wandered. The equator has apparently moved. The continents, perched on their plates, are thought to have been carried so very far and to be going in so many directions that it seems an act of almost pure hubris to assert that some landmark of our world is fixed at 73 degrees 57 minutes and 53 seconds west longitude and 40 degrees 51 minutes and 14 seconds north latitude—a temporary description, at any rate, as if for a boat on the sea. Nevertheless, these coordinates will, for what is generally described as the foreseeable future, bring you with absolute precision to the west apron of the George Washington Bridge. Nine A.M. A weekday morning. The traffic is some gross demonstration in particle physics. It bursts from its confining source, aimed at Chicago, Cheyenne, Sacramento, through the high dark roadcuts of the Palisades Sill. A young woman, on foot, is being pressed up against the rockwall by the wind booms of the big semis—Con Weimar Bulk Transportation, Fruehauf Long Ranger. Her face is Nordic, her eyes dark brown and Latin—the bequests of grandparents from the extremes of Europe. She wears mountain boots, blue jeans. She carries a single-jack sledgehammer. What the truckers seem to notice, though, is her youth, her long bright Norwegian hair; and they flirt by air horn, driving needles into her ears. Her name is Karen Kleinspehn. She is a geologist, a graduate student nearing her Ph.D., and

Unconformity at Jedburgh, borders, by John Clerk, 1787, courtesy Scottish Academic Press, Ltd., Edinburgh

there is little doubt in her mind that she and the road and the rock before her, and the big bridge and its awesome city—in fact, nearly the whole of the continental United States and Canada and Mexico to boot—are in stately manner moving in the direction of the trucks. She has not come here, however, to ponder global tectonics, although goodness knows she could, the sill being, in theory, a signature of the events that created the Atlantic. In the Triassic, when New Jersey and Mauretania were of a piece, the region is said to have begun literally to pull itself apart, straining to spread out, to break into great crustal blocks. Valleys in effect competed. One of them would open deep enough to admit ocean water, and so for some years would resemble the present Red Sea. The mantle below the crust—exciting and excited by these events—would send up fillings of fluid rock, and with such pressure behind them that they could intrude between horizontal layers of, say, shale and sandstone and lift the country a thousand feet. The intrusion could spread laterally through hundreds of square miles, becoming a broad new layer—a sill—within the country rock.

This particular sill came into the earth about two miles below the surface, Kleinspehn remarks, and she smacks it with the sledge. An air horn blasts. The passing tires, in their numbers, sound like heavy surf. She has to shout to be heard. She pounds again. The rock is competent. The wall of the cut is sheer. She hits it again and again—until a chunk of some poundage falls free. Its fresh surface is asparkle with crystals—free-form, asymmetrical, improvisational plagioclase crystals, bestrewn against a field of dark pyroxene. The rock as a whole is called diabase. It is salt-and-peppery charcoal-tweed savings-bank rock. It came to be that way by cooling slowly, at depth, and forming these beautiful crystals.

"It pays to put your nose on the outcrop," she says, turning the sample in her hand. With a smaller hammer, she tidies it up, like a butcher trimming a roast. With a felt-tip pen, she marks it "1." Moving along the cut, she points out xenoliths—blobs of the country rock that fell into the magma and became encased there like raisins in bread. She points to flow patterns, to swirls in the diabase where solidifying segments were rolled over, to layers of coarse-grained crystals that settled, like sediments, in beds. The Palisades Sill—in its chemistry and its texture—is a standard example of homogeneous magma resulting in multiple expressions of rock. It tilts westward. The sill came into a crustal block whose western extremity—known in New Jersey as the Border Fault—is thirty miles away. As the block's western end went down, it formed the Newark Basin. The high eastern end gradually eroded, shedding sediments into the basin, and the sill was ultimately revealed—a process assisted by the creation and development of the Hudson, which eventually cut out the cliffside panorama of New Jersey as seen across the river from Manhattan: the broad sill, which had cracked, while cooling, into slender columns so upright and uniform that inevitably they would be likened to palisades.

In the many fractures of these big roadcuts, there is some suggestion of columns, but actually the cracks running through the cuts are too various to be explained by columnar jointing, let alone by the impudence of dynamite. The sill may have been stressed pretty severely by the tilting of the fault block, Kleinspehn says, or it may have cracked in response to the release of weight as the load above it was eroded away. Solid-earth tides could break it up, too. The sea is not all that responds to the moon. Twice a day the solid earth bobs up and down, as much as a foot. That kind of force and that kind of distance are more than enough to break hard rock. Wells will flow faster during lunar high tides.

For that matter, geologists have done their share to bust up these roadcuts. "They've really been *through* here!" They have fungoed so much rock off the walls they may have set them back a foot. And everywhere, in profusion along this half mile of diabase, there are small, neatly cored holes, in no way resembling the shot holes and guide holes of the roadblasters, which are larger and vertical, but small horizontal borings that would be snug to a roll of coins. They were made by geologists taking paleomagnetic samples. As the magma crystallized and turned solid, certain iron minerals within it lined themselves up like compasses, pointing toward the magnetic pole. As it happened, the direction in those years was northerly. The earth's magnetic field has reversed itself a number of hundreds of times, switching from north to south, south to north, at intervals that have varied in length. Geologists have figured out just when the reversals occurred, and have thus developed a distinct arrhythmic yardstick through time. There are many other chronological frames, of course, and if from other indicators, such as fossils, one knows the age of a rock unit within several million years, a look at the mineral compasses inside it can narrow the age toward precision. Paleomagnetic insights have contributed greatly to the study of the travels of the continents, helping to show where they may have been with respect to one another. In the argot of geology, paleomagnetic specialists are sometimes called paleomagicians. Enough paleomagicians have been up and down the big roadcuts of the Palisades Sill to prepare what appears to be a Hilton for wrens and purple martins. Birds have shown no interest.

Near the end of the highway's groove in the sill, there opens a broad, forgettable view of the valley of the Hackensack. The road is descending toward the river. At an even greater angle, the sill—tilting westward—dives into the earth. Accordingly, as Karen Kleinspehn continues to move downhill she is going "upsection" through the diabase toward the top of the tilting sill. The texture of the rock becomes smoother, the crystals smaller, and soon she finds the contact where the magma—at 2000 degrees Fahrenheit—touched the country rock. The country rock was a shale, which had earlier been the deep muck of some Triassic lake, where the labyrinthodont amphibians lived, and paleoniscid fish. The diabase below the contact now is a smooth and uniform hard dark rock, no tweed—its crystals too small to be discernible, having had so little time to grow in the chill zone. The contact is a straight, clear line. She rests her hand across it. The heat of the magma penetrated about a hundred feet into the shale, enough to cook it, to metamorphose it, to turn it into spotted slate. Sampling the slate with her sledgehammer, she has to pound with even more persistence than before. "Some weird, wild minerals turn up in this stuff," she comments between swings. "The metamorphic aureole of this formation is about the hardest rock in New Jersey."

She moves a few hundred feet farther on, near the end of the series of cuts. Pin oaks, sycamores, aspens, cottonwoods have come in on the wind with milkweed and wisteria to seize living space between the rock and the road, although the environment appears to be less welcoming than the center of Carson Sink. There are fossil burrows in the slate—long stringers where Triassic animals travelled through the quiet mud, not far below the surface of the shallow lake.

There is a huge rubber sandal by the road, a crate of broken eggs, three golf balls. Two are very cheap but one is an Acushnet Titleist. A soda can comes clinking down the interstate, moving ten miles an hour before the easterly winds of the traffic. The screen of trees damps the truck noise. Karen sits down to rest, to talk, with her back against a cottonwood. "Roadcuts can be a godsend. There's a series of roadcuts near Pikeville, Kentucky—very big ones—where you can see distributary channels in a river-delta system, with natural levees, and with splay deposits going out from the levees into overbank deposits of shales and coal. It's a face-on view of the fingers of a delta, coming at you—the Pocahontas delta system, shed off the Appalachians in Mississippian-Pennsylvanian time. You see river channels that migrated back and forth across a valley and were superposed vertically on one another through time. You see it all there in one series of exposures, instead of having to fit together many smaller pieces of the puzzle."

Geologists on the whole are inconsistent drivers. When a roadcut presents itself, they tend to lurch and weave. To them, the roadcut is a portal, a fragment of a regional story, a proscenium arch that leads their imaginations into the earth and through the surrounding terrane. In the rock itself are the essential clues to the scenes in which the rock began to form—a lake in Wyoming, about as large as Huron; a shallow ocean reaching westward from Washington Crossing; big rivers that rose in Nevada and fell through California to the sea. Unfortunately, highway departments tend to obscure such scenes. They scatter seed wherever they think it will grow. They "hair everything over"—as geologists around the country will typically complain.

"We think rocks are beautiful. Highway departments think rocks are obscene."

"In the North it's vetch."

"In the South it's the god-damned kudzu. You need a howitzer to blast through it. It covers the mountainsides, too."

"Almost all our stops on field trips are at roadcuts. In areas where structure is not well exposed, roadcuts are essential to do geology."

"Without some roadcuts, all you could do is drill a hole, or find natural streamcuts, which are few and far between."

"We as geologists are fortunate to live in a period of great road building."

"It's a way of sampling fresh rock. The road builders slice through indiscriminately, and no little rocks, no softer units are allowed to hide."

"A roadcut is to a geologist as a stethoscope is to a doctor."

"An X-ray to a dentist."

"The Rosetta Stone to an Egyptologist."

"A twenty-dollar bill to a hungry man."

"If I'm going to drive safely, I can't do geology."

In moist climates, where vegetation veils the earth, streamcuts are about the only natural places where geologists can see exposures of rock, and geologists have walked hundreds of thousands of miles in and beside streams. If roadcuts in the moist world are a kind of gift, they are equally so in other places. Rocks are not easy to read where natural outcrops are so deeply weathered that a hammer will virtually sink out of sight—for example, in piedmont Georgia. Make a fresh roadcut almost anywhere at all and geologists will close in swiftly, like missionaries racing anthropologists to a tribe just discovered up the Xingu.

"I studied roadcuts and outcrops as a kid, on long trips with my family," Karen says. "I was probably doomed to be a geologist from the beginning." She grew up in the Genesee Valley, and most of the long trips were down through Pennsylvania and the Virginias to see her father's parents, in North Carolina. On such a journey, it would have been difficult not to notice all the sheets of rock that had been bent, tortured, folded, faulted, crumpled—and to wonder how that happened, since the sheets of rock would have started out as flat as a pad of paper. "I am mainly interested in sedimentology, in sedimentary structures. It allows me to do a lot of field work. I'm not too interested in theories of what happens *x* kilometres down in the earth at certain temperatures and pressures. You seldom do field work if you're interested in the mantle. There's a little bit of the humanities that creeps into geology, and that's why I am in it. You can't prove things as rigorously as physicists or chemists do. There are no white coats in a geology lab, although geology is going that way. Under the Newark Basin are worn-down remains of the Appalachians—below us here, and under that valley, and so on over to the Border Fault. In the West, for my thesis, I am working on a basin that also formed on top of a preexisting deformed belt. I can't say that the basin formed just like this one, but what absorbs me are the mechanics of these successor basins, superposed on mountain belts. The Great Valley in California is probably an example of a late-stage compressional basin—formed as plates came together. We think the Newark Basin is an extensional basin—formed as plates moved apart. In the geologic record, how do we recognize the differences between the two? I am trying to get the picture of the basin as a whole, and what is the history that you can read in these cuts. I can't synthesize all this in one morning on a field trip, but I can look at the rock here and then evaluate someone else's interpretation." She pauses. She looks back along the rockwall. "This interstate is like a knife wound all across the country," she remarks. "Sure—you could do this sort of thing from here to California. Anyone who wants to, though, had better hurry. Before long, to go all the way across by yourself will be a fossil experience. A person or two. One car. Coast to coast. People do it now without thinking much about it. Yet it's a most unusual kind of personal freedom—particular to this time span, the one we happen to be in. It's an amazing, temporary phenomenon that will end. We have the best highway system in the world. It lets us do what people in no other country can do. And it is also an ecological disaster."

In June, every year, students and professors from eastern colleges—with their hydrochloric-acid phials and their hammers and their Brunton compasses—head west. To be sure, there is plenty of absorbing geology under the shag of eastern America, galvanic conundrums in Appalachian structure and intricate puzzles in history and stratigraphy. In no manner would one wish to mitigate the importance of the eastern scene. Undeniably, though, the West is where the rocks are—the vastnesses of exposed rock—and of eastern geologists who do any kind of summer field work about seventy-five per cent go west. They carry state geological maps and the regional geological highway maps that are published by the American Association of Petroleum Geologists—maps as prodigally colored as drip paintings and equally formless in their worm-trail-and-paramecium depictions of the country's uppermost rock. The maps give two dimensions but more than suggest the third. They tell the general age and story of the banks of the asphalt stream. Kleinspehn has been doing this for some years, getting into her Minibago, old and overloaded, a two-door Ford, heavy-duty springs, with odd pieces of the Rockies under the front seat and a mountain tent in the gear behind, to cross the Triassic lowlands and the Border Fault and to rise into the Ridge and Valley Province, the folded-and-faulted, deformed Appalachians—the beginnings of a journey that above all else is physiographic, a journey that tends to mock the idea of a nation, of a political state, as an unnatural subdivision of the globe, as a metaphor of the human ego sketched on paper and framed in straight lines and in riparian boundaries between unalterable coasts. The United States: really a quartering of a continent, a drawer in North America. Pull it out and prairie dogs would spill off one side, alligators off the other—a terrain crisscrossed with geological boundaries, mammalian boundaries, amphibian boundaries: the limits of the world of the river frog, the extent of the Nugget formation, the range of the mountain cougar. The range of the cougar is the cougar's natural state, overlying segments of tens of thousands of other states, a few of them proclaimed a nation. The United States of America, with its capital city on the Atlantic Coastal Plain. The change is generally dramatic as one province gives way to another; and halfway across Pennsylvania, as you leave the quartzite ridges and carbonate valleys of the folded-and-faulted mountains, you drop for a moment into Cambrian rock near the base of a long climb, a ten-mile gradient upsection in time from the Cambrian into the Ordovician into the Silurian into the Devonian into the Mississippian (generally through the same chapters of the earth represented in the walls of the Grand Canyon) and finally out onto the Pennsylvanian itself, the upper deck, the capstone rock, of the Allegheny Plateau. Now even the Exxon map shows a new geology, roads running every which way like shatter lines in glass, following the crazed geometries of this deeply dissected country, whereas, before, the roads had no choice but to run northeast-southwest among the long ropy trends of the deformed mountains, following the endless ridges. On these transcontinental trips, Karen has driven as much as a thousand miles in a day at speeds that she has come to regard as dangerous and no less emphatically immoral. She has almost never slept under a roof, nor can she imagine why anyone on such a journey would want or need to; she "scopes out" her campsites in the late-failing light with strong affection for national forests and less for the three-dollar campgrounds where you roll out your Ensolite between two trailers, where gregarious trains honk like Buicks, and Harleys on instruments climb escarpments in the night. The physiographic boundary is indistinct where you shade off the Allegheny Plateau and onto the stable craton, the continent's enduring core, its heartland, immemorially unstrained, the steady, predictable hedreocraton—the Stable Interior Craton. There are old mountains to the east, maturing mountains to the west, adolescent mountains beyond. The craton has participated on its edges in the violent creation of the mountains. But it remains intact within, and half a nation wide—the lasting, stolid craton, slowly, slowly downwasting. It has lost five centimetres since the birth of Christ. In much of Canada and parts of Minnesota and Wisconsin, the surface of the craton is Precambrian—earth-basement rock, the continental shield. Ohio, Indiana, Illinois, and so forth—the greater part of the Midwest—is shield rock covered with a sedimentary veneer that has never been metamorphosed, never been ground into tectonic hash—sandstones, siltstones, limestones, dolomites, flatter than the ground above them, the silent floors of departed oceans, of epicratonic seas. Iowa. Nebraska. Now with each westward township the country thickens, rises—a thousand, two thousand, five thousand feet—on crumbs shed off the Rockies and generously served to the craton. At last the Front Range comes to view—the chevroned mural of the mountains, sparkling white on gray, and on its outfanning sediments you are lifted into the Rockies and you plunge through a canyon to the Laramie Plains. "You go from one major geologic province to another and—whoa!—you really know you're doing it." There are mountains now behind you, mountains before you, mountains that are set on top of mountains, a complex score of underthrust, upthrust, overthrust mountains, at the conclusion of which, through another canyon, you come into the Basin and Range. Brigham Young, when he came through a neighboring canyon and saw rivers flowing out on alluvial fans from the wall of the Wasatch to the flats beyond, made a quick decision and said, "This is the place." The scene suggested settling for it. The alternative was to press on beside a saline sea and then across salt barrens so vast and flat that when microwave relays would be set there they would not require towers. There are mountains, to be sure—off to one side and the other: the Oquirrhs, the Stansburys, the Promontories, the Silver Island Mountains. And with Nevada these high, discrete, austere new ranges begin to come in waves, range after range after north-south range, consistently in rhythm with wide flat valleys: basin, range; basin, range; a mile of height between basin and range. Beside the Humboldt you wind around the noses of the mountains, the Humboldt, framed in cottonwood —a sound, substantial, year-round-flowing river, among the largest in the world that fail to reach the sea. It sinks, it disappears, in an evaporite plain, near the bottom of a series of fault blocks that have broken out to form a kind of stairway that you climb to go out of the Basin and Range. On one step is Reno, and at the top is Donner Summit of the uplifting Sierra Nevada, which has gone above fourteen thousand feet but seems by no means to have finished its invasion of the sky. The Sierra is rising on its east side and is hinged on the west, so the slope is long to the Sacramento Valley—the physiographic province of the Great Valley—flat and sea-level and utterly incongruous within its flanking mountains. It was not eroded out in the normal way of valleys. Mountains came up around it. Across the fertile flatland, beyond the avocados, stand the Coast Ranges, the ultimate province of the present, the berm of the ocean—the Coast Ranges, with their dry and straw-brown Spanish demeanor, their shadows of the live oaks on the ground.

If you were to make that trip in the Triassic—New York to San Francisco, Interstate 80, say roughly at the end of Triassic time—you would move west from the nonexistent Hudson River with the Palisades Sill ten thousand feet down. The motions that will open the Atlantic are well under way (as things appear in present theory), but the brine has not yet come in. Behind you, in fact, where the ocean will be, are several thousand miles of land—a contiguous landmass, fragments of which will be Africa, Antarctica, India, Australia. You cross the Newark Basin. It is for the most part filled with red mud. In the mud are tracks that seem to have been made by a two-ton newt. You come to a long, low, north-south-trending, black, steaming hill. It is a flow of lava that has come out over the mud and has cooled quickly in the air to form the dense smooth textures of basalt. Someday, towns and landmarks of this extruded hill will in one way or another take from it their names: Montclair, Mountainside, Great Notch, Glen Ridge. You top the rise, and now you can see across the rest of the basin to the Border Fault, and—where Whippany and Parsippany will be, some thirty miles west of New York—there is a mountain front perhaps seven thousand feet high. You climb this range and see more and more mountains beyond, and they are the folded-and-faulted Appalachians, but middle-aged and a little rough still at the edges, not caterpillar furry and worn-down smooth. Numbers do not seem to work well with regard to deep time. Any number above a couple of thousand years—fifty thousand, fifty million—will with nearly equal effect awe the imagination to the point of paralysis. This Triassic journey, anyway, is happening two hundred and ten million years ago, or five per cent back into the existence of the earth. From the subalpine peaks of New Jersey, the descent is long and gradual to the lowlands of western Pennsylvania, where flat-lying sedimentary rocks begin to reach out across the craton—coals and sandstones, shales and limestones, slowly downwasting, Ohio, Indiana, Illinois, Iowa, erosionally losing an inch every thousand years. Where the Missouri will flow, past Council Bluffs, you come into a world of ruddy hills, Permian red, that continue to the far end of Nebraska, where you descend to the Wyoming flats. Sandy in places, silty, muddy, they run on and on, near sea level, all the way across Wyoming and into Utah. They are as red as brick. They will become the red cliffs and red canyons of Wyoming, the walls of Flaming Gorge. Triassic rock is not exclusively red, but much of it is red all over the world—red in the shales of New Jersey, red in the sandstones of Yunan, red in the banks of the Volga, red by the Solway Firth. Triassic red beds, as they are called, are in the dry valleys of Antarctica, the red marls of Worcestershire, the hills of Alsace-Lorraine. The Petrified Forest. The Painted Desert. The South African red beds of the Great Karroo. Triassic red rock is red through and through, and not merely weathered red on the surface, like the great Redwall limestone of the Grand Canyon, which is actually gray. There may have been a superabundance of oxygen in the atmosphere from late Pennsylvanian through Permian and Triassic time. As sea level changed and changed again all through the Pennsylvanian, tremendous quantities of vegetation grew and then were drowned and buried, grew and then were drowned and buried—to become, eventually, seam upon seam of coal, interlayered with sandstones and shales. Living plants take in carbon dioxide, keep the carbon in their carbohydrates, and give up the oxygen to the atmosphere. Animals, from bacteria upward, then eat the plants and reoxidize the carbon. This cycle would go awry if a great many plants were buried. Their carbon would be buried with them—isolated in rock—and so the amount of oxygen in the atmosphere would build up. All over the world, so much carbon was buried in Pennsylvanian time that the oxygen pressure in the atmosphere quite possibly doubled. There is more speculation than hypothesis in this, but what could the oxygen do? Where could it go? After carbon, the one other thing it could oxidize in great quantity was iron—abundant, pale-green ferrous iron, which exists everywhere, in fully five per cent of crustal rock; and when ferrous iron takes on oxygen, it turns a ferric red. That may have been what happened—in time that followed the Pennsylvanian. Permian rock is generally red. Red beds on an epic scale are the signs of the Triassic, when the earth in its rutilance may have outdone Mars.

As you come off the red flats to cross western Utah, two hundred and ten million years before the present, you travel in the dark, there being not one grain of evidence to suggest its Triassic appearance, no paleoenvironmental clue. Ahead, though, in eastern Nevada, is a line of mountains that are much of an age with the peaks of New Jersey—a little rounded, beginning to show age—and after you climb them and go down off their western slopes you discern before you the white summits of alpine fresh terrain, of new rough mountains rammed into thin air, with snow banners flying off the matterhorns, ridges, crests, and spurs. You are in central Nevada, about four hundred miles east of San Francisco, and after you have climbed these mountains you look out upon (as it appears in present theory) open sea. You drop swiftly to the coast, and then move on across moderately profound water full of pelagic squid, water that is quietly accumulating the sediments which—ages in the future—will become the roof rock of the rising Sierra. Tall volcanoes are standing in the sea. Then, at roughly the point where the Sierran foothills will end and the Great Valley will begin—at Auburn, California—you move beyond the shelf and over deep ocean. There are probably some islands out there somewhere, but fundamentally you are crossing above ocean crustal floor that reaches to the China Sea. Below you there is no hint of North America, no hint of the valley or the hills where Sacramento and San Francisco will be.

I used to sit in class and listen to the terms come floating down the room like paper airplanes. Geology was called a descriptive science, and with its pitted outwash plains and drowned rivers, its hanging tributaries and starved coastlines, it was nothing if not descriptive. It was a fountain of metaphor—of isostatic adjustments and degraded channels, of angular unconformities and shifting divides, of rootless mountains and bitter lakes. Streams eroded headward, digging from two sides into mountain or hill, avidly struggling toward each other until the divide between them broke down, and the two rivers that did the breaking now became confluent (one yielding to the other, giving up its direction of flow and going the opposite way) to become a single stream. Stream capture. In the Sierra Nevada, the Yuba had captured the Bear. The Macho member of a formation in New Mexico was derived in large part from the solution and collapse of another formation. There was fatigued rock and incompetent rock and inequigranular fabric in rock. If you bent or folded rock, the inside of the curve was in a state of compression, the outside of the curve was under great tension, and somewhere in the middle was the surface of no strain. Thrust fault, reverse fault, normal fault—the two sides were active in every fault. The inclination of a slope on which boulders would stay put was the angle of repose. There seemed, indeed, to be more than a little of the humanities in this subject. Geologists communicated in English; and they could name things in a manner that sent shivers through the bones. They had roof pendants in their discordant batholiths, mosaic conglomerates in desert pavement. There was ultrabasic, deep-ocean, mottled green-and-black rock—or serpentine. There was the slip face of the barchan dune. In 1841, a paleontologist had decided that the big creatures of the Mesozoic were "fearfully great lizards," and had therefore named them dinosaurs. There were festooned crossbeds and limestone sinks, pillow lavas and petrified trees, incised meanders and defeated streams. There were dike swarms and slickensides, explosion pits, volcanic bombs. Pulsating glaciers. Hogbacks. Radiolarian ooze. There was almost enough resonance in some terms to stir the adolescent groin. The swelling up of mountains was described as an orogeny. Ontogeny, phylogeny, orogeny—accent syllable two. The Antler Orogeny, the Avalonian Orogeny, the Taconic, Acadian, Alleghenian orogenies. The Laramide Orogeny. The center of the United States had had a dull geologic history—nothing much being accumulated, nothing much being eroded away. It was just sitting there conservatively. The East had once been radical—had been unstable, reformist, revolutionary, in the Paleozoic pulses of three or four orogenies. Now, for the last hundred and fifty million years, the East had been stable and conservative. The far-out stuff was in the Far West of the country—wild, weirdsma, a leather-jacket geology in mirrored shades, with its welded tuffs and Franciscan mélange (internally deformed, complex beyond analysis), its strike-slip faults and falling buildings, its boiling springs and fresh volcanics, its extensional disassembling of the earth.

There was, to be sure, another side of the page—full of geological language of the sort that would have attracted Gilbert and Sullivan. Rock that stayed put was called autochthonous, and if it had moved it was allochthonous. "Normal" meant "at right angles." "Normal" also meant a fault with a depressed hanging wall. There was a Green River Basin in Wyoming that was not to be confused with the Green River Basin in Wyoming. One was topographical and was *on* Wyoming. The other was structural and was *under* Wyoming. The Great Basin, which is centered in Utah and Nevada, was not to be confused with the Basin and Range, which is centered in Utah and Nevada. The Great Basin was topographical, and extraordinary in the world as a vastness of land that had no drainage to the sea. The Basin and Range was a realm of related mountains that coincided with the Great Basin, spilling over slightly to the north and considerably to the south. To anyone with a smoothly functioning bifocal mind, there was no lack of clarity about Iowa in the Pennsylvanian, Missouri in the Mississippian, Nevada in Nebraskan, Indiana in Illinoian, Vermont in Kansan, Texas in Wisconsinan time.

Meteoric water, with study, turned out to be rain. It ran downhill in consequent, subsequent, obsequent, resequent, and not a few insequent streams.

As years went by, such verbal deposits would thicken. Someone developed enough effrontery to call a piece of our earth an epieugeosyncline. There were those who said interfluve when they meant between two streams, and a perfectly good word like mesopotamian would do. A cactolith, according to the American Geological Institute's *Glossary of Geology and Related Sciences*, was "a quasi-horizontal chonolith composed of anastomosing ductoliths, whose distal ends curl like a harpolith, thin like a sphenolith, or bulge discordantly like an akmolith or ethmolith." The same class of people who called one rock serpentine called another jacupirangite. Clinoptilolite, eclogite, migmatite, tincalconite, szaibelyite, pumpellyite. Meyerhofferite. The same class of people who called one rock paracelsian called another despujolsite. Metakirchheimerite, phlogopite, katzenbuckelite, mboziite, noselite, neighborite, samsonite, pigeonite, muskoxite, pabstite, aenigmatite. Joesmithite. With the X-ray diffractometer and the X-ray fluorescence spectrometer, which came into general use in geology laboratories in the late nineteen-fifties, and then with the electron probe (around 1970), geologists obtained ever closer examinations of the components of rock. What they had long seen through magnifying lenses as specimens held in the hand—or in thin slices under microscopes—did not always register identically in the eyes of these machines. Andesite, for example, had been given its name for being the predominant rock of the high mountains of South America. According to the machines, there is surprisingly little andesite in the Andes. The Sierra Nevada is renowned throughout the world for its relatively young and absolutely beautiful granite. There is precious little granite in the Sierra. Yosemite Falls, Half Dome, El Capitan—for the most part the "granite" of the Sierra is granodiorite. It has always been difficult enough to hold in the mind that a magma which hardens in the earth as granite will—if it should flow out upon the earth—harden as rhyolite, that what hardens within the earth as diorite will harden upon the earth as andesite, that what hardens within the earth as gabbro will harden upon the earth as basalt, the difference from pair to pair being a matter of chemical composition and the differences within each pair being a matter of texture and of crystalline form, with the darker rock at the gabbro end and the lighter rock the granite. All of that—not to mention such wee appendixes as the fact that diabase is a special texture of gabbro—was difficult enough for the layman to remember before the diffractometers and the spectrometers and the electron probes came along to present their multiplex cavils. What had previously been described as the granite of the world turned out to be a large family of rock that included granodiorite, monzonite, syenite, adamellite, trondhjemite, alaskite, and a modest amount of true granite. A great deal of rhyolite, under scrutiny, became dacite, rhyodacite, quartz latite. Andesite was found to contain enough silica, potassium, sodium, and aluminum to be the fraternal twin of granodiorite. These points are pretty fine. The home terms still apply. The enthusiasm geologists show for adding new words to their conversation is, if anything, exceeded by their affection for the old. They are not about to drop granite. They say granodiorite when they are in church and granite the rest of the week.

When I was seventeen and staring up the skirts of eastern valleys, I was taught the rudiments of what is now referred to as the Old Geology. The New Geology is the package phrase for the effects of the revolution that occurred in earth science in the nineteen-sixties, when geologists clambered onto seafloor spreading, when people began to discuss continents in terms of their velocities, and when the interactions of some twenty parts of the globe became known as plate tectonics. There were few hints of all that when I was seventeen, and now, a shake later, middle-aged and fading, I wanted to learn some geology again, to feel the difference between the Old and the New, to sense if possible how the science had settled down a decade after its great upheaval, but less in megapictures than in day-to-day contact with country rock, seeing what had not changed as well as what had changed. The thought occurred to me that if you were to walk a series of roadcuts with a geologist something illuminating would in all likelihood occur. This was long before I met Karen Kleinspehn, or, for that matter, David Love, of the United States Geological Survey, or Anita Harris, also of the Survey, or Eldridge Moores, of the University of California at Davis, all of whom would eventually take me with them through various stretches of the continent. What I did first off was what anyone would do. I called my local geologist. I live in Princeton, New Jersey, and the man I got in touch with was Kenneth Deffeyes, a senior professor who teaches introductory geology at Princeton University. It is an assignment that is angled wide. Students who have little aptitude for the sciences are required to take a course or two in the sciences en route to some cerebral Valhalla dangled high by the designers of curriculum. Deffeyes' course is one that such students are drawn to select. He calls it Earth and Its Resources. They call it Rocks for Jocks.

Deffeyes is a big man with a tenured waistline. His hair flies behind him like Ludwig van Beethoven's. He lectures in sneakers. His voice is syllabic, elocutionary, operatic. He has been described by a colleague as "an intellectual roving shortstop, with more ideas per square metre than anyone else in the department—they just tumble out." His surname rhymes with "the maze." He has been a geological engineer, a chemical oceanographer, a sedimentary petrologist. As he lectures, his eyes search the hall. He is careful to be clear but also to bring forth the full promise of his topic, for he knows that while the odd jock and the pale poet are the white of his target the bull's-eye is the future geologist. Undergraduates do not come to Princeton intending to study geology. When freshmen fill out cards stating their three principal interests, no one includes rocks. Those who will make the subject their field of major study become interested after they arrive. It is up to Deffeyes to interest them—and not a few of them—or his department goes into a subduction zone. So his eyes search the hall. People out of his course have been drafted by the Sacramento Kings and have set records in distance running. They have also become professors of geological geophysics at Caltech and of petrology at Harvard.

Deffeyes' own research has gone from Basin and Range sediments to the floor of the deep sea to unimaginable events in the mantle, but his enthusiasms are catholic and he appears to be less attached to any one part of the story than to the entire narrative of geology in its four-dimensional recapitulations of space and time. His goals as a teacher are ambitious to the point of irrationality: At the very least, he seems to expect a hundred mint geologists to emerge from his course—expects perhaps to turn on his television and see a certified igneous petrographer up front with the starting

Kings. I came to know Deffeyes when I wondered how gold gets into mountains. I knew that most old-time hard-rock prospectors had little to go on but an association of gold with quartz. And I knew the erosional details of how gold comes out of mountains and into the rubble of streams. What I wanted to learn was what put the gold in the mountains in the first place. I asked a historical geologist and a geomorphologist. They both recommended Deffeyes. He explained that gold is not merely rare. It can be said to love itself. It is, with platinum, the noblest of the noble metals—those which resist combination with other elements. Gold wants to be free. In cool crust rock, it generally is free. At very high temperatures, however, it will go into compounds; and the gold that is among the magmatic fluids in certain pockets of interior earth may be combined, for example, with chlorine. Gold chloride is "modestly" soluble, and will dissolve in water that comes down and circulates in the magma. The water picks up many other elements, too: potassium, sodium, silicon. Heated, the solution rises into fissures in hard crust rock, where the cooling gold breaks away from the chlorine and—in specks, in flakes, in nuggets even larger than the eggs of geese—falls out of the water as metal. Silicon precipitates, too, filling up the fissures and enveloping the gold with veins of silicon dioxide, which is quartz.

When I asked Deffeyes what one might expect from a close inspection of roadcuts, he said they were windows into the world as it was in other times. We made plans to take samples of highway rock. I suggested going north up some new interstate to see what the blasting had disclosed. He said if you go north, in most places on this continent, the geology does not greatly vary. You should proceed in the direction of the continent itself. Go west. I had been thinking of a weekend trip to Whiteface Mountain, or something like it, but now, suddenly, a vaulting alternative came to mind. What about Interstate 80, I asked him. It goes the distance. How would it be? "Absorbing," he said. And he mused aloud: After 80 crosses the Border Fault, it pussyfoots along on morainal till that levelled up the fingers of the foldbelt hills. It does a similar dance with glacial debris in parts of Pennsylvania. It needs no assistance on the craton. It climbs a ramp to the Rockies and a fault-block staircase up the front of the Sierra. It is geologically shrewd. It was the route of animal migrations, and of human history that followed. It avoids melodrama, avoids the Grand Canyons, the Jackson Holes, the geologic operas of the country, but it would surely be a sound experience of the big picture, of the history, the construction, the components of the continent. And in all likelihood it would display in its roadcuts rock from every epoch and era.

In seasons that followed, I would go back and forth across the interstate like some sort of shuttle working out on a loom, accompanying geologists on purposes of their own or being accompanied by them from cut to cut and coast to coast. At any location on earth, as the rock record goes down into time and out into earlier geographies it touches upon tens of hundreds of stories, wherein the face of the earth often changed, changed utterly, and changed again, like the face of a crackling fire. The rock beside the road exposes one or two levels of the column of time and generally implies what went on immediately below and what occurred (or never occurred) above. To tell all the stories would be to tell pretty much the whole of geology in many volumes across a fifty-foot shelf, a task for which I am in every conceivable way unqualified. I am a layman who has travelled with a small core sampling of academic and government geologists ranging in experience from a student to an *éminence grise*. I wish to make no attempt to speak for all geology or to sweep in every fact that came along. I want to choose some things that interested me and through them to suggest the general history of the continent by describing events and landscapes that geologists see written in rocks.

To poke around in a preliminary way, Deffeyes and I went up to the Palisades Sill, where I was to return with Karen Kleinspehn, borrowed some diabase with a ten-pound sledge, and then began to travel westward, traversing the Hackensack Valley. It was morning. Small airplanes engorged with businessmen were settling into Teterboro. Deffeyes pointed out that if this were near the end of Wisconsinan time, when the ice was in retreat, those airplanes would have been settling down through several hundred feet of water, with the runway at the bottom of a lake. Glacial Lake Hackensack was the size of Lake Geneva and was host to many islands. It had the Palisades Sill for an eastern shoreline, and on the west the lava hill that is now known as the First Watchung Mountain. The glacier had stopped at Perth Amboy, leaving its moraine there to block the foot of the lake, which the glacier fed with meltwater as it retreated to the north. Some two hundred million years earlier, the runway would have been laid out on a baking red flat beside the first, cooling Watchung—glowing from cracks, from lava fountains, but generally black as carbon. Basalt flows don't light up the sky. Three hundred million years before that, the airplanes would have been settling down toward the same site through water—in this instance, salt water—on the eastern shelf of a broad low continent, where an almost pure limestone was forming, because virtually nothing from the worn-away continent was eroding into the shallow sea. Three random moments from the upper ninth of time.

In Paterson, I-80 chops the Watchung lava. Walking the cut from end to end, Deffeyes picked up some peripheral shale—Triassic red shale. He put it in his mouth and chewed it. "If it's gritty it's a silt bed, and if it's creamy it's a shale," he said. "This is creamy. Try it." I would not have thought to put it in coffee. In the blocky basaltic wall of the road, there were many small pockets, caves the size of peas, caves the size of lemons. As magma approaches the surface of the earth, it is so perfused with gases that it fizzes like ginger ale. In cooling basalt, gas bubbles remain, and form these minicaves. For a century and more, nothing much fills them. Slowly, though, over a minimum of about a million years, they can fill with zeolite crystals. Until well after the Second World War, not a whole lot was known about the potential uses of zeolite crystals. Nor was it known where they could be found in abundance. Deffeyes did important early work in the field. His doctoral dissertation, which dealt with two basins and two ranges in Nevada, included an appendix that started the zeolite industry. Certain zeolites (there are about thirty kinds) have become the predominant catalysts in use in oil refineries, doing a job that is otherwise assigned to platinum. Now, in Paterson, Deffeyes searched the roadcut vugs (as the minute caves are actually called) looking for zeolites. Some vugs were large enough to suggest the holes that lobsters hide in. They did indeed contain a number of white fibrous zeolite crystals—smooth and soapy, of a type that resembled talc or asbestos—but the cut had been almost entirely cleaned out by professional and amateur collectors, undeterred by the lethal traffic not many inches away. Nearly all the vugs were now as empty as they had been in their first hundred years. In the shale beyond the lava we saw the burrows of Triassic creatures. An ambulance from Totowa flew by with its siren wailing.

We moved on a few miles into the Great Piece Meadows of the Passaic River Valley, flat as a lake floor, poorly drained land. A meadow in New Jersey is any wet spongy acreage where you don't sink in above your chin. Great Piece Meadows, Troy Meadows, Black Meadows, the Great Swamp—Whippany, Parsippany, Madison, and Morristown are strewn among the reeds. The whole region, very evidently, was the bottom of a lake, for a lake itself is by definition a sign of poor drainage, an aneurysm in a river, a highly temporary feature on the land. Some lakes dry up. Others disappear after the outlet stream, deepening its valley and eroding headward into the outlet, empties the water. This one—Glacial Lake Passaic—vanished about ten thousand years ago, after the retreating glacier exposed what is now the Passaic Valley. The lake drained gradually into the new Passaic River, which fell a hundred feet into Glacial Lake Hackensack, and, en route, went over a waterfall that would one day in effect found the city of Paterson by turning its first mill wheel. At the time of its greatest extent, Lake Passaic was two hundred feet deep, thirty miles long, and ten miles wide, and seems to have been a scene of great beauty. Its margins are still decorated with sand spits and offshore bars, wave-cut cliffs and stream deltas, set in suburban towns. The lake's west shore was the worn-low escarpment of the Border Fault, and its most arresting feature was a hook-shaped basaltic peninsula that is now known to geologists as a part of the Third Watchung Lava Flow and to the people of New Jersey as Hook Mountain.

Deffeyes became excited as we approached Hook Mountain. The interstate had blasted into one toe of the former peninsula, exposing its interior to view. Deffeyes said, "Maybe someone will have left some zeolites here. I want them so bad I can taste them." He jumped the curb with his high-slung Geology Department vehicle, got out his hammers, and walked the cut. It was steep and competent, with brown oxides of iron over the felt-textured black basalt, and in it were tens of thousands of tiny vugs, a high percentage of them filled with pearl-lustred crystals of zeolite. To take a close look, he opened his hand lens—a small-diameter, ten-power

Hastings Triplet. "You can do a nice act in a jewelry store," he suggested. "You whip this thing out and you say the price is too high. These are beautiful crystals. Beautiful crystals imply slow growth. You don't get in a hurry and make something that nice." He picked up the sledge and pounded the cut, necessarily smashing many crystals as he broke their matrix free. "These crystals are like Vietnamese villages," he went on. "You have to destroy them in order to preserve them. They contain aluminum, silicon, calcium, sodium, and an incredible amount of imprisoned water. 'Zeolite' means 'the stone that boils.' If you take one small zeolite crystal, of scarcely more than a pinhead's diameter, and heat it until the water has come out, the crystal will have an internal surface area equivalent to a bedspread. Zeolites are often used to separate one kind of molecule from another. They can, for example, sort out molecules for detergents, choosing the ones that are biodegradable. They love water. In refrigerators, they are used to adsorb water that accidentally gets into the Freon. They could be used in automobile gas tanks to adsorb water. A zeolite called clinoptilolite is the strongest adsorber of strontium and cesium from radioactive wastes. The clinoptilolite will adsorb a great deal of lethal material, which you can then store in a small space. When William Wyler made *The Big Country*, there was a climactic chase scene in which the bad guy was shot and came clattering down a canyon wall in what appeared to be a shower of clinoptilolite. Geologists were on the phone to Wyler at once. 'Loved your movie. Where was that canyon?' There are a lot of zeolites in the Alps, in Nova Scotia, and in North Table Mountain in Colorado. When I was at the School of Mines, I used to go up to North Table Mountain just to wham around. Some of the best zeolites in the world are in this part of New Jersey."

There were oaks and maples on top of Hook Mountain, and, in the wall of the roadcut, basal rosettes of woolly mullein, growing in the rock. The Romans drenched stalks of mullein with suet and used them for funeral torches. American Indians taught the early pioneers to use the long flannel leaves of this plant as innersoles. Only three miles west of us was the Border Fault, where the basin had touched the range, where the stubby remnants of the fault scarp are now under glacial debris. Deffeyes said that the displacement along the fault—the eventual difference between two points that had been adjacent when the faulting began—exceeded fifteen thousand feet. Of course, this happened over several millions of years, and the mountains fronting the basin were all the while eroding, so they were never anything like fifteen thousand feet high. Generally, though, in the late Triassic, there would have been about a mile of difference, a mile of relief, between basin and range. In flash floods, boulders came raining off the mountains and piled in fans at the edge of the basin, ultimately to be filled in with sands and muds and to form conglomerate, New Jersey's so-called Hammer Creek Conglomerate—multicircled, polka-dotted headcheese rock, sometimes known as puddingstone. Here where the basin met the range, the sediments piled up so much that after all of the erosion of two hundred million years what remains is three miles thick. "I was in a bar once in Austin, Nevada," Deffeyes said, "and there was a sudden torrential downpour. The bartender began nailing plywood over the door. I wondered why he was doing that, until boulders came tumbling down the main street of the town. When you start pulling a continent apart, you have a lot of consequences of the same event. Faulting produced this basin. Sediments filled it in. Pull things apart and you produce a surface vacancy, which is faulting, and a subsurface vacancy, which causes upwelling of hot mantle that intrudes as sills or comes out as lava flows. In the Old Geology, you might have seen a sill within the country rock and said, 'Ah, the sill came much later.' With the New Geology, you see that all this was happening more or less at one time. The continent was splitting apart and the ultimate event was the opening of the Atlantic. If you look at the foldbelt in northwest Africa, you see the other side of the New Jersey story. The folding there is of the same age as the Appalachians, and the subsequent faulting is Triassic. Put the two continents together on a map and you will see what I mean. Fault blocks like this one are still in evidence, but discontinuously, from the Connecticut Valley to South Carolina. They are all parts of the suite that opened the Atlantic seaway. The story is very similar in the Great Basin—in the West, in the Basin and Range. The earth is splitting apart there, quite possibly opening a seaway. It is not something that happened a couple of hundred million years ago. It only began in the Miocene, and it is going on today. What we are looking at here in New Jersey is not just some little geologic feature, like a zeolite crystal. This is the opening of the Atlantic. If you want to see happening right now what happened here two hundred million years ago, you can see it all in Nevada."

Basin. Fault. Range. Basin. Fault. Range. A mile of relief between basin and range. Stillwater Range. Pleasant Valley. Tobin Range. Jersey Valley. Sonoma Range. Pumpernickel Valley. Shoshone Range. Reese River Valley. Pequop Mountains. Steptoe Valley. Ondographic rhythms of the Basin and Range. We are maybe forty miles off the interstate, in the Pleasant Valley basin, looking up at the Tobin Range. At the nine-thousand-foot level, there is a stratum of cloud against the shoulders of the mountains, hanging like a ring of Saturn. The summit of Mt. Tobin stands clear, above the cloud. When we crossed the range, we came through a ranch on the ridgeline where sheep were fenced around a running brook and bales of hay were bright green. Junipers in the mountains were thickly hung with berries, and the air was unadulterated gin. This country from afar is synopsized and dismissed as "desert"—the home of the coyote and the pocket mouse, the side-blotched lizard and the vagrant shrew, the MX rocket and the pallid bat. There are minks and river otters in the Basin and Range. There are deer and antelope, porcupines and cougars, pelicans, cormorants, and common loons. There are Bonaparte's gulls and marbled godwits, American coots and Virginia rails. Pheasants. Grouse. Sandhill cranes. Ferruginous hawks and flammulated owls. Snow geese. This Nevada terrain is not corrugated, like the folded Appalachians, like a tubal air mattress, like a rippled potato chip. This is not—in that compressive manner—a ridge-and-valley situation. Each range here is like a warship standing on its own, and the Great Basin is an ocean of loose sediment with these mountain ranges standing in it as if they were members of a fleet without precedent, assembled at Guam to assault Japan. Some of the ranges are forty miles long, others a hundred, a hundred and fifty. They point generally north. The basins that sep-

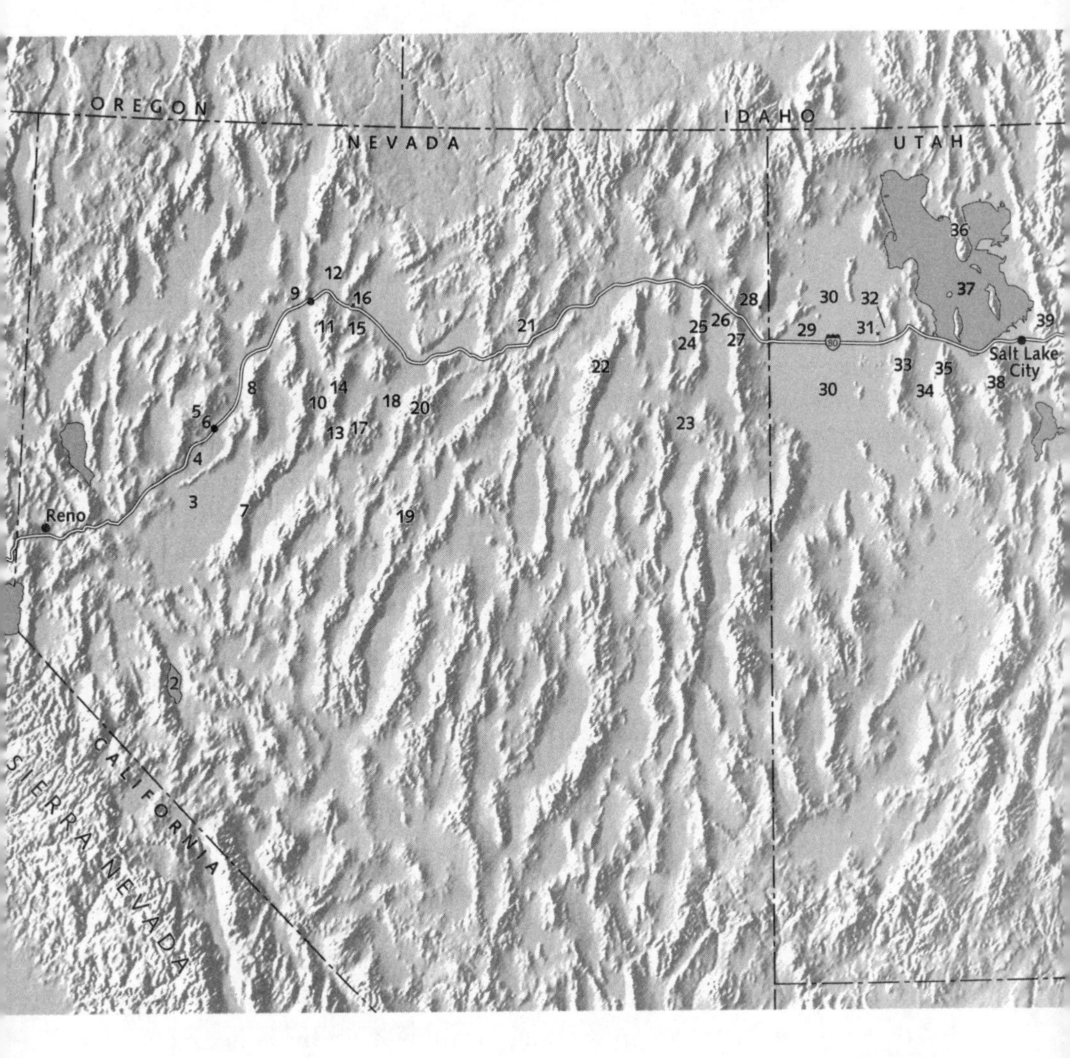

1. Donner Pass
2. Walker Lake
3. Carson Sink
4. Humboldt Sink
5. Trinity Range
6. Lovelock
7. Stillwater Range
8. Humboldt Range
9. Winnemucca
0. Pleasant Valley
11. Sonoma Range
12. Paradise Valley
13. Jersey Valley
14. Tobin Range
15. Pumpernickel Valley
16. Golconda Summit
17. Fish Creek Mountains
18. Reese River Valley
19. Toiyabe Range
20. Shoshone Range
21. Carlin Canyon
22. Ruby Mountains
23. Steptoe Valley
24. Independence Valley
25. Pequop Mountains
26. Goshute Valley
27. Toano Range
28. Pilot Peak
29. Bonneville Salt Flats
30. Great Salt Lake Desert
31. Grayback Mountain
32. Ripple Valley
33. Cedar Mountains
34. Skull Valley
35. Stansbury Mountains
36. Promontory Mountains
37. Great Salt Lake
38. Oquirrh Mountains
39. Wasatch Range

arate them—ten and fifteen miles wide—will run on for fifty, a hundred, two hundred and fifty miles with lone, daisy-petalled windmills standing over sage and wild rye. Animals tend to be content with their home ranges and not to venture out across the big dry valleys. "Imagine a chipmunk hiking across one of these basins," Deffeyes remarks. "The faunas in the high ranges here are quite distinct from one to another. Animals are isolated like Darwin's finches in the Galápagos. These ranges are truly islands."

Supreme over all is silence. Discounting the cry of the occasional bird, the wailing of a pack of coyotes, silence—a great spatial silence—is pure in the Basin and Range. It is a soundless immensity with mountains in it. You stand, as we do now, and look up at a high mountain front, and turn your head and look fifty miles down the valley, and there is utter silence. It is the silence of the winter forests of the Yukon, here carried high to the ridgelines of the ranges. "It is a soul-shattering silence," the physicist Freeman Dyson wrote of southern Nevada in *Disturbing the Universe*. "You hold your breath and hear absolutely nothing. No rustling of leaves in the wind, no rumbling of distant traffic, no chatter of birds or insects or children. You are alone with God in that silence. There in the white flat silence I began for the first time to feel a slight sense of shame for what we were proposing to do. Did we really intend to invade this silence with our trucks and bulldozers and after a few years leave it a radioactive junkyard?"

What Deffeyes finds pleasant here in Pleasant Valley is the aromatic sage. Deffeyes grew up all over the West, his father a petroleum engineer, and he says without apparent irony that the smell of sagebrush is one of two odors that will unfailingly bring upon him an attack of nostalgia, the other being the scent of an oil refinery. Flash floods have caused boulders the size of human heads to come tumbling off the range. With alluvial materials of finer size, they have piled up in fans at the edge of the basin. ("The cloudburst is the dominant sculptor here.") The fans are unconsolidated. In time to come, they will pile up to such enormous thicknesses that they will sink deep and be heated and compressed to form conglomerate. Erosion, which provides the material to build the fans, is tearing down the mountains even as they rise. Mountains are not somehow created whole and subsequently worn away. They wear down as they come up, and these mountains have been rising and eroding in fairly even ratio for millions of years—rising and shedding sediment steadily through time, always the same, never the same, like row upon row of fountains. In the southern part of the province, in the Mojave, the ranges have stopped rising and are gradually wearing away. The Shadow Mountains. The Dead Mountains, Old Dad Mountains, Cowhole Mountains, Bullion, Mule, and Chocolate mountains. They are inselberge now, buried ever deeper in their own waste. For the most part, though, the ranges are rising, and there can be no doubt of it here, hundreds of miles north of the Mojave, for we are looking at a new seismic scar that runs as far as we can see. It runs along the foot of the mountains, along the fault where the basin meets the range. From out in the valley, it looks like a long, buff-painted, essentially horizontal stripe. Up close, it is a gap in the vegetation, where plants growing side by side were suddenly separated by several metres, where, one October evening, the basin and the range —Pleasant Valley, Tobin Range—moved, all in an instant, apart. They jumped sixteen feet. The erosion rate at which the mountains were coming down was an inch a century. So in the mountains' contest with erosion they gained in one moment about twenty thousand years. These mountains do not rise like bread. They sit still for a long time and build up tension, and then suddenly jump. Passively, they are eroded for millennia, and then they jump again. They have been doing this for about eight million years. This fault, which jumped in 1915, opened like a zipper far up the valley, and, exploding into the silence, tore along the mountain base for upward of twenty miles with a sound that suggested a runaway locomotive.

"This is the sort of place where you really do not put a nuclear plant," says Deffeyes. "There was other action in the neighborhood at the same time—in the Stillwater Range, the Sonoma Range, Pumpernickel Valley. Actually, this is not a particularly spectacular scarp. The lesson is that the whole thing—the whole Basin and Range, or most of it—is alive. The earth is moving. The faults are moving. There are hot springs all over the province. There are young volcanic rocks. Fault scars everywhere. The world is splitting open and coming apart. You see a sudden break in the sage like this and it says to you that a fault is there and a fault block is coming up. This is a gorgeous, fresh, young, active fault scarp. It's growing. The range is lifting up. This Nevada topography is what you see *during* mountain building. There are no foothills. It is all too young. It is live country. This is the tectonic, active, spreading, mountain-building world. To a nongeologist, it's just ranges, ranges, ranges."

Most mountain ranges around the world are the result of compression, of segments of the earth's crust being brought together, bent, mashed, thrust and folded, squeezed up into the sky—the Himalaya, the Appalachians, the Alps, the Urals, the Andes. The ranges of the Basin and Range came up another way. The crust—in this region between the Rockies and the Sierra—is spreading out, being stretched, being thinned, being literally pulled to pieces. The sites of Reno and Salt Lake City, on opposite sides of the province, have moved apart sixty miles. The crust of the Great Basin has broken into blocks. The blocks are not, except for simplicity's sake, analogous to dominoes. They are irregular in shape. They more truly suggest stretch marks. Which they are. They trend nearly north-south because the direction of the stretching is roughly east-west. The breaks, or faults, between them are not vertical but dive into the earth at angles that average sixty degrees, and this, from the outset, affected the centers of gravity of the great blocks in a way that caused them to tilt. Classically, the high edge of one touched the low edge of another and formed a kind of trough, or basin. The high edge—sculpted, eroded, serrated by weather—turned into mountains. The detritus of the mountains rolled into the basin. The basin filled with water—at first, it was fresh blue water—and accepted layer upon layer of sediment from the mountains, accumulating weight, and thus unbalancing the block even further. Its tilt became more pronounced. In the manner of a seesaw, the high, mountain side of the block went higher and the low, basin side went lower until the block as a whole reached a state of precarious and temporary truce with God, physics, and mechanical and chemical erosion, not to mention, far below, the agitated mantle, which was running a temperature hotter than normal, and was, almost surely, controlling the action. Basin and range. Integral fault blocks: low side the basin, high side the range. For five hundred miles they nudged one another across the province of the Basin and Range. With extra faulting, and whatnot, they took care of their own irregularities. Some had their high sides on the west, some on the east.

The escarpment of the Wasatch Mountains—easternmost expression of this immense suite of mountains—faced west. The Sierra—the westernmost, the highest, the predominant range, with Donner Pass only halfway up it—presented its escarpment to the east. As the developing Sierra made its skyward climb—as it went on up past ten and twelve and fourteen thousand feet—it became so predominant that it cut off the incoming Pacific rain, cast a rain shadow (as the phenomenon is called) over lush, warm, Floridian and verdant Nevada. Cut it off and kept it dry.

We move on (we're in a pickup) into dusk—north up Pleasant Valley, with its single telephone line on sticks too skinny to qualify as poles. The big flanking ranges are in alpenglow. Into the cold clear sky come the ranking stars. Jackrabbits appear, and crisscross the road. We pass the darkening shapes of cattle. An eerie trail of vapor traverses the basin, sent up by a clear, hot stream. It is only a couple of feet wide, but it is running swiftly and has multiple sets of hot white rapids. In the source springs, there is a thumping sound of boiling and rage. Beside the springs are lucid green pools, rimmed with accumulated travertine, like the travertine walls of Lincoln Center, the travertine pools of Havasu Canyon, but these pools are too hot to touch. Fall in there and you are Brunswick stew. "This is a direct result of the crustal spreading," Deffeyes says. "It brings hot mantle up near the surface. There is probably a fracture here, through which the water is coming up to this row of springs. The water is rich in dissolved minerals. Hot springs like these are the source of vein-type ore deposits. It's the same story that I told you about the hydrothermal transport of gold. When rainwater gets down into hot rock, it brings up what it happens to find there—silver, tungsten, copper, gold. An ore-deposit map and a hot-springs map will look much the same. Seismic waves move slowly through hot rock. The hotter the rock, the slower the waves. Nowhere in the continental United States do seismic waves move more slowly than they do beneath the Basin and Range. So we're not woofing when we say there's hot mantle down there. We've measured the heat."

The basin-range fault blocks in a sense are floating on the mantle. In fact, the earth's crust everywhere in a sense is floating on the mantle. Add weight to the crust and it rides deeper, remove cargo and it rides higher, exactly like a vessel at a pier. Slowly disassemble the Rocky Mountains and carry the material in small fragments to the Mississippi Delta. The delta builds down. It presses ever deeper on the mantle. Its depth at the moment exceeds twenty-five thousand feet. The heat and the pressure are so great down there that the silt is turning into siltstone, the sand into sandstone, the mud into shale. For another example, the last Pleistocene ice sheet loaded two miles of ice onto Scotland, and that dunked Scotland in the mantle. After the ice melted, Scotland came up again, lifting its beaches high into the air. Isostatic adjustment. Let go a block of wood that you hold underwater and it adjusts itself to the surface isostatically. A frog sits on the wood. It goes down. He vomits. It goes up a little. He jumps. It adjusts. Wherever landscape is eroded away, what remains will rise in adjustment. Older rock is lifted to view. When, for whatever reason, crust becomes thicker, it adjusts downward. All of this—with the central image of the basin-range fault blocks floating in the mantle—may suggest that the mantle is molten, which it is not. The mantle is solid. Only in certain pockets near the surface does it turn into magma and squirt upward. The temperature of the mantle varies widely, as would the temperature of anything that is two thousand miles thick. Under the craton, it is described as chilled. By surface standards, though, it is generally white hot, everywhere around the world—white hot and solid but magisterially viscous, permitting the crust above it to "float." Deffeyes was in his bathtub one Saturday afternoon thinking about the viscosity of the mantle. Suddenly he stood up and reached for a towel. "Piano wire!" he said to himself, and he dressed quickly and went to the library to look up a book on piano tuning and to calculate the viscosity of the wire. Just what he guessed—10^{22} poises. Piano wire. Look under the hood of a well-tuned Steinway and you are looking at strings that could float a small continent. They are rigid, but ever so slowly they will sag, will slacken, will deform and give way, with the exact viscosity of the earth's mantle. "And that," says Deffeyes, "is what keeps the piano tuner in business." More miles, and there appears ahead of us something like a Christmas tree alone in the night. It is Winnemucca, there being no other possibility. Neon looks good in Nevada. The tawdriness is refined out of it in so much wide black space. We drive on and on toward the glow of colors. It is still far away and it has not increased in size. We pass nothing. Deffeyes says, "On these roads, it's ten to the minus five that anyone will come along." The better part of an hour later, we come to the beginnings of the casino-flashing town. The news this year is that dollar slot machines are outdrawing nickel slot machines for the first time, ever.

Deffeyes' purposes in coming to Nevada are pure and noble. His considerable energies appear to be about equally divided between the pursuit of pure science and the pursuit of noble metal. In order to enloft humanity's understanding of the basins, he has been taking paleomagnetic samples of basin sediments. He seeks insight into the way in which the rifting earth comes apart. He wants to perceive the subtle differences in the histories of one fault block and another. His ideas about silver, on the other hand, may send his children to college. This is, after all, Nevada, whose geology bought the tickets for the Spanish-American War. George Hearst found his fortune in the ground here. There were silver ores of such concentration that certain miners did nothing more to the heavy gray rocks than pack them up and ship them to Europe. To be sure, those days and those rocks—those supergene enrichments—are gone, but it has crossed the mind of Deffeyes that there may be something left for Deffeyes. Banqueting Sybarites surely did not lick their plates.

We rented the pickup in Salt Lake City—a white Ford. "If we had a bale of hay in here, we'd be Nevada authentic," Deffeyes remarked, and he swept snow off the truck with a broom. November. Three inches on the ground and more falling, slanting in to us from the west. We squinted, and rubbed the insides of the windows, and passed low commercial buildings that drifted in and out of sight. WILD DUCKS & PHEASANTS PROCESSED. DEER CUT & WRAPPED. DRIVE-IN WINDOW. 7:00 TILL MIDNIGHT. Behind us we could not see, of course, the wall of the Wasatch, its triangles and pinnacles white, but westward of the city visibility improved, and soon other mountains—the Oquirrh Mountains—came looming out of the blankness, their strata steeply dipping and as distinct as the stripes of an awning. "Those are Pennsylvanian and Permian sandstones and limestones," Deffeyes said. "There was glaciation in the Southern Hemisphere at the time. The ice came and went. Sea level kept flapping up and down. So the deposition has a striped look."

When a mountain range comes up into the air, a whole lot comes up with it. The event that had lifted the Oquirrhs—the stretching of the crust until it broke into blocks—was only among the latest of many episodes that have adjusted dramatically the appearance of central Utah. As we could plainly see from the interstate, the rock now residing in that striped mountainside had once been brutally shoved around—shoved, not pulled, and with such force that a large part of it had been tipped up more than ninety degrees, to and well beyond the vertical. Overturned. Such violence can happen on an epic scale. There is an entire nation in Europe that is upside down. It is not a superpower, but it is a whole country nonetheless—San Marino, overturned. Basin and Range faulting, on its own, has never overturned anything. The great fault blocks have a maximum tilt of thirty degrees. The event that so deformed the rock in the Oquirrhs took place roughly sixty million years ago—fifty-two million years before the Oquirrhs came into existence—and it was an event that made alpine fresh compressional mountains, which had their time here under the sun and were disassembled by erosion, taken down and washed away; and now those crazily upended stripes within the Oquirrhs are the evidence and fragmental remains of those ancestral mountains, brought up out of the earth and put on view as a component of new mountains. The new mountains—the mountains of the Basin and Range—are packages variously containing rock that formed at one time or another during some five hundred and fifty million years, or an eighth of the earth's total time. It was thought until recently that older rock was in certain of the ranges, but improved techniques of dating have shown that not to be true. Seven-eighths of the earth's time is lost here, gone without evidence—rock that disintegrated and went off to be recycled. One-eighth, for all that, is no small amount of earth history, and as the great crustal blocks of the Basin and Range have tipped their mountains into the air, with individual faults offset as much as twenty thousand feet, they have brought to the surface and have randomly exposed former seafloors and basaltic dikes, entombed rivers and veins of gold, volcanic spewings and dunal sands—chaotic, concatenated shards of time. In the Basin and Range are the well-washed limestones of clear and sparkling shallow Devonian seas. There are dark, hard, cherty siltstones from some deep ocean trench full of rapidly accumulating Pennsylvanian guck. There are Triassic sediments rich in fossils, scattered pods of Cretaceous granite, Oligocene welded tuffs. There is not much layer-cake geology. The layers have too often been tortured by successive convulsive events.

The welded tuffs were the regional surface when basin-range faulting began. And for more than twenty million previous years they had been the surface, the uppermost rock, with scant relief in the topography of these vast volcanic plains, whose great size and barren aspect are commensurate with the magnitude of the holocaust that brought the rock onto the land. Up through perhaps a hundred fissures, dikes, chimneys, vents, fractures came a furiously expanding, exploding mixture of steam and rhyolite glass, and, in enormous incandescent clouds, heavier than air, it scudded across the landscape like a dust storm. The volcanic ash that would someday settle down on Herculaneum and Pompeii was a light powder compared with this stuff, and as the great ground-covering clouds oozed into the contours of the existing landscape they sent streams hissing to extinction, and covered the streambeds and then the valleys, and—with wave after wave of additional cloud—obliterated entire drainages like plaster filling a mold. They filled in every gully and gulch, cave, swale, and draw until almost nothing stuck above a blazing level plain, and then more clouds came exploding from below and in unimpeded waves spread out across the plain. Needless to say, every living creature in the region died. Single outpourings settled upon areas the size of Massachusetts, and before the heavy ash stopped flowing it had covered twenty times that. Moreover, it was hot enough to weld. As the great clouds collapsed and condensed, they formed a compact rock in large part consisting of volcanic glass. It was so thick—as much as three hundred metres thick—that crystals formed slowly in the cooling glass. "When you bury a countryside in that much rock so hot it welds, that is the ultimate environmental catastrophe," Deffeyes remarked. "I'm glad there hasn't been one recently."

The province, stung like that, sat still here for twenty-two million years, with volcanism continuing only on its periphery, while erosion worked on the tuff, making draws and gulches, modest valleys and unspectacular hills, but not extensively altering the essentially level plain. There was no repetition of the foaming, frothing outpourings that had completely changed many tens of thousands of square miles of the face of the earth, but so much disturbance arising from and within the underlying earth was obviously precursive of disturbances to follow, when the plains of welded tuff and some thousands of feet below them began to rift into crustal blocks and become the Basin and Range.

The basins filled immediately with water, and life came into the lakes. "Late-Miocene fossils are the earliest we have wherever we have found fossils in those lakebeds," Deffeyes said. "So Basin and Range faulting can be dated to the late Miocene—about eight million years ago." Gradually, as the rain shadow lengthened, the lakes "turned chemical"—became saline or alkaline (bitter)—and eventually they dried up. There are basalt flows in the Basin and Range that are also post-Miocene—lavas that poured out on the surface well after the block faulting had begun, like the Watchungs of New Jersey. There are ruins of cinder cones—evidence of fairly recent local action—and, in the basins and on the ranges, widespread falls of light ash from volcanoes beyond the province. You see, too, the stream deltas, shoreline terraces, and wave-cut cliffs of big lakes that came into the Great Basin after Pleistocene glaciation began. The change in world climate that made ice in the north temporarily preempted the rain shadow and dropped into the Great Basin torrents from the sky. In a region where evaporation had greatly exceeded precipitation, the reverse was now the case, and the big lakes in time connected the basins and made islands of the ranges—Lake Manlius (its bed is now, in part, Death Valley); Lake Lahontan, near Reno (its bed is now, in part, the Humboldt and Carson sinks); and Lake Bonneville. Lake Bonneville grew until it was the size of Lake Erie. Then it grew some more. At Red Rock Pass, in Idaho, it spilled over the brim of the Great Basin and into the Snake River Plain. By now it was as large as Lake Michigan. It was not a glacial lake, just a sort of side effect of the distant glaciation, and it sat there for thousands of years with limestone terraces forming and waves cutting benches at the shoreline. Eventually, it began to drop, in stages, pausing wherever evaporation and precipitation were in temporary equilibrium, and more benches were cut and more terraces were made, and then as the rain shadow took over again, the water shrank back past Erie size and kept on shrinking and turning more and more chemical and getting smaller and shallower and shallower and smaller and near the end of its days became the Great Salt Lake.

The Great Salt Lake reached out to our right and disappeared in snow. In a sense, there was no beach. The basin flatness just ran to the lake and kept on going, wet. The angle formed at the shoreline appeared to be about 179.9 degrees. There were dark shapes of islands, firmaments in the swirling snow—elongate, north-south-trending islands, the engulfed summits of buried ranges. "Chemically, this is one of the toughest environments in the world," Deffeyes said. "You swing from the saltiest to the most dilute waters on the planet in a matter of hours. Some of the most primitive things living are all that can take that. The brine is nearly saturated with sodium chloride. For a short period each year, so much water comes down out of the Wasatch that large parts of the lake surface are relatively fresh. Any creature living there gets an osmotic shock that amounts to hundreds of pounds per square inch. No higher plants can take that, no higher animals—no multicelled organisms. Few bacteria. Few algae. Brine shrimp, which do live there, die by the millions from the shock."

I have seen the salt lake incredibly beautiful in winter dusk under snow-streamer curtains of cloud moving fast through the sky, with the wall of the Wasatch a deep rose and the lake islands rising from what seemed to be rippled slate. All of that was now implied by the mysterious shapes in the foreshortening snow. I didn't mind the snow. One June day, moreover, with Karen Kleinspehn—on her way west for summer field work—I stopped in the Wasatch for a picnic of fruit and cheese beside a clear Pyrenean stream rushing white over cobbles of quartzite and sandstone through a green upland meadow—cattle in the meadow, cottonwoods along the banks of this clear, fresh, suggestively confident, vitally ignorant river, talking so profusely on its way to its fate, which was to move among paradisal mountain landscapes until, through a terminal canyon, the Great Basin drew it in. No outlet. Three such rivers feed the Great Salt Lake. It does indeed consume them. Descending, we ourselves went through a canyon so narrow that the Union Pacific Railroad was in the median of the interstate and on into an even steeper canyon laid out as if for skiing in a hypnotizing rhythm of christiania turns under high walls of rose-brick Nugget sandstone and brittle shattered marine limestone covered with scrub oaks. "Good God, we are dropping out of the sky," said Kleinspehn, hands on the wheel, plunging through the big sheer roadcuts, one of which suddenly opened to distance, presented the Basin and Range.

" 'This is the place.' "

"You can imagine how he felt."

In the foreground was the alabaster city, with its expensive neighborhoods strung out along the Wasatch Fault, getting ready to jump fifteen feet. In the distance were the Oquirrhs, the Stansburys, the lake. Sunday afternoon and the Mormons were out on the flats by the water in folding chairs at collapsible tables, end to end like refectory tables, twenty people down to dinner, with acres of beachflat all to themselves and seagulls around them like sacred cows. To go swimming, we had to walk first—several hundred yards straight out, until the water was ankle-deep. Then we lay down on our backs and floated. I have never been able to float. When I took the Red Cross tests, age nine to fifteen, my feet went down and I hung in the water with my chin wrenched up like something off Owl Creek Bridge. I kicked, slyly kicked to push my mouth above the surface and breathe. I could not truly float. Now I tried a backstroke and, like some sort of hydrofoil, went a couple of thousand feet on out over the lake. Only my heels, rump, and shoulder blades seemed to be wet. I rolled over and crawled. I could all but crawl on my hands and knees. And this was June, at the south end—the least salty season, the least salty place in the whole of the Great Salt Lake.

Rolling up on one side, and propped on an elbow, I could see the Promontory Mountains across the water to the north, an apparent island but actually a peninsula, reaching southward into the lake. In 1869, a golden spike was carried into the Promontories and driven into a tie there to symbolize the completion of the first railroad to cross the North American continent—exactly one century before the first footprint on the moon, a span of time during which Salt Lake City and Reno would move apart by one human stride. In that time, also, the railroad twice became dissatisfied with the local arrangements of its roadbed—losing affection for the way of the golden spike (over the mountains) and building a causeway and wooden trestle across the lake itself, barely touching the Promontory peninsula at its southern tip. In the late nineteen-fifties, the trestle section was replaced by rock. The causeway traverses the lake like a solid breakwater, dividing it into halves. The principal rivers that flow into the Great Salt Lake all feed the southern half. The water on the north side of the causeway is generally a foot or two lower and considerably saltier than the water on the other side. Evaporate one cupful of Great Salt Lake North and you have upward of a third of a cup of salt. Evaporate a cupful of Great Salt Lake South and you have about a quarter of a cup of salt, or—nonetheless—eight times as much as from a cup of the ocean. As the lake drew at our bodies, trying to pull fresh water through our skins, it closed our pores tight and our lips swelled and became slightly numb. The water stung savagely at the slightest scratch and felt bitter as strep in the back of the throat.

We filled a bag with eggstones from the bottom, with oolites, the Salt Lake sand. It was by no means ordinary sand—not the small, smoothed-off ruins of mountains, carried down and dumped by rivers. It was sand that had formed in the lake. Just as raindrops are created around motes of dust, oolites form around bits of rock so tiny that in wave-tossed water they will stir up and move. They move, and settle, move, and settle. And while they are up in the water calcium carbonate forms around them in layer after layer, building something like a pearl. Slice one in half with a diamond saw and you reveal a perfect bull's-eye, or, as its namer obviously imagined it, a stone egg, white and yolk—an oolite. Underwater on the Bahama Banks are sweeping oolitic dunes. When a geologist finds oolites embedded in rock—in, say, some Cambrian outcrop in the Lehigh Valley—the Bahamas come to mind, and the Great Salt Lake, and, by inference, a shallow, lime-rich Cambrian sea. Our sample bag was like a ten-pound sack of sugar. I rolled over on my back, set it on my stomach, and, floating a little lower, kicked in to shore.

On the firm flat beach of the Great Salt Lake were many hundreds of thousands of brine flies—broad dark patches of them hopping and buzzing a steady collective electrical hum. A sacred gull made short bursts through the brine flies, its bill clapping. Three years before gulls ate crickets and saved the Mormons, Kit Carson shot gulls to feed the starving emigrants. Gulls, though, and brine flies are natural survivors. Now, at the end of spring runoff, dead creatures were everywhere. Osmotic shock had killed shrimp outnumbering the flies. Corpses, a couple of centimetres each, lay in hydrogen-sulphide decaying stink. Interlayered with the oolites on the bottom of the lake was a kind of galantine of brine shrimp, the greasy black muck of trillions dead.

Salt crystals clung like snow to our hair, and were spread on our faces like powder. In man-made ponds near the shore, the sun was making Morton's salt. Spaced along the beach were water towers, courtesy of the State of Utah. You pulled a rope and took a shower.

And now in the autumn snow, Deffeyes and I could see shoreline terraces of Lake Bonneville a thousand feet above us on mountain slopes. That a lake so deep had been brought down to a present average depth of thirteen feet was food for melancholia. Still shrinking, it had long since become the world's second-deadest body of water. In a couple of hundred years, it could match the Dead Sea.

"Mother of God, that's nice," said Deffeyes suddenly, braking down the pickup on the shoulder of the road. The tip of the nose of the Stansbury Mountains had been sliced off by the interstate to reveal a sheer and massive section of handsome blue rock, thinly bedded, evenly bedded, forty metres high. Its parallel planes were tilting, dipping, gently to the east, with the exception of some confused and crumpled material that suggested a snowball splatted against glass, or a broken-down doorway in an otherwise undamaged wall. Deffeyes said, "Let's Richter the situation," and he got out and crossed the road. With his hammer, he chipped at the rock, puzzled the cut. He scraped the rock and dropped acid on the scrapings. Tilted by the western breeze, the snow was dipping sixty degrees east. The bedding planes were dipping twenty degrees east; and the stripes of Deffeyes' knitted cap were dipping fifty degrees north. The cap had a big tassel, and with his gray-wisped hair coming out from under in a curly mélange he looked like an exaggerated elf. He said he thought he knew what had caused "that big goober" in the rock, and it was almost certainly not a manifestation of some major tectonic event—merely local violence, a cashier shot in a grab raid, an item for an inside page. The cut was mainly limestone, which had collected as lime mud in an Ordovician sea. The goober was dolomite.

Limestone is calcium carbonate. Dolomite is calcium carbonate with magnesium added. Together they are known as the carbonate rocks. Deffeyes was taught in college that while it seemed obvious to infer that magnesium precipitating out of water changes limestone into dolomite there was no way to check this out empirically because dolomite was forming nowhere in the world. Deffeyes found that impossible to believe. Deffeyes was already a uniformitarian—a geologist who believes that the present is the key to the past, that if you want to understand how a rock is formed you go watch it forming now. Watch basalt flows at Kilauea. Watch the festooned cross-beddings of future sandstones being sketched by the currents of Hatteras. Watch a flooding river blanket the tracks of a bear. Surely, somewhere, he thought, limestone must be changing into dolomite now. Not long after graduate school, he and two others went to Bonaire, in the Netherlands Antilles, where they found a lagoon that was concentrating under the sun and "making a juice very rich in magnesium." The juice was flowing through the limestone below and changing it into dolomite. They presented the news in *Science*. When the rock of this big Utah roadcut had been the limy bottom of the Ordovician sea, the water had been so shallow that the lime mud had occasionally been above the surface and had dried out and cracked into chips, and then the water rose and the chips became embedded in more lime mud, and the process happened again and again so that the limestone now is a self-containing breccia studded with imprisoned chips—an accident so lovely to the eye you want to slice the rock and frame it.

In age, the blue stone approached five hundred million years. Captain Howard Stansbury, USA, whose name would rest upon the mountains of which the rock was a component, was approaching fifty when he came into the Great Basin in 1849. He had been making lighthouses in Florida. The government preferred that he survey the salt lake. With sixteen mules, a water keg, and some India-rubber bags, he circumambulated the lake, and then some. People told him not to try it. He ran out of water but not of luck. And he came back with a story of having seen—far out on the westward flats—scattered books, clothing, trunks, tools, chains, yokes, dead oxen, and abandoned wagons. The Donner party went around the nose of the Stansburys in late August, 1846, rock on their left, lake marshes on their right. This huge blue roadcut, in its supranatural way, would have frightened them to death. They must have filed along just about where Deffeyes had parked the pickup, on the outside shoulder of the interstate. Deffeyes and I went back across the road, waiting first for a three-unit seven-axle tractor-trailer to pass. Deffeyes described it as "a freaking train."

Stansbury Mountains, Skull Valley . . . The Donner party found good grass in Skull Valley, and good water, and a note by a post at a spring. It had been torn to shreds by birds. The emigrants pieced it together. "Two days—two nights—hard driving—cross desert—reach water." They went out of Skull Valley over the Cedar Mountains into Ripple Valley and over Grayback Mountain to the Great Salt Lake Desert. Grayback Mountain was basalt, like the Watchungs of New Jersey. The New Jersey basalt flowed about two hundred million years ago. The Grayback Mountain basalt flowed thirty-eight million years ago. Well into this century, it was possible to find among the dark-gray outcrops of Grayback Mountain pieces of wagons and of oxhorn, discarded earthenware jugs. The snow suddenly gone now, and in cold sunshine, Deffeyes and I passed Grayback Mountain and then had the Great Salt Lake Desert before us—the dry bed of Bonneville—broader than the periphery of vision. The interstate runs close to but not parallel to the wagon trail, which trends a little more northwesterly. The wagon trail aims directly at Pilot Peak of the Pilot Range, which we could see clearly, upward of fifty miles away—a pyramidal summit with cloud coming off it in the wind like a banner unfurling. Across the dry lakebed, the emigrants homed on Pilot Peak, standing in what is now Nevada, above ten thousand feet. Along the fault scarp, at the base of Pilot Peak, are cold springs. When the emigrants arrived at the springs, their tongues were bloody and black.

"Imagine those poor sons of bitches out here with their animals, getting thirsty," Deffeyes said. "It's a wonder they didn't string the guy that invented this route up by his thumbs."

The flats for the most part were alkaline, a leather-colored mud superficially dry. Dig down two inches and it was damp and greasy. Come a little rain and an ox could go in to its knees. The emigrants made no intended stops on the Great Salt Lake Desert. They drove day and night for the Pilot Range. In the day, they saw mirages—towers and towns and shimmering lakes. Sometimes the lakes were real—playa lakes, temporary waters after a storm. Under a wind, playa lakes move like puddles of mercury in motion on a floor—two or three hundred square miles of water on the move, here today, there tomorrow, gone before long like a mirage, leaving wagons mired in unimagined mud. Very few emigrants chose to cross the Bonneville flats, although the route was promoted as a shortcut—"a nigher route"—rejoining the main migration four basins into Nevada. It was the invention of Lansford Hastings and was known as the Hastings Cutoff. Hastings wrote the helpful note in Skull Valley. His route was geologically unfavorable, but this escaped his knowledge and notice. His preoccupations were with politics. He wished to become President of California. He saw California—for the moment undefendably Mexican—as a new nation, under God, conceived at liberty and dedicated to the proposition that anything can be accomplished through promotion: President Lansford Hastings, in residence in a western White House. His strategy for achieving high office was to create a new shortcut on the way west, to promote both the route and the destination through recruiting and pamphleteering, to attract emigrants by the thousands year after year, and as their counsellor and deliverer to use them as constituent soldiers in the promised heaven. He camped beside the trail farther east. He attracted the Donners. He attracted Reeds, Kesebergs, Murphys, McCutchens, drew them southward away from the main trek and into the detentive scrub oak made fertile by the limestones of the Wasatch. The Donners were straight off the craton—solid and trusting, from Springfield, Illinois. Weeks were used hacking a path through the scrub oaks, which were living barbed wire. Equipment was abandoned on the Bonneville flats to lighten up loads in the race against thirst. Even in miles, the nigher route proved longer than the one it was shortcutting, on the way to a sierra that was named for snow.

Deffeyes and I passed graffiti on the Bonneville flats. There being nothing to carve in and no medium substantial enough for sprayed paint, the graffitists had lugged cobbles out onto the hard mud—stones as big as grapefruit, ballast from the interstate—and in large dotted letters had written their names: ROSS, DAWN, DON, JUDY, MARK, MOON, ERIC, fifty or sixty miles of names. YARD SALE. Eric's lithography was in basalt and dolomite, pieces of Grayback Mountain, apparently, pieces of the Stansburys. His name, if it sits there a century or so, will eventually explode. Salt will work into the stones along the grain boundaries. When this happens, water evaporates out of the salt, and salt crystals keep collecting and expanding until they explode the rock. In Death Valley are thousands of little heaps of crumbs that were once granite boulders. Salt exploded them. Salt gets into fence posts and explodes them at the base.

Near the far side of Utah, the flats turned blinding white, corn-snow white, and revolving winds were making devils out of salt. Over the whiteness you could see the salt go off the curve of the earth. When the drivers of jet cars move at Mach .9 over the Bonneville Salt Flats, they feel that they are always about to crest a hill. Dig into the salt and it turns out to be a crusty white veneer, like cake icing, more than an inch thick—an almost pure sodium chloride. Below it are a few inches of sand-size salt particles, and below them a sort of creamy yogurt mud that is the color of blond coffee. In much the manner in which these salts were left behind by the shrinking outline of the saline lake, there were times around the edges of North America when the shrinking ocean stranded bays that gradually dried up and left plains of salt. When the ocean came back, came up again, it spread inland over the salt, which was not so much dissolved as buried, under layers of sediment washing in from the continent. With the weight of more and more sediment, the layers of salt went deep. Salt has a low specific gravity and is very plastic. Pile eight thousand feet of sediment on it and it starts to move. Slowly, blobularly, it collects itself and moves. It shoves apart layers of rock. It mounds upon itself, and, breaking its way upward, rises in mushroom shape—a salt dome. Still rising into more shales and sandstones, it bends them into graceful arches and then bursts through them like a bullet shooting upward through a splintering floor. A plastic body moving like this is known as a diapir. The shape becomes a reverse teardrop. Generally, after the breakthrough, there will be some big layers of sandstone leaning on the salt dome like boards leaning up against a wall. The sandstone is permeable and probably has a layer of shale above it, which is not permeable. Any fluid in the sandstone will not only be trapped under the shale but will also be trapped by the impermeable salt. Enter the strange companionship of oil and salt. Oil also moves after it forms. You never find it where God put it. It moves great distances through permeable rock. Unless something traps it, it will move on upward until it reaches daylight and turns into tar. You don't run a limousine on tar, let alone a military-industrial complex. If, however, the oil moves upward through inclined sandstone and then hits a wall of salt, it stops, and stays—trapped. Run a little drill down the side of a salt dome and when you hit "sand" it may be full of oil. In the Gulf of Mexico were many of the bays that dried up covered with salt. Where the domes are now, there are towers in the Gulf. A number of salt domes are embedded in the Mississippi Delta, and have been mined. There are rooms inside them with ceilings a hundred feet high—room after room after room, like convention halls, with walls, floors, and ceilings of salt, above ninety-nine per cent pure.

Deffeyes was saying, "It's likely that in under this salt flat are mountain structures just as complicated as any of the ranges. They're just buried."

We picked up some shattered limestones and welded tuffs close by the Nevada state line. The tuff was hard, heavy, crystalline rock, freckled with feldspars and quartz. You would never dig a city out of that. The ranges now were anything but buried, and Pilot Peak reached above the shadowed basin and high into sunlight, a mile above its valleys. Soon we were climbing the Toano Range. "Here comes another roadcut," said Deffeyes near the summit. "You can feel them coming on. The Taconic Parkway would drive you nuts. I-80 gives you one when you're ready for it." What it gave in the Toanos was granite—not some sibling, son, or cousin but granite himself: sparkling black hornblendes evenly spaced through a snowy field of feldspars and quartz. It was of much the same age as the celebrated rock of the Sierra. Its presence here suggested that the great crustal meltings in the tectonic drama farther west put out enough heat even in eastern Nevada to cook up this batch of fresh granite.

In this manner we moved along from roadcut to roadcut, range to range, like barnyard poultry pecking up rock, seeing what the fault blocks had lifted from below. We crossed the Goshute Valley and went up into the Pequops into red Devonian shales, Devonian siltstones, Devonian limestones—a great many millions of years older than the granite, and from another world. These were marine rocks (by and large), full of crinoids and other marine fossils. Nothing about their appearance differed from sediment that might have collected over Illinois or Iowa in midcontinental, epicratonic seas. They provided not so much as a hint that they were actually from the continental shelf, that Pequop Summit is more or less where North America ended in Devonian time. The first attempt to move covered wagons directly across the continent to California ended at the Pequops, too. The wagons were abandoned at a spring by the eastern base of the mountains, a short hike off the interstate. Later emigrants made cooking fires with the wood of the wagons. Deffeyes was spitting out the siltstones but chewing happily on the shales.

The oolites of the Great Salt Lake were forming in the present. The dolomite of the Stansbury Mountains was almost five hundred million years old. The tuff had been welded for thirty million years. The age of the granite was a hundred million years. The rock of Pequop Summit was four times as old as that. On a scale of zero to five hundred, those samplings were bunched toward the extremes, with nothing representing the middle three hundred million years. That was just chance, though—just what the faults had happened to throw up—and farther down the road, at Golconda, would come a full-dress two-hundred-and-fifty-million-year-old Triassic show.

Geologists mention at times something they call the Picture. In an absolutely unidiomatic way, they have often said to me, "You don't get the Picture." The oolites and dolomite—tuff and granite, the Pequop siltstones and shales—are pieces of the Picture. The stories that go with them—the creatures and the chemistry, the motions of the crust, the paleoenvironmental scenes—may well, as stories, stand on their own, but all are fragments of the Picture.

The foremost problem with the Picture is that ninety-nine per cent of it is missing—melted or dissolved, torn down, washed away, broken to bits, to become something else in the Picture. The geologist discovers lingering remains, and connects them with dotted lines. The Picture is enhanced by filling in the lines—in many instances with stratigraphy: the rock types and ages of strata, the scenes at the times of deposition. The lines themselves to geologists represent structure—folds, faults, flat-lying planes. Ultimately, they will infer why, how, and when a structure came to be—for example, why, how, and when certain strata were folded—and that they call tectonics. Stratigraphy, structure, tectonics. "First you read ze Kafka," I overheard someone say once in a library elevator. "Ond zen you read ze Turgenev. Ond zen ond only zen are—you—ready—for—ze Tolstoy."

And when you have memorized Tolstoy, you may be ready to take on the Picture. Multidimensional, worldwide in scope and in motion through time, it is sometimes called the Big Picture. The Megapicture. You are cautioned not to worry if at first you do not wholly see it. Geologists don't see it, either. Not all of it. The modest ones will sometimes scuff a boot and describe themselves and their colleagues as scientific versions of the characters in John Godfrey Saxe's version of the Hindu fable of the blind men and the elephant. "We are blind men feeling the elephant," David Love, of the Geological Survey, has said to me at least fifty times. It is not unknown for a geological textbook to include snatches of the poem.

It was six men of Indostan
To learning much inclined,
Who went to see the Elephant
(Though all of them were blind),
That each by observation
Might satisfy his mind.

The first man of Indostan touches the animal's side and thinks it must be some sort of living wall. The second touches a tusk and thinks an elephant is like a spear. The others, in turn, touch the trunk, an ear, the tail, a knee—"snake," "fan," "rope," "tree."

And so these men of Indostan
Disputed loud and long,
Each in his own opinion
Exceeding stiff and strong,

Though each was partly in the right,
And all were in the wrong!

The blind men and the elephant are kept close at hand mainly to slow down what some graduate students refer to as "arm waving"—the delivery, with pumping elbows, of hypotheses so breathtakingly original that the science seems for the moment more imaginative than descriptive. Where it is solid, it is imaginative enough. Geologists are famous for picking up two or three bones and sketching an entire and previously unheard-of creature into a landscape long established in the Picture. They look at mud and see mountains, in mountains oceans, in oceans mountains to be. They go up to some rock and figure out a story, another rock, another story, and as the stories compile through time they connect—and long case histories are constructed and written from interpreted patterns of clues. This is detective work on a scale unimaginable to most detectives, with the notable exception of Sherlock Holmes, who was, with his discoveries and interpretations of little bits of grit from Blackheath or Hampstead, the first forensic geologist, acknowledged as such by geologists to this day. Holmes was a fiction, but he started a branch of a science; and the science, with careful inference, carries fact beyond the competence of invention. Geologists, in their all but closed conversation, inhabit scenes that no one ever saw, scenes of global sweep, gone and gone again, including seas, mountains, rivers, forests, and archipelagoes of aching beauty rising in volcanic violence to settle down quietly and then forever disappear—*almost* disappear. If some fragment has remained in the crust somewhere and something has lifted the fragment to view, the geologist in his tweed cap goes out with his hammer and his sandwich, his magnifying glass and his imagination, and rebuilds the archipelago.

I once dreamed about a great fire that broke out at night at Nasser Aftab's House of Carpets. In Aftab's showroom under the queen-post trusses were layer upon layer and pile after pile of shags and broadlooms, hooks and throws, para-Persians and polyesters. The intense and shrivelling heat consumed or melted most of what was there. The roof gave way. It was a night of cyclonic winds, stabs of unseasonal lightning. Flaming debris fell on the carpets. Layers of ash descended, alighted, swirled in the wind, and drifted. Molten polyester hardened on the cellar stairs. Almost simultaneously there occurred a major accident in the ice-cream factory next door. As yet no people had arrived. Dead of night. Distant city. And before long the west wall of the House of Carpets fell in under the pressure and weight of a broad, braided ooze of six admixing flavors, which slowly entered Nasser Aftab's showroom and folded and double-folded and covered what was left of his carpets, moving them, as well, some distance across the room. Snow began to fall. It turned to sleet, and soon to freezing rain. In heavy winds under clearing skies, the temperature fell to six below zero. Celsius. Representatives of two warring insurance companies showed up just in front of the fire engines. The insurance companies needed to know precisely what had happened, and in what order, and to what extent it was Aftab's fault. If not a hundred per cent, then to what extent was it the ice-cream factory's fault? And how much fault must be—regrettably—assigned to God? The problem was obviously too tough for the Chicken Valley Police Department, or, for that matter, for any ordinary detective. It was a problem, naturally, for a field geologist. One shuffled in eventually. Scratched-up boots. A puzzled look. He picked up bits of wall and ceiling, looked under the carpets, tasted the ice cream. He felt the risers of the cellar stairs. Looking up, he told Hartford everything it wanted to know. For him this was so simple it was a five-minute job.

From the high ridges right down to the level of the road, there was snow all over the Ruby Mountains. "Ugh," said Deffeyes—his comment on the snow.

"Spoken like a skier," I said.

He said, "I'm a retired skier."

He skied for the School of Mines. In other Rocky Mountain colleges and universities at the time, the best skiers in the United States were duly enrolled and trying to look scholarly and masquerading as amateurs to polish their credentials for the 1952 Olympic Games. Deffeyes was outclassed even on his own team, but there came a day when a great whiteout sent the superstars sprawling on the mountain. Deffeyes' turn for the slalom came late in the afternoon, and just as he was moving toward the gate the whiteout turned to alpenglow, suddenly bringing into focus the well-compacted snow. He shoved off, and was soon bombing. He was not hurting for weight even then. He went down the mountain like an object dropped from a tower. In the end, his time placed him high among the ranking stars.

Now, in the early evening, crossing Independence Valley, Deffeyes seemed scarcely to notice that the white summits of the Ruby Range—above eleven thousand feet, and the highest mountains in this part of the Great Basin—were themselves being reddened with alpenglow. He was musing aloud, for reasons unapparent to me, about the melting points of tin and lead. He was saying that as a general rule material will flow rather than fracture if it is hotter than half of its melting point measured from absolute zero. At room temperature, you can bend tin and lead. They are solid but they flow. Room temperature is more than halfway between absolute zero and the melting points of tin and lead. At room temperature, you cannot bend glass or cast iron. Room temperature is less than halfway from absolute zero to the melting points of iron and glass. "If you go down into the earth here to a depth that about equals the width of one of these fault blocks, the temperature is halfway between absolute zero and the melting point of the rock. The crust is brittle above that point and plastic below it. Where the brittleness ends is the bottom of the tilting fault block, which rests—floats, if you like—in the hot and plastic, slowly flowing lower crust and upper mantle. I think this is why the ranges are so rhythmic. The spacing between them seems to be governed by their depth—the depth of the cold brittle part of the crust. As you cross these valleys from one range to the next, you can sense how deep the blocks are. If they were a lot deeper than their width—if the temperature gradient were different and the cold brittle zone went down, say, five times the surface width—the blocks would not have mechanical freedom. They could not tilt enough to make these mountains. So I suspect the blocks are shallow—about as deep as they are wide. Earthquake history supports this. Only shallow earthquakes have been recorded in the Basin and Range. At the western edge of Death Valley, there are great convex mountain faces that are called turtlebacks. To me they are more suggestive of whales. You look at them and you see that they were once plastically deformed. I think the mountains have tilted up enough there to be giving us a peek at the original bottom of a block. Death Valley is below sea level. I would bet that if we could scrape away six thousand feet of gravel from these mile-high basins up here what we would see at the base of these mountains would look like the edge of Death Valley. I haven't published this hypothesis. I think it sounds right. I haven't done any field work in Death Valley. I was just lucky enough to be there in 1961 with the guy who first mapped the geology. I have been lucky all through the years to work in the Basin and Range. The Basin and Range impresses me in terms of geology as does no other place in North America. It's not at all easy, anywhere in the province, to say just what happened and when. Range after range—it is mysterious to me. A lot of geology is mysterious to me."

Interstate 80, in its complete traverse of the North American continent, goes through much open space and three tunnels. As it happens, one tunnel passes through young rock, another through middle-aged rock, and the third through rock that is fairly old, at least with respect to the rock now on earth which has not long since been recycled. At Green River, Wyoming, the road goes under a remnant of the bed of a good-sized Cenozoic lake. The tunnel through Yerba Buena Island, in San Francisco Bay, is in sandstones and shales of the Mesozoic. And in Carlin Canyon, in Nevada, the road makes a neat pair of holes in Paleozoic rock. This all but leaves the false impression that an academic geologist chose the sites—and now, as we approached the tunnel at Carlin Canyon, Deffeyes became so evidently excited that one might have thought he had done so himself. "Yewee zink bogawa!" he said as the pickup rounded a curve and the tunnel appeared in view. I glanced at him, and then followed his gaze to the slope above the tunnel, and failed to see there in the junipers and the rubble what it was that could cause this professor to break out in such language. He did not slow up. He had been here before. He drove through the westbound tube, came out into daylight, and, pointing to the right, said, "Shazam!" He stopped on the shoulder, and we admired the scene. The Humboldt River, blue and full, was flowing toward us, with panes of white ice at its edges, sage and green meadow beside it, and dry russet uplands rising behind. I said I thought that was lovely. He said yes, it was lovely indeed, it was one of the loveliest angular unconformities I was ever likely to see.

The river turned in our direction after bending by a wall of its canyon, and the wall had eroded so unevenly that a prominent remnant now stood on its own as a steep six-hundred-foot hill. It made a mammary silhouette against the sky. My mind worked its way through that image, but still I was not seeing what Deffeyes was seeing. Finally, I took it in. More junipers and rubble and minor creases of erosion had helped withhold the story from my eye. The hill, structurally, consisted of two distinct rock formations, awry to each other, awry to the gyroscope of the earth—just stuck together there like two artistic impulses in a pointedly haphazard collage. Both formations were of stratified rock, sedimentary rock, put down originally in and beside the sea, where they had lain, initially, flat. But now the strata of the upper part of the hill were dipping more than sixty degrees, and the strata of the lower part of the hill were standing almost straight up on end. It was as if, through an error in demolition, one urban building had collapsed upon another. In order to account for that hillside, Deffeyes was saying, you had to build a mountain range, destroy it, and then build a second set of mountains in the same place, and then for the most part destroy them. You would first have had the rock of the lower strata lying flat—a conglomerate with small bright pebbles like effervescent bubbles in a matrix red as wine. Then the forces that had compressed the region and produced mountains would have tilted the red conglomerate, not to the vertical, where it stood now, but to something like forty-five degrees. That mountain range wore away—from peaks to hills to nubbins and on down to nothing much but a horizontal line, the bevelled surface of slanting strata, eventually covered by a sea. In the water, the new sediment of the upper formation would have accumulated gradually upon that surface, and, later, the forces building a fresh mountain range would have shoved, lifted, and rotated the whole package to something close to its present position, with its lower strata nearly vertical and its upper strata aslant. Here in Carlin Canyon, basin-and-range faulting, when it eventually came along, had not much affected the local structure, further tilting the package only two or three degrees.

Clearly, if you were going to change a scene, and change it again and again, you would need adequate time. To make the rock of that lower formation and then tilt it up and wear it down and deposit sediment on it to form the rock above would require an immense quantity of time, an amount that was expressed in the clean, sharp line that divided the formations—the angular unconformity itself. You could place a finger on that line and touch forty million years. The lower formation, called Tonka, formed in middle Mississippian time. The upper formation, called Strathearn, was deposited forty million years afterward, in late Pennsylvanian time. Cambrian, Ordovician, Silurian, Devonian, Mississippian, Pennsylvanian, Permian, Triassic, Jurassic, Cretaceous, Paleocene, Eocene, Oligocene, Miocene, Pliocene, Pleistocene . . . In the long roll call of the geologic systems and series, those formations—those discrete depositional events, those forty million years—were next-door neighbors on the scale of time. The rock of the lower half of that hill dated to three hundred and forty million years ago, in the Mississippian, and the rock above the unconformity dated to three hundred million years ago, in the Pennsylvanian. If you were to lift your arms and spread them wide and hold them straight out to either side and think of the distance from fingertips to fingertips as representing the earth's entire history, then you would have all the principal events in that hillside in the middle of the palm of one hand.

It was an angular unconformity in Scotland—exposed in a riverbank at Jedburgh, near the border, exposed as well in a wave-scoured headland where the Lammermuir Hills intersect the North Sea—that helped to bring the history of the earth, as people had understood it, out of theological metaphor and into the perspectives of actual time. This happened toward the end of the eighteenth century, signalling a revolution that would be quieter, slower, and of another order than the ones that were contemporary in America and France. According to conventional wisdom at the time, the earth was between five thousand and six thousand years old. An Irish archbishop (James Ussher), counting generations in his favorite book, figured this out in the century before. Ussher actually dated the earth, saying that it was created in 4004 B.C., "upon the entrance of the night preceding the twenty-third day of October."

It was also conventional wisdom toward the end of the eighteenth century that sedimentary rock had been laid down in Noah's Flood. Marine fossils in mountains were creatures that had got there during the Flood. To be sure, not everyone had always believed this. Leonardo, for example, had noticed fossil clams in the Apennines and, taking into account the distance to the Adriatic Sea, had said, in effect, that it must have been a talented clam that could travel a hundred miles in forty days. Herodotus had seen the Nile Delta—and he had seen in its accumulation unguessable millennia. G. L. L. de Buffon, in 1749 (the year of *Tom Jones*), began publishing his forty-four-volume *Histoire Naturelle*, in which he said that the earth had emerged hot from the sun seventy-five thousand years before. There had been, in short, assorted versions of the Big Picture. But the scientific hypothesis that overwhelmingly prevailed at the time of Bunker Hill was neptunism—the aqueous origins of the visible world. Neptunism had become a systematized physiognomy of the earth, carried forward to the *n*th degree by a German academic mineralogist who published very little but whose teaching was so renowned that his interpretation of the earth was taught as received fact at Oxford and Cambridge, Turin and Leyden, Harvard, Princeton, and Yale. His name was Abraham Gottlob Werner. He taught at Freiberg Mining Academy. He had never been outside Saxony. Extrapolation was his means of world travel. He believed in "universal formations." The rock of Saxony was, beyond a doubt, by extension the rock of Peru. He believed that rock of every kind—all of what is now classified as igneous, sedimentary, and metamorphic—had precipitated out of solution in a globe-engulfing sea. Granite and serpentine, schist and gneiss had precipitated first and were thus "primitive" rocks, the cores and summits of mountains. "Transitional" rocks (slate, for example) had been deposited underwater on high mountain slopes in tilting beds. As the great sea fell and the mountains dried in the sun, "secondary" rocks (sandstone, coal, basalt, and more) were deposited flat in waters above the piedmont. And while the sea kept withdrawing, "alluvial" rock—the "tertiary," as it was sometimes called—was established on what now are coastal plains. That was the earth's surface as it was formed and had remained. There was no hint of where the water went. Werner was gifted with such rhetorical grace that he could successfully omit such details. He could gesture toward the Saxon hills—toward great pyramids of basalt that held castles in the air—and say, without immediate fear of contradiction, "I hold that no basalt is volcanic." He could dismiss volcanism itself as the surface effect of spontaneous combustion of coal. His ideas may now seem risible in direct proportion to their amazing circulation, but that is characteristic more often than not of the lurching progress of science. Those who laugh loudest laugh next. And some contemporary geologists discern in Werner the lineal antecedence of what has come to be known as black-box geology—people in white coats spending summer days in basements watching million-dollar consoles that flash like northern lights—for Werner's "first sketch of a classification of rocks shows by its meagreness how slender at that time was his practical acquaintance with rocks in the field." The words are Sir Archibald Geikie's, and they appeared in 1905 in a book called *The Founders of Geology*. Geikie, director general of the Geological Survey of Great Britain and Ireland, was an accomplished geologist who seems to have dipped in ink the sharp end of his hammer. In summary, he said of Werner, "Through the loyal devotion of his pupils, he was elevated even in his lifetime into the position of a kind of scientific pope, whose decisions were final on any subject regarding which he chose to pronounce them. . . . Tracing in the arrangement of the rocks of the earth's crust the history of an original oceanic envelope, finding in the masses of granite, gneiss, and mica-schist the earliest precipitations from that ocean, and recognising the successive alterations in the constitution of the water as witnessed by the series of geological formations, Werner launched upon the world a bold conception which might well fascinate many a listener to whom the laws of chemistry and physics, even as then understood, were but little known." Moreover, Werner's earth was compatible with Genesis and was thus not unpleasing to the Pope himself. When Werner's pupils, as they spread through the world, encountered reasoning that ran contrary to Werner's, pictures that failed to resemble his picture, they described all these heresies as "visionary fabrics"—including James Hutton's *Theory of the Earth; or, an Investigation of the Laws Observable in the Composition, Dissolution, and Restoration of Land*

Upon the Globe, which was first presented before the Royal Society of Edinburgh at its March and April meetings in 1785.

Hutton was a medical doctor who gave up medicine when he was twenty-four and became a farmer who at the age of forty-two retired from the farm. Wherever he had been, he had found himself drawn to riverbeds and cutbanks, ditches and borrow pits, coastal outcrops and upland cliffs; and if he saw black shining cherts in the white chalks of Norfolk, fossil clams in the Cheviot Hills, he wondered why they were there. He had become preoccupied with the operations of the earth, and he was beginning to discern a gradual and repetitive process measured out in dynamic cycles. Instead of attempting to imagine how the earth may have appeared at its vague and unobservable beginning, Hutton thought about the earth as it was; and what he did permit his imagination to do was to work its way from the present moment backward and forward through time. By studying rock as it existed, he thought he could see what it had once been and what it might become. He moved to Edinburgh, with its geologically dramatic setting, and lived below Arthur's Seat and the Salisbury Crags, remnants of what had once been molten rock. It was impossible to accept those battlement hills precipitating in a sea. Hutton had a small fortune, and did not have to distract himself for food. He increased his comfort when he invested in a company that made sal ammoniac from collected soot of the city. He performed experiments—in chemistry, mainly. He extracted table salt from a zeolite. But for the most part—over something like fifteen years—he concentrated his daily study on the building of his theory.

Growing barley on his farm in Berwickshire, he had perceived slow destruction watching streams carry soil to the sea. It occurred to him that if streams were to do that through enough time, there would be no land on which to farm. So there must be in the world a source of new soil. It would come from above—that was to say, from high terrain—and be made by rain and frost slowly reducing mountains, which in stages would be ground down from boulders to cobbles to pebbles to sand to silt to mud by a ridge-to-ocean system of dendritic streams. Rivers would carry their burden to the sea, but along the way they would set it down, as fertile plains. The Amazon had brought off the Andes half a continent of plains. Rivers, especially in flood, again and again would pick up the load, to give it up ultimately in depths of still water. There, in layers, the mud, silt, sand, and pebbles would pile up until they reached a depth where heat and pressure could cause them to become consolidated, fused, indurated, lithified—rock. The story could hardly end there. If it did, then the surface of the earth would have long since worn smooth and be some sort of global swamp. "Old continents are wearing away," he decided, "and new continents forming in the bottom of the sea." There were fossil marine creatures in high places. They had not got up there in a flood. Something had lifted the rock out of the sea and folded it up as mountains. One had only to ponder volcanoes and hot springs to sense that there was a great deal of heat within the earth—much exceeding what could ever be produced by an odd seam of spontaneously burning coal—and that not only could high heat soften up rock and change it into other forms of rock, it could apparently move whole regions of the crustal package and bend them and break them and elevate them far above the sea.

Granite also seemed to Hutton to be a product of great heat and in no sense a precipitate that somehow grew in water. Granite was not, in a sequential sense, primitive rock. It appeared to him to have come bursting upward in a hot fluid state to lift the country above it and to squirt itself thick and thin into preexisting formations. No one had so much as imagined this before. Basalt was no precipitate, either. In Hutton's description, it had once been molten, exhibiting "the liquefying power and expansive force of subterranean fire." Hutton's insight was phenomenal but not infallible. He saw marble as having once been lava, when in fact it is limestone cooked under pressure in place.

Item by item, as the picture coalesced, Hutton did not keep it entirely to himself. He routinely spent his evenings in conversation with friends, among them Joseph Black, the chemist, whose responses may have served as a sort of fixed foot to the wide-swinging arcs of Hutton's speculations—about the probable effect on certain materials of varying ratios of temperature and pressure, about the story of the forming of rock. Hutton was an impulsive, highly creative thinker. Black was deliberate and critical. Black had a judgmental look, a lean and somber look. Hutton had dark eyes that flashed with humor under a far-gone hairline and an oolitic forehead full of stored information. Black is regarded as the discoverer of carbon dioxide. He is one of the great figures in the history of chemistry. Hutton and Black were among the founders of an institution called the Oyster Club, where they whiled away an evening a week with their preferred companions—Adam Smith, David Hume, John Playfair, John Clerk, Robert Adam, Adam Ferguson, and, when they were in town, visitors from near and far such as James Watt and Benjamin Franklin. Franklin called these people "a set of as truly great men . . . as have ever appeared in any Age or Country." The period has since been described as the Scottish Enlightenment, but for the moment it was only described as the Oyster Club. Hutton, who drank nothing, was a veritable cup running over with enthusiasm for the achievements of his friends. When Watt came to town to report distinct progress with his steam engine, Hutton reacted with so much pleasure that one might have thought he was building the thing himself. While the others busied themselves with their economics, their architecture, art, mathematics, and physics, their naval tactics and ranging philosophies, Hutton shared with them the developing fragments of his picture of the earth, which, in years to come, would gradually remove the human world from a specious position in time in much the way that Copernicus had removed us from a specious position in the universe.

A century after Hutton, a historian would note that "the direct antagonism between science and theology which appeared in Catholicism at the time of the discoveries of Copernicus and Galileo was not seriously felt in Protestantism till geologists began to impugn the Mosaic account of the creation." The date of the effective beginning of the antagonism was the seventh of March, 1785, when Hutton's theory was addressed to the Royal Society in a reading that in all likelihood began with these words: "The purpose of this Dissertation is to form some estimate with regard to the time the globe of this Earth has existed." The presentation was more or less off the cuff, and ten years would pass before the theory would appear (at great length) in book form. Meanwhile, the Society required that Hutton get together a synopsis of what was read on March 7th and finished on April 4, 1785. The present quotations are from that abstract.

We find reason to conclude, *1st*, That the land on which we rest is not simple and original, but that it is a composition, and had been formed by the operation of second causes. *2dly*, That before the present land was made there had subsisted a world composed of sea and land, in which were tides and currents, with such operations at the bottom of the sea as now take place. And, *Lastly*, That while the present land was forming at the bottom of the ocean, the former land maintained plants and animals . . . in a similar manner as it is at present. Hence we are led to conclude that the greater part of our land, if not the whole, had been produced by operations natural to this globe; but that in order to make this land a permanent body resisting the operations of the waters two things had been required; *1st*, The consolidation of masses formed by collections of loose or incoherent materials; *2dly*, The elevation of those consolidated masses from the bottom of the sea, the place where they were collected, to the stations in which they now remain above the level of the ocean. . . .

Having found strata consolidated with every species of substance, it is concluded that strata in general have not been consolidated by means of aqueous solution. . . .

It is supposed that the same power of extreme heat by which every different mineral substance had been brought into a melted state might be capable of producing an expansive force sufficient for elevating the land from the bottom of the ocean to the place it now occupies above the surface of the sea. . . .

A theory is thus formed with regard to a mineral system. In this system, hard and solid bodies are to be formed from soft bodies, from loose or incoherent materials, collected together at the bottom of the sea; and the bottom of the ocean is to be made to change its place . . . to be formed into land. . . .

Having thus ascertained a regular system in which the present land of the globe had been first formed at the bottom of the ocean and then raised above the surface of the sea, a question naturally occurs with regard to time; what had been the space of time necessary for accomplishing this great work? . . .

We shall be warranted in drawing the following conclusions; *1st*, That it had required an indefinite space of time to have produced the land which now appears; *2dly*, That an equal space had been employed upon the construction of that former land from whence the materials of the present came; *Lastly*, That there is presently laying at the bottom of the ocean the foundation of future land. . . .

As things appear from the perspective of the twentieth century, James Hutton in those readings became the founder of modern geology. As things appeared to Hutton at the time, he had constructed a theory that to him made eminent sense, he had put himself on the line by agreeing to confide it to the world at large, he had provoked not a few hornets into flight, and now—like the experimental physicists who would one day go off to check on Einstein by photographing the edges of solar eclipses—he had best do some additional travelling to see if he was right. As he would express all this in a chapter heading when he ultimately wrote his book, he needed to see his "Theory confirmed from Observations made on purpose to elucidate the Subject." He went to Galloway. He went to Banffshire. He went to Saltcoats, Skelmorlie, Rumbling Bridge. He went to the Isle of Arran, the Isle of Man, Inchkeith Island in the Firth of Forth. His friend John Clerk sometimes went with him and made line drawings and watercolors of scenes that arrested Hutton's attention. In 1968, a John Clerk with a name too old for Roman numerals found a leather portfolio at his Midlothian estate containing seventy of those drawings, among them some cross sections of mountains with granite cores. Since it was Hutton's idea that granite was not a "primary" rock but something that had come up into Scotland from below, molten, to intrude itself into the existing schist, there ought to be pieces of schist embedded here and there in the granite. There were. "We may now conclude," Hutton wrote later, "that without seeing granite actually in a fluid state we have every demonstration possible of this fact; that is to say, of granite having been forced to flow in a state of fusion among the strata broken by a subterraneous force, and distorted in every manner and degree."

What called most for demonstration was Hutton's essentially novel and all but incomprehensible sense of time. In 4004 + 1785 years, you would scarcely find the time to make a Ben Nevis, let alone a Gibraltar or the domes of Wales. Hutton had seen Hadrian's Wall running across moor and fen after sixteen hundred winters in Northumberland. Not a great deal had happened to it. The geologic process was evidently slow. To accommodate his theory, all that was required was time, adequate time, time in quantities no mind had yet conceived; and what Hutton needed now was a statement in rock, a graphic example, a breath-stopping view of deep time. There was a formation of "schistus" running through southern Scotland in general propinquity to another formation called Old Red Sandstone. The schistus had obviously been pushed around, and the sandstone was essentially flat. If one could see, somewhere, the two formations touching each other with strata awry, one could not help but see that below the disassembling world lie the ruins of a disassembled world below which lie the ruins of still another world. Having figured out inductively what would one day be called an angular unconformity, Hutton went out to look for one. In a damp country covered with heather, with gorse and bracken, with larches and pines, textbook examples of exposed rock were extremely hard to find. As Hutton would write later, in the prototypical lament of the field geologist, "To a naturalist nothing is indifferent; the humble moss that creeps upon the stone is equally interesting as the lofty pine which so beautifully adorns the valley or the mountain: but to a naturalist who is reading in the face of rocks the annals of a former world, the mossy covering which obstructs his view, and renders undistinguishable the different species of stone, is no less than a serious subject of regret." Hutton's perseverance, though, was more than equal to the irksome vegetation. Near Jedburgh, in the border country, he found his first very good example of an angular unconformity. He was roaming about the region on a visit to a friend when he came upon a stream cutbank where high water had laid bare the flat-lying sandstone and, below it, beds of schistus that were standing straight on end. His friend John Clerk later went out and sketched for Hutton this clear conjunction of three worlds—the oldest at the bottom, its remains tilted upward, the intermediate one a flat collection of indurated sand, and the youngest a landscape full of fences and trees with a phaeton-and-two on a road above the rivercut, driver whipping the steeds, rushing through a moment in the there and then. "I was soon satisfied with regard to this phenomenon," Hutton wrote later, "and rejoiced at my good fortune in stumbling upon an object so interesting to the natural history of the earth, and which I had been long looking for in vain."

What was of interest to the natural history of the earth was that, for all the time they represented, these two unconforming formations, these two levels of history, were neighboring steps on a ladder of uncountable rungs. Alive in a world that thought of itself as six thousand years old, a society which had placed in that number the outer limits of its grasp of time, Hutton had no way of knowing that there were seventy million years just in the line that separated the two kinds of rock, and many millions more in the story of each formation—but he sensed something like it, sensed the awesome truth, and as he stood there staring at the riverbank he was seeing it for all humankind.

To confirm what he had observed and to involve further witnesses, he got into a boat the following spring and went along the coast of Berwickshire with John Playfair and young James Hall, of Dunglass. Hutton had surmised from the regional geology that they would come to a place among the terminal cliffs of the Lammermuir Hills where the same formations would touch. They touched, as it turned out, in a headland called Siccar Point, where the strata of the lower formation had been upturned to become vertical columns, on which rested the Old Red Sandstone, like the top of a weather-beaten table. Hutton, when he eventually described the scene, was both gratified and succinct—"a beautiful picture . . . washed bare by the sea." Playfair was lyrical:

> On us who saw these phenomena for the first time, the impression made will not easily be forgotten. The palpable evidence presented to us, of one of the most extraordinary and important facts in the natural history of the earth, gave a reality and substance to those theoretical speculations, which, however probable, had never till now been directly authenticated by the testimony of the senses. We often said to ourselves, What clearer evidence could we have had of the different formation of these rocks, and of the long interval which separated their formation, had we actually seen them emerging from the bosom of the deep? We felt ourselves necessarily carried back to the time when the schistus on which we stood was yet at the bottom of the sea, and when the sandstone before us was only beginning to be deposited, in the shape of sand or mud, from the waters of a superincumbent ocean. An epocha still more remote presented itself, when even the most ancient of these rocks, instead of standing upright in vertical beds, lay in horizontal planes at the bottom of the sea, and was not yet disturbed

by that immeasurable force which has burst asunder the solid pavement of the globe. Revolutions still more remote appeared in the distance of this extraordinary perspective. The mind seemed to grow giddy by looking so far into the abyss of time.

Hutton had told the Royal Society that it was his purpose to "form some estimate with regard to the time the globe of this Earth has existed." But after Jedburgh and Siccar Point what estimate could there be? "The world which we inhabit is composed of the materials not of the earth which was the immediate predecessor of the present but of the earth which . . . had preceded the land that was above the surface of the sea while our present land was yet beneath the water of the ocean," he wrote. "Here are three distinct successive periods of existence, and each of these is, in our measurement of time, a thing of indefinite duration. . . . The result, therefore, of this physical inquiry is, that we find no vestige of a beginning, no prospect of an end."

The Old Red Sandstone was put down by rivers flowing southward to a sea where marine strata were accumulating in the region that is now called Devon. The size, speed, and direction of the rivers—their islands, pitches, and bends—are not just inferable but can almost be seen, in structures in the Old Red Sandstone: gravel bars, point bars, ripples of the riverbeds, migrating channels, "waves" that formed of sand. The sea into which those rivers spilled ran all the way to Russia, but it was in the rock of Devonshire that geologists in the eighteen-thirties found cup corals—fossilized skeletons, cornucopian in shape—that were not of an age with corals they had found before. They had found related corals that were obviously less developed than these, and they had found corals that were more so. The less developed corals had been in rock that lay under the Old Red Sandstone. The more developed corals had been in rock above the Old Red Sandstone. Therefore, it was inferred (correctly) that the Old Red Sandstone of North Britain and the marine limestone of Devon were of the same age, and that henceforth any rock of that age anywhere in the world—in downtown Iowa City; on Pequop Summit, in Nevada; in Stroudsburg, Pennsylvania; in Sandusky, Ohio—would be called Devonian. It was a name given, although they did not know it then, to forty-six million years. They still had no means of measuring the time involved. They also had no way of knowing that those forty-six million years had ended a third of a billion years ago. All they had was their new and expanding insight that they were dealing with time in quantities beyond comprehension. Devonian—408 to 362 million years before the present.

Geologists did not have to look long at the coal seams of Europe—the coals of the Ruhr, the coals of the Tyne—to decide that the coals were of an age, which they labelled Carboniferous. The coal and related strata lay on top of the Old Red Sandstone. So, in the succession of time, the Carboniferous period (eventually subdivided into Mississippian and Pennsylvanian in the United States) would follow the Devonian, coupling on, as the science would eventually determine, another seventy-two million years—362 to 290 million years before the present.

In this manner—with their fossil assemblages and faunal successions, their hammers decoding rock—geologists in the first eighty years of the nineteenth century constructed their scale of time. It was based on the irreversible history of life. Crossing the century, it both anticipated and confirmed Darwin. When the Devonian was defined in the light of the changes in corals, Darwin was obscure and not long off the Beagle, with twenty years to go before *The Origin of Species*. Meanwhile, the geologists were out correlating strata and reading there a record less of rock than of life. The rock had been recycled, and sandstones of one era could be indistinguishable from the sandstones of another, but evolution had not occurred in cycles, so it was through the antiquity of fossils that geologists worked out the comparative ages of the rock in which the fossils were preserved. Some creatures were more useful than others. Oysters and horseshoe crabs, for example, were of marginal assistance. Oysters had appeared in the Triassic, horseshoe crabs in the Cambrian. Both had evolved minimally and had obviously avoided extinction. Some creatures, on the other hand, had appeared suddenly, had evolved quickly, had become both abundant and geographically widespread, and then had died out, or died down, abruptly. Geologists canonized them as "index fossils" and studied them in groups. Experience proved that the surest method of working out relative ages of rock was not through individual creatures but through the relating of successive strata to whole collections of creatures whose fossils were contained therein—a painstaking comparison of arrivals and extinctions that helped to characterize the divisions of the time scale and define its boundaries with precision.

Imagine an E. L. Doctorow novel in which Alfred Tennyson, William Tweed, Abner Doubleday, Jim Bridger, and Martha Jane Canary sit down to a dinner cooked by Rutherford B. Hayes. Geologists would call that a fossil assemblage. And, without further assistance from Doctorow, a geologist could quickly decide—as could anyone else—that the dinner must have occurred in the middle eighteen-seventies, because Canary was eighteen when the decade began, Tweed became extinct in 1878, and the biographies of the others do not argue with these limits. In progressive refinements, geologists with their fossil assemblages established their systems and series and stages of rock, their eras and periods and epochs of time. But, unlike Doctorow, who deals with a mere half-dozen people around a dinner table, the geologists would assemble from one set of strata hundreds and even thousands of species from all over the food chain, and by lining up their genetic histories side by side establish with near-certainty points in comparative time.

Some of these time lines were bolder than others, and none more so than the one that underlined the first appearance of megascopic fossils in abundance in the world. It marked a great and sudden explosion of life, all the major phyla having developed more or less at the same time and now acquiring skeletons and shells and teeth and other hard components that allowed them individually to be reported to the future. Because rock that held these early fossils was first studied on Harlech Dome and adjacent Welsh terrains, geologists named the system Cambrian, after the Roman name for Wales. They then named the Silurian for a Welsh tribe that bitterly defied the Romans. After some years and more comparative study, an argument broke out over the Cambro-Silurian line, a scientific battle royal in which the Cambrian forces tried to move their banner forward through time and the Silurian proponents attempted to push theirs back. The disputed block of time became a sort of demilitarized zone. Friendships came unstuck. The standoff lasted for decades, until some genius in scientific diplomacy suggested that the disputed time had enough characteristics of its own to be given the status of a discrete period, an appropriate name for which—in honor of another tribe of intractable Welsh belligerents—would be Ordovician. There was a lot of room for generosity. There was plenty of time for all. Cambrian—544 to 490. Ordovician—490 to 439. Silurian—439 to 408 million years before the present.

A British geologist went to Russia and after a season or two's tapping at the Urals named still another period in time, and system of rock, for the upland oblast of Perm. There were formations in Perm with a fossil story distinctly their own that were superimposed—as they happen to be in Pennsylvania, as they happen to be at the rim of the Grand Canyon—upon the Carboniferous. What was distinct about the character of the Permian assemblages was not only the forms to which they had evolved but also their absence in great numbers from higher, younger strata. There had evidently been a wave of death, in which thousands of species had vanished from the world. No one has explained what happened—at least not to the general satisfaction. A drastic retreat of shallow seas may have destroyed innumerable environments. The cause may have been extraterrestrial—lethal radiation from a supernova dying nearby. The wave of death occurred 250.1 million years before the present, and exactly that long ago flood basalts emerged in Siberia and quickly covered about a million and a half square kilometres with incandescent lava. The brief, intense greenhouse effect, the surge of carbon-dioxide emissions, would have stopped the upwelling of the oceans and the associated growth of nutrients. None of these hypotheses has attracted enough concurrence to be dressed out in full as a theory, but, whatever the cause, no one argues that at least half the fish and invertebrates and three-quarters of all amphibians—perhaps as much as ninety-six per cent of all marine faunal species—disappeared from the world in what has come to be known as the Permian Extinction.

It was an extinction of a magnitude that would be approached only once in subsequent history, or—to express that more gravely

—only once before the present day. The sharp line of creation at the outset of the Cambrian had an antiphonal parallel in the Permian Extinction, and the whole long stretch between the one and the other was set apart in history as the Paleozoic era. It was a unit—well below the surface but far above the bottom—just hanging there suspended in the formless pelagics of time. The Paleozoic—544 to 250 million years before the present, a fifteenth of the history of the earth. Cambrian, Ordovician, Silurian, Devonian, Mississippian, Pennsylvanian, Permian. When I was seventeen, I used to accordion-pleat those words, mnemonically capturing the vanished worlds of "Cosdmpp," the order of the periods, the sequence of the systems. It was either that or write them in the palm of one hand.

Lyell, Cuvier, Conybeare, Phillips, von Alberti, von Humboldt, Desnoyers, d'Halloy, Sedgwick, Murchison, Lapworth, Smith (William "Strata" Smith): the geologists who extended Hutton's insight and built this time scale conjoined their names in the history of the science in a way that would not be repeated for more than a hundred years, until a roster of comparable length—Hess, Heezen, McKenzie, Morgan, Wilson, Matthews, Vine, Parker, Sykes, Ewing, Le Pichon, Cox, Menard—would effect the plate-tectonics revolution. The system of rock immediately above the Paleozoic, in which all that Permian life failed to reappear, was typified by three formations in Germany—certain sandstones, limestones, and marly shales—that ran like a striped flag through the Black Forest, the Rhine Valley, and lent the name Triassic to forty-two million years. In the Triassic, the earliest subdivision of the Mesozoic era, two families of reptiles that had survived the Permian Extinction began to show patterns of unprecedented growth. This would continue for a hundred and fifty million years—through the Jurassic and out to the end of Cretaceous time, when the "fearfully great lizards," on the point of disappearance, would reach their greatest size, not to be surpassed until epochs that followed the Eocene development of whales. European geologists studying the massive limestones of the Jura—the gentle mountains of the western cantons of Switzerland and of Franche-Comté—related the copious displays of ancient life there to comparable assemblages elsewhere in the world, and called them all Jurassic. A primordial bird appeared in the Jurassic. It had claws on its wings and teeth in its bill and a reptile's long tail sprouting feathers. Its complete performance envelope as a flier was to climb a tree and jump.

Physicists, chemists, and mathematicians, taking note of all the nomenclatural inconsistencies—of time named for mountain ranges, time named for savage tribes, time named for a country here, a county there, an oblast in the Urals—have politely, gently, suggested that, in this one sense only, the time scale seems archaic, seems, if one may say so, out of date. Geology might be better served by a straightforward system of numbers. The reaction of geologists, by and large, has been to look upon this suggestion as if it had come over a bridge that exists between two cultures. A Continental geologist, in 1822, named eighty million years for the white cliffs of Dover, for the downs of Kent and Sussex, for the chalky ground of Cognac and Champagne. Related strata were spread out through Holland, Sweden, Denmark, Germany, and Poland. He called it Le Terrain Crétacé. If that name was apt, his own was irresistible. He was J. J. d'Omalius d'Halloy. Triassic, Jurassic, Cretaceous. When the Cretaceous ended, the big marine reptiles had disappeared, the flying reptiles, the dinosaurs, the rudistid clams, and many species of fish, not to mention the total elimination or severe reduction of countless smaller species from the sea. At the same point in geologic time, the flood basalts now known as the Deccan Traps came out of the mantle and quickly covered at least a million square kilometres in India, effectively stopping the upwelling of the ocean. An ocean gone stagnant would kill phytoplankton, which prosper in the currents of mixed-up seas. Break the food chain and creatures die out above the break. Phytoplankton are the base of the food chain. The Arctic Ocean, surrounded by continents that had drifted together, might have become in the Cretaceous the greatest lake in all eternity, and when the North Atlantic opened up enough to let the water flood the southern seas the life in them would have suffered a cold osmotic shock. Drastic fluctuations of sea level—also related, perhaps, to the separation of continents—might have caused changes in air temperature and ocean circulation that were enough to sunder the food chain. At the end of 1979, a small group at the Lawrence Radiation Laboratory, in Berkeley—among them the physicist Luis Alvarez, winner of a Nobel Prize, and his son Walter, who is a geologist—brought forth a piece of science in which they presented the catastrophe as the effect of an Apollo Object colliding with the earth. An Apollo Object is an "earth-orbit-crossing" asteroid that is at least a kilometre in diameter and is in the category of asteroids that have pockmarked the surface of Mercury, Mars, and the moon, and the surface of the earth as well, although most of the evidence has been obscured here by erosion. Like the general run of meteorites, an Apollo Object could be expected to contain a percentage of iridium and other platinum-like metals at least a thousand times greater than the concentration of the same metals in the crust of the earth. In widely separated parts of the world—Italy, Denmark, New Zealand—the Berkeley researchers found a thin depositional band, often just a centimetre thick, that contains unearthly concentrations of iridium. Below that sharp line are abundant Cretaceous fossils, and above it they are gone. It marks precisely the end of Cretaceous time. The Berkeley calculations suggested an asteroid about six miles in diameter hitting the earth with a punch of a hundred million megatons, making a crater a hundred miles wide. Such an occurrence—which could repeat itself tomorrow afternoon, there being several hundred big asteroids out there in threatening orbits —would have sent up a mushroom cloud containing some thirty thousand cubic kilometres of pulverized asteroid and terrestrial crust, part of which would have gone into the stratosphere and spread quickly over the earth, keeping sunlight off the lands and seas and suppressing photosynthesis. A decade after the publication of the Berkeley hypothesis, Chicxulub Crater was discovered, buried five hundred metres under Yucatán. Evidently made by an Apollo Object, it is a hundred and ten miles wide. On August 26 and 27, 1883, when the island Krakatoa, in the Sunda Strait, exploded with great violence, it sent less than twenty cubic kilometres of material into the air, but within a few days dust had spread above the whole earth, turning daylight into dusk. It made exceptionally brilliant sunsets for two and a half years. Edmund Halley, who died when James Hutton was fifteen, once wrote a paper suggesting that the way God started Noah's Flood was by directing a big comet into collision with the earth. The Cretaceous Extinction, whatever its cause, was one of the two most awesome annihilations of life in the history of the world. With the Permian Extinction before it, it framed the Mesozoic, an era of burgeoning creation within deadly brackets of time.

For establishing our bearings through time, we obviously owe an incalculable debt to vanished and endangered species, and if the condor, the kit fox, the human being, the black-footed ferret, and the three-toed sloth are at the head of the line to go next, there is less cause for dismay than for placid acceptance of the march of prodigious tradition. The opossum may be Cretaceous, certain clams Devonian, and oysters Triassic, but for each and every oyster in the sea, it seems, there is a species gone forever. Be a possum is the message, and you may outlive God. The Cenozoic era—coming just after the Cretaceous Extinction, and extending as it does to the latest tick of time—was subdivided in the eighteen-thirties according to percentages of molluscan species that have survived into the present. From the Eocene, for example, which ended some thirty-five million years ago, roughly three and a half per cent have survived. Eocene means "dawn of the recent." The first horse appeared in the Eocene. Looking something like a toy collie, it stood three hands high. From the Miocene ("moderately recent"), some fifteen per cent of molluscan species survive; from the Pliocene ("more recent"), the number approaches half. As creatures go, mollusks have been particularly hardy. Many species of mammals fell in the Pliocene as prairie grassland turned to tundra and ice advanced from the north. From the Pleistocene ("most recent"), more than ninety per cent of molluscan species live on. The Pleistocene has also been traditionally defined by four great glacial pulsations, spread across a million years—the Nebraskan ice sheet, the Kansan ice sheet, the Illinoian and Wisconsinan ice sheets. It now appears that these were the last of many glacial pulsations that have occurred in relatively recent epochs, beginning probably in the Miocene and reaching a climax in the ice sheets of Pleistocene time. The names of the Cenozoic epochs were proposed by Charles Lyell, whose *Principles of Geology* was the standard text through much of the nineteenth century. To settle problems here and there, the Oligocene ("but a little recent") was inserted in the list, and the Paleocene ("old recent") was sliced off the beginning. Paleocene, Eocene, Oligocene, Miocene, Pliocene, Pleistocene—sixty-five million to ten thousand years before the present. Divisions grew shorter in the Cenozoic—the epochs range from twenty-one million years to less than two million—because so much remains on earth of Cenozoic worlds.

Ignoring its geology, I guess I don't know a paragraph in literature that I prefer to the one Joseph Conrad begins by saying, "Going up that river was like travelling back to the earliest beginnings of the world, when vegetation rioted on the earth and the big trees were kings." He says, moments later, "This stillness of life did not in the least resemble a peace. It was the stillness of an implacable force brooding over an inscrutable intention. It looked at you with a vengeful aspect. I got used to it afterwards; I did not see it anymore; I had no time. I had to keep guessing at the channel; I had to discern, mostly by inspiration, the signs of hidden banks; I watched for sunken stones." Metaphorically, he travelled back to the Carboniferous, when the vegetal riot occurred, but scarcely was that the beginning of the world. The first plants to appear on land, ever, appeared in the Silurian. Through the Ordovician and the Cambrian, there had been no terrestrial vegetation at all. And in the deep shadow below the Cambrian were seven years for every one in all subsequent time. There were four billion years back there—since the earliest beginnings of the world. There were scant to nonexistent fossils. There were the cores of the cratons, the rock of the continental shields, the rock of the surface of the moon. There were the reefs of the Witwatersrand. There was the rock that would become the Adirondack Mountains, the Wind River summits, the Seward Peninsula, Manhattan Island. But so little is known of this seven-eighths of all history that in a typical two-pound geological textbook there are fourteen pages on Precambrian time. The Precambrian has attracted geologists of exceptional imagination, who see families of mountains in folded schists. Uranium-lead, rubidium-strontium, and potassium-argon radiometric dating have helped them to sort out their Kenoran, Hudsonian, Grenvillean orogenies, their Aphebian, Hadrynian, Paleohelikian time. Isolating the first two billion years of the life of the earth, they called it the Archean Eon. In the middle Archean, photosynthesis began. Much later in the Precambrian, somewhere in Helikian or Hadrynian time, aerobic life appeared. There is no younger rock in the United States than the travertine that is forming in Thermopolis, Wyoming. A 2.7-billion-year-old outcrop of the core of the continent is at the head of Wind River Canyon, twenty miles away. Precambrian—4,560 to 544 million years before the present.

At the other end of the scale is the Holocene, the past ten

thousand years, also called the Recent—Cro-Magnon brooding beside the melting ice. (The Primitive and Secondary eras of eighteenth-century geology are long since gone from the vocabulary, but oddly enough the Tertiary remains. The term, which is in general use, embraces nearly all of the Cenozoic, from the Cretaceous Extinction to the end of the Pliocene, while the relatively short time that follows—the Pleistocene plus the Holocene—has come to be called the Quaternary. The moraines left by ice sheets are Quaternary, as are the uppermost basin fillings in the Basin and Range.) It was at some moment in the Pleistocene that humanity crossed what the geologist-theologian Pierre Teilhard de Chardin called the Threshold of Reflection, when something in people "turned back on itself and so to speak took an infinite leap forward. Outwardly, almost nothing in the organs had changed. But in depth, a great revolution had taken place: consciousness was now leaping and boiling in a space of super-sensory relationships and representations; and simultaneously consciousness was capable of perceiving itself in the concentrated simplicity of its faculties. And all this happened for the first time." Friars of another sort—evangelists of the environmental movement—have often made use of the geologic time scale to place in perspective that great "leap forward" and to suggest what our reflective capacities may have meant to Mother Earth. David Brower, for example, the founder of Friends of the Earth and emeritus hero of the Sierra Club, has tirelessly travelled the United States delivering what he himself refers to as "the sermon," and sooner or later in every talk he invites his listeners to consider the six days of Genesis as a figure of speech for what has in fact been four and a half billion years. In this adjustment, a day equals something like seven hundred and fifty million years, and thus "all day Monday and until Tuesday noon creation was busy getting the earth going." Life began Tuesday noon, and "the beautiful, organic wholeness of it" developed over the next four days. "At 4 P.M. Saturday, the big reptiles came on. Five hours later, when the redwoods appeared, there were no more big reptiles. At three minutes before midnight, the human race appeared. At one-fourth of a second before midnight, Christ arrived. At one-fortieth of a second before midnight, the Industrial Revolution began. We are surrounded with people who think that what we have been doing for that one-fortieth of a second can go on indefinitely. They are considered normal, but they are stark raving mad." Brower holds up a photograph of the world —blue, green, and swirling white. "This is the sudden insight from Apollo," he says. "There it is. That's all. We see through the eyes of the astronauts how fragile our life really is." Brower has computed that we are driving through the earth's resources at a rate comparable to an automobile going a hundred and twenty-eight miles an hour—and he says that we are accelerating.

In like manner, geologists will sometimes use the calendar year as a unit to represent the time scale, and in such terms the Precambrian runs from New Year's Day until well after Halloween. Dinosaurs appear in the middle of December and are gone the day after Christmas. The last ice sheet melts on December 31st at one minute before midnight, and the Roman Empire lasts five seconds. With your arms spread wide again to represent all time on earth, look at one hand with its line of life. The Cambrian begins in the wrist, and the Permian Extinction is at the outer end of the palm. All of the Cenozoic is in a fingerprint, and in a single stroke with a medium-grained nail file you could eradicate human history. Geologists live with the geologic scale. Individually, they may or may not be alarmed by the rate of exploitation of the things they discover, but, like the environmentalists, they use these repetitive analogies to place the human record in perspective—to see the Age of Reflection, the last few thousand years, as a small bright sparkle at the end of time. They often liken humanity's presence on earth to a brief visitation from elsewhere in space, its luminous, explosive characteristics consisting not merely of the burst of population in the twentieth century but of the whole residence of people on earth—a single detonation, resembling nothing so much as a nuclear implosion with its successive neutron generations, whole generations following one another once every hundred-millionth of a second, temperatures building up into the millions of degrees and stripping atoms until bare nuclei are wandering in electron seas, pressures building up to a hundred million atmospheres, the core expanding at five million miles an hour, expanding in a way that is quite different from all else in the universe, unless there are others who also make bombs.

The human consciousness may have begun to leap and boil some sunny day in the Pleistocene, but the race by and large has retained the essence of its animal sense of time. People think in five generations—two ahead, two behind—with heavy concentration on the one in the middle. Possibly that is tragic, and possibly there is no choice. The human mind may not have evolved enough to be able to comprehend deep time. It may only be able to measure it. At least, that is what geologists wonder sometimes, and they have imparted the questions to me. They wonder to what extent they truly sense the passage of millions of years. They wonder to what extent it is possible to absorb a set of facts and move with them, in a sensory manner, beyond the recording intellect and into the abyssal eons. Primordial inhibition may stand in the way. On the geologic time scale, a human lifetime is reduced to a brevity that is too inhibiting to think about. The mind blocks the information. Geologists, dealing always with deep time, find that it seeps into their beings and affects them in various ways. They see the unbelievable swiftness with which one evolving species on the earth has learned to reach into the dirt of some tropical island and fling 747s into the sky. They see the thin band in which are the all but indiscernible stratifications of Cro-Magnon, Moses, Leonardo, and now. Seeing a race unaware of its own instantaneousness in time, they can reel off all the species that have come and gone, with emphasis on those that have specialized themselves to death.

In geologists' own lives, the least effect of time is that they think in two languages, function on two different scales.

"You care less about civilization. Half of me gets upset with civilization. The other half does not get upset. I shrug and think, So let the cockroaches take over."

"Mammalian species last, typically, two million years. We've about used up ours. Every time Leakey finds something older, I say, 'Oh! We're overdue.' We will be handing the dominant-species-on-earth position to some other group. We'll have to be clever not to."

"A sense of geologic time is the most important thing to suggest to the nongeologist: the slow rate of geologic processes, centimetres per year, with huge effects, if continued for enough years."

"A million years is a short time—the shortest worth messing with for most problems. You begin tuning your mind to a time scale that is the planet's time scale. For me, it is almost unconscious now and is a kind of companionship with the earth."

"It didn't take very long for those mountains to come up, to be deroofed, and to be thrust eastward. Then the motion stopped. That happened in maybe ten million years, and to a geologist that's really fast."

"If you free yourself from the conventional reaction to a quantity like a million years, you free yourself a bit from the boundaries of human time. And then in a way you do not live at all, but in another way you live forever."

One is tempted to condense time, somewhat glibly—to say, for example, that the faulting which lifted up the mountains of the Basin and Range began "only" eight million years ago. The late Miocene was "a mere" eight million years ago. That the Rocky Mountains were building seventy million years ago and the Appalachians were folding four hundred million years ago does not impose brevity on eight million years. What is to be avoided is an abridgment of deep time in a manner that tends to veil its already obscure dimensions. The periods are so long—the eighty million years of the Cretaceous, the forty-six million years of the Devonian—that each has acquired its own internal time scale, intricately constructed and elaborately named. I will not attempt to reproduce this amazing list but only to suggest its profusion. The stages and ages, as they are called—the subdivisions of all of the epochs and eras—read like a roll call in a district council somewhere in Armenia. Berriasian, Valanginian, Hauterivian, Barremian, Bedoulian, Gargasian, Aptian, Albian, Cenomanian, Turonian, Coniacian, Santonian, Campanian, and Maastrichtian, reading upward, are chambers of Cretaceous time. Actually, the Cretaceous has been cut even finer, with about fifty clear time lines now, subdivisions of the subdivisions of its eighty million years. The Triassic consists of the Scythian, the Anisian, the Ladinian, the Carnian, the Norian, and the Rhaetian, averaging seven million years. What survived the Rhaetian lived on into the Liassic. The Liassic, an epoch, comes just after the Triassic and is the early part of the Jurassic. Kazanian, Couvinean, Kopaninian, Kimmeridgian, Tremadocian, Tournaisian, Tatarian, Tiffanian . . . When geologists choose to ignore these names, as they frequently do, they resort to terms that are undecipherably simple, and will note, typically, that an event which occurred in some flooded summer 341.27 million years ago took place in the "early late-middle Mississippian." To say "middle Mississippian" might do, but with millions of years in the middle Mississippian there is an evident compunction to be more precise. "Late" and "early" always refer to time. "Upper" and "lower" refer to rock. "Upper Devonian" and "lower Jurassic" are slices of time expressed in rock.

In the middle Mississippian, there was an age called Meramecian, of about eight million years, and it was during the Meramecian that the Tonka—the older of the formations in the angular unconformity in Carlin Canyon, Nevada—was accumulating along an island coast. The wine-red sandstone and its pebbles may have been sand and pebbles of the beach. The island was of considerable size, apparently, and stood off North America in much the way that Taiwan now reposes near the coast of China. Where there were swamps, they were full of awkward amphibians, not entirely masking in their appearance the human race they would become. They struggled along on stumpy legs. The strait separating the Meramecian island from the North American mainland was about four hundred miles wide and contained crossopterygian fish, from which the amphibians had evolved. There were shell-crushing sharks, horn corals, meadows of sea lilies, and spiral bryozoans that looked like screws. The strait was warm and equatorial. The equator ran through the present site of San Diego, up through Colorado and Nebraska, and on through the site of Lake Superior. The lake would not be dug for nearly three hundred and forty million years. If in the Meramecian you were to have followed the present route of Interstate 80 moving east, you would have raised the coast of North America near the Wyoming border, and landed on a red beach. Gradually, you would have ascended through equatorial fern forests, in red soil, to a high point somewhere near Laramie, to begin there a long general downgrade among low hills to Grand Island, Nebraska, where you would have come to an arm of the sea. The far shore was four hundred miles to the east, where the Mississippi River is now, and beyond it was a low, wet, humid, flat terrain, dense with ferns and fern trees—Illinois, Indiana, Ohio. Halfway across Ohio, you would have come to a second epicratonic sea, its far shore in central Pennsylvania. In New Jersey, you would have begun to ascend mountains and ever higher mountains, their summits girt with ice and capped with snow, not unlike Mt. Kenya, not unlike the present peaks of New Guinea and Ecuador, with their snowfields and glaciers in the equatorial tropics. Reaching the site of the George Washington Bridge, you would have been at considerable altitude, looking at mountains and more mountains before you in future Africa.

If you had turned around and gone back to Nevada a million years later, still in Meramecian time, there would have been few variations to note along the way. The west coast would have moved east, but only a bit, and would still be approximately at the western end of Wyoming. There would have been a significant alteration, however, in the demeanor of the island over the strait. At a little over two inches a year, it would have moved forty miles or so eastward, compressing the floor of the strait and pushing up high mountains, like the present mountains of Timor, which have come up in much the same way to stand ten thousand feet above the Banda Sea. Up from the sea and within those Meramecian Nevada mountains came the wine-red pebbly sandstone of the Tonka formation.

Forty million years after that, when the Tonka mountains had been worn flat and the Strathearn limestones were forming over their roots, the American scene was very different. It was now the Missourian age of late Pennsylvanian time (about three hundred million years ago), and the Appalachians were still high but they were no longer alpine. Travelling west, and coming down from the mountains around Du Bois, Pennsylvania, you would have descended into a densely vegetal swamp. This was Pennsylvania in the Pennsylvanian, when vegetation rioted on the earth and the big trees were kings. They were not huge by our standards but they were big trees, some with diamond patterns precisioned in their bark. They had thick boles and were about a hundred feet high. Other trees had bark like the bark of hemlocks and leaves like flat straps. Others had the fluted, swollen bases of cypress. In and out among the trunks flew dragonflies with the wingspans of great horned owls. Amphibians not only were walking around easily but some of them had become reptiles. Through the high meshing crowns of the trees not a whole lot of light filtered down. The understory was all but woven—of rushlike woody plants and seed ferns. There were luxuriant tree ferns as much as fifty feet high. The scene suggests a tropical rain forest but was more akin to the Everglades, the Dismal Swamp, the Atchafalaya basin—a hummocky spongy landscape ending in a ragged coast. All through Pennsylvanian time, ice sheets had been advancing over the southern continents, advancing and retreating, forming and melting, lowering and raising the level of the sea, and as the sea came up and over the land in places like the swamps of Du Bois it buried them, first under beach sand and later—as the seawater deepened—under lime muds. With enough burial, the muds became limestone, the sands became sandstone, the vegetation coal. When the sea fell, erosion wore away some of that, but then the sea would rise again to bury new generations of ferns and trees under successive layers of rock. These cyclothems, as they are called, contain the coals of Pennsylvania, and similar ones the coals of Iowa and Illinois. The shallow sea that reached into western Pennsylvania and eastern Ohio was a hundred miles wide in the Missourian age of Pennsylvanian time, and after crossing the water you would have reached a beach and another coal swamp and then, in light-gray soil, a low lush tropical forest that went on through Indiana to eastern Illinois, where it ended with more coal swamps, another sea. The far shore was where the Mississippi River is now, and beyond that was an equatorial rain forest, which ended in central Iowa with another swamp, another sea. The water here was clear and sparkling, with almost no land-derived sediments settling into it, just accumulating skeletons—clean deep beds of lime. Five hundred miles over the water, you would have raised the rose-colored beaches of eastern Wyoming. Mountains stood out to the south. They were the Ancestral Rockies, and time would bevel them to stumps. Skirting them, in Pennsylvanian Wyoming, you would have traversed what seem to have been Saharan sands, wave after wave of dunal sands, five hundred miles of rose and amber pastel sands, ending at the west coast of North America, in Salt Lake City. As the Pennsylvanian sea level moved up and down here, it left alternating beds of lime and sand, which, two eras later, the nascent Oquirrh Mountains would lift to view. Two hundred miles out to sea was the site of Carlin Canyon, where muds of clean lime were settling. The Strathearn, the younger formation of the two in the Carlin unconformity, is an almost pure limestone.

The two formations, conjoined, were driven upward, according to present theory, in a collision of crustal plates that occurred in the early Triassic. The result was yet another set of new mountains—alpine mountains which erosion brought down before the end of the Jurassic, but not enough to obliterate the story that is told in Carlin Canyon. Still gazing at the Carlin unconformity, Ken Deffeyes said, "Profound as all the time is to build and destroy those mountain ranges, it is just a one-acter in the history of the Basin and Range—small potatoes, weak beer, just a little piece of time, a little piece of the action, lost in all the welter of all the other history." There had been two complete cycles of erosion and deposition and mountain building in this one place in one-fortieth of the time scale. That is what made John Playfair's mind grow giddy when James Hutton took him in 1788 to see the angular unconformity at Siccar Point. It was especially fortunate that Playfair was there, and that Playfair knew Hutton and Hutton's geology equally well, for when Hutton finally wrote his book most readers were trampled by the prose. Hutton was at best a difficult writer. Insights came to him but phrases did not. James Hall, who was twenty-seven when he went with Hutton and Playfair to Siccar Point, would say of Hutton years later, "I must own that on reading Dr. Hutton's first geological publication I was induced to reject his system entirely, and should probably have continued still to do so, with the great majority of the world, but for my habits of intimacy with the author, the vivacity and perspicuity of whose conversation formed a striking contrast to the obscurity of his writings." Hall, incidentally, melted rock in crucibles and saw how crystals formed as it cooled. He is regarded as the founder of experimental geology. John Playfair likewise assessed Hutton's literary style as containing "a degree of obscurity astonishing to those who knew him, and who heard him every day converse with no less clearness and precision than animation and force." One can imagine what Playfair thought when he read something like this in Hutton's two-volume *Theory of the Earth*:

If, in examining our land, we shall find a mass of matter which had been evidently formed originally in the ordinary manner of stratification, but which is now extremely distorted in its structure and displaced in its position,—which is also extremely consolidated in its mass and variously changed in its composition,—which therefore has the marks of its original or marine composition extremely obliterated, and many subsequent veins of melted mineral matter interjected; we should then have reason to suppose that here were masses of matter which, though not different in their origin from those that are gradually deposited at the bottom of the ocean, have been more acted upon by subterranean heat and the expanding power, that is to say, have been changed in a greater degree by the operations of the mineral region.

In that long sentence lies the discovery of metamorphic rock. But just as metamorphism will turn shale into slate, sandstone into quartzite, and granite into gneiss, Hutton had turned words into pumice. Unsurprisingly, his insights did not at once spread far and wide. They received a scattered following and much abuse. The attacks were theological, in the main, but, needless to say, geological as well—particularly with regard to his elastic sense of time. Even when people began to agree that the earth must be a great deal older than six thousand years, calculations were conservative and failed to yield the reach of time that Hutton's theory required. Lord Kelvin, as late as 1899, figured that twenty-five million years was the approximate age of the earth. Kelvin was the most august figure in contemporary science, and no one stepped up to argue. Hutton published his *Theory of the Earth* in 1795, when almost no one doubted the historical authenticity of Noah's Flood, and all species on earth were thought to have been created individually, each looking at the moment of its creation almost exactly as it did in modern times. Hutton disagreed with that, too. Writing a treatise on agriculture, he brought up the matter of variety in animals and noted, "In the infinite variation of the breed, that form best adapted to the exercise of the instinctive arts, by which the species is to live, will most certainly be continued in the propagation of this animal, and will be always tending more and more to perfect itself by the natural variation which is continually taking place. Thus, for example, where dogs are to live by the swiftness of their feet and the sharpness of their sight, the form best adapted to that end will be the most certain of remaining, while those forms that are less adapted to this manner of chase will be the first to perish; and, the same will hold with regard to all the other forms and faculties of the species, by which the instinctive arts of procuring its means of substance may be pursued." When he died, in 1797, Hutton was working on that manuscript, no part of which was published for a hundred and fifty years.

People who admired Hutton's theory of the earth became known—because of the theory's igneous aspects, its molten basalts and intruding granites—as vulcanists or plutonists, and they quickly grew to be the intellectual enemies of the Wernerian neptunists, and others who believed that God had made the world through a series of catastrophes, notably the Noachian flood. The schism between these two groups would carry well into the nineteenth and even into the twentieth century, the ratio gradually reversing. In 1800, the Huttonians were outnumbered at least ten to one. In fact, a Werner-trained neptunist took over the chair of natural history at the University of Edinburgh and for many years neptunism was official in Hutton's own city.

All this can be presumed to have bestirred John Playfair, a handsome, life-loving, and generous man of "mild majesty and considerate enthusiasm," as a contemporary described him. Never mind that the contemporary was his nephew. With all those neptunists and men of the cloth on the one side and his friend's prose on the other, the battle to Playfair must have seemed unjust, and he betook himself to alter the situation. The least of his many verbal gifts was a slow-cooled lucidity, a sense of the revealing phrase, and his *Illustrations of the Huttonian Theory of the Earth*, published in 1802, was the first fully clear and persuasive statement of what the theory was about. It is testimony to Playfair's efficacity that the opposition stiffened. "According to the conclusions of Dr. Hutton, and of many other geologists, our continents are of indefinite antiquity, they have been peopled we know not how, and mankind are wholly unacquainted with their origin," wrote the Calvinist geologist Jean André Deluc in 1809. "According to my conclusions, drawn from the same source, that of *facts*, our continents are of such small antiquity, that the memory of the revolution which gave them birth must still be preserved among men; and thus we are led to seek in the book of Genesis the record of the history of the human race from its origin. Can any object of importance superior to this be found throughout the circle of natural science?"

As geologists built the time scale, their research and accumulating data imparted to Hutton's theory an obviously increasing glow. And in the early eighteen-thirties Charles Lyell, who said in so many words that his mission in geology was "freeing the science from Moses," gave Hutton's theory and his sense of deep time their largest advance toward universality. In three volumes, he published a work whose full title was *Principles of Geology, Being an Attempt to Explain the Former Changes of the Earth's Surface, by Reference to Causes Now in Operation*. Lyell was so anti-neptunist, so anti-catastrophist that he out-Huttoned Hutton both in manner and in form. He not only subscribed to the uniformitarian process—the topographical earth building and destroying and rebuilding itself through time—but was finicky in insisting that all processes had been going on at exactly the same rate through all ages. *Principles of Geology* was to be the most enduring and effective geological text ever published. The first volume was eighteen months off the press when H.M.S. Beagle set sail from Devonport with Charles Darwin aboard. "I had brought with me the first volume of Lyell's *Principles of Geology*, which I studied attentively; and the book was of the highest service to me in many ways. The very first place which I examined, namely St. Jago in the Cape de Verde islands, showed me clearly the wonderful superiority of Lyell's manner of treating geology, compared with that of any other author whose works I had with me or ever afterwards read." When Darwin had first studied geology, he had heard lectures in Wernerian neptunism at Edinburgh, and they had very nearly put him to sleep. Nevertheless, the degree he later took at Cambridge University was in geology. He referred to himself as a geologist. His field identifications of the rocks he collected on his travels, and of the minerals within the rocks, were essentially without error. The rocks are in Cambridge, where contemporary geologists have thin-sectioned them, confirming Darwin's petrology. Voyaging on the Beagle, he was enhancing his sense of the slow and repetitive cycles of the earth and the giddying depths of time, with Lyell's book in his hand and Hutton's theory in his head. In six thousand years, you could never grow wings on a reptile. With sixty million, however, you could have feathers, too.

According to present theory, many exotic terranes moved in from the western ocean and collected against North America during a span of about three hundred million years which ended roughly forty million years ago, increasing the continent to something like its present size. Three of these assembled at the latitude of Interstate 80. It was the first of these collisions that crunched and folded the wine-red sandstone near Carlin. The second, in the early Triassic, is what apparently caused the whole Carlin unconformity to revolve quite close to its present position. Sonomia, as the second terrane has been named, included much of what is now western Nevada and eastern California, and is said to have come into the continent with such force—notwithstanding that it was moving an inch or so a year—that it overlapped its predecessor by as much as eighty kilometres before it finally stopped. The evidence of this event is known locally as the Golconda Thrust, and both its upper and lower components are exposed in a big roadcut on the western flank of Golconda Summit, where the interstate, coming up out of Pumpernickel Valley, crosses a spur of the Sonoma Range. Small wonder that Deffeyes pulled over when we came to it and said, "Let's stick our eyeballs on this one."

It was dawn at the summit. We had been awake for hours and had eaten a roadhouse breakfast sitting by a window in which the interior of the room was reflected against the black of the morning outside while a television mounted on a wall behind us resounded with the hoofbeats of the great horse Silver. *The Lone Ranger.* Five A.M. CBS's good morning to Nevada. Waiting for bacon and eggs, I put two nickels in a slot machine and got two nickels back. The result was a certain radiance of mood. Deffeyes, for his part, was thinking today in troy ounces. It would take a whole lot more than two nickels to produce a similar effect on him. Out for silver, he was heading into the hills, but first, in his curiosity, he walked the interstate roadcut, now and again kicking a can. The November air was in frost. He seemed to be smoking his breath. He remarked that the mean distance between beer cans across the United States along I-80 seemed to be about one metre. Westward, tens of hundreds of square miles were etched out by the early light: basins, ranges, and—below us in the deep foreground—Paradise Valley, the village Golconda, sinuous stands of cottonwood at once marking and concealing the Humboldt. The whole country seemed to be steaming, vapors rising from warm ponds and hot springs. The roadcut was long, high, and benched. It was sandstone, for the most part, but at its lower, westernmost end the blasting had exposed a dark shale that had been much deformed and somewhat metamorphosed, the once even bedding now wrinkled and mashed—rock folded up like wet laundry. "You can spend hours doping out one of these shattered places, just milling around trying to find out what's going on," Deffeyes said cautiously, but he was fairly sure he knew what had happened, for the sandstone that lay above contained many volcanic fragments and was full of sharp-edged grains of chert and quartz, highly varied in texture, implying to him a volcanic source and swift deposition into the sea (almost no opportunity for streams to have rounded off the grains), implying, therefore, an island arc standing in deep water on a continental margin—an Aleutian chain, a Bismarck Archipelago, a Lesser Antilles, a New Zealand, a Japan, thrust upon and overlapping the established continent, a piece of which was that mashed-up shale. Deffeyes mused his way along the cut. "There is complexity here because you have not only the upper and lower plates of the Golconda Thrust, which happened in the early Triassic; you also have basin-range faulting scarcely a hundred yards away—enormously complicating the regional picture. If you look at a geologic map of western Canada and Alaska, you can see the distinct bands of terrane that successively attached themselves to the continent. Here the pattern has been all broken up and obscured by the block faulting of the Basin and Range, not to mention the great outpouring of Oligocene welded tuff. So this place is a handsome mess. If you ever want to study this sort of collision more straightforwardly, go to the Alps, where you had a continent-to-continent collision and that was it."

So much for theory. This roadcut contained both extremities of Deffeyes' wide interests in geology, and his attention was now drawn to a large gap in the sandstone, faulted open probably six or seven million years ago and now filled with rock crumbs, as if a bomb had gone off there in the ground. The material was gradated outward from a very obvious core. In a country full of living hot springs, this was a dead one. Sectioned by the road builders, it remembered in its swirls and convolutions the commotion of water raging hot in rock. The dead hot spring had developed cracks, and they had been filled in by a couple of generations of calcite veins. Deffeyes was busy with his hammer, pinging, chipping samples of the calcite. "This stuff is too handsome to leave out here," he said, filling a canvas bag. "There was a lot of thermal action here. Most of this material is not even respectable rock anymore. It's like soil. In 1903, a mining geologist named Waldemar Lindgren found cinnabar in crud like this at Steamboat, near Reno. Cinnabar is mercury sulphide. He also found cinnabar in the fissures through which water had come up from deep in the crust. He thought, Aha! Mercury deposits are hot-spring deposits! And he applied that idea to ore deposits generally. He started classifying them according to the temperature of the water from which they were deposited—warm, hot, hotter, and so on. We know now that not all metal deposits are hydrothermal in origin, but more than half of them are. As you know, the hot water, circulating deep, picks up whatever is there—gold, silver, molybdenum, mercury, tin, uranium—and brings it up and precipitates it out near the surface. A vein of ore is the filling of a fissure. A map of former hot springs is remarkably close to a map of metal discoveries. Old hot springs like this one brought up the silver of Nevada. It would do my heart good to find silver right here in this roadcut and put it to the local highway engineer."

He took some samples, which eventually proved to be innocent of silver, and we got back into the pickup. We soon left the interstate for a secondary road heading north—up a pastel valley, tan, with a pale-green river course, fields of cattle and hay. It was a valley that had been as special to the Paiutes as the Black Hills were to the Sioux. The Paiutes gave it up slowly, killing whites in desperation to keep it, and thus bringing death on themselves. The first pioneers to settle in this "desert" were farmers—an indication of how lush and beautiful the basin must have appeared to them, ten miles wide and seventy miles long, framed in serrated ridges of north-south-trending mountains: range, basin, range. Magpies, looking like scale-model jets, kept rising into flight from the side of the road and gaining altitude over the hood of the pickup. Deffeyes said they were underdeveloped and reminded him of *Archaeopteryx*, the Jurassic bird. We crossed cattle guards that were nothing more than stripes painted on the road, indicating that Nevada cattle may be underdeveloped, too, with I.Q.s in one digit, slightly lower than the national norm.

For eight million years, Deffeyes was saying, as the crustal blocks inexorably pulled apart here and springs boiled up along the faults, silver had been deposited throughout the Basin and Range. The continually growing mountains sometimes fractured their own ore deposits, greatly complicating the sequence of events and confusing the picture for anyone who might come prospecting for ores. There was another phenomenon, however, that had once made prospecting dead simple. Erosion, breaking into hot-spring and vein deposits, concentrated the silver. Rainwater converted silver sulphides to silver chloride, heavy stuff that stayed right where it was and—through thousands of millennia—increased in concentration as more rain fell. These were the deposits, richer than an Aztec dream, that were known to geologists as supergene enrichments. Miners called them surface bonanzas. In the eighteen-sixties, and particularly in the eighteen-seventies, they were discovered in range after range. A big supergene enrichment might be tens of yards wide and a mile long, lying at or near the surface. Instant cities appeared beside them, with false-front saloons and tent ghettos, houses of sod, shanties made of barrels. The records of these communities suggest uneven success in the settling of disputes between partners over claims: "Davison shot Butler through the left elbow, breaking the bone, and in turn had one of his toes cut off with an axe." They were places with names like Hardscrabble, Gouge Eye, Battle Mountain, Treasure Hill. By the eighteen-nineties, the boom was largely over and gone. During those thirty years, there were more communities in Nevada than there are now. "Silver is our most depleted resource, because it gave itself away," said Deffeyes, looking mournful. "You didn't need a Ph.D. in geology to find a supergene enrichment."

All you needed was Silver Jim. Silver Jim was a Paiute, and he, or a facsimile, took you up some valley or range and showed you grayish rock with touches of green that had a dull waxy lustre like the shine on the horn of a cow. Horn silver. It was just lying there, difficult to lift. Silver Jim could show you horn silver worth twenty-seven thousand dollars a ton. Those were eighteen-sixties dollars and an uninflatable ton. You could fill a wheelbarrow and go down the hill with five thousand dollars' worth of silver. Three or four years ago, a miner friend of Deffeyes who lives in Tombstone, Arizona, happened to find on his own property an overlooked fragment of a supergene enrichment, a narrow band no more than a few inches thick, six feet below the cactus. Knocking off some volcanic overburden with a front-end loader, the miner went after this nineteenth-century antique and fondly dug it out by hand. He said to his children, "Pay attention to what I'm doing here. Look closely at the rock. We will never see this stuff again." In a couple of hours of a weekend afternoon, he took twenty thousand dollars from the ground.

We were off on dirt roads now with a cone of dust behind us, which Deffeyes characterized as the local doorbell. He preferred not to ring it. This talkative and generous professor—who ordinarily shares his ideas as rapidly as they come to him, spilling them out in bunches like grapes—was narrow-eyed with secrecy today. He had stopped at a courthouse briefly, and—an antic figure, with his bagging sweater and his Beethoven hair—had revealed three digits to a county clerk in requesting to see a registry of claims. The claims were coded in six digits. Deffeyes kept the fourth, fifth, and sixth to himself like cards face down on a table. He found what he sought in the book of claims. Now, fifty miles up the valley, we had long since left behind us its only town, with its Odd Fellows Hall, its mercantile company, its cottonwoods and Lombardy poplars; and there were no houses, no structures, no cones of dust anywhere around us. The valley was narrowing. It ended where ranges joined. Some thousands of feet up the high face of a distant and treeless mountain we saw an unnaturally level line.

"Is that a road?" I asked him.

"That's where we're going," he said, and I wished he hadn't told me.

Looking up there, I took comfort in the reflection that I would scarcely be the first journalist to crawl out on a ledge in the hope of seeing someone else get rich. In 1869, the editor of the *New York Herald*, looking over his pool of available reporters, must have had no difficulty in choosing Tom Cash to report on supergene enrichments. Cash roved Nevada. He reported from one place that he took out his pocketknife and cut into the wall of a shaft, removing an ore of such obviously high assay that he could roll it in his fingers and it would not crumble. Cash told the mine owner that he feared being accused of exaggeration—"of making false statements, puffing"—with resulting damage to his journalistic reputation. There was a way to avoid this, he confided to the miner. "I would like to take a sample with me of some of the richest portions." The miner handed him a fourteen-pound rock containing about a hundred and fifty troy ounces of silver (seventy-three per cent). In the same year, Albert S. Evans, writing in the San Francisco *Alta California*, described a visit with a couple of bankers and a geologist to a claim in Nevada where he was lowered on a rope into a mine. "The light of our candles disclosed great black sparkling masses of silver on every side. The walls were of silver, the roof over our heads was of silver, and the very dust that filled our lungs and covered our boots and clothing with a gray coating was of fine silver. We were told that in this chamber a million dollars' worth of silver lies exposed to the naked eye and our observations confirm the statement. How much lies back of it, Heaven only knows."

Heaven knew exactly. For while the supergene enrichments—in their prodigal dispersal through the Basin and Range—were some of the richest silver deposits ever discovered in the world, they were also the shallowest. There was just so much lying there, and it was truly bonanzan—to print money would take more time than to pick up this silver—but when it was gone it was gone, and it went quickly. Sometimes—as in the Comstock Lode in Virginia City—there were "true veins" in fissures below, containing silver of considerable value if more modest assay, but more often than not there was nothing below the enrichment. Mining and milling towns developed and died in less than a decade.

We were on our way to a nineteenth-century mine, and were now turning switchbacks and climbing the high mountainside. Deffeyes, in order to consult maps, had turned over the wheel to me. He said his interest in the secondary recovery of silver had been one result of certain computer models that had been given wide circulation in the early nineteen-seventies, using differential equations to link such things as world population, pollution, resources, and food, and allow them to swim forward through time, with a resulting prediction that the world was more or less going to come to an end by the year 2000, because it would run out of resources. "We have found all obvious deposits, and, true enough, we've got to pay the price," he said. "But they did not take into account reserves or future discoveries or picking over once again what the old-timers left behind." Seeking commissions from, for example, the Department of Energy, he began doing studies of expectable discoveries of petroleum and uranium. He sort of slid inadvertently from uranium into silver after a syndicate of New York businessmen came to him to ask for his help in their quest for gold. The group was called Eocene and was interested in scavenging old mines. Deffeyes pointed out to them that while new gold strikes were still occurring in the world and new gold mines were still being developed, no major silver mine had been discovered since 1915. The pressure for silver was immense. Dentistry and photography used two-thirds of what there was, and there were no commercial substitutes. "We've been wiped out. We've gone through it, just as we have gone through magnesium and bromine. You can raise the price of silver all you want to but you won't have a new mine." He predicted that as prices went up silver would probably outperform gold. The potentialities in the secondary recovery of silver appeared to him to be a lot more alluring than working through tailings for gold. Eocene engaged him as a consultant, to help them scavenge silver.

Now far above the basin, we were on the thin line we had seen from below, a track no wider than the truck itself, crossing the face of the mountain. It curved into reentrants and out around noses and back into reentrants and out to more noses. I was on the inboard side, and every once in a while as we went around a nose I looked across the hood and saw nothing but sky—sky and the summits of a distant range. We could see sixty, seventy miles down the valley and three thousand feet down the mountain. The declivity was by no means sheer, just steep—a steepness, I judged, that would have caused the vehicle, had it slipped off the road, to go end over end enveloped in flame at a hundred yards a bounce. My hands slid on the wheel. They were filmed in their own grease.

The equanimous Deffeyes seemed to be enjoying the view. He said, "Where did you learn to drive a truck?"

"Not that it's so god-damned difficult," I told him, "but this is about the first time."

Before 1900, the method used in this country to extract silver from most ores was to stamp the rock to powder in small stamp mills, then stir the powder into hot salt water and mercury, and, after the mercury had attracted the silver, distill the mercury. In 1887, a more thorough extraction process had been developed in England whereby silver ores were dissolved in cyanide. The method moved quickly to South Africa and eventually to the United States. An obvious application was to run cyanide through old tailings piles to see what others had missed, and a fair amount of such work was done, in particular during the Depression. There had been so many nineteenth-century mines in Nevada, however, that Deffeyes was sure that some had been ignored. He meant to look for them, and the first basin he prospected was the C Floor of Firestone Library, up the hill from his office in Princeton. There he ran through books and journals and began compiling a catalogue of mines and mills in the Basin and Range that had produced more than a certain number of dollars' worth of silver between 1860 and 1900. He prefers not to bandy the number. He found them in many places, from barrel-cactus country near the Colorado River to ranges near the Oregon line, from the Oquirrh Mountains of Utah to the eastern rampart of the Sierra Nevada. In all, he listed twenty-five. The larger ones, like the Comstock, had been worked and reworked and cyanided to death, and "tourists were all over them like ants." A scavenger had best consider lesser mines, out-of-the-way mines—the quick-shot enrichments, the small-fissure lodes, where towns grew and died in six years. He figured that any mine worth, say, a million dollars a hundred years ago would still be worth a million dollars, because the old mills at best extracted ninety per cent of the silver in the ores, and the ten per cent remaining would be worth about what the ninety per cent had been worth then. Pulling more books and journals off the shelves, he sought to learn if and where attention had been paid to various old mines in the nineteen-thirties, and wherever he discovered activity at that time he crossed off those mines.

His next move was to buy aerial photographs from the United States Geological Survey. The pictures were in overlapping pairs, and each pair covered sixteen square miles. "You look at them with stereo equipment and you are a giant with eyeballs a mile apart and forty thousand feet in the air. God, do you have stereovision! Things jump off the earth. You look for tailings. You look for dumps. You look for the faint scars of roads. The environmentalists are right. A scar in this climate will last. It takes a long time for the terrain to erase a road. You try to reason like a miner. If this was a mine, now where would I go for water? If this was a mill here, by this stream, then where is the mine? I was looking for mines that were not marked on maps. I could see dumps in some places. They stood out light gray. The old miners made dumps of rock that either contained no silver at all or did not contain enough silver to be worth their while at the time. I tried to guess roughly the volume of the dumps. Mill tailings made unnatural light-gray smudges on the pictures. Some of the tailings and dumps I found in these mountains appear on no maps I've seen."

He flew to Nevada, chartered a light plane, and went over the country a thousand feet above the ground, taking fresh private pictures with a telephoto lens. When he flew over places where other scavengers looked up and waved, he crossed those places off his list. He went in on the ground then, to a number of sites, and collected samples. He had machines at home that could deal with the samples in ways unheard of just a few years before, let alone in the nineteenth century. Kicking at old timbers, he looked at the nails. Wire nails came into use in 1900 and are convenient index fossils of the Age of Cyanide. He hoped for square nails.

Deffeyes was on his own now. His relationship with Eocene had faded out after they had chosen, on various points, to follow counsel other than his, and they transferred their scrutiny to Arizona, preferring not to cope with winter. One day in Princeton, his wife, Nancy Deffeyes, was looking through a stack of hundred-year-old *Engineering & Mining Journals* when she found a two-line reference to certain mining efforts in the eighteen-seventies that eventually assumed prominence on her husband's list, and that was what had brought him here and why we were crawling like a Japanese beetle across the face of this mountain.

We turned a last corner, with our inner wheels resting firmly on the road and the two others supported by Deffeyes' expectations. Now we were moving along one wall of a big V-shaped canyon that eventually became a gulch, a draw, a crease in the country, under cottonwoods. In the upper canyon, some hundreds of acres of very steep mountainside were filled with holes and shafts, hand-forged ore buckets, and old dry timbers. There were square nails in the timbers. An ore bucket was filled with square nails. "Good litter," Deffeyes said, and we walked uphill past the mine and along a small stream into the cottonwoods. The stream was nearly dry. Under the cottonwoods were the outlines of cabins almost a century gone. Here at seven thousand feet in this narrow mountain draw had lived a hundred people, who had held their last election a hundred years earlier. They had a restaurant, a brewery, a bookstore. They had seven saloons. And now there was not so much as one dilapidated structure. There were only the old unhappy cottonwoods, looking alien and discontented over the moist bed of the creek. Sixteen stood there, twisted, surviving—most of them over four feet thick. "Those cottonwoods try an environmentalist's soul," Deffeyes said. "They transpire water like running fountains. If you were to cut them down, the creek would run. Cottonwoods drink the Humboldt. Some of the tension in this country is that miners need water. Getting rid of trees would preserve water. By the old brine-and-mercury method, it took three tons of water to mill one ton of ore. There was nothing like that in this creek. They had to take the ore from here to a big enough stream, and that, as it happens, was a twelve-mile journey using mules. They would have gone out of here with only the very best ore. There was probably a supergene enrichment here over a pretty good set of veins. They took what they took and were gone in six years."

We walked back down to the mine, below which the stream—in flash flood once or twice a century over several million years—had cut the deep sharp V of its remarkably plunging valley. A number of acres of one side had been used as a dump, and Deffeyes began to sample this unused ore. "They must have depended on what they could see in the rock," he said. "If it was easy to see, they got it all. If it was complicated and gradational, they couldn't differentiate as well, and I think they threw it here." The material was crumbly, loose, weathered, unstable underfoot, a pyramid side of decomposing shards. Filling small canvas bags at intervals of six feet, he worked his way across it. With each step, he sank in above his ankles. He was about two hundred feet above the stream. Given the steepness of the ground and the proximity of all the loose material to the critical angle of repose, I had no trouble imagining that he was about to avalanche, and that he would end up in an algal pool of the trickling stream below us, buried under megatons of unextracted silver. The little stream was a jumble of boulders, testimony of the floods, with phreatophytes around the boulders like implanted spears. Deffeyes obviously was happy and without a fear in the world. When a swift-rising wind blew dust in his face, he mooed. Working in cold sunshine with his orange-and-black conical cap on his head, he appeared to be the Gnome of Princeton, with evident ambition to escalate to Zurich.

To make a recovery operation worthwhile, he said, he would have to get five ounces of silver per ton. The figures would turn out to be better than that. Before long, he would have a little plastic-lined pond of weak cyanide, looked after by a couple of technicians, down where the ore from this mine had been milled. A blue streak in the tailings there would come in at fifty-eight ounces a ton—richer than any tailings he had ever found in Nevada. "You put cyanide on that ore, the silver leaps out of it," he would say. "I have enough cyanide there to kill Cincinnati. People have a love-hate relationship with cyanide. Abelson showed that lightning acts on carbon dioxide and other atmospheric components to make hydrogen cyanide, and hydrogen cyanide polymerizes and later reacts with water to form amino acids, which are the components of proteins—and that may be how life began. Phil Abelson is an editor at *Science*. He's a geochemist, and he worked on the Manhattan Project. To get the silver out of here at an acceptable price, you need small-scale technology. You need miniaturized equipment, simple techniques. In the nineteenth century, they made sagebrush fires to heat the brine to dissolve the silver chloride. When mercury picked up the silver, they knew they had 'the real stuff' from the squeak. A mercury-and-silver mixture is what the dentist uses, and when he mashes it into your tooth it makes the same squeak."

Deffeyes' methodology would depend on more than sagebrush and sound. In time, he would have a portable laboratory there, size of a two-hole privy, and in it would be, among other things, a silver single-ion electrode and an atomic-absorption spectrophotometer. He could turn on a flame, close two switches, and see at once the amount of silver in a sample. For a short while, he would have a five-pound ingot of raw silver on the floor, propping open the door. When he was finished with his pond, he would withdraw the cyanide and turn it into a marketable compound known as Prussian blue. He would cover his pond with dirt and sow it with crested wheat.

And now, finishing up his sampling at the mine in the mountains, he filled a large burlap bag with ore he would take home to improve his technique of extraction. The smaller samples he had taken were for assays of silver in various parts of the slope. "I'm nothing but a ragpicker," he said. "A scavenger armed with a forty-thousand-dollar X-ray machine." The wind picked up another cloud of dust off the dump and blew it into his face. He mooed. "That may feel like dirt to you, but it feels like money to me," he said.

"How much money would you say that felt like?" I asked him.

He took out a Magic Marker and began to do metric conversions, geometry, and arithmetic on the side of a new canvas bag. "Well, this section of the dump is at least fifteen thousand cubic metres," he said. "That is the most conservative figure. At two hundred dollars a ton, that works out to about three million dollars, left here in the side of the hill."

"What are those red stakes up there?"

"Somebody seems to think they're finding new ore. I'm interested in the old stuff, down here."

"If you've got good silver in those bags, what about Eocene? What if they decide they still own you? What if they go to the sheriff?"

"Eocene doesn't own me, and Eocene doesn't own the contents of my head. The law has long since decided that. But if anybody comes after me I want you to go to jail cheerfully rather than surrender your notes."

As we wound down the mountain at the end of the day, we

stopped to regard the silent valley—the seventy miles of basin under a rouge sky, the circumvallate mountains, and, the better part of a hundred miles away, Sonoma Peak, of the Sonoma Range. Deffeyes said, "If you reduced the earth to the size of a baseball, you couldn't feel that mountain. With a telephoto lens, you could convince someone it was Everest." Even at this altitude, the air was scented powerfully with sage. There was coyote scat at our feet. In the dark, we drove back the way we had come, over the painted cattle guards and past jackrabbits dancing in the road, pitch-dark, and suddenly a Black Angus was there, standing broadside, middle of the road. With a scream of brakes, we stopped. The animal stood still, thinking, its eyes unmoving—a wall of beef. We moved slowly after that, and even more slowly when a white sphere materialized on our right in the moonless sky. It expanded some, like a cloud. Its light became so bright that we stopped finally and got out and looked up in awe. A smaller object, also spherical, moved out from within the large one, possibly from behind it. There was a Saturn-like ring around the smaller sphere. It moved here and there beside the large one for a few minutes and then went back inside. The story would be all over the papers the following day. The *Nevada State Journal* would describe a "Mysterious Ball of Light" that had been reported by various people at least a hundred miles in every direction from the place where we had been. "By this time we decided to get the hell out of there," a couple of hunters reported, "and hopped in our pickup and took off. As we looked back at it, we saw a smaller craft come out of the right lower corner. This smaller craft had a dome in the middle of it and two wings on either side, but the whole thing was oval-shaped." Someone else had said, "I thought it was an optical illusion at first, but it just kept coming closer and closer so that I could see it wasn't an illusion. Then something started coming out of the side of it. It looked like a star, and then a ring formed around it. A kind of ring like you'd see around Saturn. It didn't make any noises, and then it vanished."

"Now we're both believers," said one of the hunters. "And I don't ever want to see another one. We're pretty good-sized men and ain't scared of nothing except for snakes and now flying saucers."

After the small sphere disappeared, the large one rapidly faded and also disappeared. Deffeyes and I were left on the roadside among the starlighted eyes of dark and motionless cattle. "Copernicus took the world out of the center of the universe," he said. "Hutton took us out of a special place somewhere near the beginning of things and left us awash in the middle of the immensity of time. An extraterrestrial civilization could show us where we are with regard to the creation of life."

We also went to Jersey Valley, between the Fish Creek Mountains and the Tobin Range, where Deffeyes had once spent a couple of field seasons collecting data for his doctoral thesis. He had lived in a tent in the oven weather, and had chugalugged water in quart draughts while examining the rising mountains and the sediments the mountains had shed. The thick welded tuff of the Oligocene catastrophe, having been the regional surface when the faulting started, was the first material to break into grains that washed and rolled downhill. When erosion wore through the tuff and into the older rock below, it sent the older rock also in fragments to the basins. Reading up through a basin was like reading down through a range. Deffeyes had locally described this record, and now he wished to relate its timing to the development of the province as a whole. Forty miles off the interstate and with a lot of dust settling behind, he paused on the brow of a small hill at the head of Jersey Valley. It was intimate, compared with others in the Basin and Range. For perhaps twenty miles, it ran on south between snow-covered mountains and was filled with a delirium of sage. Deffeyes let out a cowboy yell. There were cinder cones standing in the valley, young and basaltic—enormous black anthills of the Pleistocene. Here and there was a minor butte, an erosional remnant, kept intact by sandstone at the top, but approaching complete disintegration, and, like a melting sugar lump, soon to be absorbed into the basin plain. "In a lot of valleys in Nevada all you will see is sagebrush, and not know that eight feet below you is a hell of an interesting story,"

Deffeyes said. "I found late-Miocene horse teeth over there in the Tobin Range."

"How did you know they were late Miocene?"

"I didn't. I sent them to a horse-teeth expert. I also found beaver teeth, fish, a camel skeleton, and the jaw of a rhinoceros not so far from here. The jaw was late Miocene, too. Early- and middle-Miocene fossils are absent from the province. You'll remember that wherever we have found fossils in the basin sediments the oldest have been late Miocene. So, if the vertebrate paleontologists have their heads screwed on right, the beginning of the faulting of the Basin and Range can be dated to the late Miocene. Vertebrate paleontology is an important old sport, like tossing the caber."

We left the dirt road and drove a mile or so up a pair of ruts, then continued on foot across a rough cobbly slope. We went down into a dry gulch, climbed out of it, and walked along the contour of another slope. These declivities were not discrete hills but fragments of great alluvial fans that were spilling off the mountains and were creased by streams that were as dry as cracks in leather. In their intermittent way, these streams had exposed successive layers of sediment, all of which happened to be dark in hue, with the pronounced exception of light-gray layers of ash. The ash was from elsewhere, from outside the province, punctuation brought in on the wind. It had come from volcanoes standing, probably, in what is now the Snake River Plain, two hundred miles away. Settling into the basin lakes—long-gone Miocene, Pliocene lakes—much of it had turned into zeolites. Among the zeolites, Deffeyes had found in Jersey Valley three million tons of a variety called erionite, which is named for wool, and is fibrous, and when it gets into the linings of human lungs causes mesothelioma. There are more millions of tons of erionite throughout the Basin and Range, passively causing nothing. But if twenty-five valleys of the province were to be filled up with forty-six hundred concrete shelters for MX missiles, as the Defense Department had proposed, wind would present extraordinary hazards during the process of construction. It would be difficult to overestimate the amount of fine material that can be borne long distances by the wind. The largest single layer of ash that Deffeyes found in Jersey Valley was ten feet thick. He once showed it to Howel Williams, of the University of California at Berkeley, whom he regarded as "the greatest of volcanologists." Deffeyes asked Williams what might have been the size of a volcano that from two hundred miles away could send out such an explosion of ash. Williams just stood there impressed, shaking his head.

On a shelf above us was a pile of sticks of a size that in moister country could well have been collected by a beaver. "Hawk," Deffeyes said. "Note the southern exposure. The hawk went solar long ago. The sun incubates the eggs and the hawk is free to soar." Running his eye over the sequence of sediments revealed in the slope before us, he decided to begin his work right there. He was carrying in his hand a device of his own invention with which he hoped to accomplish the delicate operation of removing paleomagnetic samples from unconsolidated lake sediments. Less delicately, he had equipped me with a military shovel. He asked me to go along the slope digging foxholes a couple of feet deep in order to get rid of the weathered surface and prepare the way for him. As the mountains had given up grains and the grains had come down into the basin, any that had magnetite in them would have settled in a uniform manner, pointing like compasses toward the earth's magnetic pole. Since the late Miocene, the earth's magnetic field had reversed itself twenty times—from north to south, from south back to north—and the dates of those reversals had by now become well established. If Deffeyes could somehow collect unconsolidated but firmly compacted sediment and keep it from falling apart and destroying its own evidence while he carried it to a paleomagnetic lab, he might be able to compare what he already knew from his vertebrate time scale—his expertized horse teeth, his rhinoceros jaw—with the paleomagnetic time scale as expressed by the magnetite in the successive basin sediments. He would thus improve his knowledge of what occurred when—in this basin, this range. Later, he could correlate the ash falls and other stratigraphy of Jersey Valley with other valleys in the region, and make clearer the story of how it all took shape, adding polish to chapters of the Basin and Range. And so he had invented and machined a corer that would tap clear-plastic tubing gingerly into the earth with a micropiledriver made of non-magnetic aluminum. As I began the crude initial digging, Deffeyes said, "There are ten thousand feet of sediment here, and all of it has been deposited in eight million years. I have high hopes for the success of these endeavors. For each sample, I would prefer to go twenty feet into the slope instead of two. I would like to have a bulldozer as a substitute for you. But one has to settle for what one can get."

The first time I put my foot to the ground, the shovel broke in half. It was decapitated. After that, I had to hold its head in my hands and scrape as with an awkward trowel.

"There's more to this paleomagnetism game than reversals," Deffeyes said, "more than just determining when, and whether, the magnetic pole was in the north or south. The earth's magnetic field is such that a compass needle at the equator will lie flat, while a compass needle at the poles will want to stand straight up on end—with all possible gradations of that in the latitudes between. So by looking at the paleomagnetic compasses in rock you can tell not only whether the magnetic pole was in the north or south when the rock formed but also—from the more subtle positions of the needles—the latitude of the rock at the time it formed."

On the striated pavement of Algeria lies the till of polar glaciers. There are tropical atolls in Canada, tropical limestones in Siberia, tropical limestones in Antarctica. From fossils, from climates preserved in stone, such facts were known long before paleomagnetism was discovered; but they were, to say the least, imperfectly understood. Paleomagnetism, first perceived in 1906, eventually confirmed what the paleoclimatologists and paleontologists had been saying about the latitudes of origins of rocks, but it did not resolve the mystery of the phenomenon, because there seemed to be two equally reasonable explanations. Either the rock had moved (and continents with it) or the whole earth had rolled, like a child's top slowly turning on its side, and the poles and equator had wandered. Either the equator had gone to Minnesota or Minnesota had gone to the equator.

As early as the sixteenth century, the specific movements of the earth's surface that eventually became known as continental drift and plate tectonics had been hypothesized. The Flemish geographer Abraham Ortelius, in the third edition of his *Thesaurus Geographicus* (Antwerp, 1596), postulated that the American continents were "torn away from Europe and Africa" by earthquakes and other catastrophic events. "The vestiges of the rupture reveal themselves," he continued, "if someone brings forward a map of the world and considers carefully the . . . projecting parts of Europe and Africa . . .

along with the recesses of America." In centuries that followed, various writers called attention to the suggestive shapes of landmasses, but almost no one else imagined that the landmasses had been driven apart, let alone by what mechanism. In 1838, the Scottish philosopher Thomas Dick, of County Angus, published his *Celestial Scenery; or, the Wonders of the Planetary System Displayed: Illustrating the Perfections of Deity and a Plurality of Worlds*, in which he noted how neatly western Africa could lock itself tight around the horn of Brazil, "and Nova Scotia and Newfoundland would block up a portion of the Bay of Biscay and the English Channel, while Great Britain and Ireland would block up the entrance to Davis's Straits." Such an assembly would "form one compact continent." And "a consideration of these circumstances renders it not altogether improbable that these continents were originally conjoined, and that, at some former physical revolution or catastrophe, they may have been rent asunder by some tremendous power, when the waters of the ocean rushed in between them, and left them separated as we now behold them." I am indebted to Alan Goodacre, of the Geological Survey of Canada, for this high-assay nugget, and to James Romm, of Bard College, for the quotations from Ortelius, which they separately reported in the British journal *Nature* in 1991 and 1994, backdating by three centuries the continental-drift hypothesis attributed in textbooks to the meteorologist Alfred Wegener, of Graz in the Styrian Alps. Ortelius and Dick fared better than Wegener, for while their propositions achieved no significant attention, Wegener's won a considerable fame that rapidly decayed into notoriety.

In an address to the German Geological Association in 1912, and three years later in his essay *Die Enstehung der Kontinente und Ozeane*, Wegener based his concept not only on the jigsaw fit of Africa and the Americas but also on the likeness of certain rocks on the two sides of the ocean, and on comparisons of living and fossil creatures. He knew nothing of paleomagnetism, which was in its infancy and was many years away from yielding insight to the problem, but he was the promulgator of the hypothesis of continental drift. Unfortunately, he attempted to explain how the continents moved. He envisioned them plowing like icebreakers through solid basalt. Almost no one believed his hypothesis, any more than Benjamin Franklin had been believed when, in 1782, possibly after a visit to Edinburgh, he said he thought that the surface parts of the earth were floating about on a liquid interior. Wegener had received fame as a record-setting balloonist, an Arctic explorer, and now he was making an assertion for which his name would live in mockery for about fifty years. In life and in death, he was a target of scorn. His idea provoked gibes, jeers, sneers, derision, raillery, burlesque, mockery, irony, satire, and sarcasm, but it could not be ignored. In 1928, the American Association of Petroleum Geologists published a symposium on continental drift. It included a paper called "Some of the Objections to Wegener's Theory," by Rollin T. Chamberlin, of the University of Chicago, who expressed what was then the prevailing attitude among geologists and would continue to be until the nineteen-seventies, after which it would cease to prevail but not to survive:

> Wegener's hypothesis in general is of the foot-loose type, in that it takes considerable liberty with our globe, and is less bound by restrictions or tied down by awkward, ugly facts than most of its rival theories. Its appeal seems to lie in the fact that it plays a game in which there are few restrictive rules and no sharply drawn code of conduct. So a lot of things go easily. But taking the situation as it now is, we must either modify radically most of the present rules of the geological game or else pass the hypothesis by. The best characterization of the hypothesis which I have heard was the remark made at the 1922 meeting of the Geological Society of America at Ann Arbor. It was this: "If we are to believe Wegener's hypothesis we must forget everything which has been learned in the last seventy years and start over again."

Through the nineteen-thirties, and particularly after the Second World War, paleomagnetic data accrued, and, as it presented its story of kaleidoscopic environments changing through time in any given place, academic geologists sketched on globes and maps their curves of apparent polar wander. Here is where the poles were at the end of the Silurian; this is where they went from there. Rocks of identical age, sampled in various parts of the world, indicated as much in their imprisoned compasses.

Some geologists—little cells of them in South Africa, the odd don or two at Cambridge—preferred the other explanation, but they were few, and in geology departments around the world everybody would annually crowd in to hear Lucius P. Aenigmatite, Regius Professor of Historical Geology, give his world-renowned lecture ridiculing continental drift. Oil geologists, when they had found what they were looking for in deep sandstones put down by ancient rivers, naturally yearned to know in what direction those rivers had flowed. They had long since learned empirically that if you wanted to find the direction of the stream you had to use different pole positions for well cores of different ages. Whether this was the result of polar wander or continental drift did not much matter to the flying red horse. Other geologists satisfied themselves by deciding that the paleomagnetic compasses were unreliable, notwithstanding that oil companies were using them to make money. Certain English geologists produced confusion by embracing continental drift and then drawing up narratives and maps that showed continents moving all over the earth with respect to a fixed and undriftable England.

By the late nineteen-fifties, paleomagnetic evidence had piled up so high that it demanded improved explication. India, for example, yielded data that put it out of harmony with the rest of the world with respect to polar wander. Either there was an inexplicable series of anomalies in the data or India itself had moved, coming up from the Southern Hemisphere and completely crossing the equator, rapidly, and at a rate of speed (as much as twenty-two centimetres a year) completely out of synchronization with the rate at which the equator's position had differed in other terrains. More data, and increased sophistication in the analysis of data, began to show that polar-wander curves—once thought to be in agreement worldwide—could differ some from continent to continent. Curves based on Paleozoic and Triassic rock in North America and in Europe looked much alike but, oddly, stood separate in the way that a single line will appear to be double in inebriate vision. The gap corresponds to the present width of the Atlantic Ocean. The opening of the Atlantic began in the Triassic.

If the hypothesis of continental drift had long been overshadowed by the hypothesis of polar wander, the reverse would before long be true. Researchers in paleomagnetism at Cambridge University concluded that their data were showing them that both hypotheses could be correct, as later research at Princeton would confirm. The poles indeed had wandered. The continents had moved as well. The phenomenon of "apparent polar wander" had been caused, right enough, by the movement of masses of land, but concomitantly the earth had rolled—and patterns of "true polar wander" were seen to be superimposed on all the other motions of the shifting surface of the world. But what motions? If the continents had drifted, then in what manner were they drifting? Where had they come from and where were they going? What would happen if two should collide? Since they obviously were not plowing through solid basalt, how in fact did they move? It was all within a decade—1960–1968—that these questions were given answers of startling cohesion, as not only paleomagnetists but seismologists and oceanographers, geologists and geophysicists, whose specialties had been diverging through time, suddenly drew together around new outpourings of information and produced a chain of scientific papers whose interlocking insights would for most geologists fundamentally adjust their understanding of the dynamics of the earth.

"It was a change as profound as when we gave up the Biblical story," Deffeyes said as he tapped his collector into the ground. "It was a change as profound as Darwinian evolution, or Newtonian or Einsteinian physics."

The papers themselves had straightforward scientific titles, some of which—perhaps only in the afterlight of their great effect—seem to resound with the magnitude of the subject: "History of Ocean Basins," "Rises, Trenches, Great Faults, and Crustal Blocks," "Sea-Floor Spreading and Continental Drift," "Seismology and the New Global Tectonics." From Berkeley, Princeton, San Diego, New York, Canberra, Cambridge (England), there were about twenty primal contributions which, taken together, can be said to have constituted the plate-tectonics revolution.

Now plate tectonics is widely taken for granted. When I was in high school, there was essentially no television in America, and four years later television had replaced flypaper. When I was in high school, in the nineteen-forties, the term "plate tectonics" did not exist—albeit there was one remarkably prescient paragraph in our physical-geology textbook about the motions and mechanisms of continental drift. Today, children in schoolrooms just assume that the story being taught them is as old as the hills, and was told by God himself to their teacher in 4004 B.C.

The story is that everything is moving, that the outlines of continents by and large have nothing to do with these motions, that "continental drift" is actually a misnomer, that only the world picture according to Marco Polo makes much sense in the old-time browns and greens and Rand McNally blues. The earth is at present divided into some twenty crustal segments called plates. Plate boundaries miscellaneously run through continents, around continents, along the edges of continents, and down the middle of oceans. The plates are thin and rigid, like pieces of eggshell. In miles, sixty deep by nine thousand by eight thousand are the dimensions of the Pacific Plate. "Pacific Plate" is not synonymous with "Pacific Ocean," which wholly or partly covers many other plates. There are virtually no landmasses associated with some plates—the Cocos Plate, the Nazca Plate. Some plates are almost entirely land—the Arabian Plate, the Iranian Plate, the Eurasian Plate, a large part of which used to be known as the (heaven help us) China Plate. (Jokes may be invisible to some geologists. Harry Hess, who in 1960 opened out the new story with his "History of Ocean Basins," began it with these words: "The birth of the oceans is a matter of conjecture, the subsequent history is obscure, and the present structure is just beginning to be understood. Fascinating speculation on these subjects has been plentiful, but not much of it predating the last decade holds water.") Certain major plates are about half covered with ocean—the South American Plate, the African Plate, the North American Plate. Australia and India are parts of the same plate. It is shaped like a boomerang, with the landmasses at either end. It may be in the early stages of separating itself into two plates. The northern section has a slightly different motion but there is no sharp boundary. In Africa, the terrane east of the Great Rift Valley is far enough along in the act of separation to be called the Somalian Plate, but the boundary is not yet continuous. Continents in themselves are not drifting, are not cruise ships travelling the sea. Continents are high parts of plates. East-west, the North American Plate starts in the middle of the Atlantic Ocean and ends in San Francisco. West-east, the Eurasian Plate begins in the middle of the Atlantic Ocean and ends in the sea of Okhotsk.

It is the plates that move. They all move. They move in varying directions and at different speeds. The Adriatic Plate is moving north. The African Plate came up behind it and drove it into Europe—drove Italy like a nail into Europe—and thus created the Alps. The South American Plate is moving west. The Nazca Plate is moving east. The Antarctic Plate is spinning, like pan ice in a river.

As has happened only twice before in geology—with Abraham

Werner's neptunist system and James Hutton's *Theory of the Earth* —the theory of plate tectonics has assembled numerous disparate phenomena into a single narrative. Where plates separate, they produce oceans. Where they collide, they make mountains. As oceans grow, and the two sides move apart, new seafloor comes into the middle. New seafloor is continuously forming at the trailing edge of the plate. Old seafloor, at the leading edge of a plate, dives into deep ocean trenches—the Kuril Trench, the Aleutian Trench, the Marianas Trench, the Java Trench, the Japan Trench, the Philippine Trench, the Peru-Chile Trench. The seafloor goes down four hundred miles after it goes into the trenches. On the way down, some of it melts, loses density, and—white-hot and turbulent—rises toward the surface of the earth, where it emerges as volcanoes, or stops below as stocks and batholiths, laccoliths and sills. Most of the volcanoes of the world are lined up behind the ocean trenches. Almost all earthquakes are movements of the boundaries of plates —shallow earthquakes at the trailing edges, where the plates are separating and new material is coming in, shallow earthquakes along the sides, where one plate is ruggedly sliding past another (the San Andreas Fault), and earthquakes of any depth down to four hundred miles below and beyond the trenches where plates are consumed (Japan, 1923; Chile, 1960; Alaska, 1964; Mexico, 1985). A seismologist discovered that deep earthquakes under a trench had occurred on a plane that was inclined forty-five degrees into the earth. As ocean floors reach trenches and move on down into the depths to be consumed, the average angle is something like that. Take a knife and cut into an orange at forty-five degrees. To cut straight down would be to produce a straight incision in the orange. If the blade is tilted forty-five degrees, the incision becomes an arc on the surface of the orange. If the knife blade melts inside, little volcanoes will come up through the pores of the skin, and together they will form arcs, island arcs—Japan, New Zealand, the Philippines, the New Hebrides, the Lesser Antilles, the Kurils, the Aleutians.

Where a trench happens to run along the edge of a continent and subducting seafloor dives under the land, the marginal terrain will rise. The two plates, pressing, will create mountains, and volcanoes will appear as well. The Peru-Chile Trench is right up against the west coast of South America. The Nazca Plate, moving east, is

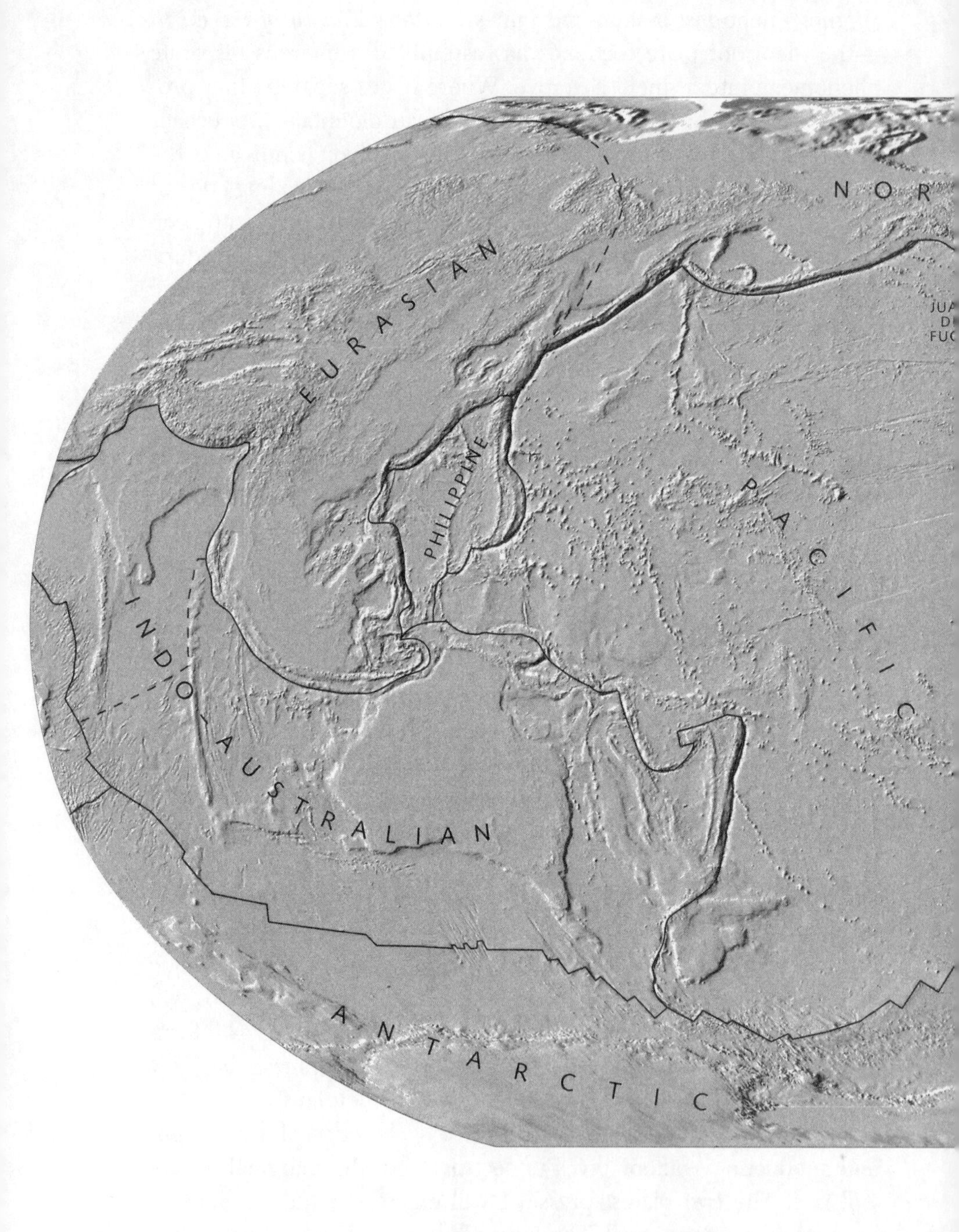
EURASIAN
PHILIPPINE
PACIFIC
INDO-AUSTRALIAN
ANTARCTIC

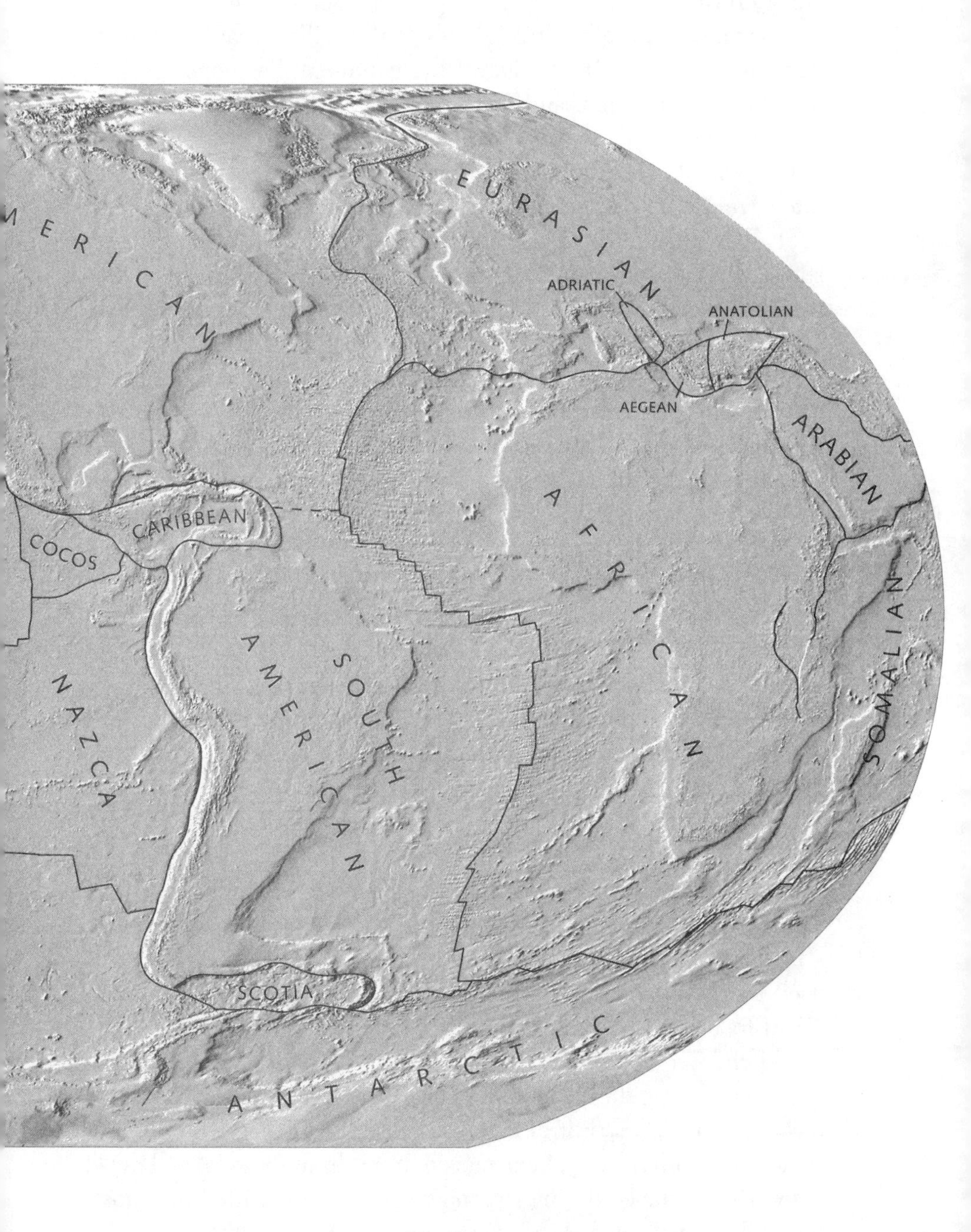

MAJOR LITHOSPHERIC PLATES

AND SOME MINOR ONES

going down into the trench. Interspersed among the uplifted Andes are four thousand miles of volcanoes. The Pacific Ocean floor, going down to melt below that edge of the continent, has done much to help lift it twenty thousand feet.

Seafloor—ocean crust—is dense enough to go down a trench, but continents are too light, too buoyant. When a continent comes into a trench, it will become stuck there, causing havoc. Even if part of it goes down some dozens of kilometres, it will eventually get stuck. Australia is such a continent, and where it has jammed a trench it has buckled up the earth to make the mountains of New Guinea, sixteen thousand five hundred feet.

When two continental masses happen to move on a collision course, they gradually close out the sea between them—barging over trenches, shutting them off—and when they hit they drive their leading edges together as a high and sutured welt, resulting in a new and larger continental mass. The Urals are such a welt. So is the Himalaya. The Himalaya is the crowning achievement of the Indo-Australian Plate. India, in the Oligocene, crashed head on into Tibet, hit so hard that it not only folded and buckled the plate boundaries but also plowed in under the newly created Tibetan Plateau and drove the Himalaya five and a half miles into the sky. The mountains are in some trouble. India has not stopped pushing them, and they are still going up. Their height and volume are already so great they are beginning to melt in their own self-generated radioactive heat. When the climbers in 1953 planted their flags on the highest mountain, they set them in snow over the skeletons of creatures that had lived in the warm clear ocean that India, moving north, blanked out. Possibly as much as twenty thousand feet below the seafloor, the skeletal remains had formed into rock. This one fact is a treatise in itself on the movements of the surface of the earth. If by some fiat I had to restrict all this writing to one sentence, this is the one I would choose: The summit of Mt. Everest is marine limestone.

Plates grow, shrink, combine, disappear, their number changing through time. They shift direction. Before the Pliocene, there was a trench off California. Seafloor moved into it from the west and dived eastward into the earth. Big volcanoes came up. Under the volcanoes, the melted crust cooled in huge volumes as new granite batholiths. Basin-range faulting has elevated the batholiths to fourteen thousand feet, and weather has sketched them out as the Sierra Nevada.

When seafloor goes into a trench, there can be a certain untidiness as segments are shaved off the top. They end up sitting on the other plate, large hunks of ocean crust that formed as much as a few thousand miles away and are now emplaced strangely among the formations of the continent. The California Coast Ranges—the hills of Vallejo, the hills of San Simeon, the hills of San Francisco—are a kind of berm that was pushed up out of the water by the incoming plate, including large slices of the seafloor and a jumble of oceanic and continental materials known to geologists as the Franciscan mélange. Geologists used to earn doctorates piecing together the stratigraphy of the Franciscan mélange, finding bedding planes in rock masses strewn here and there, and connecting them with dotted lines. Plate tectonics reveals that there is no stratigraphy in the mélange, no consecutive story of deposition—just mountains of bulldozed hash. The eastward motion of the ocean plate stopped soon after basin-range faulting began. The plate started moving in another direction. The trench, ceasing to be a trench then, was replaced by the San Andreas Fault.

Mountain building, in the Old Geology, had been seen as a series of orogenies rhythmically spaced through time, in part resulting from isostatic adjustments, and in part the work of "earth forces" that were not extensively explained. As mountains were disassembled, their materials were deposited in huge troughs, depressions, downbendings of the crust that were known as geosynclines. Earth forces made the geosynclines. As sediment accumulated in them, its weight pressed ever farther down into the mantle until the mantle would take no more, and then there came a trampoline effect, an isostatic bounce, that caused the material to rise. The Gulf of Mexico was a good example of a geosyncline, with a large part of the Rocky Mountains sitting in it as more than twenty-five thousand feet of silt, sand, and mud, siltstone, sandstone, and shale. "The South will rise again!" Deffeyes used to say. The huge body of sediment would one day be lifted far above sea level and dissected by weather and wrinkled into mountains in the way that the skin of an apple wrinkles as the apple grows old and dry. The steady rhythm of these orogenies was known as "the symphony of the earth"—the Avalonian Orogeny in latest Precambrian time, the Taconic Orogeny in late Ordovician time, the Acadian Orogeny in late Devonian time, the Antler Orogeny in Mississippian time, the Alleghenian Orogeny in Pennsylvanian-Permian time, the Laramide Orogeny in Cretaceous-Tertiary time. It was a slow march of global uplifting effects—predictable—proceeding through history in stately order. By the end of the nineteen-sixties, the symphony had come to the last groove, and was up in the attic with the old Aeolian. Mountain building had become a story of random collisions, unpredictable, whims of the motions of the plates, which, when continents collided or trenches otherwise jammed, could give up going one way and move in another. The Avalonian, Taconic, Acadian, and Alleghenian orogenies were now seen, in plate theory, not as distinct events but as successive parts of the same event, which involved the closing of an ocean called Iapetus that existed more or less where the Atlantic is today. The continents on either side of Iapetus came together not head-on but like scissors closing from the north, folding and faulting their conjoining boundaries to make the Atlas Mountains and the Appalachian chain. It was a Paleozoic story, and the motions finally stopped. In the Mesozoic, an entirely new dynamic developed and the crust in the same region began to pull apart, to break into blocks that formed a new province, a Eurafrican-American basin and range. The blocks kept on separating until a new plate boundary formed, and eventually a new marine basin, which looked for a while like the Red Sea before widening to become an ocean.

It was on the mechanics of the seafloor that geology's revolutionary inquiries were primarily focussed in the early days. Harry Hess, a mineralogist who taught at Princeton, was the skipper of an attack transport during the Second World War, and he carried troops to landings—against the furious defenses of Iwo Jima, for example, and through rockets off the beaches of Lingayen Gulf. Loud noises above the surface scarcely distracted him. He had brought along a new kind of instrument called a Fathometer, and, battle or no battle, he never turned it off. Its stylus was drawing pictures of the floor of the sea. Among the many things he discerned there were dead volcanoes, spread out around the Pacific bottom like Hershey's Kisses on a tray. They had the arresting feature that their tops had been cut off, evidently the work of waves. Most of them were covered with thousands of feet of water. He did not know what to make of them. He named them guyots, for a nineteenth-century geologist at Princeton, and sailed on.

The Second World War was a technological piñata, and, with their new Fathometers and proton-precession magnetometers, oceanographers of the nineteen-fifties—most notably Bruce Heezen and Marie Tharp at Columbia University—mapped the seafloor in such extraordinary detail that in a sense they were seeing it for the first time. (Today, the very best maps are classified, because they reveal the places where submarines hide.) What stood out even more prominently than the deep trenches were mountain ranges that rose some six thousand feet above the general seafloor and ran like seams through every ocean and all around the globe. They became known as rises, or ridges—the Mid-Atlantic Ridge, the Southeast Indian Ocean Ridge, the East Pacific Rise. They fell away gently from their central ridgelines, and the slopes extended outward hundreds of miles, to the edges of abyssal plains—the Hatteras Abyssal Plain, the Demerara Abyssal Plain, the Tasman Abyssal Plain. Right down the spines of most of the submarine cordilleras ran high axial valleys, grooves that marked the summit line. These eventually came to be regarded as rift valleys, for they proved to be the boundaries between separating plates. As early as 1956, oceanographers at Columbia had assembled seismological data suggesting that a remarkable percentage of all earthquakes were occurring in the mid-ocean rifts—a finding that was supported, and then some, after a worldwide system of more than a hundred seismological monitoring stations was established in anticipation of the nuclear-test-ban treaty of 1963. If there was to be underground testing, one had to be able to detect someone else's tests, so a by-product of the Cold War was seismological data on a scale unapproached before. The whole of plate tectonics, a story of steady-state violence along boundaries, was being brought to light largely as a result of the development of instruments of war. Earthquakes "focus" where earth begins to move, and along transform faults like the San Andreas the focusses were shallow. At the ocean trenches they could be very deep. The facts accrued. Global maps of the new seismological data showed earthquakes not only clustered all along the ridges of the seafloor mountains but also in the trenches and transform faults, with the result that the seismology was sketching the earth's crustal plates.

To Rear Admiral Hess, as he had become in the U.S. Naval Reserve, it now seemed apparent that seafloors were spreading away from mid-ocean ridges, where new seafloor was continuously being created in deep cracks, and, thinking through as many related phenomena as he was able to discern at the time, he marshalled his own research and the published work of others up to 1960 and wrote in that year his "History of Ocean Basins." In the nineteen-forties, a professor at Delft had written a book called *The Pulse of the Earth*, in which he asserted with mild cynicism that where gaps exist among the facts of geology the space between is often filled with things "geopoetical," and now Hess, with good-humored candor, adopted the term and announced in his first paragraph that while he meant "not to travel any further into the realm of fantasy than is absolutely necessary," he nonetheless looked upon what he was about to present as "an essay in geopoetry." He could not be sure which of his suppositions might be empty conjecture and which might in retrospect be regarded as precocious insights. His criterion could only have been that they seemed compelling to him. His guyots, he had by now decided, were volcanoes that grew at spreading centers, where they protruded above the ocean surface and were attacked by waves. With the moving ocean floor they travelled slowly down to the abyssal plains and went on eventually to "ride down into the jaw crusher" of the deep trenches, where they were consumed. "The earth is a dynamic body with its surface constantly changing," he wrote, and he agreed with others that the force driving it all must be heat from deep in the mantle, moving in huge revolving cells (an idea that had been around in one form or another since 1839 and is still the prevailing guess in answer to the unresolved question: What is the engine of plate tectonics?). Hess reasoned also that the heat involved in the making of new seafloor is what keeps the ocean rises high, and that moving outward the new material gradually cools and subsides. The rises seemed to be impermanent features, the seafloor altogether "ephemeral." "The whole ocean is virtually swept clean (replaced by new mantle material) every three hundred to four hundred million years," he wrote, not then suspecting that ocean crust is actually consumed in half that time. "This accounts for the relatively thin veneer of sediments on the ocean floor, the relatively small number of volcanic seamounts, and the present absence of evidence of rocks older than Cretaceous in the oceans." In ending, he said, "The writer has attempted to invent an evolution for ocean basins. It is hardly likely that all of the numerous assumptions made are correct. Nevertheless it appears to be a useful framework for testing various and sundry groups of hypotheses relating to the oceans. It is hoped that the framework with necessary patching and repair may eventually form the basis for a new and sounder structure."

In 1963, Drummond Matthews and Fred Vine, of Cambridge University, published an extraordinary piece of science that gave to Hess's structure much added strength. Magnetometers dragged back and forth across the seas had recorded magnetism of two quite different intensities. Plotted on a map, these magnetic differences ran in stripes that were parallel to the mid-ocean ridges. The magnetism over the centers of the ridges themselves was uniformly strong. Moving away from the ridges, the strong and weak stripes varied in width from a few kilometres to as many as eighty. Vine and Matthews, chatting over tea in Cambridge, thought of using this data to connect Harry Hess's spreading seafloor to the time scale of paleomagnetic reversals. The match would turn out to be exact. The weaker stripes matched times when the earth's magnetic field had been reversed, and the strong ones matched times when the magnetic pole was in the north. Moreover, the two sets of stripes—calendars, in effect, moving away from the ridge—seemed to be symmetrical. The seafloor was not only spreading. It was documenting its age. L. W. Morley, a Canadian, independently had reached the same conclusions. Vine and Matthews' paper was published in *Nature* in September, 1963, and became salient in the development of plate tectonics. In January of the same year, Morley had submitted almost identical ideas to the editors of *Nature*, but they were not yet prepared to accept them, so Morley then submitted the paper in the United States to the *Journal of Geophysical Research*, which rejected it summarily. Morley's paper came back with a note telling him that his ideas were suitable for a cocktail party but not for a serious publication.

Data confirming the Vine-Matthews hypothesis began to accumulate, nowhere more emphatically than in a magnetic profile of the seafloor made by the National Science Foundation's ship Eltanin crossing the East Pacific Rise. The Eltanin's data showed that the seafloor became older and older with distance from the spreading center, and with perfect symmetry for two thousand kilometres on either side. All through the nineteen-sixties, ships continued to cruise the oceans dragging magnetometers behind, and eventually computers were programmed to correlate the benthic data with the surface wanderings of the ships. Potassium-argon dating had timed the earth's magnetic reversals to apparent perfection for the last three and a half million years. Geologists at Columbia calculated the rate of seafloor spreading for those years and then assumed the rate to have been constant through earlier time. On that assumption, they extrapolated a much more extensive paleomagnetic time scale. (Improved radiometric dating later endorsed the accuracy of the method.) And with that scale they swiftly mapped the history of ocean basins. Compared with a geologic map of a continent, it was a picture handsome and spare. As the paleomagnetist Allan Cox, of Stanford University, would describe it in a book called *Plate Tectonics and Geomagnetic Reversals*, "The structure of the seafloor is as simple as a set of tree rings, and like a modern bank check it carries an easily decipherable magnetic signature."

Meanwhile, geophysicists at Toronto, Columbia, Princeton, and the Scripps Institution of Oceanography were filling in the last major components of the plate-tectonics paradigm. They figured out the geometry of moving segments on a sphere, showed that deformation happens only at the margins of plates, charted the relative motions of the plates, and mapped for the first time the plate boundaries of the world.

If it was altogether true, as Hess had claimed, that with relative frequency "the whole ocean is virtually swept clean," then old rock should be absent from deep ocean floors. Since 1968, the drill ships Glomar Challenger and JOIDES Resolution have successively travelled the world looking for, among other things, the oldest ocean rocks. The oldest ever found is Jurassic. In a world that is 4.56 billion years old, with continental-shield rock that has been dated to 3.96, it is indeed astonishing that the oldest rock that human beings have ever removed from a seafloor has an age of a hundred and eighty-five million years—that the earth is twenty-five times as old as the oldest rock of the oceans. In 1969, it seemed likely that the oldest ocean floor would be found in the Northwest Pacific. The Glomar Challenger went there to see. Two Russians were aboard who believed that rock older than Jurassic—rock of the Paleozoic, in all likelihood—would be discovered. They took vodka with them to toast the first trilobite to appear on deck. Trilobites, index fossils of the Paleozoic, came into the world at the base of the Cambrian and went out forever in the Permian Extinction—sixty-five million years before the age of the oldest rock ever found in modern oceans. As expected, the oceanic basement became older and older as the ship drilled westward from Hawaii. But even at the edge of the Marianas Trench, the Russians were disappointed. No vodka. Ah, but there might be older rock on the other side of the trench, in the floor of the Philippine Sea. The ship pulled up its drill pipe and moved across the trench. This time the rock was Miocene, more or less a tenth as old as the Jurassic floor. The Russians broke out the vodka. A toast! Neil Armstrong and Edwin Aldrin were walking around on the moon.

"In the old days we would have called this North America," Deffeyes said, sinking another clear tube into the ground. "We now think of plates. The plate-tectonics revolution came as a surprise, with very little buildup to it. There was none of that cloud that precedes a political revolution. In the nineteen-fifties, when I was a graduate student, nearly all the faculty at Princeton thought continental drift was sheer baloney. A couple of years later, Harry Hess broke it open. I had thought I would go through my career without anything like it. Oil and mining seemed enough of a contribution to keep one going. But now something had come along that was so profound that it took the whole science with it. We used to think that continents grew like onions around old rock. That was overturned by plate tectonics. And we could see now how amazingly fast you could put up a mountain range. A continent-to-continent collision was a hell of an episode at a limited place. After the Appalachians and the Urals were recognized as continent-to-continent sutures, people said, 'O.K., where's the suture in California?' Geologists kept jumping up and saying, 'I've got the suture! I've got the suture!' It turned out, of course, that there were at least three sutures. In each instance, a great island had closed up a sea and hit into America—just as India hit Tibet, just as Kodiak Island, which is a mini-India, is about to plow into Alaska. Fossils from the mid-Pacific have been found here in the West, and limestones that lithified a thousand miles south of the equator. Formations in California have alien fossils with cousins in the rock of New Guinea. For a while, people were going around naming a defunct ocean for every suture. The first piece, coming in from the west, was the one that rode up onto North America about forty miles, not a trivial distance, in Mississippian time. That was the action that first tipped the rock in the Carlin unconformity. The old name for it was the Antler Orogeny. In the early Triassic, the second one arrived—the Golconda Thrust—and rode fifty miles over the trailing edge of the first one, and in the Jurassic the third one came in, sutured on somewhere near Sacramento, and more or less completed California. I have read that two geologists have found in Siberia a displaced terrane that was taken off of North America. The Lord giveth and the Lord taketh away."

I mentioned that I had read in *Geology* that one out of eight geologists does not accept plate-tectonic theory.

He said, "There are still a few people dragging their feet. They don't want to come into the story."

I asked him if he thought the Uinta Mountains could be explained in terms of plate theory. The Uintas are a range in the Rockies, seven hundred miles from the sea, and they run east-west, unlike virtually all other ranges for thousands of miles around them. If the Western cordilleras were raised by colliding plates, how did the Uintas happen to come up at right angles to the other mountains?

He said, "You must have been talking to a Rocky Mountain geologist." He said nothing else for a time, while he tapped at the earth I had uncovered and captured a perfect sample. Then he said, "The north side of the Uintas is a spectacular mountain wall. Glorious. You come upon it and suddenly you see structurally the boundary of the range. But you don't see what put it there. The Uintas are mysterious. They are not a basin-range fault block, yet they have come up nearly vertically, with almost no compression evident. You just stand there and watch them go up into the sky. They don't fit our idea of plate tectonics. The Rockies in general will be one of the last places in the world to be deciphered in terms of how many hits created them, and just when, and from where."

The article in *Geology* was based on a questionnaire that was circulated toward the end of the nineteen-seventies. The results indicated that forty per cent of geologists had come to feel that plate theory was "essentially established," while a roughly equal number preferred to qualify a bit and say that it was "fairly well established." Eleven per cent felt that the theory was "inadequately proven." Seven per cent said they had accepted continental drift before 1940. Six per cent thought plate tectonics would be "still in doubt" in the late nineteen-eighties. And one geologist predicted that the theory would eventually be rejected.

"At any given moment, no two geologists are going to have in their heads exactly the same levels of acceptance of all hypotheses and theories that are floating around," Deffeyes said. "There are always many ideas in various stages of acceptance. That is how science works. Ideas range from the solidly accepted to the literally half-baked—those in the process of forming, the sorts of things about which people call each other up in the middle of the night. All science involves speculation, and few sciences include as much speculation as geology. Is the Delaware Water Gap the outlet of a huge lake all other traces of which have since disappeared? A geomorphologist will tell you that, in principle, the idea is O.K. You have to deal with partial information. In oil drilling, you had better be ready to act shrewdly on partial information. Do physicists do that? Hell, no. They want to have it to seven decimal places on their Hewlett-Packards. The geologist has to choose the course of action with the best statistical chance. As a result, the style of geology is full of inferences, and they change. No one has ever seen a geosyncline. No one has ever seen the welding of tuff. No one has ever seen a granite batholith intrude."

Since I was digging his sample pits, I felt enfranchised to remark on what I took to be the literary timbre of his science.

"There's an essential difference," he said. "The authors of literary works may not have intended all the subtleties, complexities, undertones, and overtones that are attributed to them by critics and by students writing doctoral theses."

"That is what God says about geologists," I told him, chipping into the sediment with his broken shovel.

"You may recall Archelaus's explanation of earthquakes," he said cryptically. "Earthquakes were caused by air trapped in underground caves. It shook the earth in its effort to escape. Everyone knew then that the earth was flatulent."

Deffeyes said he had asked his friend Jason Morgan—whose paper "Rises, Trenches, Great Faults, and Crustal Blocks" defined the boundaries of the plates—what he was going to do for an encore. Morgan said he didn't know, but possibly the most exciting thing to do next would be to prove the theory wrong.

That would be a reversal comparable to the debunking of Genesis. I remembered Eldridge Moores, of the University of California at Davis, telling me what it had been like to be in graduate school at the height of the plate-tectonics revolution, and how he had imagined that the fervor and causal excitement of it was something like landing on Guadalcanal in the middle of the action of "a noble war." Tanya Atwater, a marine geologist who eventually joined the faculty of the Massachusetts Institute of Technology, was then a graduate student at the Scripps Institution of Oceanography. In a letter written to Allan Cox at Stanford, she re-creates the milieu of the time. "Seafloor spreading was a wonderful concept because it could explain so much of what we knew, but plate tectonics really set us free and flying. It gave us some firm rules so that we could predict what we should find in unknown places. . . . From the moment the plate concept was introduced, the geometry of the San Andreas system was an obviously interesting example. The night Dan McKenzie and Bob Parker told me the idea, a bunch of us were drinking beer at the Little Bavaria in La Jolla. Dan sketched it on a napkin. 'Aha!' said I, 'but what about the Mendocino trend?' 'Easy!,' and he showed me three plates. As simple as that! The simplicity and power of the geometry of those three plates captured my mind that night and has never let go since. It is a wondrous thing to have the random facts in one's head suddenly fall into the slots of an orderly framework.

It is like an explosion inside. That is what happened to me that night and that is what I often felt happen to me and to others as I was working out (and talking out) the geometry of the western U.S. . . . The best part of the plate business is that it has made us all start communicating. People who squeeze rocks and people who identify deep-ocean nannofossils and people who map faults in Montana suddenly all care about each other's work. I think I spend half my time just talking and listening to people from many fields, searching together for how it might all fit together. And when something does fall into place, there is that mental explosion and the wondrous excitement. I think the human brain must love order."

Deffeyes, meanwhile, had joined Shell before the excitement developed. Growing up in oil fields, he had grown to like them, to admire the skill and independence of the crews, the competent manner in which they lived with danger. "Like a bullfighter, you are careful. So danger is not an overwhelming risk. But it is always there. And you can be crushed, burned, asphyxiated, destroyed by an explosion. A crew on a rig floor runs pipe in the hole with swift precision, and any piece of equipment can take your hand off just as fast." As a small boy, he often went into oil fields with his father, whose assignments changed many times—Oklahoma City, Hutchinson, Great Bend, Midland, Hobbs, Casper. As a teen-ager, Deffeyes played the French horn in the Casper Civic Symphony. He debated on the high-school team. He became—as he has remained—a forensic marvel, the final syllables of his participles and gerunds ringing like Buddha's gongs. In the way that others collected stamps, he collected rocks. For counsel, he took his specimens to the geologists in town, of whom there were plenty, including Paul Walton, who in 1948 had suggested to J. Paul Getty that he go to Kuwait. High-school summers, Deffeyes worked as an assistant shooter with seismic crews and as a roustabout maintaining wells. When he finished graduate school and moved on to Houston with Shell, he was ignorant not only of the imminent revolution in geologic theory but also of the approaching atrophy in successful exploration for oil. M. King Hubbert, an outstanding geological geophysicist, was with Shell at the time, and Deffeyes had only settled in when Hubbert happened to predict (with amazing accuracy) the approaching date when more oil would be coming out of American ground than geologists would be discovering. He predicted the energy crisis that would inevitably follow. When Deffeyes saw Hubbert's evanescing figures, he saw disappearing with them what had looked to be his most productive years. He resigned from Shell to go into teaching and was soon on the faculty at Oregon State, where he set himself up as a chemical oceanographer, because the ocean was where things were happening. The university had bought from the government a small ship left over from the Second World War and had converted it for oceanographic research. "Working for an oil company had suddenly become like working for a railroad—a dying industry. Now in this new field new equipment was being improvised, and the problems were the same as they were in the oil fields when I was a kid. In the ocean, we used bottom-hole pressure gauges and other oil-field equipment. I could feel the same sort of excitement I had felt years before in the oil fields, and with the same sorts of people—roustabouts and roughnecks—in the crew."

Unfortunately, Deffeyes had a signal defect as an oceanographer. He got terribly seasick. His enthusiasm grew moist, and he began to contrive to remain ashore. Then in October, 1965, J. Tuzo Wilson, of the University of Toronto, and Fred Vine, of Cambridge, published a paper in which they defined an oddly isolated piece of mid-ocean ridge off the coasts of Oregon and Washington. It was the spreading center of what would eventually become known as the Juan de Fuca Plate, one of the smallest of all the crustal plates in the world. The volcanoes of the Cascades—Mt. Hood, Mt. Rainier, Mt. St. Helens, Glacier Peak—were lined up behind its trench. "Continental drift is one hypothesis I'll get seasick for," Deffeyes decided, and he signed up for a week's use of the ship. He had no program in mind. To ask for a suggestion, he picked up a telephone and called Harry Hess. Instantly, Hess said, "Go to the ridge and dredge some rock from the axial valley. It better not be old." Hess's hypothesis that new seafloor forms at ocean rises had scarcely been tested. This was before the Eltanin profile, and before the voyages of the Glomar Challenger. Hess's immediate response to Deffeyes was to suggest a test that could have shelved his hypothesis then and there.

Deffeyes went out to dredge the rock, but first he had to find the ridge, so he made a long pass with his echo sounder tracing the profile of the bottom. The ridge-axis rock, when he dredged it up, was extremely young. But what in the end interested Deffeyes at least as much was the benthic profile that had been traced by the stylus of the sounder. The profile of the spreading center in the ocean bottom off Oregon seemed remarkably familiar to someone who had done his thesis field work in Nevada. It appeared to be, in miniature, a cross section of the Basin and Range. The new crust, spreading out, had broken into fault blocks and had become a microcosm of the Basin and Range, because both were expressions of the same cause. It was a microcosm, too, of the Triassic lowlands of the East two hundred and ten million years ago—Triassic Connecticut, Triassic New Jersey—with their border faults and basalt flows, their basins and ranges, gradually extending, pulling apart, to open the Atlantic. The Red Sea of today was what the Atlantic and its two sides had looked like about twenty million years after the Atlantic began to open. The Red Sea today was what the Basin and Range would probably look like at some time in the future.

In December, 1972, the astronaut Harrison Schmitt, riding in Apollo 17, looked down at the Red Sea and the Gulf of Aden—at a simple geometry that seemed to have been made with a jigsaw barely separating Africa and the Arabian peninsula. He told the people at Mission Control, "I didn't grow up with the idea of drifting continents and seafloor spreading. But I tell you, when you look at the way pieces of the northeastern portion of the African continent seem to fit together, separated by a narrow gulf, you could make a believer out of anybody." Schmitt was one of the eighty per cent who were changing their minds about the new global theory. In addition to his astronaut's training, he had a Ph.D. in geology, and he would bring back from the moon a hundred kilos of rock.

Twenty miles out of Winnemucca, and the interstate is dropping south toward the Humboldt Range. A coyote runs along beside the road. It is out of its element, tongue out, outclassed, under minimum speed. Deffeyes says that most ranges in the Basin and Range had one or two silver deposits in them, if any, but the Humboldts had five. We have also entered the bottomlands of the former Lake Lahontan. The hot-springs map shows more activity in this part of the province. Extension of the earth's crust has been somewhat more pronounced here, Deffeyes explains, and hence there are more ore deposits. He feels that when a seaway opens up, the spreading center will be somewhere nearby. Or possibly back in Utah, in the bed of Lake Bonneville. "But this one has better connections."

"Connections?"

"Death Valley. Walker Lake. Carson Sink." An Exxon map of the western United States is spread open on the seat between us. He runs his finger from Death Valley to Carson Sink and on northward to cross the interstate at Lovelock. "The ocean will open here," he repeats. "Or in the Bonneville basin. I think here."

A few miles off the road is the site of a planned community dating from the nineteen-sixties. It was to have wide streets and a fountained square, but construction was delayed and then indefinitely postponed. Ghostless ghost town, it had been named Neptune City.

With the river on our right, we round the nose of the Humboldt Range, as did the Donner party and roughly a hundred and sixty-five thousand other people, in a seventeen-year period, heading in their wagons toward Humboldt Sink, Carson Sink, and the terror of days without water. But first, as we do now, they came into broad green flats abundantly fertile with grass, knee-high grass, a fill for the oxen, the last gesture of the river before it vanished into the air. The emigrants called this place the Big Meadows of the Humboldt, and something like two hundred and fifty wagons would be resting here at any given time.

"There was a sea here in the Triassic," Deffeyes remarks. "At least until the Sonomia terrane came in and sutured on. The sea was full of pelagic squid, and was not abyssal, but it was deep enough so the bottom received no sunlight, and bottom life was not dominant."

"How do you know it was not dominant?"

"Because I have looked at the siltstones and the ammonites in them, and that is what I see there."

Visions of oceans before and behind us in time, we roll on into Lovelock. SLOW—DUST HAZARD. Lovelock, Nevada 89419. There are cumulus snow clouds overhead and big bays of blue in the cold sky, with snow coming down in curtains over the Trinity Range, snow pluming upward over the valley like smoke from a runaway fire. Lovelock was a station of the Overland Stage. It became known throughout Nevada as "a good town with a bad water supply." An editor of the *Lovelock Review-Miner* wrote in 1915, "There is little use in trying to induce people to locate here until the water question is settled. . . . Maybe the water does not kill anyone, but it certainly drives people away." In 1917, Lovelock was incorporated as a third-class city, and one of its first acts was to enforce a ban on houses of prostitution within twelve hundred feet of the Methodist Episcopal Church. Another was a curfew. Another ordered all city lights turned off when there was enough moon.

JAX CASINO LIBERAL SLOTS

TWO STIFFS SELLING GAS AND MOTEL

WATER SUPPLY FROM PRIVATE WELL

LOVELOCK SEED COMPANY
GRAINS AND FEED

Here in the Big Meadows of the Humboldt, the principal employer is the co-op seed mill on the edge of town, which sends alfalfa all over the world.

On the sidewalks are men in Stetsons, men in three-piece suits, men in windbreakers, tall gaunt overalled men with beards. There are women in Stetsons, boots, and jeans. A thin young man climbs out of a pickup that is painted in glossy swirls of yellow and purple, and has a roll bar, balloon tires, headphones, and seventeen lights.

There are terraces of Lake Lahontan above the ballfield of the Lovelock Mustangs. Cattle graze beside the field. The Ten Commandments are carved in a large piece of metamorphosed granite outside the county courthouse.

NO. 10:
THOU SHALT NOT COVET THY NEIGHBOR'S WIFE,
NOR HIS MANSERVANT, NOR HIS MAIDSERVANT,
NOR HIS CATTLE

BRAZEN ONAGER—BAR—BUD—PIZZA

WHOO-O-A MOTEL

"Lovelock was a person's name," Deffeyes cautions.

LOVELOCK MERCANTILE

The name is fading on the cornice of Lovelock Mercantile. It was built in 1905, expanded in 1907, is the bus stop now, liquor store, clothing store, grocery store, real-estate office, bakery, Western Union office—all in one room. There is a sign on one of the columns that hold up the room:

WE CANNOT ACCEPT
GOVERNMENT MEAL TICKETS

Across the valley is a huge whitewash "L" on a rock above the fault scar of the Humboldt Range.

We go into Sturgeon's Log Cabin restaurant and sit down for coffee against a backdrop of rolling cherries, watermelons, and bells. A mountain lion in a glass case. Six feet to the tip of the tail. Shot by Daniel (Bill) Milich, in the Tobin Range.

I hand Deffeyes the Exxon map and ask him to sketch in for me the opening of the new seaway, the spreading center as he sees it coming. "Of course, all the valleys in the Great Basin are to a greater or lesser extent competing," he says. "But I'd put it where I said—right here." With a pencil he begins to rough in a double line, a swath, about fifteen miles wide. He sketches it through the axis of Death Valley and up into Nevada, and then north by northwest through Basalt and Coaldale before bending due north through Walker Lake, Fallon, and Lovelock. "The spreading center would connect with a transform fault coming in from Cape Mendocino," he adds, and he sketches such a line from the California coast to a point a little north of Lovelock. He is sketching the creation of a crustal plate, and he seems confident of that edge, for the Mendocino transform fault—the Mendocino trend—is in place now, ready to go. He is less certain about the southern edge of the new plate, because he has two choices. The Garlock Fault runs east-west just above Los Angeles, and that could become a side of the new plate; or the spreading center could continue south through the Mojave Desert and the Salton Sea to meet the Pacific Plate in the Gulf of California. "The Mojave sits in there with discontinued basin-and-range faulting," Deffeyes says, almost to himself, a substitute for whistling, as he sketches in the alternative lines. "There has to be a transform fault at the south end of the live, expanding rift. The sea has got to get through somewhere."

Now he places his hands on the map so that they frame the Garlock and Mendocino faults and hold between them a large piece of California—from Bakersfield to Redding, roughly, and including San Francisco, Sacramento, and Fresno—not to mention the whole of the High Sierra, Reno, and ten million acres of Nevada. "You create a California Plate," he says. "And the only question is: Is it this size, or the larger one? How much goes out to sea?" British Columbia is to his left and Mexico is to his right, beside his coffee cup on the oak Formica. The coast is against his belly. He moves his hands as if to pull all of central California out to sea. "Does this much go?" he says. "Or do the Mojave and Baja go with it?" A train of flatcars pounds through town carrying aircraft engines.

My mind has drifted outside the building. I am wondering what these people in this dry basin—a mile above sea level—would think if they knew what Deffeyes was doing, if they were confronted with the news that an ocean may open in their town. I will soon find out.

"What?"

"Are you stoned?"

"The way I see it, I won't be here, so the hell with it."

"It's a little doubtful. It could be, but it's a little doubtful."

"If it happens real quick, I guess a couple of people will die, but if it's like most other things they'll find out about it hundreds of years before and move people out of here. The whole world will probably go to hell before that happens anyway."

"You mean salt water, crests, troughs, big splash, and all that? Don't sweat it. You're safe here—as long as Pluto's out there."

"We got a boat."

"That's the best news I've heard in a couple of years. When I go bye-bye to the place below, why, that water will be there to cool me. I hope it's Saturday night. I won't have to take an extra bath."

"It may be a good thing, there's so many politicians; but they may get an extra boat. I used to be a miner. Oh, I've been all over. But now they've got machines and all the miners have died."

"The entire history of Nevada is one of plant life, animal life, and human life adapting to very difficult conditions. People here are the most individualistic you can find. As district attorney, I see examples of it every day. They want to live free from government interference. They don't fit into a structured way of life. This area was settled by people who shun progress. Their way of life would be totally unattractive to most, but they chose it. They have chosen conditions that would be considered intolerable elsewhere. So they would adapt, easily, to the strangest of situations."

"I've been here thirty-three years, almost half of that as mayor. I can't quite imagine the sea coming in—although most of us know that this was all underwater at one time. I know there's quite a fault that runs to the east of us here. It may not be active. But it leaves a mark on your mind."

"Everybody's entitled to an opinion. Everybody's entitled to ask a question. If I didn't think your question was valid, I wouldn't have to answer you. I'd hope the fishing was good. I wouldn't mind having some beach-front property. If it was absolutely certified that it was going to happen, we should take steps to keep people out of the area. But as chief of police I'm not going to be alarmed."

"It'll be a change to have water here instead of desert. By God, we could use it. I say that as fire chief. We get seventy fire calls a year, which ain't much, but then we have to go a hundred miles to put out those damned ranch fires. We can't save much, but we can at least put out the heat. I got a ten-thousand-gallon tank there, which is really something for a place with no water. I guess I won't still be here to see the ocean come, and I'm glad of it, because I can't swim."

Meanwhile, Deffeyes, in Sturgeon's Log Cabin, applies the last refining strokes to his sketchings on the map. "The Salton Sea and Death Valley are below sea level now, and the ocean would be there if it were not for pieces of this and that between," he says. "We are extending the continental crust here. It is exactly analogous to the East African Rift, the Red Sea, the Atlantic. California will be an island. It is just a matter of time."

Book 2

In Suspect Terrain

The paragraph that follows is an encapsulated history of the eastern United States, according to plate-tectonic theory and glacial geology.

About a thousand million years ago, a continent of unknown dimensions was rifted apart, creating an ancestral ocean more or less where the Atlantic is now. The older ocean has been called Iapetus, because Iapetus was the father of Atlas, for whom the Atlantic is named. Some geologists, who may feel that their science is dangerously clever, are snappish about Iapetus. They prefer to say proto-Atlantic. The ancestral ocean existed a great deal longer than the Atlantic has, but gradually, across some two hundred and fifty million years in the Paleozoic era, it closed. Moving toward each other, the great landmasses on either side buckled and downwarped the continental shelves and then came together in a crash no less brutal than slow—a continent-to-continent collision marked by an alpine welt, which has reached its old age as the Appalachian Mountains. In the Mesozoic era, two hundred and ten million years ago, rifting began again, pulling apart certain segments of the mountain chain, creating fault-block basins—remnants of which are the Connecticut River Valley, central New Jersey, the Gettysburg battlefields, the Culpeper Basin—and eventually parting the earth's crust enough to start a new ocean, which is now three thousand miles wide and is

Detail from "Delaware Water Gap," by George Innes, 1859. Collection of the Montclair Art Museum

still growing. Meanwhile, a rhythm of glaciation has been established in what is essentially the geologic present. Ice sheets have been forming on either side of Hudson Bay and have spread in every direction to cover virtually all of Canada, New England, New York, and much of New Jersey, Pennsylvania, and the Middle West. The ice has come and gone at least a dozen times, in cycles that seem to require about a hundred thousand years, and, judging by other periods of glaciation in the earlier history of the earth, the contemporary cycles have only begun. About fifty more advances can be expected. Some geologists have attempted to isolate the time in all time that runs ten thousand years from the Cro-Magnons beside the melting ice to the maternity wards of the here and now by calling it the Holocene epoch, with the implication that this is our time and place, and the Pleistocene—the "Ice Age"—is all behind us. The Holocene appears to be nothing more than a relatively deglaciated interval. It will last until a glacier two miles thick plucks up Toronto and deposits it in Tennessee. If that seems unlikely, it is only because the most southerly reach of the Pleistocene ice fields to date stopped seventy-five miles shy of Tennessee.

Anita Harris is a geologist who does not accept all that is written in that paragraph. She is cool toward aspects of plate tectonics, the novel theory of the earth that explains mountain belts and volcanic islands, ocean ridges and abyssal plains, the deep earthquakes of Alaska and the shallow earthquakes of a fault like the San Andreas as components of a unified narrative, wherein the shell of the earth is divided into segments of varying size, which separate to form oceans, collide to make mountains, and slide by one another causing buildings to fall. In a revolutionary manner, plate-tectonic theory burst forth in the nineteen-sixties, and Anita Harris is worried now that the theory is taught perhaps too glibly in schools. In her words: "It's important for people to know that not everybody believes in it. In many colleges, it's all they teach. The plate-tectonics boys move continents around like crazy. They publish papers every year revising their conclusions. They say that a continental landmass up against the eastern edge of North America produced the Appalachians. I know about some of the geology there, and what they say about it is wrong. I don't say they're wrong everywhere. I'm open-minded. Too often, though, plate tectonics is oversimplified and overapplied.

I get all heated up when some sweet young thing with three geology courses tells me about global tectonics, never having gone on a field trip to look at a rock."

As she made these comments, she was travelling west on Interstate 80, approaching Indiana on a gray April morning. She had brought me along to "do geology," as geologists like to say—to see the countryside as she discerned it. Across New Jersey, Pennsylvania, and Ohio, she had been collecting, among other things, limestones and dolomites for their contained conodonts, index fossils from the Paleozoic, whose extraordinary utility in oil and gas exploration had been her discovery, with the result that Mobil and Chevron, Amoco and Arco, Chinese and Norwegians had appeared at her door. She was driving, and she wore a railroad engineer's striped hat, a wool shirt, blue jeans, and old split hiking boots—hydrochloric acid for testing limestones and dolomites in a phial in a case on her hip. With her high cheekbones, her assertive brown eyes, her long dark hair in twin ponytails, she somehow suggested an American aborigine. Of middle height, early middle age, she had been married twice—first to a northern-Appalachian geologist, and now to a southern-Appalachian geologist. She was born on Coney Island and grew up in a tenement in Williamsburg Brooklyn. There was not a little Flatbush in her manner, soul, and speech. Her father was Russian, and his name in the old country was Herschel Litvak. In Brooklyn, he called himself Harry Fishman, and sometimes Harry Block. According to his daughter, English names meant nothing to Russian Jews in Brooklyn. She grew up Fishman and became in marriage Epstein and Harris, signing her geology with her various names and imparting some difficulty to followers of her professional papers. With her permission, I will call her Anita, and let the rest of the baggage go. Straightforwardly, as a student, she went into geology because geology was a means of escaping the ghetto. "I knew that if I went into geology I would never have to live in New York City," she once said to me. "It was a way to get out." She was nineteen years old when she was graduated from Brooklyn College. She remembers how pleased and astounded she was to learn that she could be paid "for walking around in mountains." Paid now by the United States Geological Survey, she has walked uncounted mountains.

After the level farmlands of northwestern Ohio, the interstate climbed into surprising terrain—surprising enough to cause Anita to suspend her attack on plate tectonics. Hills appeared. They were steep in pitch. The country resembled New England, a confused and thus beautiful topography of forested ridges and natural lakes, stone fences, bunkers and bogs, cobbles and boulders under maples and oaks: Indiana. Rough and semi-mountainous, this corner of Indiana was giving the hummocky lie to the reputed flatness of the Middle West. Set firmly on the craton—the Stable Interior Craton, unstirring core of the continent—the whole of Middle America is structurally becalmed. Its basement is coated with layers of rock that are virtually flat and have never experienced folding, let alone upheaval. All the more exotic, then, were these abrupt disordered hills. Evidently superimposed, they almost seemed to have been created by the state legislature to relieve Indiana. Not until the nineteenth century did people figure out whence such terrain had come, and how and why. "Look close at those boulders and you'll see a lot of strangers," Anita remarked. "Red jasper conglomerates. Granite gneiss. Basalt. None of those are from anywhere near here. They're Canadian. They have been transported hundreds of miles."

The ice sheets of the present era, in their successive spreadings overland, have borne immense freight—rock they pluck up, shear off, rip from the country as they move. They grind much of it into gravel, sand, silt, and clay. When the ice melts, it gives up its cargo, dumping it by the trillions of tons. The most recent advance has been called the Wisconsinan ice sheet, because its effects are well displayed in Wisconsin. Its effects, for all that, are not unimpressive in New York. The glacier dumped Long Island where it is (nearly a hundred per cent of Long Island), and Nantucket, and Cape Cod, and all but the west end of Martha's Vineyard. Wherever the ice stopped and began to melt back, it signed its retreat with terminal moraines—huge accumulations of undifferentiated rock, sand, gravel, and clay. The ice stopped at Perth Amboy, Metuchen, North Plainfield, Madison, Morristown—leaving a sinuous, morainal, lobate line that not only connects these New Jersey towns but keeps on going to the Rocky Mountains. West of Morristown, old crystalline rock from the earth's basement—long ago compressed, distorted, and partially melted, driven upward and westward in the Appalachian upheavals—stands now in successive ridges, which are called the New Jersey Highlands. They trend northeast-southwest. With a notable exception, they have discouraged east-west construction of roads. When the last ice sheet set down its terminal moraine, it built causeways from one ridge to another, on which Interstate 80 rides west. Over the continent, the ice had spread southward about as evenly as spilled milk, and there is great irregularity in its line of maximum advance. South of Buffalo, it failed to reach Pennsylvania, but it plunged deep into Ohio, Indiana, Illinois. The ice sheets set up and started Niagara Falls. They moved the Ohio River. They dug the Great Lakes. The ice melted back in stages. Pausing here and there in temporary equilibrium, it sometimes readvanced before continuing its retreat to the north. Wherever these pauses occurred, as in northeastern Indiana, boulders and cobbles and sand and gravel piled up in prodigal quantity—a cadence of recessional moraines, hills of rock debris. The material, heterogeneous and unsorted, has its own style of fabric, in which geologists can see the moves and hesitations of the ice, not to mention its weight and velocity. Scottish farmers, long before they had any idea what had laid such material upon Scotland, called it till, by which they meant to convey a sense of "ungenial subsoil," of coarse obdurate land.

"This would be a good place for a golf course," Anita remarked, and scarcely had she uttered the words than—after driving two thousand yards on down the road with a dogleg to the left—we were running parallel to the fairways of a clonic Gleneagles, a duplicated Dumfries, a faxed Blairgowrie, four thousand miles from Dumfriesshire and Perthshire, but with natural bunkers and traps of glacial sand, with hummocky roughs and undulating fairways, with kettle depressions, kettle lakes, and other chaotic hazards. "If you want a golf course, go to a glacier" is the message according to Anita Harris. "Golf was invented on the moraines, the eskers, the pitted outwash plains—the glacial topography—of Scotland," she explained. "All over the world, when people make golf courses they are copying glacial landscapes. They are trying to make countryside that looks like this. I've seen bulldozers copying Scottish moraines in places like Louisiana. It's laughable."

On warm afternoons in summer, the meltwater rivers that pour from modern glaciers become ferocious and unfordable, like the Suiattle, in Washington, coming down from Glacier Peak, like the

Yentna, in Alaska, falling in tumult from the McKinley massif. Off the big ice sheets of the Pleistocene have come many hundreds of Suiattles and Yentnas, most of which are gone now, leaving their works behind. The rivers have built outwash plains beyond the glacial fronts, sorting and smoothing miscellaneous sizes of rock—moving cobbles farther than boulders, and gravels farther than cobbles, and sands farther than gravels, and silt grains farther than sands—then gradually losing power, and filling up interstices with groutings of clay. Enormous chunks of ice frequently broke off the retreating glaciers and were left behind. The rivers built around them containments of gravel and clay. Like big, buried Easter eggs, the ice sat there and slowly melted. When it was gone, depressions were left in the ground, pitting the outwash plains. The depressions have the shapes of kettles, or at least have been so described, and "kettle" is a term in geology. All kettles contained water for a time, and some contain water still. Rivers that developed under glaciers ran in sinuous grooves. Rocks and boulders coming out of the ice fell into the rivers, building thick beds contained between walls of ice. When the glacier was gone, the riverbeds were left as winding hills. The early Irish called them eskers, meaning pathways, because they used them as means of travel above detentive bogs. Where debris had been concentrated in glacial crevasses, melting ice left hillocks, monticles, hummocks, knolls, braes—collections of lumpy hills known generically to the Scots as kames. In Indiana as in Scotland—in La Bresse and Estonia as in New England and Quebec—the sort of country left behind after all these features have been created is known as kame-and-kettle topography.

The interstate was waltzing with the glacier—now on the outwash plain, now on moraine, among the kettles and kames of Scottish Indiana. Roadcuts were green with vetch covering glacial till. We left 80 for a time, the closer to inspect the rough country. The glacier had been away from Indiana some twelve thousand years. There were many beds of dried-up lakes, filled with forest. In the Boundary Waters Area of northern Minnesota, the ice went back ten thousand years ago, possibly less, and most of the lakes it left behind are still there. The Boundary Waters Area is the scene of a contemporary conservation battle over the use and fate of the lakes. "Another five thousand years and there won't be much to fight about," Anita said, with a shrug and a smile. "Most of those Minnesota lakes will probably be as dry as these in Indiana." Some of the larger and deeper ones endure. We made our way around the shores of Lake James, Bingham Lake, Lake of the Woods, Loon Lake. Like Walden Pond, in Massachusetts, they were kettles.

The woods around them were bestrewn with boulders, each an alien, a few quite large. If a boulder rests above bedrock of another type, it has obviously been carried some distance and is known as an erratic. In Alaska, I have come upon glacial erratics as big as office buildings, with soil developing on their tops and trees growing out of them like hair. In Pokagon State Park, Indiana, handsome buildings looked out on Lake James—fieldstone structures, red and gray, made of Canadian rocks. The red jasper conglomerates were from the north shore of Lake Huron. The banded gray gneisses were from central Ontario. The sources of smaller items brought to Indiana by the ice sheets have been less easy to trace—for example, diamonds and gold. During the Great Depression, one way to survive in Indiana was to become a pick-and-shovel miner and earn as much as five dollars a day panning gold from glacial drift—as all glacial deposits, sorted and unsorted, are collectively called. There were no nuggets, nothing much heavier than a quarter of an ounce. But the drift could be fairly rich in fine gold. It had been scattered forth from virtually untraceable sources in eastern Canada. One of the oddities of the modern episodes of glaciation is that while three-fifths of all the ice in the world covered North America and extended south of Springfield, Illinois, the valley of the Yukon River in and near Alaska was never glaciated, and as a result the gold in the Yukon drainage—the gold of the richest placer streams ever discovered in the world—was left where it lay, and was not plucked up and similarly scattered by overriding ice. Miners in Indiana learned to look in their pans for menaccanite—beanlike pebbles of iron and titanium that signalled with some consistency the propinquity of gold. The menaccanite had come out of the exposed Precambrian core of Canada—the Canadian craton, also known as the Canadian Shield. There were garnets in the gold pans, too—and magnetite, amphibole, corundum, jasper, kyanite. Nothing in that list is native to Indiana, and all are in the Canadian Shield. There is Canadian copper in the drift of Indiana, and there are diamonds that are evidently

Canadian, too. Hundreds have been discovered—pink almond-shaped hexoctahedrons, blue rhombic dodecahedrons. Weights have approached five carats, and while that is modest compared with twenty-carat diamonds found in Wisconsin, these Indiana diamonds have nonetheless been accorded the stature of individual appellations: the Young Diamond (1898), the Stanley Diamond (1900).

The source of a diamond is a kimberlite pipe, a form of diatreme—a relatively small hole bored through the crust of the earth by an expanding combination of carbon dioxide and water which rises from within the earth's mantle and moves so fast driving magma to the surface that it breaks into the atmosphere at supersonic speeds. Such events have occurred at random through the history of the earth, and a kimberlite pipe could explode in any number of places next year. Rising so rapidly and from so deep a source, a kimberlite pipe brings up exotic materials the like of which could never appear in the shallow slow explosion of a Mt. St. Helens or the flows of Kilauea. Among the materials are diamonds. Evidently, there are no diamond pipes, as they are also called, in or near Indiana. Like the huge red jasper boulders and the tiny flecks of gold, Indiana's diamonds are glacial erratics. They were transported from Canada, and by reading the fabric of the till and taking bearings from striations and grooves in the underlying rock—and by noting the compass orientation of drumlin hills, which look like sculptured whales and face in the direction from which their maker came—anybody can plainly see that the direction from which the ice arrived in this region was something extremely close to 045°, northeast. At least one pipe containing gem diamonds must exist somewhere near a line between Indianapolis and the Otish Mountains of Quebec, because the ice that covered Indiana did not come from Kimberley—it formed and grew and, like an opening flower, spread out from the Otish Mountains. With rock it carried and on rock it traversed, it narrated its own journey, but it did not reveal where it got the diamonds.

There is a layer in the mantle, averaging about sixty miles below the earth's surface, through which seismic tremors pass slowly. The softer the rock, the slower the tremor—so it is inferred that the low-velocity zone, as it is called, is close to its melting point. In the otherwise rigid mantle, it is a level of lubricity upon which the plates of the earth can slide, interacting at their borders to produce the effects known as plate tectonics. The so-termed lithospheric plates, in other words, consist of crust and uppermost mantle and can be as much as ninety miles thick. Diamond pipes are believed to originate a good deal deeper than that—and in a manner which, as most geologists would put it, "is not well understood." After drawing fuel from surrounding mantle rock—compressed water from mica, in all likelihood, and carbon dioxide from other minerals—the material is thought to work slowly upward into the overlying plate. Slow it may be at the start, but a hundred and twenty miles later it comes out of the ground at Mach 2. The result is a modest crater, like a bullet hole between the eyes.

No one has ever drilled a hundred and twenty miles into the earth, or is likely to. Diamond pipes, meanwhile, have brought up samples of what is there. It is spewed all over the landscape, but it also remains stuck in the throat, like rich dense fruitcake. For the most part, it is peridotite, which is the lowest layer of the subcontinental package and is believed to be the essence of the mantle. There is high-pressure recrystallized basalt, full of garnets and jade. There are olivine crystals of incomparable size. The whole of it is known as kimberlite, the matrix rock of diamonds.

The odds against diamonds appearing in any given pipe are about a hundred to one. Carbon will crystallize in its densest form only under conditions of considerable heat and pressure—pressures of the sort that exist deep below the thickest parts of the plates, pressures of at least a hundred thousand pounds per square inch. The thickest parts of the plates are the continental cores, the cratons. All diamond-bearing kimberlites ever found have been in pipes that came up through cratons. Down where diamonds form, they are stable, but as they travel upward they pass through regions of lower pressure, where they will swiftly turn into graphite. Only by passing through such regions at tremendous speed can diamonds reach the earth's surface as diamonds, where they cool suddenly and enter a state of precarious preservation that somehow betokens to human beings a touching sense of "forever." Diamonds shoot like bullets through the earth's crust. Nonetheless, they are often found within rinds of graphite. Countless quantities turn into graphite altogether or disappear into the air as carbon dioxide. At room temperature and surface pressure, diamonds are in repose on an extremely narrow thermodynamic shelf. They want to be graphite, and with a relatively modest boost of heat graphite is what they would become, if atmospheric oxygen did not incinerate them first. They are, in this sense, unstable—these finger-flashing symbols of the eternity of vows, yearning to become fresh pencil lead. Except for particles that are sometimes found in meteorites, diamonds present themselves in nature in no other way.

Kimberlite is easily eroded. A boy playing jacks in South Africa in 1867 picked up an alluvial diamond that led to the discovery of a number of pipes, one of which became the Kimberley Mine. From that pipe alone, fourteen million carats followed. The rock source of diamonds had never before been known. The Regent, the Koh-i-noor, the Great Mogul had been eroded out by streams. As the ice walls of the Pleistocene moved across Quebec, resculpting mountains, digging lakes, they apparently dozed through kimberlite pipes, scattering the contents southwest. The ice that plucked up the diamonds not only brought questions with it but also obscured the answers. How many pipes are there? Where are they? How rich are they in diamonds? If one ten-millionth of their content is gem diamond, they would be worth mining. They are somewhere northeast of Indiana. They are in all likelihood less than a quarter of a mile wide. They may be under glacial drift. They may be under lakes. A few have been discovered—none of value. Presumably, there are others, relatively studded with diamonds. Many people have searched. No one has found them.

"In Siberia, a few years ago, a couple of diamond pipes were located after diamonds were discovered in glacial drift," Anita told me.

I said, "Possibly some Russian geologists could be helpful here."

Looking out across the water of Lake James at a line of morainal hills, she chose to ignore the suggestion. The hills screened the outwash plain beyond. After some moments, she said, "Rocks remember. They may not be able to tell you exactly where in Canada to look for a diamond pipe, but when you have diamonds in this drift you'd better believe it is telling you that diamond pipes are there. Rocks are the record of events that took place at the time they formed. They are books. They have a different vocabulary, a different alphabet, but you learn how to read them. Igneous rocks tell you the temperature at which they changed from the molten to the solid state, and they tell you the date when that happened, and hence they give you a picture of the earth at that time, whether they formed three thousand million years ago or flowed out of the ground yesterday. In sedimentary rock, the colors, the grain sizes, the ripples, the crossbedding give you clues to the energy of the environment of deposition—for example, the force and direction and nature of the rivers that laid down the sediments. Tracks and trails left by organisms—and hard parts of their bodies, and flora in the rock—tell whether the material came together in the ocean or on the continent, and possibly the depth and temperature of the water, and the temperature on the land. Metamorphic rocks have been heated, compressed, and recrystallized. Their mineral composition tells you if they were originally igneous or sedimentary. Then they tell you what happened later on. They tell you the temperatures when they changed. At one point, I wanted to major in history. My teachers steered me into science, but I really majored in history. I grew up in topography like this, believe it or not. Looking at these lakes and hills, you'd never think of Brooklyn. For that matter, you'd never think of Indiana. I didn't know what bedrock meant. I remember how amazed I was to discover, in learning to read rocks, how much history there was. All the glacial stuff arrived just yesterday and is sitting on the surface. Most of Brooklyn is a pitted outwash plain. Brooklyn means broken land."

A day would come when I would pick up Anita Harris at the home of a cousin of hers in Morganville, New Jersey, and drive across the Narrows Bridge to Brooklyn. She had not seen her neighborhood for twenty-five years. Her cousin, Murray Srebrenick, who gave us coffee before we left, was more than a little solicitous toward us, and even somewhat embarrassed, as if he were in the presence of people with an uncorrectable defect. He, too, had grown up in

Brooklyn, and now, as an owner and operator of trucks, he supported his suburban life hauling clothes to Seventh Avenue. On runs through the city to various warehouses, he and his drivers knew what routes to avoid, but often enough they literally ran into trouble. Crime was part of his overhead, and as he rinsed the coffee cups he finally came out with what he was thinking and pronounced us insane. He spoke with animation, waving a pair of arms that could bring down game. Old neighborhood or no old neighborhood, he said, he would not go near Williamsburg, or for that matter a good many other places in Brooklyn; and he reeled off stories of open carnage that might have tested the stomach of the television news. I wondered what it might be like to die defending myself with a geologist's rock hammer. Anita, for her part, seemed nervous as we left for the city. Twenty-five years away, she seemed afraid to go home.

It was an August day already hot at sunrise. "In Williamsburg, I lived at 381 Berry Street," she said as we crossed the big bridge. "It was the worst slum in the world, but the building did have indoor plumbing. Our first apartment there was a sixth-floor walkup. The building was from the turn of the century and was faced with red Triassic sandstone." Brooklyn was spread out before us, and Manhattan stood off to the north, with its two sets of skyscrapers three miles apart—the ecclesiastical spires of Wall Street, and beyond them the midtown massif. Anita asked me if I had ever wondered why there was a low saddle in the city between the stands of tall buildings.

I said I had always assumed that the skyline was shaped by human considerations—commercial, historical, ethnic. Who could imagine a Little Italy in a skyscraper, a linoleum warehouse up in the clouds?

The towers of midtown, as one might imagine, were emplaced in substantial rock, Anita said—rock that once had been heated near the point of melting, had recrystallized, had been heated again, had recrystallized, and, while not particularly competent, was more than adequate to hold up those buildings. Most important, it was right at the surface. You could see it, in all its micaceous glitter, shining like silver in the outcrops of Central Park. Four hundred and fifty million years in age, it was called Manhattan schist. All through midtown, it was at or near the surface, but in the region south of Thirtieth Street it began to fall away, and at Washington Square it descended abruptly. The whole saddle between midtown and Wall Street would be underwater, were it not filled with many tens of fathoms of glacial till. So there sat Greenwich Village, SoHo, Chinatown, on material that could not hold up a great deal more than a golf tee—on the ground-up wreckage of the Ramapos, on crushed Catskill, on odd bits of Nyack and Tenafly. In the Wall Street area, the bedrock does not return to the surface, but it comes within forty feet and is accessible for the footings of the tallest things in town. New York grew high on the advantage of its hard rock, and, New York being what it is, cities all over the world have attempted to resemble it. The skyline of nuclear Houston, for example, is a simulacrum of Manhattan's. Houston rests on twelve thousand feet of montmorillonitic clay, a substance that, when moist, turns into mobile jelly. After taking so much money out of the ground, the oil companies of Houston have put hundreds of millions back in. Houston is the world's foremost city in fat basements. Its tall buildings are magnified duckpins, bobbing in their own mire.

We skirted Brooklyn on the Belt Parkway, heading first for Coney Island, where Anita had spent many a day as a child, and where, somewhat impatiently, she had been born. Her mother, seven months pregnant, took a subway to the beach one day, and Anita first drew breath in Coney Island Hospital.

"Cropsey Avenue," she said now, reading a sign. "Keep right, we're going off here."

I went into the right lane, signals blinking, but the exit was chocked with halted traffic. There were police. There were flashing lights. Against the side of an abused Pontiac, a young man was leaning palms flat, like a runner stretching, while a cop addressed him with a drawn pistol. "Welcome home, Anita," said Anita.

The broad beach was silent, so early in the morning, where people in ten thousands had been the day before, and where numbers just as great would soon return. The Parachute Jump stood high in relief. The Cyclone was in shadow and touched by slanting light. Reminiscently, Anita ran her eye from the one to the other and to the elevated railways beyond. When a fossil impression is left in sand by the outside of an organic structure, it is known in geology as an external mold. One would not have to be a sedimentologist to read this beach, with its colonies of giant bivalves. We walked to the strandline, the edge of the water, where the play of waves had concentrated heavy dark sands—hematite, magnetite, small garnets broken out by the glacier from their matrix of Manhattan schist.

The beach itself, with its erratic sands, was the extremity of the outwash plain. The Wisconsinan ice sheet, arriving from the north, had come over the city not from New England, as one might guess, but primarily from New Jersey, whose Hudson River counties lie due north of Manhattan. Big boulders from the New Jersey Palisades are strewn about in Central Park, and more of the same diabase is scattered through Brooklyn. The ice wholly covered the Bronx and Manhattan, and its broad snout moved across Astoria, Maspeth, Williamsburg, and Bedford-Stuyvesant before sliding to a stop in Flatbush. Flatbush was the end of the line, the point of return for the Ice Age, the locus of the terminal moraine. Water poured in white tumult from the melting ice, carrying and sorting its freight of sands and gravels, building the outwash plain: Bensonhurst, Canarsie, the Flatlands, Coney Island. When Anita was a child, she would ride the D train out to Coney Island, with an old window screen leaning against her knees. She sifted the beach sand for lost jewelry. In the beach sand now, she saw tens of thousands of garnets. There is a lot of iron in the Coney Island beach as well, which makes it tawny from oxidation, and not a lot of quartz, which would make it white. The straw-colored sand sparkled with black and silver micas—biotite, muscovite—from Fifth Avenue or thereabouts, broken out of Manhattan schist. A beach represents the rock it came from. Most of Coney Island is New Jersey diabase, Fordham gneiss, Inwood marble, Manhattan schist. Anita picked up some sand and looked at it through a hand lens. The individual grains are characteristically angular and sharp, she said, because the source rock was so recently crushed by the glacier. To make a well-rounded grain, you need a lot more time. Weather and waves had been working on this sand for fifteen thousand years.

If the gneissic grains and garnets were erratics, so in their way were the Schenley bottles, the Pepsi-Cola cans, the Manhattan Schlitz, the sand-coated pickles and used paper plates.

"Colonial as penguins, dirtier than mud daubers," I observed of the creatures of the beach.

"We rank with bats, starlings, and Pleistocene sloths as the great messmakers of the world," said Anita, and we left Coney Island for Williamsburg.

North over the outwash plain we followed Ocean Parkway five miles—broad, tree-lined Ocean Parkway, with neat houses in trim neighborhoods, reaching into shaded streets. Ahead, all the while, loomed the terminal moraine, suggesting, from a distance, an escarpment, but actually just a fairly steep hill. Eastern Parkway defines its summit, two hundred feet high. Two hundred feet of till. Near Prospect Park you begin to climb. One moment you are level on the plain and the next you are nose up, gaining altitude. There are cemeteries in every direction: Evergreens Cemetery, Lutheran Cemetery, Mt. Carmel, Cypress Hills, Greenwood Cemetery—some of the great necropolises of all time, with three million under sod, moved into the ultimate neighborhood, the terminal moraine. "In glacial country, all you have to do is look for cemeteries if you want to find the moraine," Anita said. "A moraine is poor farmland—steep and hummocky, with erratics and boulders. Yet it's easy ground to dig in, and well drained. An outwash plain is boggy. There's a cemetery over near Utica Avenue that's in the outwash. Most people prefer moraine. I would say it's kind of distasteful to put your mother down into a swamp."

Ebbets Field, where they buried the old Brooklyn Dodgers, was also on the terminal moraine. When a long-ball hitter hit a long ball, it would land on Bedford Avenue and bounce down the morainal front to roll toward Coney on the outwash plain. No one in Los Angeles would ever hit a homer like that.

We detoured through Prospect Park, which is nestled into the morainal front and is studded with big erratics on raucously irregular ground. It looks much like Pokagon Park, in Indiana, with the difference that the erratics there are from the Canadian Shield and these were from the New Jersey Palisades. Pieces of the Adirondacks have been found in Pennsylvania, pieces of Sweden on the north German plains, and no doubt there is Ticonderoga dolomite, Schenectady sandstone, and Peekskill granite in the gravels of Canarsie and the sands of Coney Island. But such distant transport, while it characterizes continental ice sheets wherever they have moved, accounts for a low percentage of the rock in glacial drift. The glacier cuts and fills. Continuously, it plucks up material and sets it down, plucks it up, sets it down. It taketh away, and then it giveth. A diamond may travel from Quebec to Indiana, some dolomite from Lake George to the sea, but most of what is lifted is dropped nearby—boulders from New Jersey in Prospect Park.

"Glacial geology is simple to deal with," Anita said, "because so much of what the glacier created is preserved. Also, you can go places and see the same processes working. You can go to Antarctica and see continental glaciation. There's alpine glaciation in Alaska."

This warm clear summer day was now approaching noon, and Prospect Park was quiet and unpeopled. It was all but deserted. Anita as a child had come here often. She remembered people and picnics everywhere she looked, none of this ominous silence. "I suppose it isn't safe," she said, and we moved on toward Williamsburg.

As we drew close, she became even more obviously nervous. "They tell me it's just the worst slum in the world now," she said. "I don't know if I should tell you to roll up all the windows and lock the doors."

"We would die of the heat."

"This is a completely unnatural place," she went on. "It's a totally artificial environment. Cockroaches, rats, human beings, and pigeons are all that survive. At Brooklyn College, my instructors had difficulty relating geology to the lives of people in this artificial world. In the winter, maybe you froze your ass off waiting for the subway. Maybe that was a way to begin discussing glaciation. In the city, let me tell you, no one knows from geology."

We went first to her high school. It appeared to be abandoned and was not. It was a besooted fortress with battlements. Inside were tall cool hallways that smelled of polish and belied the forbidding exterior. She had walked the halls four years with A's on her report cards and been graduated with high distinction at the age of fifteen. We went to P.S. 37, her grade school. It was taller than wide and looked like an old brick church. It was abandoned, beyond a doubt—glassless and crumbling. Trees of heaven, rooted in the classroom floors, were growing out the windows. Anita said,

"At least I'm glad I saw my school, I think, before they take it away."

We came to Broadway and Berry Street, and now she had before her for the first time in twenty-five years the old building where she had lived. It was a six-story cubical tenement, with so many fire escapes that it seemed to be faced more with iron than with the red Triassic stone. Anita looked at the building in silence. Usually quick to fill the air with words, she said nothing for long moments. Then she said, "It doesn't look as bad as it did when I lived here."

She stared on at the building for a while before speaking again, and when she did speak the nervousness of the morning was completely gone from her voice. "It's been sandblasted," she said. "They've cleaned it up. They've put a new facing on the lower stories, and they've sandblasted the whole building. People are wrong. They're wrong in what they tell me. This place looks cleaner than when I lived here. The whole neighborhood still looks all right. It hasn't changed. I used to play stickball here in the street. This is my neighborhood. This is the same old neighborhood I grew up in. I'm not afraid of this. I'm getting my confidence up. I'm not afraid."

We moved along slowly from one block to another. A young woman crossed the street in front of us, pushing a baby carriage. "She's wearing a wig, I promise you," Anita said. "Her head may be shaved." Singling out another woman among the heterogeneous people of the neighborhood, she said, "Look. See that woman with the turban? She has her hair covered on purpose. They're Chassidic Jews. Their hair is shaved off or concealed so they will not be attractive to passing men." There was a passing man with long curls hanging down either side of his head—in compliance with a dictum of the Pentateuch. "Just to be in the streets here is like stepping into the Middle Ages," Anita said. "Fortunately, my parents were not religious. I would have thought these people would have moved out of here long ago. Chassidic Jews are not all poor, I promise you. Their houses may not look like much, but you should see them inside. They're diamond cutters. They handle money. And they're still here. People are wrong. They are wrong in what they have told me."

We went out of the noon sun into deep shade under the Williamsburg Bridge, whose immense stone piers and vaulting arches seemed Egyptian. She had played handball under there when she was a girl. "There were no tennis courts in this part of the world, let me tell you." When the boys went off to swim in the river, she went back to Berry Street. "Me? In the river? Not me. The boys swam nude."

In the worst parts of summer, when the air was heavy and the streets were soft, Anita went up onto the bridge, climbing to a high point over the river, where there was always a breeze. Seven, eight years old, she sat on the pedestrian walk, with her feet dangling, and looked down into the Brooklyn Navy Yard. The Second World War was in full momentum. U.S.S. Missouri, U.S.S. Bennington, U.S.S. Kearsarge—she saw keels going down and watched battleships and carriers grow. It was a remarkable form of entertainment, but static. Increasingly, she wondered what lay beyond the bridge. One day, she got up the courage to walk all the way across. She set foot on Manhattan and immediately retreated. "I wanted to go up Delancey Street, but I was too scared."

Next time, she went up Delancey Street three blocks before she turned around and hurried home. In this manner, through time, she expanded her horizons. In the main, she just looked, but sometimes she had a little money and went into Manhattan stores. About the only money she ever had she earned returning bottles for neighbors, who gave her a percentage of the deposit. Her idea of exceptional affluence was a family that could afford fresh flowers. Her mother was a secretary whose income covered a great deal less than the family's needs. Her father was a trucker ("with a scar on his face that would make you think twice"), and his back had been broken in an accident. He would spend three years in traction, earning nothing. Gradually, Anita's expeditions on foot into Manhattan increased in length until she was covering, round trip, as much as twelve miles. Her line of maximum advance was somewhere in Central Park. "That's as far as I ever got. I was too scared." Going up the Bowery and through the East Village, she had no more sense of the geology than did the men who were lying in the doorways. When she looked up at the Empire State Building, she was unaware that it owed its elevation to the formation that outcropped in Central Park; and when she saw the outcrops there, she did not wonder why, in the moist atmosphere of the American East, those great bare shelves of sparkling rock were not covered with soil and vegetation. In Wyoming, wind might have stripped them bare, but Wyoming is miles high and drier than the oceans of the moon. Here in the East, a river could wash rock clean, but this rock was on the high ground of an island, far above flood and tide. She never thought to wonder why the rock was scratched and grooved, and elsewhere polished like the foyer of a bank. She didn't know from geology.

In Brooklyn College, from age fifteen onward, she read physics, mineralogy, structural geology, igneous and metamorphic petrology. She took extra courses to the extent permitted. To attend the college she had to pay six dollars a semester, and she meant to get everything out of the investment she could. There were also lab fees and breakage fees. Breakage fees, in geology, were not a great problem. Among undergraduate colleges in the United States, this one was relatively small, about the size of Harvard, which it resembled, with its brick-and-white-trim sedate Colonial buildings, its symmetrical courtyards and enclosed lawns; and like Harvard it stood on outwash. Brooklyn College is in south Flatbush, seaward of the terminal moraine. When Anita was there, in the middle nineteen-fifties, there were so many leftists present that the college was known as the Little Red Schoolhouse. She did not know from politics, either. She was in a world of roof pendants and discordant batholiths, elastic collisions and neutron scatteration, and she branched out into mineral deposits, field mapping, geophysics, and historical geology, adding such things to the skills she had established earlier in accounting, bookkeeping, typing, and shorthand. It had been assumed in her family that she would be a secretary, like her mother.

Now when she goes up Fifth Avenue—as she did with me that summer day—she addresses Fifth Avenue as the axis of the trough of a syncline. She knows what is underfoot. She is aware of the structure of the island. The structure of Manhattan is one of those paradoxes in spatial relations which give geologists especial delight and are about as intelligible to everyone else as punch lines delivered in Latin. There is a passage in the oeuvre of William F. Buckley, Jr., in which he remarks that no writer in the history of the world has ever successfully made clear to the layman the principles of celestial navigation. Then Buckley announces that celestial navigation is dead simple, and that he will pause in the development of his present narrative to redress forever the failure of the literary class to elucidate this abecedarian technology. There and then—and with intrepid, awesome courage—he begins his explication; and before he is through, the oceans are in orbit, their barren shoals are bright with shipwrecked stars. With that preamble, I wish to announce that I am about to make perfectly clear how Fifth Avenue, which runs along the high middle of a loaf of rock that lies between two rivers, runs also up the center of the trough of a syncline. When rock is compressed and folded, the folds are anticlines and synclines. They are much like the components of the letter S. Roll an S forward on its nose and you have to the left a syncline and to the right an anticline. Each is a part of the other. Such configurations in rock compose the structure of a region, but will not necessarily shape the surface of the land. Erosion is the principal agent that shapes the surface of the land; and erosion—particularly when it packs the violence of a moving glacier—can cut through structure as it pleases. A carrot sliced the long way and set flat side up is composed of a synclinal fold. Manhattan, embarrassingly referred to as the Big Apple, might at least instructively be called the Big Carrot. River to river, erosion has worn down the sides, and given the island its superficial camber. Fifth Avenue, up on the high ground, is running up the center of a synclinal trough.

On the upper West Side that afternoon, Anita drew her rock hammer and relieved Manhattan of some dolomite marble, which she took from an outcrop for its relevance to her research in conodonts. She found the marble "overcooked." She said, "To get that kind of temperature, you have to go down thirty or forty thousand feet, or have molten rock nearby, or have a high thermal gradient, which can vary from place to place on earth by a factor of four. This marble is so cooked it is almost volatilized. This—you better believe—is hot rock." At Seventy-second Street and West End Avenue, she stopped to admire a small apartment building whose façade, in mottled greens and black, was elegant with serpentine. On Sixty-eighth Street between Fifth and Madison, she was impressed by a house of gabbro, as anyone would be who had spent a childhood emplaced like a fossil in Triassic sand. It was a house of great wealth, the house of gabbro. Up the block was a house of granite, even grander than the gabbro, and beyond that was a limestone mansion so airily patrician one feared it might dissolve in rain. Anita dropped acid on it and watched it foam.

Jack Epstein, Anita's northern-Appalachian geologist, went to Brooklyn College, too, and subsequently enrolled in the master's program at the University of Wyoming. Anita tried to follow, in 1957, but the geology department in Laramie offered no fellowships for first-year graduate students. ("I needed money. I didn't have a pot to cook in.") She looked into places like Princeton, with geology departments outstanding in the world, but they were even less receptive than Wyoming. In those days, Princeton would not have admitted a woman had she been a direct descendant of Sir Charles Lyell offering as tuition her weight in gems. Anita applied to ten schools in all. The best offer came from Indiana University, in Bloomington, where her professors were soon much aware of her as an extremely bright and aggressive student with the disconcerting habit of shaking her head while they talked, as if to say no, no, no, no, you cratonic schnook, you don't know from nothing. Something of the sort was not always far from her thoughts. ("I am not a very orthodox geologist. I do buy some dogma, if I think it's common sense.")

Bloomington stood upon Salem limestone, which, in the terminology of the building trade, makes beautiful "dimension stone," and is cut to be the cladding of cities. It formed from lime mud in the Meramecian age of middle Mississippian time—between 348 and 340 million years before the present—when Bloomington was at the bottom of a shallow arm of the transgressing ocean, an epicratonic sea. "You people in New York may have your Empire State Building," a professor pointed out to Anita. "But out here we have the hole in the ground it came from."

Anita and Jack Epstein were married in 1958, and, with their newly acquired master's degrees, went to work for the United States Geological Survey. Within the profession, the Survey had particular prestige. A geologist who sought field experience was likely to obtain it in such quantity and variety nowhere else. Anita and Jack Epstein looked upon geology as "an extremely applied science" and shared a conviction that field experience was indispensable in any geological career—no less essential to a modern professor than it ever was to a pick-and-shovel prospector. ("People should go out and get experience and not just turn around and teach what they've been taught.") In their first year in the Survey—to an extent beyond anything they could ever have guessed—they would get what they sought.

Because geology is sometimes intuitive even to the point of being subjective, the sort of field experience one happens to acquire may tend to influence one's posture with regard to deep questions in the science. Geologists who grow up with young rocks are likely to subscribe strongly to the doctrine of uniformitarianism, whereby the present is seen to be the key to the past. They discern a river sandbar in a wall of young rock; they see a sandbar in a living river; and they know that each is in the process of becoming the other, cyclically through time. Whatever is also was, and ever again shall be. Geologists who grow up with very old rock tend to be impressed by the fact that it has been around since before the earliest development of life, and to imagine a progression in which the recycling of the earth's materials is a subplot in a dramatic story that begins with dark scums in motion on an otherwise featureless globe and evolves through various continental configurations toward the scenery of the earth today. They refer to the earliest part of that story as "scum tectonics." The rock cycle—with its crumbling mountains being carried to the sea to form there the rock of mountains to be —is the essence of the uniformitarian principle, which was first articulated by James Hutton, of Edinburgh, at the end of the eighteenth century. Hutton, with his depths of time—his vision of great crustal changes occurring slowly through unguessable numbers of years—opened the way to Darwin (time is the first requirement of evolution) and also placed emphasis on repetitive processes and a sense that change is largely gradual. In contemporary dress, these concepts are still at odds in geology. Some geologists seem to look upon the rock record as a frieze of catastrophes interspersed with gaps, while others prefer to regard everything from rock-slides and volcanic eruptions to rifted continents and plate collisions as dramatic passages in a quietly unfolding story. If you grow up in Brooklyn, you are free to form your prejudices where you may.

Anita Epstein's sense of the dynamics of the earth underwent considerable adjustment one night in 1959, when she and her husband were on summer field assignment in southwestern Montana. They were there to do geologic mapping and studies in structure and stratigraphy in the Madison Range and the Gallatin Range, where Montana is wrapped around a corner of Yellowstone Park. They lived in a U.S.G.S. house trailer in a grove of aspens on the Blarneystone Ranch, a lovely piece of terrain whose absent owner was Emmett J. Culligan, the softener of water. Since joining the Survey, they had worked in Pennsylvania, mapping quadrangles in the region of the Delaware Water Gap, and had spent the winter at headquarters in Washington, and now they were being given a chance to see some geology in a part of the United States where it is particularly visible—in Anita's words, "where it all hangs out."

The ranch was close by Hebgen Lake, which owed itself to a dam in the valley of the Madison River. The valley ran along the line of a fault that was thought to be inactive until that night. The air was crisp. The moon was full. The day before, a fire watcher in a tower in the Gallatins had become aware of an unnerving silence. The birds were gone, he realized. Birds of every sort had made a wholesale departure from his mountain. It would be noted by others that bears had taken off as well, while bears that remained walked preoccupied in circles. The Epsteins had no knowledge of these signs and would not have known what to make of them if they had. They were unaware then that Chinese geologists routinely watch wildlife for intimations of earthquakes. They were also unaware that David Love, of the Survey's office in Laramie, had published an abstract only weeks before called "Quaternary Faulting in and near Yellowstone Park," in which he expressed disagreement with the conventional wisdom that seismic activity on a grand scale was a thing of the past in that region. He said he thought a major shock was not unlikely. Anita was shuffling cards, 11:37 P.M., when the lantern above her began to swing, crockery fell from cabinets, and water leaped out of a basin. Jack tried to catch the swinging lantern and "it beaned him on the head." The floor of the trailer was moving in a way that reminded her of the Fun House at Coney Island. They ran outside. "Trees were toppling over. The solid earth was like a glop of jelly," she would recall later. In the moonlight, she saw soil moving like ocean waves, and for all her professed terror she was collected enough to notice that the waves were not propagating well and were cracking at their crests. She remembers something like thirty seconds of "tremendous explosive noise," an "amplified tornado." She was close to the epicenter of a shock that was felt three hundred and fifty miles away and markedly affected water wells in Hawaii and Alaska. East and west from where she stood ran an eighteen-mile rip in the surface of the earth. The fault ran straight through Culligan's ranch house, and had split its levels, raising the back twelve feet. The tornado sound had been made by eighty million tons of Precambrian mountainside, whose planes of schistosity had happened to be inclined toward the Madison River, with the result that half the mountain came falling down in one of the largest rapid landslides produced by an earthquake in North America in historical time. People were camped under it and near it. Among the dead were some who died of the air blast, after flapping like flags as they clung to trees. Automobiles rolled overland like tumbleweed. They were inundated as the river pooled up against the rockslide, and they are still at the bottom of Earthquake Lake, as it is called—a hundred and eighty feet deep.

The fault offset the water table, and the consequent release of artesian pressure sent grotesque fountains of water, sand, and gravel spurting into the air. Yet the dam at Hebgen Lake held—possibly because the lake's entire basin subsided, in places as much as twenty-two feet. Seiche waves crossed its receding surface. A seiche is a freshwater tsunami, an oscillation in a bathtub. The surface of Hebgen Lake was aslosh with them for twelve hours, but the first three or four were the large ones. Entering lakeside bungalows, they drowned people in their beds.

When a volcano lets fly or an earthquake brings down a mountainside, people look upon the event with surprise and report it to each other as news. People, in their whole history, have seen comparatively few such events; and only in the past couple of hundred years have they begun to sense the patterns the events represent. Human time, regarded in the perspective of geologic time, is much too thin to be discerned—the mark invisible at the end of a ruler. If geologic time could somehow be seen in the perspective of human time, on the other hand, sea level would be rising and falling hundreds of feet, ice would come pouring over continents and as quickly go away. Yucatáns and Floridas would be under the sun one moment and underwater the next, oceans would swing open like doors, mountains would grow like clouds and come down like melting sherbet, continents would crawl like amoebae, rivers would arrive and disappear like rainstreaks down an umbrella, lakes would go away like puddles after rain, and volcanoes would light the earth as if it were a garden full of fireflies. At the end of the program, man shows up—his ticket in his hand. Almost at once, he conceives of private property, dimension stone, and life insurance. When a Mt. St. Helens assaults his sensibilities with an ash cloud eleven miles high, he writes a letter to the *New York Times* recommending that the mountain be bombed.

As the night returned to quiet and the ground ceased to move, Anita recovered whatever composure she had lost, picked up her deck of cards, and said to herself, "That's the way it goes, folks. The earth's a very shaky mobile thing, and that's how it works. Apparently, the mountains around here are still going up." Later, she would say, "We were taught all wrong. We were taught that changes on the face of the earth come in a slow steady march. But that isn't what happens. The slow steady march of geologic time is punctuated with catastrophes. And what we see in the geologic record are the catastrophes. Look at a graded sandstone and see the bedding go from fine to coarse. That's a storm. That's one storm—when the water came up and laid the coarse material down over the fine. In the rock record, the tranquillity of time is not well represented. Instead, you have the catastrophes. In the Southwest, they live from one catastrophe to another, from one flash flood to the next. The evolution of the world does not happen a grain at a time. It happens in the hundred-year storm, the hundred-year flood. Those things do it all. That earthquake made a catastrophist of me."

No one knew where the bears went when they left the Gallatin Range. When they came back, they were covered with mud.

Catastrophism in another form presented itself that autumn when Jack Epstein was transferred to the office of the Geological Survey's Water Resources Division in Alexandria, Louisiana. There was no position for Anita, and she could not have had a job even if one had been open, for it was a rule of the Survey that spouses could not work for the same supervisor. The Alexandria office was small, and included one supervisor. Her nascent geological career was suddenly aborted. She taught physics and chemistry in a Rapides Parish high school. In the summer that followed, she worked for the state government as an interviewer in the unemployment office. She did her geology when and where she could. Driving home from work, she saw people dressed like signal flags hitting golf balls on fake moraines.

Fortunately, her husband was even less interested in the water resources of Louisiana than she was in the unemployment interviews. They decided they needed Ph.D.s to improve their chances of working somewhere else. They enrolled at Ohio State, and in eastern Pennsylvania took up the summer field work that led to their dissertations. They did geologic mapping and biostratigraphy among the ridges of the folded Appalachians—noting the directional trends of the various formations (the strike) and their angles of dip, along a narrow band of deformation from the Schuylkill Gap near Reading to the Delaware Water Gap, and on toward the elbow of the Delaware River where Pennsylvania, New Jersey, and New York conjoin. The most recent ice sheet had reached the Water Gap—where the downcutting Delaware River had sawed a mountain in two—and had filled the gap, and even overtopped the mountain, and then had stopped advancing. So the country of their dissertations was filled with fossil tundra, with kames and eskers, with periglacial boulders and the beds of vanished lakes, with erratics from the Adirondacks, with a vast imposition of terminal moraine. Like the outwash of Brooklyn and the tills of Indiana, this Pennsylvania countryside helped to give Anita her sophistication in glacial geology, which was consolidated at Ohio State, whose Institute of Polar Studies trains specialists in the field. Glacial evidence was not, however, what drew her particular attention. The Wisconsinan ice was modern, in the long roll of time, in much the way that Edward VII is modern compared with a hominid skull. The ice melted back out of the Water Gap seventeen thousand years ago. Anita was more interested in certain stratigraphic sequences in rock that protruded through the glacial debris and had existed for several hundred million years. She would crush this rock, separate out certain of its components, and under a microscope at fifty to a hundred magnifications study its contained conodonts, hard fragments of the bodies of unknown marine creatures—hard as human teeth, and of the same material. At a hundred magnifications, some of them looked like wolf jaws, others like shark teeth, arrowheads, bits of serrated lizard spine—not unpleasing to the eye, with an asymmetrical, objet-trouvé appeal. Many of them resembled conical incisors, and in 1856 this had caused a Latvian paleontologist to give them their name. Conodonts were in many formations but were most easily extracted from limestone and dolomite—the carbonate rocks. They would become useful to geologists because they were all over the earth, and because the creatures that left them behind had appeared in the world early in the Paleozoic and had vanished forever at the end of the Triassic. Yet not until the late nineteen-fifties did studies begin to be published that brought conodonts to prominence as index fossils, helping to subdivide a specific zone of time, a fifteenth of the history of the earth, running from 512 to 208 million years before the present. As the conodont-bearing creatures evolved through those years, their conodonts became increasingly complex, with apparatus extending in denticles, bars, and blades. Geologists, observing these changes, could readily assign relative ages to the places where conodonts were found.

After collecting her samples, Anita could not have been shipping them back to a better place than Ohio State. Just as Johns Hopkins has been celebrated for lacrosse, Hartwick for soccer, and Rollins for tennis, Ohio State is known for conodonts. Geologists call

Ohio State "a conodont factory." Like all the other workers there (a specialist in the field is known in the profession as "a conodont worker"), she noticed incidentally as she catalogued the evolutionary changes in her specimens that some were light and some were dark. They were white, brown, yellow, tan, and gray. Since they were coming into Columbus from all over the United States, and in fact the world, she began to notice that in a general way their colors followed geographical patterns. She wondered what that might suggest. She looked at conodonts from Kentucky and Ohio, which were of a yellow so pale it was almost white. From western Pennsylvania they were jonquil, from central Pennsylvania brown. The ones she had collected north of Schuylkill Gap were black. She thought at first there was something wrong with her samples, but her adviser told her that in all likelihood the blackness was merely the result of pressures attendant when the limestone or dolomite was being deformed. He did not encourage her to make a formal study of the matter, and she returned to her absorptions with conodont biostratigraphy. On one of her trips east, she crossed New York State, collecting dolomite and limestone all the way. From Lake Erie to the Catskills, New York State is a cake of Devonian rock, lying flat in a swath sixty miles wide. You can travel across it chipping off rock of the same approximate age, and not just any old Devonian samples—for the Devonian period covers forty-six million years—but, say, limestone and dolomite from the Gedinnian age, which is seven million years of early Devonian, or even from the Helderbergian stage of middle Gedinnian time. For as much as a hundred and fifty miles, you could follow a line of time no broader than three million years. You could cut it that fine. Anita did something of the sort, and crushed the rocks at Ohio State. She noticed that the conodonts were amber in Erie County, tan in Schuyler and Steuben. They were cordovan in Tioga and Broome. In Albany County, they were dark as pitch.

She wondered what the colors might be suggesting about the geologic history of the region.

Nothing much, her adviser assured her. The colors were the results of tectonic pressures.

It had been just a passing thought. She let it go, and went back to work on her thesis, which would be titled "Stratigraphy and Conodont Paleontology of Upper Silurian and Lower Devonian Rocks of New Jersey, Southeastern New York, and Eastern Pennsylvania." She was documenting subtle evolutionary differences in conodonts close to the Silurian-Devonian boundary, a point in time just over four hundred million years ago. And, arranging her microfossils in chronicle form, she was differentiating and cataloguing in time the units of rock from which they had been removed. This in turn would help her to understand the structure of the country in which she had picked up the rock. Conodont colors faded in her mind.

By 1966, having completed their course work at Ohio State, Anita and Jack Epstein had returned to work for the Geological Survey—he to concentrate on northern-Appalachian geology and she to take what she could get, which was a map-editing job in Washington. She would have preferred to work on conodonts, but the federal budget at that time covered only one conodont worker, and someone else had the job. Before long, she had become general editor of all geologic mapping taking place east of the Mississippi River. She dealt with hundreds of geologists. There were fifteen hundred in the Survey, and the quality of their work, their capacity for visualizing plunging synclines and recumbent folds, tended to vary. She looked upon some of them as "losers." Such people were sent to what she privately described as "penal quadrangles": the lesser bayous of Louisiana, the Okefenokee Swamp. If they did not know strike from dip, they could go where they would encounter neither. She did not feel pity. Better to be a loser in the United States, she thought, than to be a geological peasant in China. There are four hundred thousand people in the Chinese Geological Survey. "It's a hell of an outfit," in Anita's words. "If they want to see exposed rock, they don't depend on streambanks and roadcuts, as we do. If an important Chinese geologist wants to see a section of rock, the peasants dig out a mountainside."

She was a map editor for seven years, during all of which she continued her conodont research, almost wholly on her own time. Collecting rock from Maryland and Pennsylvania, she crushed it and "ran the samples" at home. Running samples was not just a matter of pushing slides past the nose of a microscope. After pulverizing the rock and dissolving most of it in acid, she had to sort its remaining components, and this could not be done chemically, so it had to be done physically. It was a problem analogous to the separation of uranium isotopes, which in the early nineteen-forties had brought any number of physicists to a halt. It was also something like sluicing gold, but you could not see the gold.

Anita primarily uses tetrabromoethane, an extremely heavy and extremely toxic fluid that costs three hundred dollars a gallon. Granite will float in tetrabromoethane. Quartz will float in tetrabromoethane. Conodonts sink without a bubble. Her hands in rubber gloves within a chemical hood, she pours the undissolved rock residue into the tetrabromoethane. The lighter materials, floating, are removed. Inconveniently, conodonts are not all that sink. Pyrite, among other things, sinks, too. With methylene iodide, a fluid even heavier than tetrabromoethane, she turns the process around. In methylene iodide, the pyrite and whatnot go to the bottom, while the conodonts, among other things, float. Electromagnetically, she further concentrates the conodonts. She can now have a look at them under a microscope, seeing "bizarre shapes that any idiot can recognize," and assign them variously to the Anisian, Ladinian, Cayugan, Osagean, Llandoverian, Ashgillian, or any other among tens of dozens of subdivisions of Cambrian, Ordovician, Silurian, Devonian, Mississippian, Pennsylvanian, Permian, and Triassic time.

While recording ages, she could not ignore colors, and the question of their possible significance returned to her mind. In the Appalachians generally, formations thickened eastward. The farther east you went, the deeper the rock had once been buried—the greater the heat had once been. Heat appeared to her to have affected the color of the conodonts in the same manner that it affects the color of butter—turning it from yellow to light brown to darker brown to black-and-ruined smoking in the pan. Oh, she thought. You could use those things as thermometers. They might help in mapping metamorphic rock. The process by which heat and pressure change one kind of rock into another is divided into grades of intensity. Maybe conodont colors, plotted on a map, could demonstrate the shadings of the grades. At work, she began saying to people, "Show me a conodont and I'll tell you where in the Appalachians it came from." With amazing accuracy she repeatedly passed the test. She imagined that color had been controlled by carbon fixing. In the presence of heat, she thought, the amount of carbon in a given conodont would have remained constant while the amounts of hydrogen and oxygen declined, which is what happens in heated butter. No one seemed to agree with her. One way to test her idea might have been to scan for individual elements with an electron probe, but this was 1967 and electron probes in those days could not pick up light elements like hydrogen and oxygen. She sought other avenues of proof—with other types of equipment that no one has at home. The Geological Survey had a question for her, however. They said, "Who needs to know this anyway?" The Survey had been established to serve the public.

"O.K., to hell with it," Anita told herself. Half a dozen years went by. With the oil embargo of 1973, the Survey felt a need to do everything possible to effect an increase in the nation's energy resources. Its Branch of Oil & Gas Resources was expanded fifteen-fold. There were new positions for about two hundred front-rank geologists. They were hired away from oil companies or brought in from elsewhere in the Survey. What attracted the people from the companies was the opportunity to do publishable research. To run the branch, Peter R. Rose gave up his position as a staff geologist of Shell. Leonard Harris, of the Survey, a southern-Appalachian geologist whose interests had moved northward from the Ozarks, came into the Oil & Gas Branch, too. One day, he mentioned to Anita that he understood she was interested in conodonts. He said he would like to have some of her rock samples analyzed for "organic maturation."

She listened to this dark-haired blue-eyed geologist as if he had come from a place a great deal more distant than the Ozarks. Just how did he propose to discern organic maturation? "Do you do that chemically?" she asked him.

"Yes," he said. "You can. And you can also do it by observing changes in organic materials such as fossil pollen and spores, where they exist."

"How do you do *that*?" she said.

"By looking at color change," he said. "You see, the pollen and spores—"

"Stop!" she said. "Stop right there. They change from pale yellow to brown to black. Am I right?"

"Right," he said. He was matter-of-fact in tone. He was, among other things, an oil geologist, while she was not. Oil companies had been using the colors of fossil pollen and fossil spores to help identify rock formations that had achieved the sorts of temperatures in which oil might form. Land-based plants, with their pollen and spores, had not developed on earth until a hundred and thirty million years after the beginning of the Paleozoic era, however. Nor would they ever be as plentiful and as nearly ubiquitous as marine fossils. Hearing Leonard Harris mention oil companies and their use of color alteration in pollen and spores, Anita realized in the instant that she had—in her words—"reinvented the wheel." And then some. She had not known that pollen and spores were used as geothermometers in the oil business, and now that she knew it she could see at once that conodonts used for the same purpose would have different geographical applications, covering greater ranges of temperature and different segments of time.

"I think I can do the assessments easier and better by using conodonts," she said to Harris. "Conodonts change color, too, and in the same way."

It was his turn to be surprised. "How come I never heard about that?" he said.

She said, "Because no one knows it."

Petroleum—the transmuted fossils of ocean algae—forms when the rock that holds the fossils becomes heated to the temperature of a cup of coffee and remains as warm or warmer for at least a million years. The minimal temperature is about fifty degrees Celsius. At lower temperatures, the algal remains will not turn into oil. At temperatures hotter than a hundred and fifty degrees, any oil or potential oil within the rock is destroyed. ("The stuff is there, throughout the Appalachians. You look at the rocks and you see all this dead oil.") The narrow "petroleum window," as it is called—between fifty degrees and a hundred and fifty degrees—is scarcely a fourteenth part of the full temperature variation of the crust of the earth, a fact that goes a long way toward explaining how the human race could have used up such a large part of the world's petroleum in one century. Not only must the marine algae have been buried for adequate time at depths where temperatures hover in the window but once oil has formed it is subject to destruction underground if for one reason or another the temperature of its host rock rises.

Natural gas is to oil as politicians are to statesmen. Any organic material whatsoever will form natural gas, and will form it rapidly, at earth-surface temperatures and on up to many hundreds of degrees. In Anita's words: "You get natural gas as soon as anything drops dead. For oil, the requisites are the organic material and the thermal window. When they look for oil, they don't know what they've got until they drill a hole." In trying to figure out where to drill, geologists have an obvious need for geothermometers. Pollen and spores are of considerable use, but only when they have fossilized in certain rocks. Moreover, they are absent altogether from early Paleozoic times, and they are extremely rare in rock from the deep sea.

Leonard Harris asked Anita how many years she had been "sitting on" her discovery about conodonts.

About ten, she told him. The last thing she had wished to do was to keep it secret, but no one had shown much interest. She gave him slides of the New York State east-west series, and told him that a comparable set could be got together for Pennsylvania, too. Harris went south and traversed the state of Tennessee, collecting carbonate rocks that were close in age, and when Anita ran the conodonts she found the color alterations quite the same as in the northern states—dark in the east, pale in the west. Leonard and Anita reported all this to Peter Rose, leader of the Oil & Gas Branch, pointing out that the variations in conodont color could lead to a cheap and rapid technique of finding rock in the petroleum window. Rose said he couldn't understand why no one in the United States had ever thought of this if it was as obvious as all that. Anita told him that for years she had been puzzled by the same question, since the procedure would be one that "any idiot ought to be able to follow, because all you need is to be not color-blind."

At Rose's request, Anita's division of the Geological Survey allowed her to work two days a week on conodonts. Weekends, she worked on them at home. Actual temperature values had not been assigned to the varying colors. She did so in a year of experiments. She began with the palest of conodonts from Kentucky and heated them at varying temperatures until they became canary and golden and amber and chocolate and cordovan, black, and gray. With enough added heat, they would turn white and then clear. At nine hundred degrees Celsius, they disintegrated. By cooking her samples in a great many variations of the ratio of time to temperature, she was able to develop a method of extrapolating laboratory findings onto the scale of geologic time. She concluded that pale-yellow conodonts could remain at about fifty degrees indefinitely without changing color. If they were to remain at sixty to ninety degrees for a million years or more, they would be amber. The earth's thermal gradient varies locally, but generally speaking the temperature of rock increases about one degree Celsius for each hundred feet of depth. A conodont would have to be lodged in rock buried three thousand to six thousand feet in order to experience temperatures of the sort that would turn it amber. At depths of nine thousand to fifteen thousand feet, she discovered, conodonts would turn light brown in roughly ten million years. If they spent ten million years at, say, eighteen thousand feet, they would be dark brown. In comparable amounts of time but at greater and greater depths, they would turn black, gray, opaque, white, clear as crystal. Anita also cooked conodonts in pressure bombs, because it had been suggested to her that the pressures of great tectonism—the big dynamic events in the crust, with mountains building and whole regions being kneaded like dough—might also affect conodont colors. Her experiments convinced her that pressure has little effect on color; heat is what primarily causes it to change.

Of course, plenty of heat is produced by deep burial during major tectonic events. Her conodonts from New Jersey were black and from Kentucky pale essentially because huge disintegrating eastern mountain ranges had buried the near ones very deep and the far ones scarcely at all. The East is for the most part the wreckage of the Ancestral Appalachians, and—as is exemplified in the Devonian rock of New York—the formations are thickest close to where the mountains stood. A continuous sedimentary deposit that is thousands of feet thick in eastern Pennsylvania may be ten feet thick in Ohio. Where oil was first discovered in western Pennsylvania, it was seeping out of rocks and running in the streams. It is of a character and purity so remarkable that people used to buy it and drink it for their health. Anita looked at conodont samples from rock that surrounded this truly exceptional oil. In the temperature range of eighty to a hundred and twenty degrees, they were in the center of the petroleum window. They were golden brown.

With a year of tests run, with Kodachrome pictures, with graphs and charts of what she called her "wind-tunnel models," she was prepared to tell her story. The Geological Society of America was to meet in Florida in November, 1974, and she arranged to deliver a paper there. "I prepared carefully—I always do—so I wouldn't phumpfer. But the G.S.A. meeting was not momentous. They were academics, and not particularly knowledgeable about exploration techniques." Five months later, scarcely knowing what to anticipate, she went to Dallas and spoke before the American Association of Petroleum Geologists. It was the same show, but this time it was playing in the right house. Requests and invitations poured upon her from oil companies wherever they might be, and from geological societies situated in oil centers like Calgary and Tulsa. "It filled a big hole in their technology," Anita has said, recalling those days. "They have to be able to assess the thermal level of deposits, and this was a simple way to do it."

Anita became a conodont specialist for the United States Geological Survey, full time. She lives in Maryland. Her home is an island in flower beds and lawn. She gets up at four-thirty and drives to work at the Survey's headquarters in Reston, Virginia. Oil companies have continued to beat the path to her door, as have oil geologists from every continent but Antarctica, including large delegations from the Chinese Geological Survey. While oil prospectors are using brown and yellow conodonts to guide them to the thermal window, mineral prospectors are using white ones in the search for copper, iron, silver, and gold. White conodonts and clear conodonts, products of the highest temperatures, suggest the remains of thermal hot spots, thermal aureoles, ancient hydrothermal springs—places where metallic minerals would have come up in solution to be precipitated out into veins.

Soon after her discovery, universities began calling her. She was pleased to appear at places like Princeton, pleased to be given an opportunity to demonstrate what could be learned elsewhere. Women students were in her audience now. In the late nineteen-seventies, she and her colleagues published a succession of scientific papers whose title pages perforce encapsulated not only their professional endeavors but something of their private lives. The "senior author" of a scientific publication is the person whose name is listed first and whose work has been of primary importance to the project, while other authors are listed more or less in diminishing order, like the ingredients on a can of stew. The benchmark paper came in 1977. Entitled "Conodont Color Alteration—an Index to Organic Metamorphism," it was "by Anita G. Epstein, Jack B. Epstein, and Leonard D. Harris." Then, in 1978, came "Oil and Gas Data from Paleozoic Rocks in the Appalachian Basin: Maps for Assessing Hydrocarbon Potential and Thermal Maturity (Conodont Color Alteration Isograds and Overburden Isopachs)"—virtually an oil-prospecting kit, a highly specialized atlas—"by Anita G. Harris, Leonard D. Harris, and Jack B. Epstein." And scarcely a year after that appeared a summary document called "Conodont Color Alteration, an Organo-Mineral Metamorphic Index, and Its Application to Appalachian Basin Geology"—"by Anita G. Harris."

Anita Harris—beginning her trip west on Interstate 80 with her rock hammer, her sledgehammer, her hydrochloric acid, and me—stopped at a lookoff near Allamuchy, New Jersey, about five miles west of Netcong. It was a cool April morning, the tint of the valley pastel green, and from our relatively high perspective, at an altitude of a thousand feet, the eye was drawn eighteen miles west across a gulf of air to the forested wall of Kittatinny Mountain, filling the skyline of two states, its apparently endless flat ridgeline broken only by one deep notch, which centered and arrested the view and was as sharply defined as a notch in a gunsight: the Delaware Water Gap, where the big river comes obliquely through the mountain, like a thief through a gap in a fence. There was a ridge or two near and below us, but the distance to the Water Gap was occupied largely by the woodlots, hedgerows, and striped fields of a broad terrain as much as seven hundred feet lower than the spot on which we stood and of such breathtaking proportions and fetching appearance that it could be mentioned in a sentence with the Shenandoah Valley. The picture of New Jersey that most people hold in their minds does not include a Shenandoah Valley. Nevertheless, this New Jersey Appalachian landscape not only looked like the Shenandoah, it actually was the Shenandoah, in the sense that it was a fragment of a valley that runs south from New Jersey to Alabama and north from New Jersey into Canada—a single valley, one continuous geology, known to science as the Great Valley of the Appalachians and to local peoples here and there as Champlain, Shenandoah, Tennessee Valley, but in New Jersey by no special name. This integral, elongate, predominantly carbonate valley disappears and reappears through the far Northeast, until in pieces it presents itself in Newfoundland and then dives under the sea. Its marbles are minable in Vermont, in Tennessee. It was the route of armies—the avenue to Antietam, the site of Chickamauga, Saratoga, Ticonderoga. It stands in the morning shadow of the Annieopsquotch Mountains, of the Green Mountains, of the New Jersey Highlands, of the Berkshire, Catoctin, and Great Smoky Mountains, which are fraternal in structure and composition and are all of Precambrian age. The lookoff where we stood was a part of that Appalachian complex. It was crystalline rock above a thousand million years old—and the rock in the valley was younger, and in the Kittatinny younger still. (Geologists avoid the word "billion" because in one part and another of the English-speaking world the quantity it refers to differs by three orders of magnitude. A billion in Great Britain is a million million.) We were looking from the New Jersey Highlands into another segment of the cordillera—the beginnings of the physiographic Ridge and Valley Province, the folded-and-faulted deformed Appalachians, the long ropy ridges of the eastern sinuous welt, which Edmund Wilson had once written off as "fairly unimportant creases in the earth covered with trees."

"Geology repeats itself," Anita remarked, and she went on to say that anyone who could understand the view before us would have come to understand in a general way the Appalachians as a whole—that what we were looking at was the fragmental evidence and low remains of alpine massifs immeasurably high and wide, massifs which for the most part had stood behind us to the east, and were now largely disintegrated and recycled into younger rock that is tens of thousands of feet deep and wedges out to the west in ever-diminishing quantity until what covers Ohio is a thin veneer.

The appearance of a country is the effect primarily of water, running off the landscape, cutting out valleys, dozing wantonly as glacier ice. The sculpturing is external. But it is influenced and can even be controlled by the rock within: by the relative strength, not to mention the solubility, of successive strata, and by the folds and faults—the structure—that the rock has been given. Figuring out the Appalachians was Problem 1 in American geology, and a difficult place to begin, for it was scarcely a matter of layer-cake legibility, like the time scale in the walls of the Grand Canyon. It was a compressed, chaotic, ropy enigma four thousand kilometres from end to apparent end, full of overturned strata and recycled rock, of steep faults and horizontal thrust sheets, of folds so tight that what had once stretched twenty miles might now fit into five. The country seemed to consist of parallel meandering belts—the Piedmont, the Precambrian highlands, the Great Valley, the folded-and-faulted deformed mountains, the Allegheny Plateau. It was high and resistant, low and vulnerable. (I have heard the Shenandoah described, if not dismissed, as "a strip of weak rock.") The early Appalachian geologists, in their horse-drawn buggies, their suits and ties, developed a sense of physiography that tuned them to the land, and when they saw long sugarloaf hills they had learned to suspect that there was dolomite within, and when they looked up at coxcomb ridges they felt the presence of Cambrian sandstones, and of Cambrian shales in the valleys beyond. The higher, harder ridges would be thick, Silurian quartzites, more often than not, while flourishing green lowlands with protruding ribs of rock would owe their shape and their fertility to limestones assembled in Ordovician seas. There were knolls in the valleys. Inside the knolls were shales. Shale breaks up easily but will not dissolve like limestone, so the shales became blisters in the limestone valleys. Of the two carbonate rocks, limestone is a good deal more soluble than dolomite, and that was why dolomite would retain itself in sugarloaves above the limestone valleys. Once the early geologists had developed this sense of the substrate, they shook the reins and moved with dispatch, filling in the first American geologic maps with a general accuracy that is impressive still.

Identifying what is there scarcely describes what happened to put it there, however. The history of the earth may be written in rock, but history is not coherent on a geologic map, which shows a region's uppermost formations in present time, while indicating little of what lies farther down and less of what is gone from above. At a given place—a given latitude and longitude—the appearance of the world will have changed too often to be recorded in a single picture, will have been, say, at one time below fresh water, at another under brine, will have been mountainous country, a quiet plain, equatorial desert, an arctic coast, a coal swamp, and a river delta, all in one Zip Code. These scenes are discernible in, among other things, the sedimentary characteristics of rock, in its chemical composition, magnetic components, interior color, hardness, fossils, and igneous, metamorphic, or depositional age. But as parts of the historical narrative these items of evidence are just phrases and clauses, often wildly disjunct. They are like odd pieces from innumerable jigsaw puzzles. The rock column—a vertical representation of the crust at some point on the earth—holds a great deal of inferable history, too. But rock columns are generalized; they are atremble with hiatuses; and they depend in large part on well borings, which are shallow, and on three-dimensional seismic studies, which are new, and far between. To this day, in other words, there remains in geology plenty of room for the creative imagination. All the more amazing is the extent to which the early geologists, who travelled the Appalachians in the eighteen-twenties and thirties, not only catalogued the evident rock but also worked out stratigraphic relationships among various formations and began to see composite structure. Starting close up, with this rock type, that mountain, this formation, that valley—with what they could see and know—they gradually began to form tentative regional pictures. Piece by piece over the next century and a half, they and their successors would put together logically sequential narratives presenting the comprehensive history of the mountain belt. As new evidence and insight came along, old logic sometimes fell into discard. When plate tectonics arrived, its revelations were embraced or accommodated but by no means universally accepted. The Appalachians, meanwhile, continued slowly to waste away. The debate about their origins did not.

Observing the valley scene, the gapped and distant ridgeline, Anita said that mountains in this region had come up and been worn down not once but a number of times: the Appalachians were the result of a series of pulses of mountain building, the last three of which had been spaced across two hundred and fifty million years —the Taconic Orogeny, the Acadian Orogeny, and the Alleghenian Orogeny. The first stirrings of the Taconic Orogeny began nearly five hundred million years ago. After the mountains it lifted had been largely eroded away, their stubs and their detritus, much of which had turned into sedimentary rock, became involved in the Acadian Orogeny; and when the Acadian Orogeny was long gone by, its mountain stubs and lithified debris were caught up in the Alleghenian Orogeny, which drove into the sky still another massif, the ruins of which lay all about us now. In such manner had each of the orogenies of the Appalachians cannibalized the products of previous pulses, and now we were left with this old mountain range, by weather almost wholly destroyed, but nonetheless containing in a traceable and unarguable way the rock of its ancestral mountains. She said the Delaware Water Gap, with its hard quartzites, represented action from the heart of the story, debris from the Taconic Orogeny: boulders, pebbles, sands, and silts carried down from bald mountains by the rapids of big braided rivers—a runoff unimpeded by vegetation, when not so much as one green leaf existed in the terrestrial world.

Long before the Taconic mountain-building pulse was felt, the scene was very different. A subdued continent, consisting of what is now the basement rock of North America, stood low with quiet streams, collecting on its margins clean accumulations of sand. One can infer the flat landscape, the slow rivers, the white beaches, in the rock that remains from those Cambrian sands. Sea level, never constant, moved generally upward all through Cambrian time. The water advanced upon the continent at an average rate of ten miles every million years, spreading across the craton successive coastal sands. Potsdam sandstone. Antietam sandstone. Waynesboro sandstone. Eau Claire sandstone. There were fifty-four million years in the long tectonic quiet of Cambrian time, 544 to 490 million years before the present. By the end of the Cambrian and the beginning of the Ordovician, the ocean had spread its great bays upon the continents to an extent that has not been equalled in five hundred million years, with the possible exception of the highest Cretaceous seas. No one knows why. There is a fixed amount of water in the world. It can rain and run, evaporate, freeze, sit in deep cold pools on abyssal plains, but it cannot leave the earth. When large amounts of it collect as ice upon the continents, the level of the sea drastically goes down. In much of Cambro-Ordovician time, glaciation was absent from the world, and almost all water was in a fluid state. But that alone will not explain the signal height of the sea. In most of the known history of the earth, glacial ice has actually been insignificant. Ice ages, such as the present one, are extremely rare. What seems likely is that ocean floors were higher in Cambro-Ordovician time, fluffed up by more than the usual amount of heat from the restless mantle—heat of the sort that has created the Hawaiian Islands in the middle of a lithospheric plate, and heat of the sort that lifts mid-ocean ridges, where plates diverge. Whatever the reason, the sea came up so far that it covered more than half of what is now the North American craton. And after the white clean beaches and shelf sands had spread their broad veneer, lime muds began to accumulate in the epicratonic seas. The lime muds were the skeletons, the macerated shells, the calcareous hard parts of marine creatures. The material turned into limestone, and where conditions were appropriate the limestone, infiltrated by magnesium, became dolomite. As the deposit grew to a general thickness of two thousand feet, these Cambro-Ordovician carbonates buried ever deeper the sandstone below them and the Precambrian rock below that, pressing it all downward like the hull of a loading ship, into the viscous mantle—but sedately, calmly, a few inches every thousand years.

Absorbing the valley scene, the gapped and distant ridgeline, the newly plowed fields where arrowheads appear in the spring, I remarked that we had entered the dominion of the Minsi, the northernmost band of the Lenape. They came into the region toward the dawn of Holocene time and lost claim to it in the beginnings of the Age of Washington. Like index fossils, they now represent this distinct historical stratum. Their home and prime hunting ground was the Minisink—over the mountain, beside the river, the country upstream from the gap. The name Delaware meant nothing to them. It belonged to a family of English peers. The Lenape named the river for themselves. I knew some of this from my grade-school days, not many miles away. The Minisink is a world of corn shocks and islands and valley mists, of trout streams and bears, today. Especially in New Jersey, it has not been mistreated, and, with respect to the epoch of the Minsi, geologically it is the same. The Indians of the Minisink were good geologists. Their trails ran great distances, not only to other hunting parks and shell-mounded beach camps but also to their quarries. They set up camps at the quarries. They cooked in vessels made of soapstone, which they cut from the ground in what is now London Britain Township, Chester County, Pennsylvania. They made adzes of granite, basalt, argillite, even siltstone, from sources closer to home. They went to Berks County, in Pennsylvania, for gray chalcedony and brown jasper. They used glacial-erratic hornfels. They made arrowheads and spear points of Deepkill flint. They made drills and scrapers of Onondaga chert. Flint, chert, and jasper are daughters of chalcedony, which in turn is a variety of quartz. The Eastern flint belt runs from Ontario across New York State and then south to the Minisink. The Indians did not have to attend the Freiberg Mining Academy to be able to tell you that. They understood empirically the uniform bonding of cryptocrystalline quartz, which cannot be separated along flat planes but fractures conchoidally, by percussion, and makes a razor's edge. The forests were alive with game on the sides of what the Lenape called the Endless Mountain. There were eels, shad, sturgeon in the river. The people lived among maize fields in osier cabins. They worshipped light and the four winds—all the elements of nature, orchestrated by the Great Manito. The burial grounds of the Minsi display the finest vistas in the Minisink. As the dead were placed in the earth, they began their final journey, through the Milky Way.

Indians first appeared in the Minisink about a hundredth of a million years ago. The carbonate rocks in the valley before us were close to five hundred million years old. They were one-ninth as old as the earth itself. For their contained conodonts, they were of special interest to Anita, and after the interstate dropped to the valley floor we stopped at roadcuts to collect them. The roadcuts suggested the ruins of blocky walls built by Hellenic masons. They were on both sides of the interstate and in the median, too. There were bluish-white and pale-gray limestones, dolomites weathered buff. Rock will respond to weather with varieties of color—rusting red in its own magnetite, turning green from trace copper. Its appearance can be deceiving. Geologists are slow to identify exposures they have not seen before. They don't just cruise around ticking off names at distances that would impress a hunter. They go up to outcrops, hit them with hammers, and look at the rock through ten-power lenses. If the possibilities include the carbonates, they try a few drops of hydrochloric acid. Limestone with hydrochloric acid on it immediately forms a head, like beer. Dolomite is less responsive to acid. With her sledgehammer, Anita took many pounds of roadcut, and not without effort. Again and again, she really had to slam the wall. Looking at the fresh surface of a piece she removed, she said she'd give odds it was dolomite. It was not responsive to acid. She scraped it with a knife and made powder. Acid on the powder foamed. "This dolomite is clean enough to produce beautiful white marble if it were heated up and recrystallized," she said. "When it became involved in the mountain building, if it had got up to five hundred degrees it would have turned into marble, like the Dolomites, in Italy. There is not a lot of dolomite in the Dolomites. Most of the rock there is marble." She pointed in the roadcut to the domal structures of algal stromatolites—fossil colonies of microorganisms that had lived in the Cambrian seas. "You know the water was shallow, because those things grew only near the light," she said. "You can see there was no mud around. The rock is so clean. And you know the water was warm, because you do not get massive carbonate deposition in cold water. The colder the water, the more soluble carbonates are. So you look at this roadcut and you know you are looking into a clear, shallow, tropical sea."

With dry land adrift and the earth prone to rolling, that Cambrian sea and New Jersey below it would have been about 20 degrees from the equator—the present latitude of Yucatán, where snorkelers kick along in transparent waters looking through their masks at limestones to be. The Yucatán peninsula is almost all carbonate and grew in its own sea. As did Florida. Under the shallow waters of the

Bahamas are wave-washed carbonate dunes, their latitude between 20 and 26 degrees. At the end of Cambrian time, the equator crossed what is now the North American continent in a direction that has become north-south. The equator came in through the Big Bend country in Texas and ran up through the Oklahoma panhandle, Nebraska, and the Dakotas. If in late Cambrian time you had followed the present route of Interstate 80, you would have crossed the equator near Kearney, Nebraska. In New Jersey, you would have been in water scarcely above your hips, wading among algal mounds and grazing gastropods. You could have waded to the equator. West of Chicago and through most of Illinois, you would have been wading on clean sand, the quiet margin of the Canadian craton, which remained above the sea. The limy bottom apparently resumed in Iowa and went on into eastern Nebraska, and then, more or less at Kearney, you would have moved up onto a blistering-hot equatorial beach and into low terrain, subdued hills, rock that had been there a thousand million years. It was barren to a vengeance with a hint of life, possibly a hint of life—rocks stained green, stained red by algae. Wyoming. Past Laramie, you would have come to a west-facing beach and, after it, tidal mudflats all the way to Utah. The waters of the shelf would now begin to deepen. A hundred miles into Nevada was the continental slope and beyond it the blue ocean.

If you had turned around and gone back after thirty million years—well into Ordovician time, say four hundred and sixty million years ago—the shelf edge would still have been near Elko, Nevada, and the gradually rising clean-lime seafloor would have reached at least to Salt Lake City. Across Wyoming, there may have been low dry land or possibly continuing sea. The evidence has almost wholly worn away, but there is one clue. In southeastern Wyoming, a diamond pipe came up about a hundred million years ago, and, in the tumult that followed the explosion, marine limestone of late Ordovician age fell into the kimberlite and was preserved. In western Nebraska, you would have crossed dry and barren Precambrian terrain and by Lincoln have reached another sea. Iowa, Illinois, Indiana. The water was clear, the bottom uneven—many shallows and deeps upon the craton. In Ohio, the sea would have begun to cloud, increasingly so as you moved on east, silts slowly falling onto the lime. In Pennsylvania, as you approached the site of the future Delaware

Water Gap, the bottom would have fallen away below you, and where it had earlier been close to the surface it would now be many tens of fathoms down.

"The carbonate platform collapsed," Anita said. "The continental shelf went down and formed a big depression. Sediments poured in." Much in the way that a sheet of paper bends downward if you move its two ends toward each other in your hands, the limestones and dolomites and the basement rock beneath them had subsided, forming a trough, which rapidly filled with dark mud. The mud became shale, and when the shale was drawn into the heat and pressure of the making of mountains its minerals realigned themselves and it turned into slate. We moved on west a couple of miles and stopped at a roadcut of ebony slate. Anita said, "Twelve thousand feet of this black mud was deposited in twelve million years. That's a big pile of rock." The formation was called Martinsburg. It had been folded and cleaved in orogenic tumult following its deposition in the sea. As a result, it resembled stacks of black folios, each of a thousand leaves. Just to tap at such rock and remove a piece of it is to create something so beautiful in its curving shape and tiered laminations that it would surely be attractive to a bonsai gardener's eye. It seems a proper setting for a six-inch tree. I put a few pieces in the car, as I am wont to do when I see some Martinsburg. Across the Delaware, in Pennsylvania, the formation presents itself in large sections that are without joints and veins, the minerals line up finely in dense flat sheets, and the foliation planes are so extensive and straight that slabs of great size can be sawed from the earth. The rock there is described as "blue-gray true unfading slate." It is strong but "soft," and will accept a polishing that makes it smoother than glass. From Memphis to St. Joe, from Joplin to River City, there is scarcely a hustler in the history of pool who has not racked up his runs over Martinsburg slate. For anybody alive who still hears corruption in the click of pocket billiards, it is worth a moment of reflection that not only did all those pool tables accumulate on the ocean floor as Ordovician guck but so did the blackboards in the schools of all America.

The accumulation of the Martinsburg—the collapsing platform, the inpouring sediments—was the first great sign of a gathering storm. Geological revolution, crustal deformation, tectonic upheaval would follow. Waves of mountains would rise. Martinsburg time in earth history is analogous to the moment in human history when Henry Hudson, of the Dutch East India Company, sailed into the bay of the Lenape River.

Completing the crossing of the Great Valley of the Appalachians, Anita and I passed more limestones, more slate. Their original bedding planes, where we could discern them, were variously atilt, vertical, and overturned, so intricate had the formations become in the thrusting and folding of the long-gone primal massifs. The road came to the river, turned north to run beside it, and presented a full view of the break in the Kittatinny ridge, still far enough away to be comprehended in context but close enough to be seen as the phenomenon it is: a mountain severed, its folds and strata and cliffs symmetrical, thirteen hundred feet of rock in close fraternal image from the skyline to the boulders of a blue-and-white river. Small wonder that painters of the Hudson River School had come to the Delaware to do their best work. George Inness painted the Delaware Water Gap many times, and he chose this perspective—downriver about four miles—more than any other. I have often thought of those canvases—with their Durham boats on the water and cows in the meadows and chuffing locomotives on the Pennsylvania side—in the light of Anita's comment that you would understand a great deal of the history of the eastern continent if you understood all that had made possible one such picture. She was suggesting, it seemed to me, a sense of total composition—not merely one surface composition visible to the eye but a whole series of preceding compositions which in the later one fragmentarily endure and are incorporated into its substance—with materials of vastly differing age drawn together in a single scene, a composite canvas not only from the Hudson River School but including everything else that had been a part of the zones of time represented by the boats, gravels, steeples, cows, trains, talus, cutbanks, and kames, below a mountain broken open by a river half its age.

The mountain touched the Martinsburg, and its rock was the younger by at least ten million years. Kittatinny Mountain is largely quartzite, the primary component of the hubs of Hell. In the post-tectonic, profoundly eroded East, quartzite has tended to stand up high. The Martinsburg is soft, and is therefore valley. There is nothing but time between the two. Where the formations meet, a touch of a finger will cover both the beginning and the end of the ten million years, which are dated at about 440 and 430 million years before the present—from latest Ordovician time to a point in the early Silurian. During that time, something apparently lifted the Martinsburg out of its depositional pit and held it above sea level until weathers wore it low enough to be ready to accept whatever might spread over it from higher ground. The quartzite—as sand—spread over it, coming down from Taconic mountains. The sand became sandstone. Upward of fifty million years later, the sand grains fused and turned into quartzite in the heat and the crush of new rising mountains, or possibly a hundred million years after that, in the heat and the crush of more mountains. The Delaware River at that time was not even a cloud in the sky. Rivers of greater size were flowing the other way, crossing at wild angles the present route of the Delaware. Rivers go wherever the country tells them to, if the country is in vertical motion. The country would not be right for the Delaware for roughly a hundred million more years, and still another hundred million years would go by before the river achieved its present relationship with Kittatinny Mountain. No one knows how the river cut through. Did it cut from above through country now gone and lying as mud in the sea? Did it work its way through the mountain as two streams, eroding headward from either side, the one finally capturing the other? Was there once a great lake spilling over the mountain and creating the gap as its outlet? The big-lake idea has attracted no support. It is looked upon less as a hypothesis than as a theoretically possible but essentially foolish guess. There was for a time an ice-defended lake between the mountain and the Wisconsinan glacier. When the ice melted, the lake ran out through the Water Gap, leaving in evidence its stream deltas and seasonally banded bottom deposits. However, Glacial Lake Sciota, as it is called, was eight miles long and two hundred feet deep and could not have cut a gap through much of anything but sugar.

The ice arrived twenty-three thousand years before the present. The terminal moraine is only ten miles south of the gap. Nonetheless, the ice front was something like two thousand feet thick, for it went over the top of the mountain. It totally plugged and must have widened the Water Gap. It gouged out the riverbed and left there afterward two hundred feet of gravel. Indians were in the Minisink when the vegetation was tundra. Ten thousand years ago, when the vegetation changed from tundra to forest, Indians in the Minisink experienced the change. The styles in which they fractured their flint—their jasper, chert, chalcedony—can be correlated to Anatolian, Sumerian, Mosaic, and Byzantine time. Henry Hudson arrived in the New World about four hundred years before the present. He was followed by Dutch traders, Dutch colonists, Dutch miners. They discovered ore-grade copper in the Minisink, or thought they did. Part fact, part folklore, it is a tradition of the region that a man named Hendrik Van Allen assessed Kittatinny Mountain and decided it was half copper. The Dutch crown ordered him to establish a mine, and to build a road on which the ore could be removed. The road ran up the Minisink and through level country to the Hudson River at Esopus Creek (Kingston, New York). A hundred miles long, it was the first constructed highway in the New World to cover so much distance. It covers it still, and is in many places scarcely changed. When Van Allen was not busy supervising the road builders, he carried on an elite flirtational minuet with the daughter of a Lenape chief. The chief was Wissinoming, his daughter Winona. One day, Van Allen went alone to hunt in the woods near the river islands of the Minisink, and he discharged his piece in the direction of a squirrel. The creature scurried through the branches of trees. Van Allen shot again. The creature scurried through the branches of other trees. Van Allen reloaded, stalked the little bugger, and, pointing his rifle upward, sighted with exceptional care. He fired. The squirrel fell to the ground. Van Allen retrieved it, and found an arrow through its heart. By the edge of the river, Winona threw him a smile from her red canoe. They fell in love. In the Minisink, there was no copper worth mentioning. Van Allen didn't care. Winona rewrote the country for him, told him the traditions of the river, told him the story of the Endless Mountain. In the words of Winona's legend as it was eventually set down, "she spoke of the old tradition of this beautiful valley having once been a deep sea of water, and the bursting asunder of the mountains at the will of the Great Spirit, to uncover for the home of her people the vale of the Minisink." In 1664, Peter Stuyvesant, without a shot, surrendered New Amsterdam and all that went with it to naval representatives of Charles,

King of England. Word was sent to Hendrik Van Allen to close his mines and go home. It was not in him to take an Indian wife to Europe. He explained these matters to Winona in a scene played out on the cliffs high above the Water Gap. She jumped to her death and he followed.

On foot at the base of the cliffs—in the gusts and shattering noise of the big tractor-trailers passing almost close enough to touch—we walked the narrow space between a concrete guard wall and the rock. Like the river, we were moving through the mountain, but in the reverse direction. Between the mirroring faces of rock, rising thirteen hundred feet above the water, the gap was so narrow that the interstate had been squeezed in without a shoulder. There was a parking lot nearby, where we had left the car—a Delaware Water Gap National Recreation Area parking lot, conveniently placed so that the citizen-traveller could see at point-blank range this celebrated natural passage through a mountain wall, never mind that it was now so full of interstate, so full of railroad track and other roadways that it suggested a convergence of tubes leading to a patient in Intensive Care. We saw painted on a storm sewer a white blaze of the Appalachian Trail, which came down from the mountain in New Jersey, crossed the river on the interstate, and returned to the ridgetop on the Pennsylvania side. There were local names for the sides of the gap in the mountain. The Pennsylvania side was Mt. Minsi, the New Jersey side Mt. Tammany. The rock of the cliffs above us was cleanly bedded, stratified, and had been not only deposited but also deformed in the course of the eastern orogenies. Regionally, it had been pushed together like cloth on a table. The particular fragment of the particular fold that erosion had left as the sustaining rock of the mountain happened to be dipping to the northwest at an angle of some forty-five degrees. As we walked in that general direction, each upended layer was somewhat younger than the last, and each, in the evidence it held, did not so much suggest as record progressive changes in Silurian worlds. "The dip always points upsection, always points toward younger rocks," Anita said. "You learn that the first day in Geology I."

"Do you ever get tired of teaching ignoramuses?" I asked her.

She said, "I haven't worked on this level since I don't know when."

Near the road and the river, at the beginning of the outcrop, great boulders of talus had obscured the contact between the mountain quartzite and the underlying slate. To move on through the gap, traversing the interior of the mountain, was to walk from early to late Silurian time, to examine an assembly of rock that had formed between 435 and 410 million years before the present. The first and oldest quartzite was conglomeratic. Its ingredients had lithified as pebbles and sand. Shouting to be heard, Anita said, "In those pebbles you can see a mountain storm. You can see the pebbles coming into a sandbar in a braided river. There is very little mud in this rock. The streams had a high enough gradient to be running fast and to carry the mud away. These sands and pebbles were coming off a mountain range, and it was young and high."

A braided river carries such an enormous burden of sand and gravel that it does not meander through its valley like most streams, making cutbanks to one side and point bars opposite. Instead, it runs in braided channels through its own broad bed. Looking at those Silurian conglomerates, I could all but hear the big braided rivers I had seen coming down from the Alaska Range, with gravels a mile wide, caribou and bears on the gravel, and channels flowing in silver plaits. If those rivers testify, as they do, to the erosional disassembling of raw young mountains, then so did the rock before us, with its clean river gravels preserved in river sand. "Geology repeats itself," Anita said, and we moved along, touching, picking at the rock. She pointed out the horse-belly curves of channel-fill deposits, and the fact that none was deeper than five feet—a result of the braiding and the shifting of the channels. Evidently, the calm earth and quiet seas that were described by the older rock we had collected up the road had been utterly revolutionized in the event that built the ancient mountains, which, bald as the djebels of Arabia, had stood to the east and shed the sand and gravel this way. In the ripple marks, the crossbedding, the manner in which the sands had come to rest,

Anita could see the westerly direction of the braided-river currents more than four hundred million years ago.

Three hundred years ago, William Penn arrived in this country and decided almost at once that the Lenape were Jews. "Their eye is little and black, not unlike a straight-look't Jew," he wrote home. "I am ready to believe them of the Jewish race. . . . A man would think himself in Dukes-place or Berry-street in London, when he seeth them." They were "generally tall, straight, well-built" people "of singular proportion." They greased themselves with clarified bear fat. Penn studied their language—the better to know them, the better to work out his treaties. "Their language is lofty, yet narrow, but like the Hebrew. . . . One word serveth in the place of three. . . . I must say that I know not a language spoken in Europe that hath words of more sweetness or greatness, in accent or emphasis, than theirs." Penn heard "grandeur" in their tribal proper names. He listed them: Tammany, Poquessin, Rancocas, Shakamaxon. He could have added Wyomissing, Wissinoming, Wyoming. He made treaties with the Lenape under the elms of Shakamaxon. Tammany was present. He was to become the most renowned chief in the history of the tribe. Many years after his death, American whites in eastern cities formed societies in his name, and called him St. Tammany, the nation's patron saint. Penn's fondness for the Lenape was the product of his admiration. Getting along with the Lenape was not difficult. They were accommodating, intelligent, and peaceful. The Indians revered Penn as well. He kept his promises, paid his way, and was fair.

Under the elms of Shakamaxon, the pledge was made that Pennsylvania and the Lenape would be friends "as long as the sun will shine and the rivers flow with water." Penn outlined his needs for land. It was agreed that he should have some country west of the Lenape River. The tracts were to be defined by the distance a man could walk in a prescribed time—typically one day, or two—at an easygoing pace, stopping for lunch, for the odd smoke, as was the Lenape manner. In camaraderie, the Penn party and the Indians gave it up somewhere in Bucks County. Penn went home to England. He died in 1718.

About fifteen years later, Penn's son Thomas, a businessman who had a lawyer's grasp of grasping, appeared from England with a copy of a deed he said his father had transacted, extending his lands to the north by a day and a half's walk. He made it known to a new generation of Lenape, who had never heard of it, and demanded that they acquiesce in the completion of—as it came to be called—the Walking Purchase. With his brothers, John and Richard, he advertised for participants. He offered five hundred acres of land for the fleetest feet in Pennsylvania. In effect, he hired three marathon runners. When the day came—September 19, 1737—the Lenape complained. They could not keep up. But they followed. Their forebears had made a bargain. The white men "walked" sixty-five miles, well into the Poconos. Even so blatant an affront might in time have been accepted by the compliant tribe. But now the brothers made an explosive mistake. Their new terrain logically required a northern boundary. Illogically, the one they drew did not run east to a point on the river close to the Water Gap but northeast on a vector that encompassed and annexed the Minisink. Massacres ensued. Buildings were burned. Up and down the river, white scalps were cut. The Lenape reached for "the French hatchet." Peaceful, accommodating they once had been, but now they were participants in the French and Indian Wars. Where they had tolerated whites in the Minisink, they burned whole settlements and destroyed the occupants. They killed John Rush. They killed his wife, his son and daughter. They killed seventeen Vanakens and Vancamps. They pursued people on the river and killed them in their boats. They killed Hans Vanfleara and Lambert Brink, Piercewell Goulding and Matthew Rue. They could not, however, kill their way backward through time. They never would regain the Minisink.

As we moved along beside the screaming trucks, we were averaging about ten thousand years per step. The progression was not uniform, of course. There might be two million years in one fossil streambed, and then the next lamination in the rock would record a single season, or a single storm—on one flaky surface, a single drop of rain. We looked above our heads at the projecting underside of a layer of sandstone patterned with polygons, impressions made as the sand came pouring down in storm-flood waters over mud cracks that had baked in the sun. From a layer of conglomerate, Anita removed a pebble with the pick end of her rock hammer. "Milky quartz," she said. "Bull quartz. We saw this rock back up the road in the Precambrian highlands. When the Taconic Orogeny came, it lifted the older rock, and erosion turned it into pebbles and sand, which is what is here in this conglomerate. It's an example of how the whole Appalachian system continually fed upon itself. These are Precambrian pebbles, in Silurian rock. You'll see Silurian pebbles in Devonian rock, Devonian pebbles in Mississippian rock. Geology repeats itself." Now and again, we came to small numbers that had been painted long ago on the outcrops. Anita said she had painted the numbers when she and Jack Epstein were working on the geology of the Water Gap. She said, "I'd hate to tell you how many months I've spent here measuring every foot of rock." Among the quartzites were occasional bands not only of sandstone but of shale. The shales were muds that had settled in a matter of days or hours and had filled in the lovely periodicity of the underlying ripples in the ancient river sand. For each picture before us in the rock, there was a corresponding picture in her mind: scenes of the early Silurian barren ground, scenes of the rivers miles wide, and, over all, a series of pictures of the big Taconic mountains to the east gradually losing their competition with erosion in the wash of Silurian rain—a general rounding down of things, with river gradients declining. There were pictures of subsiding country, pictures of rising seas. She found shale that had been the mud of an estuary, and fossil shellfish, fossil jellyfish, which had lived in the estuary. In thin dark flakes nearby she saw "a little black lagoon behind a beach." And in a massive layer of clear white lithified sand she saw the beach. "You don't see sand that light except in beaches," she said. "That is beach sand. You would have looked westward over the sea."

To travel then along the present route of Interstate 80, you would have been in need of a seaworthy shallow-draft boat. The journey could have started in mountain rapids, for the future site of the George Washington Bridge was under thousands of feet of rock. Down the huge fans of boulders and gravel that leaned against the mountains, the west-running rivers raced toward the epicontinental sea. They projected their alluvium into the water and spread it so extensively that up and down the long flanks of the Taconic sierra the alluvium coalesced, gradually building westward as an enormous collection of sediment—a deltaic complex. At the future site of the Water Gap, you would have shoved off the white beach and set a westerly course across the sea—looking back from time to time up the V-shaped creases of steep mountain valleys. That was the world in which the older rock of the Water Gap had been forming—the braided-river conglomerates, the estuary mud, the beach sand. In the Holocene epoch, the Andes would look like that, with immense fans of gravel coming off their eastern slopes—the essential difference being vegetation, of which there was virtually none in the early Silurian. The sea was shallow, with a sandy bottom, in Pennsylvania. The equator had shifted some and was running in the direction that is now northeast-southwest, through Minneapolis and Denver. There were muds of dark lime in the seafloors of Ohio, and from Indiana westward there were white-lime sands only a few feet under clear water.

If you had turned around and come back twenty-five million years later, in all likelihood you would still have been riding the sparkling waves of the limestone-platform sea, but its extent in the late Silurian is not well reported. Most of the rock is gone. There are widely scattered clues. Among the marine limestones that fell into the diamond pipe near Laramie, some are late Silurian in age. From Wyoming toward the east, there seems to have existed a vastly extrapolated sea. The extrapolation stops in Chicago. You would have come upon a huge coral reef, which is still there, which grew in Silurian time, and did not grow in a desert. It was a wave-washed atoll then, a Kwajalein, an Eniwetok, and in time it would become sugary blue dolomite packed with Silurian shells. After standing there more than four hundred million years, the dolomite would be quarried to become for many miles the concrete surface of Interstate 80 and to become as well the foundations of most of the tall buildings that now proclaim Chicago, as the atoll did in Silurian time. Interstate 80 actually crosses the atoll on bridges above the quarry, which is as close an approximation to the Grand Canyon as Chicago is likely to see, and may be its foremost attraction. Beyond the atoll, you would have come to other atolls and hypersaline seas. When water is about three times as salty as the ocean, gypsum will crystallize out. Sticking up from the bottom in central Ohio were dagger-length blades of gypsum crystal. You would have been bucking hot tropical trade winds then, blowing toward the equator, evaporating the knee-deep sea. East of Youngstown, red muds clouded the water—muds coming off the approaching shore. The beach was in central Pennsylvania now, near the future site of Bloomsburg, near the forks of the Susquehanna. The great sedimentary wedge of the delta complex had grown a hundred miles. The Taconic mountains were of humble size. The steep braided rivers were gone, their wild conglomerates buried under meandering mudbanked streams moving serenely through a low and quiet country—a rose-and-burgundy country. There were green plants in the red earth, for the first time ever.

Walking forward through time and past the tilted strata of the Water Gap, we had come to sandstones, siltstones, and shales, in various hues of burgundy and rose. In the irregular laminations of the rock—in its worm burrows, ripples, and crossbeds—Anita saw and described tidal channels, tidal flats, a river coming into an estuary, a barrier bar, a littoral sea. She saw the delta, spread out low and red, the Taconic mountains reduced to hills. We had left behind us the rough conglomerates and hard gray quartzites that had come off the Taconic mountains when they were high—the formation, known as Shawangunk, that forms the mural cliffs above the Delaware River. (The quartzites are paradisal to rappeling climbers, who refer in their vernacular to "the gap rap," a choice part of "the Gunks.") And now, half a mile up the highway and twenty million years up the time scale, we were looking at the younger of the two formations of which Kittatinny Mountain is locally composed. Generally red, the rock is named for Bloomsburg, outer reach of the deltaic plain in late Silurian time, four hundred and ten million years before the present.

Less than two hundred years before the present, when the United States was twenty-four years old, the first wagon road was achieved through the Water Gap. The dark narrow passage in rattlesnake-defended rock had seemed formidable to Colonial people, and the Water Gap had not served them as a transportational gateway but had been left aloof, mysterious, frightening, and natural.

In the hundred feet or so of transition rocks between the gray Shawangunk and the red Bloomsburg, we had seen the Silurian picture change from sea and seashore to a low alluviated coastal plain; and if we had a microscope, Anita said, we would see a few fish scales in the Bloomsburg river sands—from fish that looked like pancake spatulas, with eyes in the front corners.

In 1820, the Water Gap was discovered by tourists. They were Philadelphians with names like Binney.

Breaking away some red sandstone, Anita remarked that it was telling a story of cut-and-fill—the classic story of a meandering stream. The stream cuts on one side while it fills in on the other. Where bits and hunks of mudstone were included in the sandstone, the stream had cut into a bank so vigorously that it undermined the muddy soil above and caused it to fall. Meanwhile, from the opposite bank—from the inside of the bend in the river—a point bar had been building outward, protruding into the channel, and the point bar was preserved in clean sandstones, where curvilinear layers, the crossbeds, seemed to have been woven of rushes.

Confronted with a mountain sawed in half, a traveller would naturally speculate about how that might have happened—as had the Indians before, when they supposedly concluded that the Minisink had once been "a deep sea of water." Samuel Preston agreed with the Indians. In 1828, in a letter to *Hazard's Register of Pennsylvania*, Preston referred to the Water Gap as "the greatest natural curiosity in any part of the State." He went on to hypothesize that "from the appearance of so much alluvial or made land above the mountain, there must, in some former period of the world, have been a great dam against the mountain that formed all the settlements called Minisink into a lake, which extended and backed the water at least fifty miles." And therefore, he worked out, "from the water-made land, and distance that it appears to have backed over the falls in the river, the height must, on a moderate calculation, have been between one hundred and fifty and two hundred feet, which would have formed a cataract, in proportion to the quantity of water, similar to Niagara." Preston was a tourist, not a geologist. The first volume of Charles Lyell's *Principles of Geology*, the textbook that most adroitly explained the new science to people of the nineteenth century, would not be published for another two years, let alone cross the sea from London. All the more remarkable was Preston's Hypothesis. Like many an accomplished geologist who would follow, Preston made excellent sense even if he was wrong. Withal, he had the courage of his geology. "If any persons think my hypothesis erroneous," he concluded, "they may go and examine for themselves. . . . The Water Gap will not run away."

While sediments accumulated slowly in the easygoing lowlands of the late Silurian world, iron in the rock was oxidized, and therefore the rock turned red. Alternatively, it could have been red in the first place, if it weathered from a red rock source. There were dark-hued muds and light silts in the outcrop, settled from Silurian floods. There were balls and pillows, climbing ripples, flow rolls, and mini-dunes—multihued structures in the river sands. Maroon. Damask. Carmine. Rouge.

Artists were the Delaware Water Gap's most effective discoverers. Inadvertently, they publicized it. They almost literally put it on the map. Arrested by the symmetries of this geomorphological phenomenon, they sketched, painted, and engraved it. The earliest dated work is the Strickland Aquatint, 1830, with a long and narrow flat-bottomed Durham boat in the foreground on the river, four crewmen standing at their oars, a steersman (also standing) in the stern, and in the background the wildwoods rising up the mountain with its deep, improbable incision.

Cutting and filling, a stream would cross its own valley, gradationally leaving gravels under sands under silts under muds under fine grains that settled in overbank floods. With nothing missing, the sequence was before us now, and was many times reiterated in the rock—a history of the migrations of the stream as it spread layer upon layer through its subsiding valley, 412, 411, 410 million years ago.

In 1832, Asher B. Durand came upon the scene. Durand was one of the founders of what in time would be labelled the Hudson River School. The term was a pejorative laid by a critic on painters who went outdoors to vent their romantic spirits. They went up the Hudson, they went up the Rockies, and they went into the Water Gap unafraid. Durand painted another Durham boat. His trees looked Japanese. The picture was published after Durand himself made a copper engraving. It contributed to the axiom that where an easel had stood a hotel would follow. Kittatinny House was established in 1833, sleeping twenty-five.

Anita chipped out a piece of Bloomsburg conglomerate—evidence in itself that the stream which had made it was by no means spent. The rolling Silurian countryside must have been lovely—its river valleys velvet green. There were highland jaspers among the pebbles in the sand.

The early geologists began arriving in 1836, led by Henry Darwin Rogers. They were conducting Pennsylvania's first geological survey. In the deep marine Martinsburg slate and in the mountain strata that stood above it—in the "plication" and the "corrugation" of the sediments—Rogers saw "stupendous crust-movement and revolution," the "most momentous" of ancient times, and reported to Harrisburg what would eventually become known as the Taconic and Alleghenian orogenies. He decided that something had wrenched the mountain in New Jersey several hundred feet out of line with its counterpart in Pennsylvania. "I conceive these transverse dislocations to pervade all the great ridges and valleys of our Appalachian region," he wrote, "and to be a primary cause of most, if not all, of those deep notches which are known by the name of Water Gaps, and which cleave so many of our high mountain ridges to their very bases."

There were some thin green beds among the Bloomsburg reds. Anita said they were the *Kupferschiefer* greens that had given false hope to the Dutch. Whatever else there might be in the Bloomsburg Delta, there was not a great deal of copper. In the eighteen-forties, the mines of the Minisink were started up anew. They bankrupted out in a season. The Reverend F. F. Ellinwood delivered the "Dedication Sermon" in the Church of the Mountain, village of Delaware Water Gap, Pennsylvania, August 29, 1854, a year that Ellinwood placed in the sixth millennium after Creation. "The rude blasts of six thousand winters have howled in undaunted wildness over the consecrated spot, while yet its predicted destiny was not fulfilled," he told the congregation. "But here, at length, stands, in very deed the church firmly built upon the rock, and it is our hope and prayer that the gates of Hell shall not prevail against it. . . . For many centuries past, has Jehovah dwelt in the rocky fastnesses of this mountain. Ere there was a human ear to listen, His voice was uttered here in the sighing of the breeze and the thunder of the storms, which even then were wont to writhe in the close grapple of this narrow gorge. Ere one human footstep had invaded the wildness of the place, or the hand of art had applied the drill and blast to the silent rock, God's hand was working here alone—delving out its deep, rugged pathway for yonder river, and clothing those gigantic bluffs and terraces with undying verdure, and the far gleaming brightness of their laurel bloom." The hand of art, that very summer, was blasting the Delaware, Lackawanna & Western Railroad into the silent rock. Stagecoaches would soon leave the scene. A pathway by the river was replaced with rails. The sycamores that shaded it were felled. A telegraph wire was strung through the gap. Given a choice between utility and grandeur, people apparently wanted to have it both ways. Trains would travel in one direction carrying aristocrats and in the other carrying coal.

Anita put her fingers on fossil mud cracks, evidence not only of hours and seasons in the sun but of tranquillity in the environment in which the rock had formed. She also moved her fingers down the smooth friction streaks of slickensides (tectonic scars made by block sliding upon block, in the deforming turbulence of later times).

In the Ecological epoch, the Backpackerhaus School of photography will not so much as glance at anything within twenty-five miles of a railhead, let alone commit it to film, but in the eighteen-fifties George Inness came to the Water Gap and set up his easel in sight of the trains. The canvases would eventually hang in the Metropolitan Museum, the Tate Gallery, the National Gallery (London). Meanwhile, in 1860, Currier & Ives made a lithograph from one of them and published it far and wide. By 1866, there were two hundred and fifty beds in Kittatinny House alone, notwithstanding that the manager had killed a huge and ferocious catamount not far from the lobby. That scarcely mattered, for this was the New World, and out in the laurel there were also wolves and bears. The gap was on its way to becoming a first-class, busy summer resort.

"Note the fining-upward cycles," Anita said. "Those are cross-bedded sandstones with mud clasts at the base, rippled to unevenly bedded shaly siltstones and sandstones in the middle, and indistinctly mud-cracked bioturbated shaly siltstones with dolomite concretions at the top."

It was a lady visiting the Water Gap in the eighteen-sixties who made the once famous remark "What a most wonderful place would be the Delaware Water Gap if Niagara Falls were here."

The Aldine, in 1875, presented three wood engravings of the

Water Gap featuring in the foreground gentlemen with walking sticks and ladies with parasols, their long full dresses sweeping the quartzite. The accompanying text awarded the Delaware Water Gap an aesthetic edge over most of the alpine passes of Europe. *The Aldine* subtitled itself "The Art Journal of America" but was not shy to make dashes into other fields. "The mountains of Pennsylvania are far less known and visited than many of the American ranges at much greater distance, and even less than many of the European ranges, while they may be said to vie in beauty with any others upon earth, and to have, in many sections, features of grandeur entitling them to eminent rank," the magazine told its readers. "Not only the nature lover, by the way, has his scope for observation and thought in the Water Gap. The scientist has something to do, and is almost certain to do it, if he lingers there for any considerable period. He may not have quite decided how Niagara comes to be where it is—whether it was originally in the same place, or down at the mouth of the St. Lawrence; but he will find himself joining in the scientific speculations of the past half-century, as to whether the Water Gap changed to be what it is at the Flood; or whether some immense freshet broke through the barriers once standing across the way and let out what had been the waters of an immense inland lake."

By 1877, Kittatinny House was five stories high. *Harper's Weekly*, at the end of the season, ran a wood engraving of the Water Gap in color by Granville Perkins, who had taken enough vertical license to outstretch El Greco. Under the enlofted mountain, a woman reclined on the riverbank with a pink parasol in her hand. A man in a straw boater, dark suit, was stretched beside her like a snake in the grass.

In the crossbedding and planar bedding of the Bloomsburg rocks, as we slowly traced them forward through time, there had been evidence of what geologists call the "lower upper flow regime." That was now becoming an "upper lower flow regime."

When people were bored with the river, there were orchestras, magicians, lecturers, masquerade balls. They could read one another's blank verse:

Huge pile of Nature's majesty! how oft
The mind, in contemplation wrapt, has scann'd

Thy form serene and naked; if to tell,
That when creation from old chaos rose,
Thou wert as now thou art; or if some cause,
Some secret cause, has rent thy rocky mantle,
And hurl'd thy fragments o'er the plain below.
The pride of man may form conceptions vast,
Of all the fearful might of giant power
That rent the rampart to its very base,
Giving an exit to Lenape's stream,
And wildly mixing with woods and waters.
A mighty scene to set enchantment free,
Burst the firm barrier of eternal rock,
If by the howling of volcanic rage,
Or foaming terror of Noachian floods.
Let fancy take her strongest flight. . . .
But, as for us, let speculations go,
And be the food of geologic sons;
Who from the pebble judge the mountain's form . . .

Anita said the rock had been weakened here in this part of the mountain. The river, cutting through the formations, had found the weakness and exploited it. "Wherever a water gap or a wind gap exists, there is generally tectonic weakness in the bedrock," she went on. "The rock was very much fractured and shattered. There is particularly tight folding here."

The hotels were in Pennsylvania, and were so numerous in the eighteen-eighties and eighteen-nineties that they all but jostled one another, and suffered from the competition. Up the slope from Kittatinny House, as in a game of king-of-the-mountain, stood Water Gap House, elongate and white, with several decks of circumambient veranda under cupolas that appeared to be mansard smokestacks. All it lacked was a stern wheel. There was a fine view. On the narrow floodplain and river terraces of New Jersey, where I-80 would be, there were cultivated fields and split-rail fences, corn shocks in autumn, fresh furrows in spring.

Anita and I came to the end of the Bloomsburg, or as far as it went in the outcrops of the gap. "These are coarse basal sands," she said of one final layer. "They were deposited in channels and point bars through lateral accretion as the stream meandered." In all, there were fifteen hundred feet of the formation, reporting the disintegration of high Silurian worlds.

Ten or twelve years after the turn of the century, a Bergdoll touring car pulled into the porte cochère of Water Gap House and the chauffeur stepped out, leaving Theodore Roosevelt alone in the open back while a photograph arrested his inscrutable face, his light linen suit, his ten-gallon paunch and matching hat. This must have been a high moment for the resort community, but just as Teddy (1858–1919) was in his emeritus years, so, in a sense, was the Water Gap. A fickling clientele preferred Niagaras with falls. An intercity trolley had been added to the scene. Two miles downstream—in what had been George Inness's favorite foreground—was a new railroad bridge that looked like a Roman aqueduct. Rails penetrated the gap on both sides of the river. There was a golf course—dramatic in its glacial variations on precipitous tills pushed by the ice up the side of the broken mountain—where Walter Hagen, in 1926, won the Eastern Open Championship. Soon thereafter, the tournament was played for the last time. Walter Hagen was not coming back, and neither was the nineteenth century. The perennial Philadelphians were now in Maine. In 1931, Kittatinny House burned up like a signal fire. Freight trains wailed as they rumbled past the embers. In 1960 came the interstate—a hundred and sixty years after the first wagon road. As a unit of earth history, a hundred and sixty years could not be said to be exactly nothing—although, in the gradually accumulated red rock beside the river, ninety-four thousand such units were represented. To put it another way, in the fifteen-hundred-foot thickness of the Bloomsburg formation, there were five millimetres for each hundred and sixty years. The interstate, with its keloid configuration, was blasted into the Shawangunk quartzites, blasted into the redbeds of the Bloomsburg, along the New Jersey side. As if that was not enough for one water gap, it turned and crossed the river.

In all the rock we had walked by, the rivers and streams that carried the material had been flowing west and northwest. I looked over the bank at the inventive Delaware, going the other way. "When did the Delaware River come into existence?" I asked Anita.

She shrugged, and said, "Long ago."

I said, "Really."

She turned and looked back toward the great slot in the mountain, and said, "In the late Jurassic, maybe. Possibly the early Cretaceous. I can look it up. I didn't pay much attention to that part of geology."

In round numbers, then, the age of the river was a hundred and fifty million years. The age of the Water Gap rock was four hundred million years. Another fifty million years before that, the Taconic mountains appeared. The river 150, the rock 400, the first ancestral mountains 450 million years before the present—these dates are so unwieldy that they might as well be off a Manchu calendar unless you sense the pace of geologic change and draw an analogy between, say, a hundred million years of geology and one human century, with its upward-fining sequences, its laminations of events, its slow deteriorations and instant catastrophes. You see the rivers running east. Then you see mountains rise. Rivers run off them to the west. Mountains come up like waves. They crest, break, and spread themselves westward. When they are spent, there is an interval of time, and then again you see the rivers running eastward. You look over the shoulder of the painter and you see all that in the landscape. You see it if first you have seen it in the rock. The composition is almost infinitely less than the sum of its parts, the flickers and glimpses of a thousand million years.

Over the bridge and out of the gap, we paid twenty-five cents at the booths of the Delaware River Joint Toll Bridge Commission. The collector was a citizen so senior he appeared to have been alive for a sixty-millionth of the history of the world. "Have a nice day," he said. We were moving west and would soon be rising into what geologists refer to as "the so-called Pocono Mountains"—actually a layered flatland that has been cut up by confused streams into forested mesas with names like Mt. Pohopoco. The long continuous welt of the deformed Appalachians—the Ridge-and-Valley belt of folded mountains—is extremely narrow at this latitude. As much as eighty miles wide in the course of its run from Alabama into Canada, it is a fifteen-mile isthmus where it is crossed by Interstate 80 at the eastern end of Pennsylvania. The foldbelt is narrow there because the Poconos refused to deform. When tectonism came and rock was being corrugated left and right, the strata that would become the Poconos were somewhat compressed but did not bend. "The rocks took the shock of the tectonics and didn't buckle," Anita said. "They shattered some, but they didn't move much. They didn't have the glide planes."

Now, scarcely a mile from the toll booths and still very much in the foldbelt, we came to what road builders call a throughcut, where the road had been blasted through the tip of a ridge. We stopped, crossed the interstate, and climbed the higher side. The rock was calcareous shale, and had been seafloor mud about three hundred and ninety million years ago, possibly ten fathoms down then, and gritty with fragments of shells and corals. There were brachiopods in the rock (something like clams and scallops) and cornucopian corals. Certain categories of these lone-growing, conical corals were the index fossils that led nineteenth-century geologists working in Devonshire to recognize the relative age of the rock the corals were found in, and to call the time Devonian. "If it weren't for this roadcut, I'd never have been able to measure these rocks," Anita said. "The next exposure is halfway between Kingston and Albany. When they first made this road, we came in and mapped in a hurry, before they laid on that god-damned grass seed, all that straw and organic tar. In the East, no one knows from geology."

The sea had been in retreat here in the early Devonian, and as we made our way uphill, and pursued the dip of the strata, we hiked two or three million years through progressively shallower marine deposits and came to a conglomerate full of pearly white quartz that had been tossed and rounded by surf. Beyond the conglomerate was light, coarse-grained sandstone—a fossil Devonian beach. The sea would have been out there to the west, the equator running more or less along the boundary between Canada and Alaska. We turned and went the other way, up through woods and around the nose of the ridge. We were far above the interstate now and looking down on the tops of big rolling boxes—North American Van Lines, moving

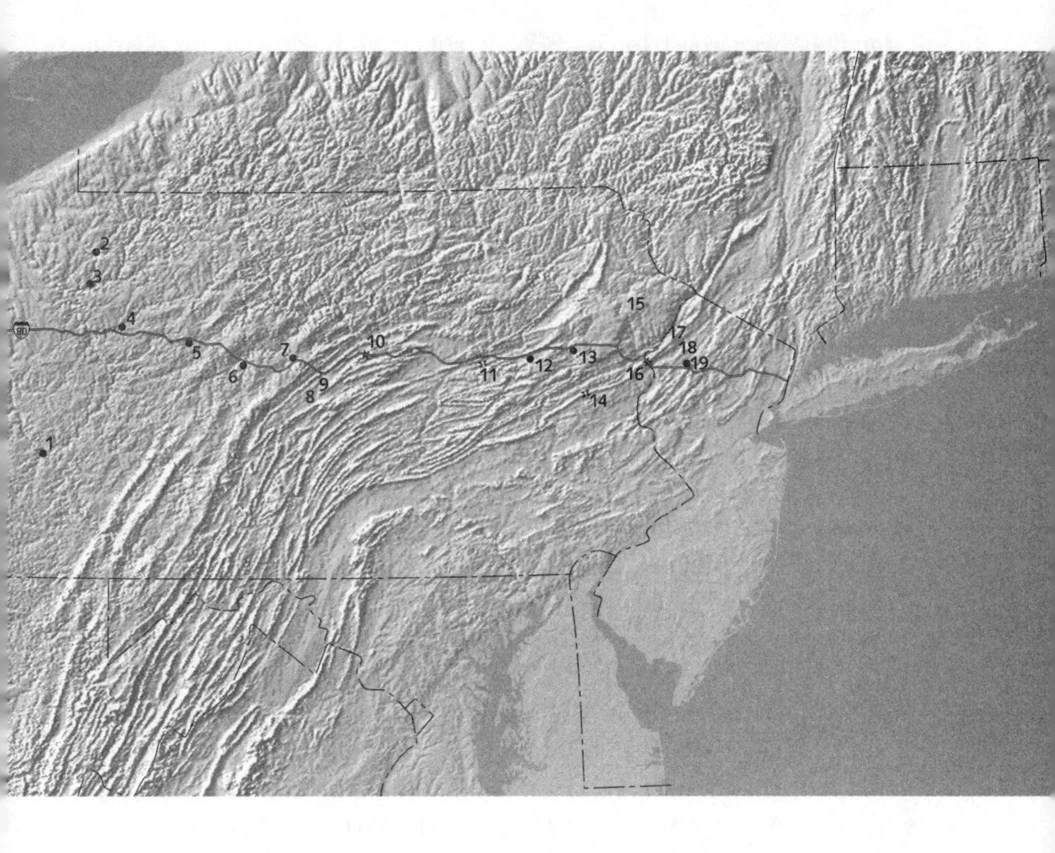

1. Pittsburgh
2. Titusville
3. Oil City
4. Clarion
5. Du Bois
6. Clearfield
7. Snow Shoe
8. Nittany Valley
9. Bald Eagle Mountain
10. Holy Toledo Cut
11. Fryingpan Gap
12. Bloomsburg
13. Hickory Run State Park
14. Lehigh Gap
15. Pocono Plateau
16. Delaware Water Gap
17. Kittatinny Mountain
18. Great Valley of the Appalachians
19. Allamuchy

families from coast to coast. We went on through more woods, in an easterly direction, against the dip, until at length, high on the far side of the ridge, we reached another beach, ten or fifteen million years older than the first one. In the comings and goings—transgressions, regressions—of the epicontinental sea, the strandline had paused here in late Silurian time. Anita said, "This was the barrier beach when the red beds of the Delaware Water Gap were paper-laminated lagoon muds behind barrier islands. Geology is predictable. If you find lagoon mud, you should find beach sand not far away." On through the woods, she walked offshore to an exposure of dark, shallow-water limestone. "This is what I've come up here for," she said. "This is as pure a limestone as you can get." She remembered the outcrop from her mapping days, and now she wanted the conodonts. With her sledgehammer, she went at the rock. It was grudging, competent. She set off sparks. Working hard, she slowly filled two canvas bags, each having a capacity approaching one cubic foot. As she had done on hundreds of similar journeys, she would carry the bags into the post office of a small town somewhere and set them on the counter with a lithic clunk while the postmaster's eyeballs moved forward over the tops of his reading glasses. There was a frank number, a printed label. "ANITA G. HARRIS, U.S.G.S. BRANCH OF PALEONTOLOGY AND STRATIGRAPHY, WASHINGTON, D.C. 20560, OFFICIAL BUSINESS." Seeing that, 18412 would develop a security clearance in the lower strata of his frown, and with solemnity accept the rocks.

It was a shelly, coastal limestone. There were cup corals in it, and a profusion of brachiopods that looked like filberts. "Farmers call these hazelnut rocks," Anita said. A little farther along the outcrop, the limestone was full of small round segments of the stems of sea lilies—tall, graceful animals with petalled heads that grew like plants on stems. "The sea lilies grew in clear, shallow water a little offshore," Anita said. "It was a coast like Fiji's, or the Philippines', or Guatemala's. The coral and the thick shells tell you the water was warm. The rock is dark because it is full of dead oil, which came in later—much later. Oil migrated into these rocks and was cooked at high temperature. The conodonts will tell me the temperature."

In the deformed, sedimentary Appalachians, the rock not only had been compressed like a carpet shoved across a floor but in places had been squeezed and shoved until the folds tumbled forward into recumbent positions. Some folds had broken. Some entire regions had been picked up and thrust many miles northwest. Dozens of other complexing events had locally affected the structures of the Ridge and Valley Province. One therefore could not know what to expect next. Whole sequences might suddenly be upside down, or repeat themselves, or stand on end reading backwards. Among such rocks, time moves in and out and up and down as well as by.

"It's a real schlemazel," Anita said. "Not by accident is geology called geology. It's named for Gaea, the daughter of Chaos."

Among the west-dipping Silurian formations of the Delaware Water Gap, one might project but could not reasonably expect Devonian rock to westward. It would be there if the stratigraphic package was intact and had not been overturned. The rock of that first big Pennsylvania roadcut was early Devonian in age. Leaving it, we moved seven miles west along the interstate and twenty million years up the time scale, where we stopped at a roadcut of middle Devonian marine siltstones and shales, so rich in organic residues that it was black as carbon, with corals that had been sliced by dynamite and resembled sections of citrus. Cambrian, Ordovician, Silurian, Devonian—for this crunched and shuffled country we were experiencing remarkable consistency in an upward voyage through time. And now in the silken muds of these Devonian seafloors we were seeing the final stages in the long tranquil interval between the Taconic and Acadian revolutions. The rock coarsened abruptly as we drove on westward. There were cobbly conglomerates. They were the first explosive belch from the new Acadian mountains, which came up in the east at a rate ten times as fast as erosion could destroy them and, with a new system of rivers, rapidly shed this downpour of rock. A few miles farther on, another ten or fifteen million years, and we were among roadcuts containing upper Devonian stream channels of a quiet country, a low alluvial plain—point bars, cutbanks, ripple marks in red river sands. We were forty million years past the Water Gap, and the geology was repeating itself on an epic scale. A new set of coalescing fans had come off the Acadian mountains, and, as the great sierra disintegrated, its detritus spread westward thick upon the country and into the sea—at least ten thousand feet thick in the east and gradually thinning to the west, this immense new clastic wedge, to be known in geology as the Catskill Delta.

It stands at the surface of a huge piece of country. Erosion working into the high eastern end has cut the shapes of the Catskill Mountains. The rock lies essentially flat there, and is flat all the way to the shores of Lake Erie. It is the uppermost rock of half of New York State. It is the rock of Chenango, the rock of Chautauqua, too. It is the rock of Seneca, Ithaca, Elmira, Oneonta. In Pennsylvania, it is largely buried, or was sliced and kneaded into the deformed mountains, but as the so-called Poconos it stands flat and high. The Poconos actually are part and parcel of the New York Devonian clastic wedge. The Poconos are a tongue of New York State penetrating Pennsylvania.

The Acadian mountains are gone. The wedge remains. The Acadians, in their Devonian prime, must have been a crowd of Kanchenjungas, to judge by their sedimentary remains, which reach almost to Indiana. As the mountains came down, they stood ever deeper in debris. At Denver, the Rocky Mountains are up to their hips in their own waste. The sedimentary wedge that has come off the Rockies is thickest there by the mountain front, and gradually thins to the east. Kansas and Nebraska are like pieces cut from a wheel of cheese—lying on their sides, thick ends to the west. Altitude in itself suggests the volume of material. Kansas and Nebraska are three thousand feet higher in the west than in the east.

We were running on the summits of the Poconos—uneven but essentially level topography, the Pocono equivalent of Alpine minarets. Where we saw stratified rock in roadcuts, it seemed level enough to stop a bubble. For the most part, it was Catskill sandstone, red as borscht, from latest Devonian and earliest Mississippian time. The summits of the Poconos were not only cragless, they ran on under the scrub oaks as far as the eye could see. There were peat bogs. There was a great deal of standing water. The landscape was bestrewn with hummocky lumps of gravel. "There's no way that streams brought all that gravel up here," Anita said. "Religious farmers say it's evidence of the Great Flood."

If so, the Great Flood was frozen. These were morainal gravels, outwash gravels. Interstate 80 marks almost exactly the Wisconsinan ice sheet's line of maximum advance in the Poconos.

We made a short digression from the interstate to see—in some Devonian siltstone—a tidal flat that was stuffed with razor clams. The surface of the rock had a Fulton Market look. It was a paisley of conglomerate clams. Three hundred and seventy-five million years old, they resembled exactly their modern counterparts. "Things haven't changed much," Anita said as she got back into the car.

We drove on into Hickory Run State Park, where we walked through heavy woods toward a clear space ahead. We seemed to be approaching a body of water. Its edge resembled a shore, and its seventeen acres were surrounded by conifers, whose jagged silhouettes invoked a northern pond. In place of water, however, the pond consisted of boulders—thousands of big boulders, some of them thirty feet long, nearly all of them red rock weathered dusty rose, and all of them accordant with a horizontal plane, causing them to seem surreally a lake of red boulders. DAD, MOM, HARRY, and GEORGE had been there in 1970 with a can of acrylic spray. JOE VIZZARD came some years later. Dozens of others had daubed the rocks on days ranging backward to 1935, when the park was established. The big red-boulder expanse was difficult to cross, however, and its sorry guestbook was confined to one corner. The boulders were stunningly beautiful—in their lacustrine tranquillity, their lovely color, their spruce-rimmed absence of all but themselves. We walked out some distance, stepping from one to another multiton red potato.

Anita lost her balance and almost went below. "What a klutz," she remarked.

I thought I might be learning a geologic term.

"These are periglacial boulders," she went on. "They're not erratics. They haven't really moved. In the climates we have now, big boulders that are not erratics just don't appear in the woods. Only a remarkable set of conditions would produce this scene. You had to start with the right bedrock. You had to have the right angle of dip, the right erosional shape for flushing, the right distance from the glacier. The terminus of the glacier was about half a mile away. The climate was arctic. Imagine the frost heaves after water in summer got into the bedrock and that kind of winter came to explode it. Gravels, sands, and clay were completely flushed away by meltwater, leaving these boulders."

Twenty-five thousand years ago—in the late Pleistocene, or, relatively speaking, the geologic present—arctic frost had broken out the boulders and had begun the weathering that rounded them. The Acadian mountains, wearing away three hundred and fifty million years ago, had provided the material of which the boulders were made. Back on the interstate and continuing across the Pocono Plateau, we ran through more flat-lying red strata of the same approximate age, and Anita said, "Remember the Bloomsburg? This rock is fifty million years later, and it looks like the Bloomsburg, and it was formed on another low, alluviated coastal plain, when the Acadian mountains were dying down. As I've been telling you, geology is predictable once you learn a few facts. Geology repeats itself all through the rock column."

On the geologic time scale, anyone could assign these events to their respective places and sense the rhythms of the cycles—rock cycles, glacial cycles, orogenic cycles: overlapping figures in the rock. Taken all together, though, they seemed to ask somewhat more than they answered, to reveal less than nature kept concealed. The evidence showed that the Acadian mountains had come down, as had the Taconics before them, and each had spread westward new worlds of debris. One could also discern that the Acadian Orogeny had folded and faulted the sedimentary rock that had formed from the grit of the earlier mountains, and metamorphosed the rock as well—changed the shales into slates, the sandstones into quartzites, the limestones and dolomites into marble. A third revolution would follow—the Alleghenian Orogeny, in Pennsylvanian-Permian time. Another mountain wave would crest, break, and send its swash to westward. It was all very repetitive, to be sure—the great ranges rising, falling, rising, falling, covering and creating landscapes, as if successive commingling waters were to rush up a beach and freeze. But why? How? You see in rock that geology repeats itself, but you do not see what started the process. In the rivers in rock you find pieces of mountains, but you do not find out why the mountains were there.

I said to Anita, "What made the mountains rise?"

"The Acadian mountains?"

"All of them—Taconic, Acadian, Alleghenian. What made them come up in the first place?"

It has been Anita's style as a geologist to begin with an outcrop and address herself to history from there—to begin with what she can touch, and then to reason her way back through time as far as she can go. A river conglomerate, as tangible rock, unarguably presents the river. The river speaks of higher ground. The volume of sediment that the river has carried can imply a range of mountains. To find Precambrian jaspers in the beds of younger rivers means that the Precambrian, the so-called basement rock, was lifted to form the mountains. These are sensible inferences drawn cleanly through an absence of alternatives. To go back in this way, retrospectively, from scene to shifting scene, is to go down the rock column, groping toward the beginning of the world. There is firm ground some of the way. Eventually, there comes a point where inference will shade into conjecture. In recesses even more remote, conjecture may usurp the original franchise of God.

By reputation, Anita is a scientist with an exceptionally practical mind, a geologist with few weaknesses, who is at home in igneous and metamorphic petrology no less than in sedimentology. She has been described as an outstanding biostratigrapher, a paleontologist who knows the rocks in the field and can go up to a problem and solve it. In my question to her, I was, I will confess, rousing her a little. I knew what her answer would be. "I don't know what made the mountains come up in the first place," she said. "I have some ideas, but I don't know. The plate-tectonics boys think they know."

When the theory of plate tectonics congealed, in the nineteen-sixties, it had been brought to light and was strongly supported by worldwide seismic data. With the coming of nuclear bombs and limitation treaties and arsenals established by a cast of inimical peoples, importance had been given to monitoring the earth for the tremors of testing. Seismographs in large numbers were salted through the world, and over a decade or so they revealed a great deal more than the range of a few explosions. A global map of earthquakes could be drawn as never before. It showed that earthquakes tend to concentrate in lines that run up the middles of oceans, through some continents, along the edges of other continents—seamlike, around the world. These patterns were seen—in the light of other data—to be the outlines of lithospheric plates: the broken shell of the earth, the twenty-odd pieces of crust-and-mantle averaging sixty miles thick and varying greatly in length and breadth. Apparently, they were moving, moving every which way at differing speeds, awkwardly disconcerting one another—pushing up alps—where they bumped. Coming apart, they very evidently had opened the Atlantic Ocean, about a hundred and eighty million years ago. Where two plates have been moving apart during the past twenty million years, they have made the Red Sea. Ocean crustal plates seemed to dive into deep ocean trenches and keep on going hundreds of miles down, to melt, with the result that magma would come to the surface as island arcs: Lesser Antilles, Aleutians, New Zealand, Japan. If ocean crust were to dive into a trench beside a continent, it could lift the edge of the continent and stitch it with volcanoes, could make the Andes and its Aconcaguas, the Cascade Range and Mt. Rainier, Mt. Hood, Mt. St. Helens. It was a worldwide theory—revolutionary, undeniably exciting. It brought disparate phenomena into a single story. It explained cohesively the physiognomy of the earth. It linked the seafloor to Fujiyama, Morocco to Maine. It cleared the mystery from long-known facts: the glacial striations in rock of the Sahara, the equator's appearances in Fairbanks and Nome. It was a theory that not only opened oceans but closed them, too. If it tore land apart, it could also suture it, in collisions that perforce built mountains. Italy had hit Europe and made the Alps. Australia had hit New Guinea and made the Pegunungan Maoke. Two continents met to make the Urals. India, at unusual speed, hit Tibet. Eras before that, South America, Africa, and Europe had, as one, hit North America and made the Appalachians. The suture was probably the Brevard Zone, a long, northeast-trending fault zone in the southern Appalachians with very different rock types on either side and no discernible matchup of offset strata. Discontinuous extensions of the Brevard Zone seemed to reach to the Catoctin Mountains and on to Staten Island. Southeastern Staten Island apparently was a piece of the Old World. Ships that sailed for Europe had arrived when they went under the bridge.

Plate theory was constructed in ten years by people with hard data who were consciously and frankly waxing "geopoetical" as well. Once the essentials of the theory were complete—after the discovery of seafloor spreading had led to an understanding of trench subduction, and after the plates and their motions had finally been outlined and described—the theory took a metaphysical leap into the sancta of the gods, flaunting its bravado in the face of Yahweh. It could make a scientist uncomfortable. Instead of reaching back in time from rock to river to mountains that must have been there—and then on to inference and cautious conjecture in the dark of imperceivable unknowns—this theory by its conception, its nature, and its definition was applying for the job of Prime Mover. The name on the door changed. There was no alternative. The theory was panterrestrial, panoceanic. It was the past and present and future of the world, sixty miles deep. It was every scene that ever was on earth. Either it worked or it didn't. Hoist it was to heaven with its own petard. "Established" there, it looked not so much backward from the known toward the unknown as forward from the invisible to its product the surface of the earth. Anita was more worried than made hostile by all this. By no means did she reject plate theory out of hand. There were applications of it with which she could not agree. Moreover, it was too fast a vehicle for its keys to be given to children.

"The plate-tectonics people have certain set patterns that they expect to see," Anita said. "They kind of lock themselves in. If something doesn't fit the theory, they'll find some sort of reason. They'll say that something is missing, or that it was subducted, or that it has not yet been found in the subsurface. They make things fit."

"Do you believe that ocean crust is subducted into trenches, that it melts and then comes up behind the trenches as volcanoes and island arcs?" I asked her.

"That is straightforward," she said. "And I have no doubt that one edge of the Pacific Plate is grinding northwest through California. What I object to is plate tectonics taken as absolute gospel. To stuff that I know about, it's been overapplied—without attention to geologic details. It's been misused terribly. It has misrepresented facts. It has oversimplified the world. The Atlantic spreading open I absolutely believe. How long it has been spreading open I don't know. I don't really believe that North America and South America were up against one another. The whole Pacific margin is thrusting from west to east, but there is no continent colliding with it. I don't see that plate tectonics explains all of these things. I think tectonics on continents is different from tectonics in oceans, and what works in oceans is often misapplied on land. As a result, there is less understanding of regional geology. The plate-tectonic model is so generalized and is used so widely that people do not get good regional pictures anymore. People come out of universities with Ph.D.s in plate tectonics and they couldn't identify a sulphide deposit if they fell over it. Plate tectonics is not a practical science. It's a lot of fun and games, but it's not how you find oil. It's a cop-out. It's what you do when you don't want to think."

Before the plate-tectonics revolution, back in the penumbra of the Old Geology, mountains (as has been noted) were thought to be driven upward from a deep-seated source known as a geosyncline—a profound downwarp of the crust, a long trough below the sea, which sediments fell into. East of North America, for example, the muds that would become Martinsburg slates first became rock in a geosyncline. The great trough trended northeast, like the mountains it would produce. How the mountains came up was not absolutely defined, but the story seemed clear, even if the authorship was somewhat moot, and it was a story of rhythmically successive orogenies, chapter headings in the biography of the earth. Some geologists preferred to liken them to punctuation marks, because mountain-building phases took up so little of all time—as little as one per cent, no more than ten per cent, depending on the geologist who was calculating the time.

Anita said, choosing North America's most eminent example, "The Gulf of Mexico is a big geosyncline, if you want. The big bird-foot delta of the Mississippi River is one hell of a sedimentary pile. Drill twenty-two thousand feet down and you're still in the Eocene. The crust will take about forty thousand feet of sediment—that's the elastic limit. Then it regurgitates the sediment, which begins to rebound. The sediment is also heated up, melted. Water, gas, and oil come out of the rock. Sedimentary layers move up with thermal drives as well as with isostasy. Sedimentary layers also move laterally, and are thus thrust sheets. In Cambro-Ordovician time, fifty million years or so before the Taconic mountains came up, the continent was to the west of us here, the coastline was in central Ohio, and to the east of us, where the Atlantic shelf is now, stood an island arc like Japan. There are volcaniclastic sediments of that age from Newfoundland to Georgia—just about the length of Japan. The present coast of Asia is the Ohio coastline in that story. Picture the sediment that is pouring off the Japanese islands into the Sea of Japan. The Martinsburg slates were shed not from the continent—not from Ohio—but mainly from the east, from the island arc offshore. You pile up forty thousand feet of sediment and it pops. The Martinsburg popped. The Taconic mountains came up. Once the process starts, it keeps itself going. You push up a mountain range, erode it into the west. The material depresses the crust. It is low-density material and it is brought down into the regime of high-density material. When enough has been piled on, the low-density material comes back up. That is how orogenic waves propagate themselves, each mountain mass being cannibalized to produce a new mountain mass to the west. But I still don't know what started the process."

Historical geologists, in the olden days, pieced together that narrative. Economic geologists, in their pragmatic way, cared less. In describing the minable Martinsburg—the blue-gray true unfading slate—C. H. Behre, Jr., wrote in 1933, "Sedimentary rocks are often compressed from the sides through what may be loosely described as shrinking of the crust of the earth; how this shrinking is brought about is, for the present purpose, beside the point. It has the well-recognized effect, however, that layers or bedding planes are wrinkled or thrown into 'folds.' "

By the nineteen-seventies, what Behre had loosely described was widely believed to be the impact of one continent colliding with another, as Iapetus, the proto-Atlantic ocean, was closed and the suture of the two continents became the spine of the Appalachians. The successive pulses of orogeny—Taconic, Acadian, Alleghenian—were attributed to the irregular shapes of shelves and coastlines of the continents. Where they bulged, the action would have an early date, and especially where some cape, point, or peninsula had a similar feature coming from the opposite side. Such headlands, in advance contact, were said to have produced the Taconic Orogeny. Great bays, eventually coming against one another, set off the Acadian Orogeny. The Alleghenian Orogeny was the final crunching scrum, completing the collision. The apparent suture was a line running through Brevard, North Carolina, more or less connecting Atlanta, Asheville, and Roanoke, not to mention Africa and America.

The Martinsburg seafloor and the underlying carbonate rocks had unquestionably been broken into thrust sheets and shuffled like cards. Uplifted with their Precambrian basement, they had, in perfect harmony with the Old Geology, become mountains that shed their sediments—shed their clastic wedges—and buried the Martinsburg deep enough to turn it into slate, buried the carbonates deep enough to turn them into marble. Thus, plate tectonics fit. Plate tectonics may have restyled the orogeny and dilapidated the geosyncline, but it fit the classical evidence.

There were, to be sure, certain anomalies, which suggested further study. If the Brevard Zone was the suture, how come it was so short? It was evident for a hundred miles, dubious for a few hundred more, and nonexistent after that. If the Taconic, Acadian, and Alleghenian orogenies were subdivisional impacts of a single intercontinental collision, how come they took so long? In plate-tectonic theory, plates move at differing speeds, the average being two inches a year. The successive orogenic pulses that resulted in what we know as the Appalachian Mountains occurred across a period of about two hundred and fifty million years. In two hundred and fifty million years, at two inches a year, you can move landmasses a third of the way around the world. Geologists ordinarily require vast stretches of time to account for their theories. In this case, they have too much time. They have two continents in the act of collision for two hundred and fifty million years.

In 1972, scarcely four years after the lithospheric plates had first been identified and the theory that described them had become news in the world, Anita and two co-authors published a set of papers offering strong evidence of plate tectonics in action, apparent proof that Sweden, or something like it, had once been in Pennsylvania. It was an inference drawn from conodont paleontology. The papers were widely cited for their support of plate-tectonic theory, and are cited still. North of Reading, in the Great Valley of the Appalachians, Anita had found early Ordovician conodonts of a type previously unknown in North America but virtually identical with early Ordovician conodonts from the rock of Scandinavia. All over North America were early Ordovician conodonts from tropical and subtropical seas. Their counterparts in Scandinavia were from cooler water, and so were these strangers Anita found in Pennsylvania. She found them in what is known as an exotic block, embedded in a far-travelled thrust sheet. They happened to be within a third of a mile of warm-water American conodonts in rock of about the same age which had moved hardly at all. The Scandinavian conodonts had apparently come to Pennsylvania with the closing of the proto-Atlantic ocean and been dropped ashore off the leading edge of the arriving plate. "Even I said, 'Oh, this piece peeled off the oncoming European-African plate and got dumped in here along with the clastics,' " Anita said, telling the story. "Everybody cited those papers. To this day they are called 'marvellous, landmark papers.' I could eat my heart out. The papers have been used as prime backup proofs of plate-tectonic applications in the northern Appalachians. Even now, a lot of the people who use the plate-tectonic model for interpreting the Appalachians are completely unaware that those papers were based on a paleontological misinterpretation."

Working in Nevada three years later, Anita had found Scandinavian-style conodonts of middle Ordovician age. Her husband, Leonard Harris, savored the discovery not for its embarrassment to his wife, needless to say, but for its air-brake effect on the theory of plate tectonics. "Now, how could that be?" he would ask. "How did this happen? Europe can't hit you in Nevada." From the Toiyabe Range she had taken the cool-water fossils, and moving east to other mountains—from basin to range—she had come to middle Ordovician carbonates that contained a mixture of conodonts of both the cool-water and the warm-water varieties, the American and the Scandinavian styles. Farther east, in limestone of the same age in Utah, she had found only warm-water conodonts. She realized now the absoluteness of her error. Utah had been pretty much the western extremity of the vast Bahama-like carbonate platform that covered North America under shallow Ordovician seas. In western Utah, the continental shelf had begun to angle down toward the floor of the Pacific, and in central Nevada the continent had ended in deep cool water. The conodont types differed as a result of water temperatures, not as a result of their geographic origins. Shallow or deep, conodonts of northern Europe were the same, because the water was cool at all depths. But here in America, with the equator running through the ocean where San Francisco would someday be, Ordovician water temperatures varied according to depth. Those apparently Scandinavian fossils were forming in deep cool water, the American ones in warm shallows.

Moving east from the Toiyabe Range and into Utah, Anita had gone from outcrop to outcrop through the Ordovician world, from ocean deeps to the rising shelf into waist-deep limestone seas. She could see now that the thrusting involved in the eastern orogenies had shoved the cool-water conodonts and their matrix rock from the deep edge of the continental rise into what would be Pennsylvania. They had travelled, to be sure—but they had more likely come from Asbury Park than from Stockholm. In the thrusting and telescoping of the strata, the transition rocks of the American continent's eastern slope had been deeply buried. In them, almost surely, would be a mixture of cool-water and warm-water conodont types. To the east of the Toiyabe Range, there had been less telescoping, and the full sequence was traceable—from the cool deep continental edge up the slope to the warm far-reaching platform. "The change had nothing to do with moving plates," Anita concluded. "Nothing to do with plate tectonics. I blew it. It was an environmental change, an environmental sequence."

More recently, working in Alaska, she had seen the sequence again, this time in tightly banded concentration, for the "American" conodonts were from reefs around Ordovician volcanic islands with steeply plunging sides and the "Scandinavian" conodonts were from cold deeps nearby.

Swerving to avoid a pothole, Anita said, "The plate-tectonics boys look at faunal lists and they go hysterical moving continents around. It's not the paleontologists doing it. It's mostly the geologists, misusing the paleontology. Think what geologists would make of the present east coast of the United States if they did not understand oceanography and the resulting distribution of modern biota. Put yourself forty or fifty million years from now trying to reconstruct the east coast of the United States by looking at the remains in the rock. God help you, you would probably have Maine connected to Labrador, and Cape Hatteras to southern Florida. You'd have a piece of Great Britain there, too, because you see the same fauna. Well, did you ever hear of ocean currents? Did you ever hear of the Gulf Stream? The Labrador Current? The Gulf Stream brings fauna north. The Labrador Current brings fauna south. I think that a lot of the faunal anomalies you see in the ancient record, and which are explained by invoking plate tectonics, can be explained by ocean currents bringing fauna into places they shouldn't be. In the early days of plate tectonics, a lot of us, including me, jumped on the bandwagon in order to explain the distribution anomalies we were seeing not only in the eastern Appalachians but in North America as a whole. When we better understood the paleoecologic controls on the animals some of us were working on, there was no reason to invoke plate tectonics."

The experience was cautionary, to say the least. It did not close her mind to plate tectonics, but it opened a line of suspicion and made her skeptical of the theory's insistent universality. Her discomfort varies with distance from the mobile ocean floors. She likes to describe herself as a "protester." The protest is not so much against the theory itself as against excesses of its application—up on the dry land. "A number of these people took very interesting ideas that apply to ocean floors and tried to apply them to everything," she remarked. "They tried to extrapolate plate tectonics through all geologic time. I don't know that that holds. My husband has blown some of their ideas apart."

Leonard Harris, sometimes known as Appalachian Harris, was very much a protester, too. Tragically, he died in 1982, a relatively early victim of cancer and related trouble. He was a genial and soft-spoken, almost laconic man with a lean figure that had walked long distances without the help of trails. He liked to build ideas on studied rock, and was not easily charmed by megapictures global in their sweep. He referred to the long deep time before the Appalachian orogenies as "the good old days." With regard to plate tectonics, he looked upon himself as a missionary of contrary opinion—not flat and rigid but selective, where he had knowledge to contribute. His wife has compared him to Martin Luther, nailing theses to the door of the castle church.

For some years he assisted oil companies in the training of geologists and geophysicists in southern-Appalachian geology, and in return the companies made available to him their proprietary data from seismic investigations of the Appalachian crust. Later on, these data were supplemented by the seismic thumpings of the U.S.G.S. and several university consortiums, whose big trucks go out with devices that literally shake the earth while vibration sensors record wave patterns reflected off the rock deep below. The technique is like computed axial tomography—the medical CAT scan. The patterns reveal structure. They reveal folds, faults, laminations, magmatic bodies both active and cooled. They report the top of the mantle. They also reveal density, and hence the types of rock. Moving cross-country, the machines make subterranean profiles known as seismic lines. Seismic shots with explosives have been used for years in the search for oil. Alaska is crisscrossed with all but indelible "seismic" disruptions of tundra. The reserves of Prudhoe Bay were discovered in this manner. Dynamite in the populous East could irritate the public, so the universities and the U.S.G.S. use a behemoth called Vibroseis to shiver the timbers of the earth. One of the first discoveries the vibrations reported was that the Brevard Zone is relatively shallow and the crust below it is American rock that does not in any vague way reflect a continent-to-continent suture. Africa was nowhere in the picture. The Brevard Zone proved to be the toboggan-like front end of a large and essentially horizontal thrust sheet.

Plate-tectonic theorists accommodated this news by moving the suture fifty miles east. The new edge of Africa was under Kings Mountain. Seismic shots took the stitches out of Kings Mountain. "When we got the data for the Brevard"—as Leonard Harris liked to tell the story—"they pushed the suture to Kings Mountain, and when we got data for that they said the suture must be under the coastal plain, and now that we are getting data for the coastal plain they say it must be in the continental shelf. Well, we've got data out there, too." Up and down the Appalachians, wherever such data were collected, thrust sheets were seen to have moved in a northwesterly direction, and much of the thrusting had never been suspected before. Conventional thought had been that the old rock of the Green Mountains, the Berkshires, the New Jersey Highlands, the Catoctin Mountains, and the northern Blue Ridge was in place, firmly rooted—autochthonous, as geologists are wont to say. It may have been crushed and pounded in the various orogenies, and metamorphosed, too, but it was nonetheless thought to be securely glued where it first had formed as rock. The belt was supposed to have been the fixed starting block from which, somehow, thrusting had proceeded northwest. The idea had come up through the Old Geology and been incorporated into the substance of plate tectonics. Then, in 1979, Vibroseis rumbled into the country and showed that from Quebec to the Blue Ridge the entire belt was deracinated. The Great Smokies and the Skyline Drive, Camp David and the Reading Prong, the Berkshires and the Green Mountains—all of it had moved, at least a few tens of miles and as much as a hundred and seventy-five miles, northwest.

Using the new data, Leonard meticulously drew a palinspastic reconstruction of North American rock, showing it as it had appeared before it was shoved and deformed. He chose a cross section that had been shot more or less from Knoxville to Charleston and out to sea. The reconstruction showed that the rock of the Ridge and Valley—the folded-and-faulted, deformed Appalachians—had been squeezed so much that its breadth had been reduced about sixty miles. The supposedly rooted Blue Ridge had been moved inland from what is now the coast. Rock of the present Piedmont had come from three hundred miles out in the present sea. This left Africa out in the cold and plate-tectonic theory in no small need of a substitute for what had been—and in many classrooms would continue to be—the world's most "classical" example of a continent-to-continent suture.

With patience geological, the believers restyled their belief, apparently according to the criterion "If at first it doesn't fit, fit, fit again." There was suggestive help from the West. A great deal of land out there had not been there when the carbonate rocks sloped away to ocean-crustal deeps in Ordovician time. California, Oregon, Washington, British Columbia had appeared where there was no continental structure of any kind. Up and down the western margin, in fact, there was an unaccounted-for swath of land averaging four hundred miles wide. There was also the whole of Alaska. How did all that country come to be where it is? What compressed the western mountains? If Europe were on the international date line, these questions would have a ready answer, but inconveniently it was not.

No one was enthusiastic enough to suggest a hit-and-run visit from China. Where, then, since Ordovician time, had the North American continent acquired nine hundred million acres of land?

There was an answer in the concept of microplates, also known as exotic terranes. New Guineas, New Zealands, New Caledonias, Madagascars, Kodiaks, Mindanaos, Fijis, Solomons, and Taiwans had come over the sea to collect like driftwood against the North American craton. The first such terrane identified was Wrangellia, named for the stratovolcanic Wrangells, some of the Fujis of Alaska. Dozens of other exotic terranes have since been named—Sonomia, Stikinia, the Smartville Block. If a piece of country is possibly exotic and possibly not—if it is so enigmatic that no one can say whether it has come from near or far—it is known as suspect terrane. I returned one time from a visit to the country north of the Tanana River, in eastern interior Alaska, where streams that resemble gin come down from mountains and into the glacial Yukon. A geologist in New Jersey welcomed me home with an article from *Nature* which described the Alaskan region of the upper Yukon. "The terrane is probably composite," said *Nature*, "with nappes of upper Palaeozoic oceanic assemblages thrust across a quartzo-feldspathic and silicic volcanic-rich protolith of probable Precambrian to known Palaeozoic age and of unknown continental affinity." I was appalled to discover that that was where I had been, and mildly disturbed to learn that terrain long familiar to me had now become suspect.

Taiwan, at this writing, is evidently on its way to the Chinese mainland. Taiwan is the vanguard of a lithospheric microplate and consists of pieces of island arc preceded by an accretionary wedge of materials coming off the Eurasian Plate and materials shedding forward from the island's rising mountains. As the plate edges buckle before it, the island has plowed up so much stuff that it has filled in all the space between the accretionary wedge and the volcanic arc, and thus its components make an integral island. It is in motion northwest. For the mainland government in Beijing to be wooing the Taiwanese to join the People's Republic of China is the ultimate inscrutable irony. Not only will Taiwan inexorably become one with Red China. It will hit into China like a fist in a belly. It will knock up big mountains from Hong Kong to Shanghai. It is only a question of time.

As an exotic terrane on the verge of collision with a continent, Taiwan is a model not only for the building of the American West but for the application of microplate-tectonic theory to the eastern orogenies and the closing of the proto-Atlantic. In this respect, a plane fare to Taiwan has been described as "a ticket to the Ordovician," a time when something or other, beyond question, produced the Taconic Orogeny, and if it was not the slamming-in of a continent against North America, then possibly it was the arrival of an exotic island like Taiwan. The analogy becomes wider. South of Taiwan are Luzon, Mindanao, Borneo, Celebes, New Guinea, Java, and hundreds of dozens of smaller islands from the Malay Peninsula to the Bismarck Archipelago. Coming up below them is Australia, palpably moving north, headed for collision with China, with a confusion of microplates lying between. According to microplate theory, as Europe, Africa, and South America closed in upon North America through Paleozoic time, there rode before them an ocean full of Javas, New Guineas, Borneos, Luzons, Taiwans, and maybe hundreds of dozens of smaller islands. The Avalon Peninsula of Newfoundland appears to have been a part of such an island, and the Carolina Slate Belt, and a piece of Rhode Island east of Providence, and Greater Boston. A schedule of arrivals of incoming exotic terranes will account—as a simple continent-to-continent collision cannot do—for the long spreads of time between one and another of the Appalachian mountain-building pulses. As someone once compacted it for me, "you sweep the New Zealands and Madagascars out of the ocean and then you close it with the Alleghenian Orogeny." Disagreeing interpreters see terranes of highly varied dimension. Nominated as the terrestrial remains of one exotic block is the whole of New England from Williamstown eastward, arriving in the Ordovician to lift the Taconic mountains.

Exotic terranes and their effects represent only one of the responses of plate-tectonic theorists to the embarrassment caused by the failure of Exhibit A among intercontinental collisions to exhibit a finite suture. Another response has been the notion that when two continents collide there is every possibility that one will split the other, like an axe blade entering cedar; if so, you would find the invaded country rock both above and below the invader. The concept is known as flake tectonics. Its message to Vibroseis is to stop shaking and go home. With a little erosion and flake tectonics, you can have the native rock reaching far under the rock from across the sea. Even so, the bunching of exotic terranes seems to solve more problems than flake tectonics does. Exotic terranes not only explain the intervals of time involved in the Taconic, Acadian, Alleghenian orogenies, they suggest as well why Taconic deformation occurred in the northern but not the southern Appalachians. Shortening collisional boundaries, they restore some dignity to the Brevard suture.

Anita turned on the windshield wipers and wiped an April shower. Beside the interstate, the Pocono Devonian roadcuts were of much the same age and character as ones we had seen before. We passed them by. "Better not to do geology in the rain," Anita said. "It's unfair to the rocks." With regard to the possibility of exotic terranes having added themselves to eastern North America, she said, "If you stretch out the overthrusts in the Appalachians, they show that—before the mountain building began—the continent was much larger than it is now, not smaller."

I remembered Leonard Harris—one day at their home in Laurel, Maryland—saying, "The Brevard Zone is the sort of fault you would see in any thrust belt. With the plate-tectonic model, anybody can write a history of an area without having been there. These people have no way to evaluate what they're doing. They just make up stories."

"Plate-tectonic interpretations often start where data stop," Anita had said. "These people will just *float* microplates around. If the West is made of microplates, where the hell was the landmass that produced the pieces?"

"They want to be science-fiction writers," Leonard said. "That's what they want to do. They really look at it in a science-fiction mode. I have never been able to do that. If you don't know what caused something, you don't know; and that's the way it is."

"Yeah, but it's a much more romantic way to look at things," she said. "And it certainly does turn students on."

"People love it."

"It allows them to play all kinds of games without the necessity and painstaking dogwork of gathering facts. It allows them to write papers without killing themselves getting data."

"People want the science-fiction story. It's easier to believe that pieces of the world move than it is to see a sand grain move. The principal problem about interpreting the Appalachians is that there have been no available subsurface data in the Blue Ridge and Piedmont. All interpretations, up until 1979, were based on what people thought was a rooted system. Their ideas were based on offshore data, where they had 3-D—you know, seismic data, magnetic data —and these data were more or less applied onshore. The concepts were developed from the ocean to the land. Now that we are beginning to get subsurface data on land, we are testing their concepts. A lot of what people have been saying is not hanging together. Some of what they have said *has* hung together."

I said to them, "One would gather from the seismic lines that for a continent-to-continent collision you'd have to go pretty far east to find the suture."

"I don't think you can go far east enough," Anita said. "The oceanic basin is out there."

"When you start working on the shore and you look offshore, you've got an immediate problem," Leonard said. "They tell us that the oldest ocean crust that has ever been found is Jurassic. Onshore, we have everything that's ever been built—from the Precambrian on up. We have a continuum. We have something that has been preserved much longer. We have rock that is nearly four billion years old. So we have a problem relating. If all the ocean crust is Jurassic, or younger, there's a lot happening here onshore that is never preserved out there. It's difficult to compare the two."

Anita said, "I believe in plate tectonics—just not in the way they're perpetrating it for places like the East Coast. It shouldn't be used as the immediate answer to every problem. That's what I object to. Now that their suture zones have disappeared, people are going to microplates."

"They seem to be saying that you don't have to see any order," Leonard said. "Because it's all chaos, and if it's chaos why worry about it?"

"What we try to do is pull the thrust plates apart and make them into some sort of recognizable geologic model," Anita said.

Leonard said, "You pull something apart to see what it might have been, not what you think it was in advance. It might have been a shelf, a basin. You work at it, and see what it was."

"The plate-tectonics boys make no attempt to do this, because they see no reason to," Anita said. "There are too many pieces missing. Each existing piece is an entity unto itself. Everything is random pieces."

"Most people have never had an opportunity to work with thrust-faulted areas. We've lived with them all our lives. If we go along a fault system far enough, we can actually see the next thrust plate. Maybe I'll have to go a hundred miles until I find out what it really looked like. You do that by making a model. You pull the thrusts apart and see what the country originally looked like. But until you've done that, and been faced with that problem, it's natural to say, 'God, these are so different. They could be microcontinents.' You can reconstruct a large flat piece out to the east as an original depositional basin. You can see volcanic terrane that was partly onshore, partly offshore. You can look at that as a basin, too, just sitting there, a continuous thing. You see the same thing from Georgia north. The Appalachian belts are almost continuous basins, showing different kinds of depositional patterns. They're not exotic pieces."

"Not at all."

"Science is not a detached, impersonal thing. People will be influenced as much by someone who is a spellbinder as by someone with a good, logical story. It is spellbinding to say that these belts are exotic and were built through time by micro or macro pieces aggregated to the continent. But the fact that you've got seismic lines without any apparent suture lines makes you wonder what really happened. Where are those Devonian and Taconic sutures? Are they just not being recognized? Or are they in fact thrust plates?"

I thought also of field trips in the company of geologists trying to puzzle out the details of plate-tectonic theory. Metamorphic details. Geophysical details. The dialogue is not without crescendos. They debate in a language exotic in itself, and shuffle like a blackjack deck the stratigraphic units of the world. In Vermont, say, walking the hard-packed dirt roads among Black Angus meadows and roll-mop hay, over plank bridges—"LEGAL LOAD LIMIT 24,000 POUNDS" —and down through the black spruce to Cambrian outcrops jutting up as ledges in fast, clear streams, they argue.

"You've got the right first approximation, but you've got to go ahead and prove that it's the correct approximation."

"We're talking about developing fabrics."

"It's pretty clear now that fabrics don't develop that way."

"For an anisotropic crystal, I don't think you can say what you just said. You've got to put in another sentence there to justify using that approximation."

"I don't see what you're saying. Anisotropic or not, that's the definition of being stressed."

"When you say the thermodynamic stability of the phase that's growing is sigma 1, sigma 2, sigma 3, divided by 3, it's proof for an isotropic crystal. For garnets growing, that's fine. For mica, that's not fine."

The hills roundabout are decidedly footloose and no one knows how far they have moved. The rock they are made of has flopped over in recumbent folds and is older than the rock it rests on. In the Old Geology, these hills were described as large pieces of the high Taconic mountains, which had slid downhill by gravity and come to rest in the westward seas. Now they are seen variously as remnants of thrust sheets or as a possible exotic terrane.

"You can perhaps picture for yourself that the allochthon was at one time more extensive. It was coming this way through a sea of black mud, and here is the record of it. This is where it touches North America—at least that's a possibility."

"But there are no remnants of the western side of the ocean."

"Evidence would be in the rest of the conglomerate, little bits of limestone debris. Evidence would be in the seismic line."

"But that evidence could be . . . You could imbricate the stuff that's coming off North America."

"Yes. Yes, you can."

"So I don't think that's definitive."

"I'm not saying it is. I'm just saying here's another possibility.

And I'm going to stick to that for the time being, as well as the Chain Lakes ophiolite."

It doesn't matter that you don't understand them. Even they are not sure if they are making sense. Their purpose is trying to. Everyone has crowded in. The science selects these people—with their jeans and boots and scuffed leather field cases and hats of railroad engineers. To them, just being out here is in no small measure what it's all about. "The three key things in this science are travel, travel, and travel," one says. "Geology is legitimized tourism." When geologists convene at an outcrop, they see their own specialties first, and sometimes last, in the rock. People listen closely for techniques applicable to areas they work in elsewhere. If someone is a specialist in little bubbles that affect cleavage planes, others will turn to the specialist for comment when cleavage is of interest in the rock. The conversation runs in links from specialty to specialty, from minutia to minutia—attempting to establish new agreement, to identify problems not under current research. From time to time, details compose. The picture vastly widens.

"Aren't we in North America?"

"You are in North America. Yes."

"And you are in Europe."

"That's one possibility. Yes."

"You are standing across the ocean."

"No. I'm not standing across the ocean. I'm transplanted here. Is the Atlantic between me and you? No."

"You are allochthonous."

"You're damned right I am."

"You are rootless."

"Not to mention recumbent."

"Only after hours."

"There may be another suture."

"There may be another suture, but this is the only one we've got."

"No, no, you've got another one, which goes up through Quebec."

"No, that's not in place. The Canadian seismic line proves it. You will remember also that Laval—way back, 1965—came out with late Ordovician fossils in the Sherbrooke anomalies. Where else do you find a continuous sedimentation from the middle Ordovician up to the Silurian—from Rangeley Lake up to New Brunswick in one belt? The Sherbrooke thing, restored, would come from where it ought to come from. So I'm suggesting either that two continents collided, and you have one basin there, receiving continuous sedimentation, or . . ."

"You may have a double arc."

"You might have a double arc."

"There's another solution."

"Sure, but I'm saying let's take the simplest configuration."

"Why not just have one arc with basins on both sides of it?"

"No. No. You have the Bronson Hill anticlinorium, and then you have the Ascot-Weedon."

"You have a volcanic arc on the stable side of the subduction zone, an expected arc above the downgoing slab."

"You have a short-lived slab going down below the Ascot-Weedon and the one of longer duration that's on the other side. I would somehow think that there has to be something in these rocks, in the limestones, that you'd be able to hopefully connect to that platform."

"The only thing I can say is . . ."

"What about the blue quartz?"

"What about the blue quartz? The stratigraphy of the Taconic rock matches unit for unit with Cambrian rocks of Avalon, and the fossils look alike. That's all I can tell you. Nowhere else do we find this sort of thing except Wales."

A structural geologist with a foot on each continent looks up and aside from this contentious scene. "While geologists argue, the rocks just sit there," he remarks. "And sometimes they seem to smile."

The car hit an erosional vacuity that almost threw it off the road. Geology versus the State of Pennsylvania. Geology wins. In eastern weather, the life expectancy of an interstate is twenty years. Mile after mile, I-80 had been heaved, split, dissolved, and cratered. A fair amount of limestone is incorporated in the road surface in Pennsylvania. Limestone is soluble in distilled water, let alone in acid rain. "Acid rains eat the surface, then water goes in and freezes, thaws, freezes again, and fractures the hell out of the road," Anita said, easing down toward minimum speed. "That, of course, is exactly how water works on bedrock. But an interstate can't be compared to bedrock. An interstate has no soil protecting it. And it's mostly carbonate. It's not very resistant stuff."

We were sixty miles into Pennsylvania and had descended from the Pocono Plateau, generally running backward through time and down through the detritus of two great ranges of mountains. Now the country was familiar—valley, ridge, valley, ridge. We were again in the deformed Appalachians. While the Delaware Water Gap had been a part of the main trunk of the foldbelt, this was an offshoot that curled around the western Poconos—a broad cul-de-sac whose long ropy ridges ended like fingers, gesturing in the direction of New York State. It was rhythmic terrain, predictable and beautiful, the quartzite ridges and carbonate valleys of the folded-and-faulted mountains, trending southwest, while the interstate negotiated with them for its passage toward Chicago. Looked at in continental scale on a physiographic map or a geologic map—on almost any map that doesn't obscure the country with exaggerated human improvements—the sinuosity of the deformed Appalachians is as consistent as the bendings of a moving serpent. In Alabama, the mountains come up from under the Gulf Coastal Plain and bend right into Georgia and then left into Tennessee and right into North Carolina and left in the Virginias and right in Pennsylvania and left in New Jersey and New York and right in Quebec and New Brunswick and left in Newfoundland. Some people believe that in this Appalachian sinuosity we are seeing the coastline of the Precambrian continent—North America in the good old days, when the Taconic Orogeny was off in the future and these big, scalloped bends were the bays of Iapetan seas. Signify what they may, their repetitive formality through two thousand miles does not suggest the random impacts of exotic terranes, nor, for that matter, does it suggest the ragged margins of crashing continents. The Great Valley, as the most prominent axial feature of the Appalachians, also seems inconvenient to the narrative of colliding lands, because the soft black slates and

shales and the dissolving carbonates that make up the valley from end to end were all moved a great many miles northwest, and if random New Zealands and the odd Madagascar were shoving at different times in different places, one would expect the formations to be considerably offset. One would imagine that miscellaneously disturbed rock—folded, faulted, shuffled, thrusted, disarranged—would be much too chaotically bulldozed to emerge through erosion as an integral, elongate, and geometrically formal valley. Similar thoughts come to mind with regard to the Precambrian highlands in their sinuous journey from the Great Smokies to the Green Mountains, and the Piedmont as well: the consistent, curving, parallel stripes of the Appalachian ensemble.

Anita, for the moment, was more interested in tangible limestone than in how it had been shoved and deformed. Eight miles west of Bloomsburg, she saw a limestone outcrop that looked good enough to sample. It was a quarter of a mile off the interstate, and we walked to it. She dropped some acid on it. Vigorously, the outcrop foamed. "It's upper Silurian limestone," she said. "I shouldn't be able to tell you that without running the conodonts, but I know."

"What if you're wrong?"

"Then I'm wrong, aren't I? They pay me to do the best I can. Geologists are detectives. You work with what you have."

She swung full force with her sledgehammer. The stone did not crack. "This profession is very physical," she complained, and belted the outcrop again.

Her knees sometimes turn black-and-blue when she carries samples down from mountains. She once handed a suitcase to a Greyhound bus driver who said, "What have you got in here, baby—rocks?" She was content to have them ride in the baggage compartment. A geologist I know in California would be unnerved by that. When he travels home from far parts of the world, he buys two airline seats—one for himself and one for his rocks.

We passed Limestoneville. We crossed Limestone Run and the West Branch of the Susquehanna, and now the road was running in a deep crease, a V with sides of about twelve hundred vertical feet: White Deer Ridge and Nittany Mountain—quartzites of early Silurian age, shed west from the Taconic Orogeny. There were quartzite boulders all through the steep woods but a notable absence of outcrops, of roadcuts, of exposures of any kind. In fact, with the exception of the limestones she had collected, we were not seeing much rock to write home about, and Anita was becoming impatient. "No wonder I never did geology in this part of Pennsylvania," she said. "There are no exposures—just colluvium lying in the woods." Multiple ridges were squeezed in close here. Characteristically, the interstate would yield to the country, to the southwestward sweep of the corrugated mountains, as it ran in a valley under a flanking ridge, biding its time for a gap. One would soon appear—not a national landmark with a history of landscape painters and lovestruck Indians, but a water gap, nonetheless—sliced clean through the ridge. Like a fullback finding a hole in the line, the road would cut right and go through. On the far side, it would break into the clear again, veering southwestward in another valley, gradually moving over toward the next long ridge. There would be another gap. Small streams had cut countless gaps. All within twenty miles of one another, for example, were Bear Gap in Buffalo Mountain, Green Gap in Nittany Mountain, Fryingpan Gap in Naked Mountain, Fourth Gap in White Deer Ridge, Third Gap, Second Gap, First Gap, Schwenks Gap, Spruce Gap, Stony Gap, Lyman Gap, Black Gap, McMurrin Gap, Frederick Gap, Bull Run Gap, and Glen Cabin Gap—among others.

In Precambrian, Cambrian, and much of Ordovician time, rivers ran southeastward off the American continent into the Iapetan ocean. Then the continental shelf bent low, and the Martinsburg muds poured into the depression from the east. Whether they were coming from Africa, Europe, or some accretionary, displaced, hapless Taiwan is completely unestablished, but what is not unestablished is the evidence preserved in the sediment—sand waves, ripple marks, crossbedded point bars—showing currents that flowed west and northwest. In later rock, such evidence is everywhere, showing eastern American rivers flowing toward what is now the middle of the continent all through the rest of Paleozoic time. As each successive orogeny produced another uplift in the east, fresh rivers would pour from it, building their depositional wedges, their minor and major deltas, but running always in a westerly direction. The last orogeny was pretty much spent about two hundred and fifty million years ago, in the Permian. For some tens of millions of years after that, the mountains were reduced by weather in a tectonically quiet world. Then, in early Mesozoic time, "earth forces" began to pull the terrain apart. According to present theory, the actual split, deep enough to admit seawater, came at some point in the Jurassic. The Atlantic opened. On the American side of the break, extremely short steep rivers flowed into the new sea, but for the most part the drainages of what is now the eastern seaboard continued to flow west. By Cretaceous time, the currents had reversed, assuming the present direction of the Penobscot, the Connecticut, the Hudson, the Delaware, the Susquehanna, the Potomac, the James.

Rivers come and go. They are younger by far than the rock on which they run. They wander all over their valleys and sometimes jump out. They reverse themselves and occasionally disappear—their behavior differentiated by textures in the solid earth below. The tightly folded Appalachians are something like the ribs of a washboard. The direction of the structure lies across the direction of scrubbing. In the Paleozoic era, when the tectonic washboard was made and repeatedly lifted from the east, falling rainwater, gathering in streams, found its way westward across the ribs.

With the coming of the Atlantic—the Mesozoic split—the principal drainages of the American East at first continued to flow toward the Midwest. A part of the plate-tectonic story is that a great deal of heat accompanies tectonic rifting and the heat lifts the two sides of the rift like trapdoors facing each other. The shores of the Red Sea look like that. On both sides are mountains, nine, ten, twelve thousand feet high. Extremely short steep rivers fall into the Red Sea. Principal drainages—the intermittent rivers of Arabia—run eastward almost from the east shore many hundreds of miles, and from near the west shore Egyptian rivers run west to the Nile. The world's mid-ocean ridges—the spreading centers of plate tectonics—are configured like the rift of the Red Sea. Typically, the two sides are of gentle pitch, and gradually rise six thousand feet higher than the flanking abyssal plains. Groovelike down the ridgelines run submarine rift valleys. Into the rift valleys of eastern Africa pour extremely short steep rivers, while long ones, like the Congo, rise close to the rift but flow away westward a thousand miles to the sea. It was the discovery and confirmation of spreading centers that opened the story of plate tectonics—and this is still the aspect of the theory that provokes the least debate. Eastern America, in Jurassic time, gradually subsided. The present explanation would be that as the ocean grew wider and the heat of the spreading center became more distant, the region cooled like a collapsing soufflé, while the weight of water and accumulating sediments also pressed down on the continental shelf. In any case, the broad package of land that had tilted northwestward for approximately three hundred million years now seesawed and began again to tilt the other way.

Rivers turned around, pooled temporarily against the ribs of the washboard, and ran over them, seeking weaknesses in the rock. Anew, the running water began to etch out the country. It was a process analogous to photoengraving, wherein acid differentially eats pictures into treated sheets of metal. The new and reversed eastern rivers differentially eroded the Appalachian structures. Where they got into the shales and the carbonates, they dug deep and wide. Where they found quartzite and other metamorphic rock, they encountered tough resistance. Sometimes, working down into the country, they came to the arching quonset roofs of anticlines, and slicing their way through quartzite found limestones within. It was like slicing into the foil around a potato and finding the soft interior. The water would remove the top of the arch, dig a valley far down inside, and leave quartzite stubs to either side as ridges flanking the carbonate valley. Streams eroding headward ate up the hillsides back into the mountainsides, digging grooves toward the nearest divide. On the other side was another stream, doing the same. Working into the mountain, the two streams drew closer to each other until the divide between them broke down and they were now confluent, one stream changing direction, captured. In this manner, some thousands of streams—consequent streams, pirate streams, beheaded streams, defeated streams—formed and re-formed, shifting valleys, making hundreds of water gaps with the general and simple objective of finding in the newly tilted landscape the shortest possible journey to the sea. A gap abandoned by its streams is called a wind gap. In the regional context, the water gap of the Delaware River is a little less phenomenal than it once appeared to be.

Until about 1970, the picture in vogue of the early Cenozoic American East was of a vast peneplain, a flat world of scant relief, with oxbowed meandering rivers heading almost nowhere. The assault of water on the ancestral mountains was thought to have worn down the whole topography close to sea level. The peneplain then rose up, according to the hypothesis, and rivers dissected it, flushing out the soft rock and leaving hard rock high, in the form of remarkably level ridges—as flat as the peneplain, of which they were thought to be remnants. Where the rivers of the peneplain had flowed across the tops of buried ridges, they cut down through them as the ridges came up—making gaps. That was the history as it was taught for three-quarters of a century. It was known as the hypothesis of the Schooley Peneplain, after Schooley Mountain, in New Jersey, which looks like an aircraft carrier. The Schooley Peneplain is out of vogue. It is an emeritus idea. It has been replaced by a story out of steady-state physics having to do with the relationship of level ridgelines to certain degrees of slope. A graduate student once remarked to me that old hypotheses never really die. He said they're like dormant volcanoes.

Under the carbonate valleys and quilted farms, the rock was buried from view. The beauty of the fields against steep-rising forests, the shimmer of April green, was not doing much for Anita. She was in need of a lithic fix. Her fingers tapped the wheel. She reminded me of a white-water fanatic on a meandering stretch of flat river. "No wonder I never did geology in this part of Pennsylvania," she said again. It had been a long time between rocks. "I'd really like to go to Iran someday," she went on, desperately. "The Zagros Mountains are another classic fold-and-thrust belt. The thing about the Zagros is that there's no vegetation. You can see everything. They're a hundred per cent outcrop."

She had scarce uttered the words when the road jumped to the right and through a nameless gap and past a roadcut twenty metres high—Bald Eagle quartzite—and then more and higher roadcuts of Juniata sandstone in red laminations dipping steeply to the west. "I take it back. This is one hell of a series, let me tell you," Anita said. More rock followed, rock in the median, rock right and left, and we ran on to scout it, to take it in whole. The road was descending now through gorges of red rock—the results of precision blasting, of instant geomorphology. Their depth increased. They shadowed the road. And in their final bend was the revealed interior of a mountain, geographically known as Big Mountain. There had been a natural gap, but it had not been large enough, and dynamite had contributed three hundred thousand years of erosion. The entire mountain had been cut through—not just a toe or a spur. "Holy Toledo! Look at that son of a bitch!" Anita cried out. "It's a hell of an exposure, a hell of a cut." More than two hundred and fifty feet high and as red as wine, it proved to be the largest man-made exposure of hard rock on Interstate 80 between New York and San Francisco. It was an accomplishment that might impress the Chinese Geological Survey. "When you're doing geology, look for the unexpected," Anita instructed me, forgetting the Zagros Mountains.

We stopped on the shoulder in the shadow of the rock. "Holy Toledo, look at that son of a bitch," Anita repeated, with her head thrown back. "Mamma mia!" The bedding was aslant in long upsweeping lines, of which a few were green. Almost due south of Lock Haven and thirty-one miles west of the Susquehanna River, it was Juniata sandstone, brought down off the Taconic uplift and spread to the west by the same system of rivers that transported the rock of the Delaware Water Gap. "This would be a beautiful place to measure the thickness of the section," Anita said. "It's completely exposed. It's consistent. There are no faults. The thin green bands are where deposition was too rapid for oxidation to take place." Evidence of geologists was everywhere. They had painted numbers and letters on the rock. They had removed countless paleomagnetic plugs. The bedding, seen close, was not monotonously even, as rock would be that formed in still water. Instead, it was full of the migrating channels, feathery crossbedding, natural levees, and overbank deposits of its thoroughly commemorated river. There were little maroon mud flakes. They were plucked off flats in a storm.

We went back a few miles and slowly reviewed the rock. When again we approached the huge roadcut, Anita said, "In Illinois, this would be a state park."

The bedding planes of the Holy Toledo cut, as I would ever after refer to those enormous walls of red stone, were dipping to the east. Over the past few miles, the rock of the country had been folded ninety degrees. To the immediate west, therefore, we would be going down in time and predictably would descend in space to a Cambro-Ordovician carbonate valley, which is what happened, as the road fell away bending left and down into Nittany Valley, where ribs of dolomite protruded here and there among rich-looking pastures turning green, gentle streamcourses, white farms. "Penn State sits on Nittany dolomite," Anita said. "It's twenty miles down this valley."

Some remnant Cambrian sandstone formed a blister in the valley. The interstate drifted around it in a westerly way and toward the foot of still another endless mountain—Bald Eagle, the last ridge of the deformed Appalachians. After the Cambrian sandstone, the Ordovician dolomite, there was Silurian quartzite in the gap that broke through the mountain. Its strata dipped steeply west. The rock had bent again, and again we were moving upward through history. Now, though, the dip of the strata would reverse no more. In a dozen miles of ever younger rock, we climbed through the Paleozoic era almost from beginning to end. We went up through time at least three hundred million years and up through the country more than a thousand vertical feet, the last ten miles uphill all the way, from Bald Eagle Creek to Snow Shoe, Pennsylvania—the longest steady grade on I-80 east of Utah—while light, wind-driven snow began to fall.

We had come to the end of the physiographic province of the folded-and-faulted mountains, and the long ascent recapitulated Paleozoic history from the clean sands of the pre-tectonic sea to the dense twilight of Carboniferous swamps. We came up through the debris of three cordilleras, through repetitive sandstones and paper shales—Silurian paper shales, Devonian paper shales, Mississippian paper shales—crumbling on their shelves like acid-paper books in libraries. The shales were so incompetent that they would long since have avalanched and buried the highway had they not been benched—terraced in the manner of Machu Picchu. In other roadcuts, Catskill Delta sandstones, beet-red and competent, were sheer. We had gone through enough hard ridges and soft valleys for me not just to sense but to see the Paleozoic pageant repeatedly played in the rock. For all the great deformity and complexity, the mountains now gone had left patterns behind. The land rising and falling, the sea receding and transgressing, the ancestral rivers losing power through time had not just obliterated much of what went before but had always imposed new scenes, and while I, for one, could not hold so many hundreds of pictures well related in my mind I felt assured beyond doubt that we were moving through more than chaos.

The strata at the foot of the ten-mile hill had been nearly vertical. Gradually, through the long climb, they levelled out. They leaned backward, relaxed, one degree every two million years, until in the end they were flat—at which moment the interstate left the deformed Appalachians and itself became level on the Allegheny Plateau.

The rock was now Pennsylvanian—massive river sandstones of Pennsylvanian time. Flat, deck-like, it was comparatively undisturbed. It had been shed, to be sure, from eastern mountains, but had not been much affected by their compressive drive. Crazed streams had disassembled the plateau, leaving half-eaten wedding cakes, failed pyramids, oddly polygonal hair-covered hills. Pittsburgh was built upon such geometries, its streets and roads faithful to the schizophrenic streams, its hills separating its people into socio-racial ethno-religious piles—up this one the snobs, up that the Jews, up this the tired, up that the poor.

A hundred miles northeast of Pittsburgh in the flurrying snow there were numerous roadcuts now, and in them were upward-fining sequences of sandstones, siltstones, shales—Allegheny black shales—underlying more levels of sandstone, siltstone, and shale. "If you were a prospector for coal, you'd go bananas when you saw these black shales," Anita said. "There ought to be coal in these roadcuts.

This is Pennsylvania in the Pennsylvanian—the home office of the rock."

Pennsylvania in the Pennsylvanian was jungle—a few degrees from the equator, like southern Indonesia and Guadalcanal. The freshwater swamp forests stood beside the nervously changing coastline of a saltwater bay, just as Sumatran swamps now stand beside the Straits of Malacca, and Bornean everglades beside the Java Sea. This was when glacial cycles elsewhere in the world were causing sea level to oscillate with geologic rapidity, and the swamps pursued the shoreline as the sea went down, and marine limestone buried the swamps as the sea returned. In just one of these cycles, the shoreline would move as much as five hundred miles—the sea transgressing and regressing through most of Pennsylvania and Ohio. There were so many such cycles at close intervals in Pennsylvanian time that Pennsylvanian rock sequences are often striped like regimental ties—the signature of glaciers half the world away. They existed three hundred million years ago, and glacial patterns of that kind have not been repeated until now, when the measure of our own brief visit to the earth is being recorded as a paper-thin stripe in time.

On both sides of the interstate, above the silhouettes of screening trees, we saw the tops of draglines—the necks and heads of industrial giraffes. They and predecessor machines had been working for fifty years, altering the topography, stripping the coal beds of Pennsylvania—in all, a mineral deposit worth a great deal more than the diamond mines of Kimberley and the goldfields of the Klondike. Coal was in the roadcuts now and would continue to be for many tens of miles—in layers that were not the dull deep gray of the Allegheny shale but truly black and shining. Layered light and dark, the roadcuts looked like Hungarian tortes. Reading up from the bottom, there was sandstone, siltstone, shale, coal, sandstone, siltstone, shale, coal. We would see limestones farther on, capping the coal where sea had covered the swamps. The present sequence was built behind a coastline—as is happening now, for example, in the bayous of the Mississippi Delta—by rivers meandering to and fro, covering with sand the matted vegetation. "These roadcuts are a textbook on the making of coal," Anita said. Buried and compressed, vegetal debris first becomes peat—a mélange of spores, seed coats, wood, bark, leaves, and roots which looks like chewing tobacco and burns about as well. Peat bears much the same relation to coal that snow does to glacier ice. As snow is ever more buried and compacted, it recrystallizes and becomes ice—on the average ten times as dense as the original snow. As peat is buried, compacted, subjected to geothermal heat, it gradually gives up much of its oxygen, hydrogen, and nitrogen, and concentrates its content of carbon. The American Geological Institute's *Glossary of Geology* defines coal as "a readily combustible rock." By weight, any rock that is half carbonaceous material is coal. Its density is roughly ten times the density of peat. In the United States, there is enough peat to keep Ireland warm for a thousand years. The United States uses almost none of it, because the United States also happens to have a great deal more coal than any other country in the world. Peat that remains near the surface will never become coal. Buried three-quarters of a mile, it becomes bituminous. With a microscope, you can see wood and bark, leaves and roots, seed coats and spores in bituminous coal—and even identify the plants they came from. Buried deeper and folded severely under pressure, it becomes anthracite. Anthracite is roughly ninety-five per cent carbon and is so hard that it fractures conchoidally, like an arrowhead. Anthracite is iridescent, and burns with a clear blue flame. Coal is a record of tectonics. In late Pennsylvanian time, when the third set of mountains came up in the east and shed still another wedge of debris, kneading it into what had gone before, the great pressure, deep burial, and severe folding produced the anthracites of eastern Pennsylvania, the pod-shaped coalfields of the folded-and-faulted mountains, which erosion and isostasy have lifted from the depths. Anthracite seams are often upside down or standing on end. Here in the Allegheny Plateau, burial was reasonably deep but tectonic pressures were minor, and the result is a lesser grade of coal.

We stopped and tried to collect some but had difficulty finding a sample that would not break up in the hand. "This is very flaky, high-ash coal," Anita said. "People take it anyway. They come out to these roadcuts with buckets and take it home to burn."

We moved on through miles of coal-streaked roadcuts, and topographically to somewhat higher ground, where the coal seams were thicker. "As you go westward and upsection, you get more coal, because the rivers, growing older, became more sluggish," Anita said. "The floodplains became broader. There was more ponded water. There was more area for vegetation to grow and accumulate—like the lower Mississippi Valley today." About five miles east of Clearfield, we stopped at a long, high throughcut full of coal. Draglines were working on both sides of the road. We chipped out some samples with rock hammers. The samples had integrity. "This is a hell of a coal," Anita remarked. "Good commercial coal. To make it, there would have been about three thousand feet of Pennsylvanian stuff on top of it, which has been removed by erosion. Three thousand feet is the amount of overburden that will produce coal of this rank."

Stirred within by all these free B.t.u.s (twelve thousand per pound), I flailed at the cut with my rock hammer and filled a bag with good commercial coal, to take home and burn in my stove. Anita commented that coal dust was blacking my face.

I wiped at it with a bandanna, and asked her, "Did I get it all?"

She said, "Good enough for government work." And we headed up the road.

When the final great pulse of mountain building folded eastern Pennsylvania, the deep burial and tectonic crush may have done wonders for the coal seams there, but all the oil in the country rock was burned black and destroyed. Conodonts were blackened, too. As Anita's many samplings would prove, conodonts become lighter in color and hue in a westward trend across the state—from black to cordovan to dusky orange to brightening levels of yellow. Running west of Du Bois and Clarion now, and less than fifty miles from Ohio, we were out of the browns and well into the gold. If the quality of coal improves eastward, the theoretical quality of petroleum goes the other way. We took a right off the interstate. Soon we were cruising on Petroleum Street, in downtown Oil City.

We continued north. In the fifteen miles between Oil City and Titusville lay the Napa Valley of early American oil. It was a

V-shaped, intimate valley, five hundred feet from rim to river, and along its floor were oil refineries so small they were almost cute. They did not suggest the starry lighted skeletal cities of Exxon's Baton Rouge Refinery or Sunoco's Marcus Hook. They suggested Christian Brothers, the Beringer Winery, the Beaulieu Vineyard. One refinery followed another. Wolf's Head, Pennzoil. They stood beside Oil Creek, which was so named in the eighteenth century because petroleum dripped out of its banks and into the water. Indians had found it, three centuries before, to judge by the age of trees that were growing in pits they had dug to collect the oil in pools. The Senecas rubbed their skins with it. They may have used it for light and heat. The use of petroleum is old in the world. Workmen laid asphalt three thousand years before Jesus Christ. The first energy crisis involving petroleum was in 1875 B.C. The first oil spills were natural, and were not so large that they could not be cleaned up by bacteria that feed on oil. In 1853, in California, a lieutenant in the Corps of Engineers reported that "the channel between Santa Barbara and the islands is sometimes covered with a film of mineral oil, giving to the surface the beautiful prismatic hues that are produced when oil is poured on water." Always, it was found in seeps. Even until a few years after the Second World War, all Iranian oil fields were associated with surface seeps. The first well in Texas—1865—was drilled near a seep. A well in Ontario had been drilled six years earlier, and in the same summer the first commercial oil well in the United States was drilled in Pennsylvania by Colonel Edwin Drake—less than a hundred steps from Oil Creek.

Colonel Drake had no record of military service. He was a sick railroad conductor—forty years old but debilitated, too fragile to remain upright in the lurching aisles of the New York & New Haven. To the Pennsylvania Rock Oil Company of Connecticut, which had bought farmland and timberland along Oil Creek, he had committed his life savings. Drake was not a geologist. He did not know that petroleum is primarily the remains of marine algae that pile up dead on the floors of shallow seas in situations that prevent oxidation. He did not know that the algal corpses slowly stew for millions of years at temperatures just high enough to crack them into crude. He did not know that oil forms in one kind of rock and moves into another—forms in, say, the lagoonal muds of epicontinental seas, and moves later into the sandstones that were once the barrier beaches between the lagoons and the open sea. He did not know that Oil Creek had cut down through Pennsylvanian and Mississippian formations and on into a Devonian coast. Drake knew none of this in 1859, and neither did the science of geology. What Drake did know was that there was negotiability in the stuff that was dripping into Oil Creek. It was even used as medicine. Fleets of red wagons had been all over eastern America selling seepage as a health-enhancing drink. "Kier's Genuine Petroleum! Or Rock Oil! A natural remedy . . . possessing wonderful curative powers in diseases of the Chest, Windpipe and Lungs, also for the care of Diarrhea, Cholera, Piles, Rheumatism, Gout, Asthma, Bronchitis, Scrofula, or King's Evil; Burns and Scalds, Neuralgia, Tetter, Ringworm, obstinate eruptions of the skin, Blotches and Pimples on the face, biles, deafness, chronic sore eyes, erysipelas . . ." Drake had, in addition, the encouragement of a Yale professor of chemistry who ran a bottle of the seepage through his lab and said, "It appears to me . . . that your company have in their possession a raw material from which, by simple and not expensive process, they may manufacture very valuable products. It is worthy of note that my experiments prove that nearly the *whole* of the raw product may be manufactured without waste." And what Drake had, above all, was the inspiration to go after the substance in its reservoir rock, not to be content to blot it up from the streambanks but to drill for it, never mind that he was making a fool of himself in the eyes of the local rubes. He would punch their tickets later. At sixty-nine and a half feet, he completed his discovery well.

There was an oil rush to Oil Creek, and frontier conditions in shantytowns, and forests of derricks on denuded hills. There was a town called Red Hot, Pennsylvania. There was Petroleum Centre. Pithole City. Babylon. In three months, the population of Pithole City went from nobody to fifteen thousand. River flatboats carried the oil to market. Their holds were divided into compartments, much as the holds of supertankers are divided now. Millers in the valley were paid royalties to release water on cue from millponds, raising the level of the creek to float the flatboats downstream. They sometimes broke and spilled.

The Dramatic Oil Company was established in the valley by John Wilkes Booth, who ruined his well trying to make it more productive. With failure, he departed, in the fall of 1864, to look for other things to do.

I am indebted for many of these facts to Ernest C. Miller, of the West Penn Oil Company, who collected them for the Pennsylvania Historical and Museum Commission.

By 1871, oil was being pumped from the ground in nine countries, but ninety-one per cent of world production still came from Pennsylvania. When it was distilled into its components—paraffin, kerosene, and so forth—the gasoline, which in those days had no commercial value, was poured off into the ground.

Petroleum is rare because it represents an extremely low percentage of the life that has lived on earth. In rock, the ratio of all organic carbon to petroleum carbon is eleven thousand to one. For petroleum carbon to turn into oil and be preserved, many conditions have to align, the most important of which is the thermal history of the source rock—the temperature through time as recorded by, among other things, the colors of conodonts. "The petroleum in this valley makes some of the best lubricating oil in the world," Anita said. "It is a very low-specific-gravity oil and needs little refining, because it has been refined to near-perfection by natural earth processes. It has been at low temperatures—around a hundred degrees Celsius—for maybe two hundred million years. You can practically take it out of the ground and put it in your car."

For a hundred and fifty miles, we had been traversing country that was free of glacial drift. Nowhere to be seen were the tills and erratics, the drumlins and kames left behind by Wisconsinan ice. Like a lifted hem, the line of maximum advance had been up in New York State somewhere, but now, in westernmost Pennsylvania, the glacial front had billowed south, and where Interstate 80 meets the eightieth meridian we again crossed the terminal moraine. Sign of the ice was everywhere—the alien boulders in the woods, the directional scratches on the country rock, the unsorted gravels, cobbles, and sands. The signature of glaciation is as bold as John Hancock's and as consistently recognizable wherever ice has moved across the solid earth. In the presence of the evidence, one has no difficulty imagining the arctic ambience, the high blue-white ice lobes thickening to the north, the white surface wide as the continent and swept by uninterrupted gales, the view in sunlight blinding, relieved only by isolated mountain summits, ice moving around them in the way that water slides past boulders in a stream.

Welcome to Ohio. A sign in the median said "STAY AWAKE! STAY ALIVE!" Ohio is not rich in roadcuts. It is a little less poor, however, than Indiana, Illinois, Iowa, and Nebraska, and before long we were running through burrowed marine shales and walls of lithified river sand. It was rippled Carboniferous sandstone. We were still in rock of that age, but gradually and imperceptibly we had been losing altitude since we climbed the Allegheny front. The eastern rim of the plateau had been more than two thousand feet above sea level, and by now we were down to half that, as we moved farther away from the ancestral mountains and their wedge of sediment thinned.

We had come into the continent's province of supreme tectonic calm, the Stable Interior Craton, where a thin veneer of sediment lies flat upon the stolid fundament, where the geology—even by geological standards—is exceptionally slow. "This is the most conservative part of the U.S.," Anita said. "I've often thought about it. The wildest, craziest people are in the most tectonically active places."

And yet the craton stirs. There is no part of the face of the earth that vertically and laterally does not move. The bedding planes in Midwestern rock, which appear to be absolutely level, do in fact dip. They will descend across a great many miles and then rise, arching over the far rim of a vast and shallow bowl, and then subtly dip again to form a similar bowl: the Findlay Arch, the Michigan Basin, the Kankakee Arch, the Illinois Basin. Anita called the arches "basement highs." She said Hudson Bay is a continental basin, slowly filling up. The basins of the Midwest are filled to the brim with level ground. They are products of the creaking motions of the craton, in

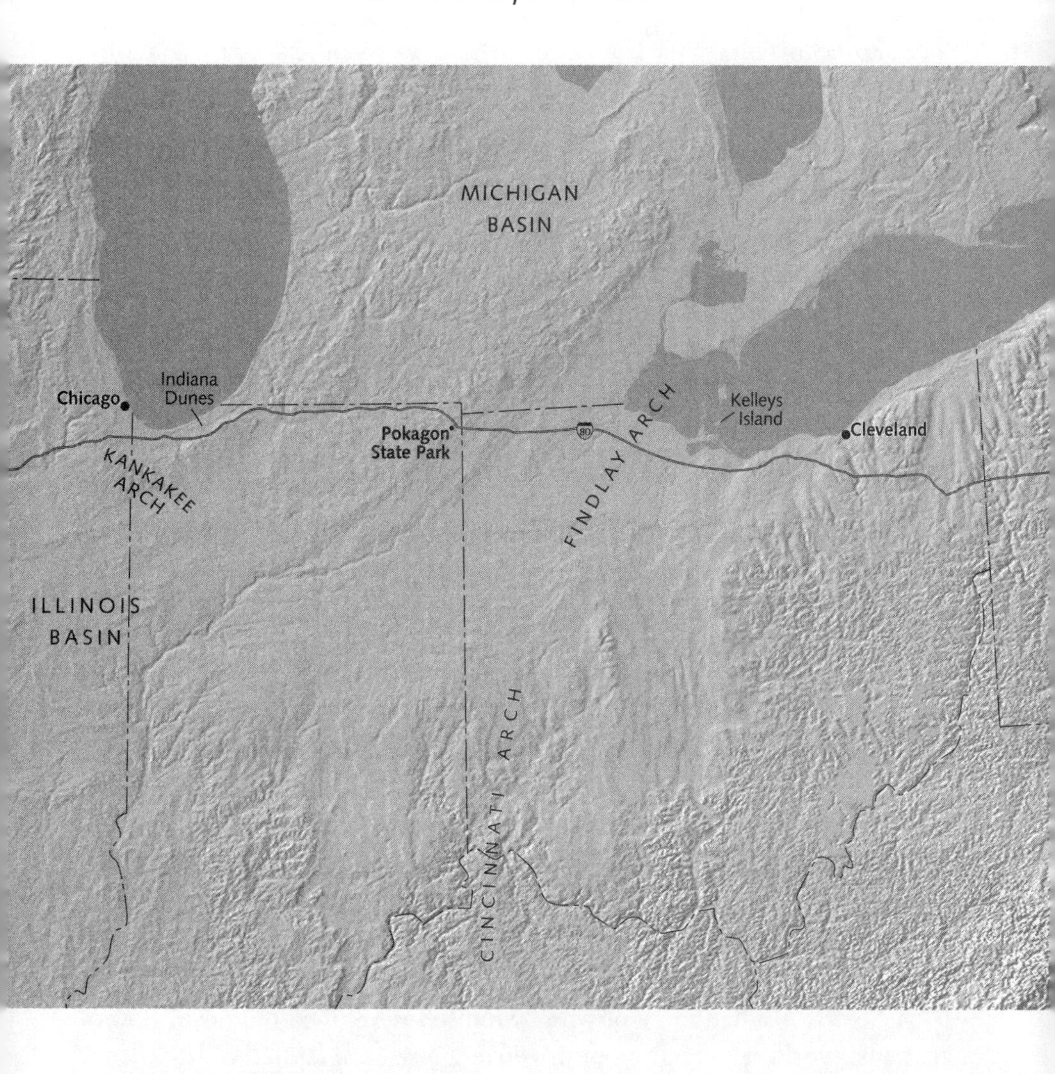

response perhaps to plays of force from deep within the mantle—a process that, in the general phrase, is "not well understood." They represent a degree of tectonic activity about as lively as the setting in of rigor mortis. This has not always been the regional story. There are roots of long-gone mountains deep in the rock of the stable craton, but it has not had an orogeny in a thousand million years. "What has the Midwest been doing since then? It's been sitting around doing nothing," Anita said. "It has just sat here ho-humming." Shallow seas may have quietly arrived and departed, and coal beds formed in the ground, but in all that time there has been no occurrence that can begin to rival in scope or total change the advent from the north of walls of marching ice.

The ice was Antarctic in breadth. The traceable episodes of recent continental glaciation have each placed about as much ice over North America as is upon Antarctica now. In Wisconsinan time, which lasted about seventy-five thousand years and ended ten thousand years ago, three-fifths of all the ice in the world was on North America, another fifth covered much of Europe, and the rest was scattered. Of all special fields within the science, glacial geology is the most evident, the least inferred. It is, for one thing, contemporary. The ice is in recess but has not gone away. In addition to the ice of Antarctica, there is ice more than two miles thick over Greenland. There are twenty-seven thousand square miles of ice on Alaska (four per cent of Alaska). In Alaska, as in Switzerland and elsewhere in the world, you can see cirque glaciers feeding into the master glaciers of alpine valleys. You can see that the cirque glaciers have dug scallops into the high ridges, and where three or four cirque glaciers have been arranged like petals they have torn away the rock until all that remains is a slender horn—the Kitzsteinhorn, the Finsteraarhorn, the Matterhorn. Not only are ice sheets, ice fields, and individual glaciers operating today with effects observable as motions occur, but wherever they once flowed their products remain in abundance and intact. They have come and gone so recently.

The evidence may seem obvious now, but not until the eighteen-thirties did anyone comprehend its significance. There had been insights, hints, and clues. James Hutton, the figure from the Scottish Enlightenment who by himself developed the novel view of the world on which modern geology rests, mentioned in his *Theory of the Earth* (1795) that the gravels and boulders of Switzerland's great valley appeared to have been put there by ancient extensions of alpine ice. But Hutton, who formed his theory among the scratched granites and drifted gravels of Scotland, never suspected that Scotland itself had been a hundred per cent covered—actually dunked into the mantle—by ten thousand feet of ice.

In 1815, in the Swiss Val de Bagnes, below the Pennine Alps, a mountaineer remarked to a geologist that all those big boulders standing around in odd places had been carried there by a glacier long since gone. The mountaineer's name was Perraudin. He was a hunter of chamois. The geologist was Jean de Charpentier. He did not believe the hunter and ignored the information. In Europe, Noah's Flood had for so long been regarded as the principal sculptor of the earth that almost no one was inclined to hazard an alternative interpretation. If boulders were out of touch with bedrock of their type, diluvian torrents had moved them, or flows of diluvian mud. In 1821, a Swiss bridge-and-highway engineer named Ignace Venetz told the Helvetic Society of Natural Sciences that he believed what the mountaineer had told Charpentier. He believed, in addition, that boulders had been scattered all over Switzerland by glaciers of "*hauteur gigantesque*" from "*une époque qui se perd dans les nuits des temps*." Venetz was ignored, too—until Charpentier decided, twelve years later, that his suppositions were probably correct. Charpentier caused Venetz's paper to be published and meanwhile went out to gather, name, and classify evidence of moving ice: erratic boulders, striations and polish on bedrock, lateral and terminal moraines. In 1834, he submitted to the Helvetic Society his "*Notice sur la Cause Probable du Transport des Blocs Erratiques de la Suisse*," which was also ignored, not to say ridiculed.

Charpentier was political in the scientific world. Great "savants" like Leopold von Buch and Alexander von Humboldt had been classmates of his at the Freiberg Mining Academy. He lived above Lake Geneva in the alpine valley of the Rhone. Savants collected in numbers at his table. In the summer of 1836, Jean Louis Rodolphe Agassiz, a professor of natural history at the College of Neuchâtel, took a house up the road. Agassiz was only twenty-nine years old, but he had done work in paleontology for which he had earned a considerable reputation. He had travelled, too. He had become a protégé of von Humboldt. He had worked in Paris for Georges Cuvier. And like von Humboldt, von Buch—like everybody else who had heard about the theory of the ice—he thought it absurd.

When von Humboldt went on field trips to look at rocks, he wore a top hat, a white cravat, and a black double-breasted frock coat that reached to his knees. He was imitated by, among others, Cuvier and von Buch. Agassiz was less formal, but in no particular did he resemble a scuffed-booted, blue-jeaned, twentieth-century field geologist when he set out with Charpentier to stroll through the valley of the upper Rhone. What Agassiz saw forever altered his life, as ice had altered the valley. When he left, he had no remaining doubt of the truth of what Perraudin, Venetz, and Charpentier believed. Wandering the Swiss countryside low and high, he found further evidence everywhere he went—grooved rock, polished rock, moraines where ice had long been gone, boulders rounded off and set where water never could have shoved them. He visited similar landscapes in enough places to spread far in his imagination the contiguity they implied, and in one spark of intuition he saw the ice covering more than the valley, the canton, the nation. The idea of continental glaciation fell into place—a stunning moment of realization that ice many thousands of feet thick had been contiguous from Ireland to Russia. When the Helvetic Society met in Neuchâtel in the summer of 1837, Louis Agassiz—as its president-elect—addressed the savants. Instead of reading an expected discourse in paleontology, he outlined at great length the evidence and chronology of glacial history as he had come to see it, announcing to the Society and to the world at large what would before long be known as the Ice Age.

He called it the Époque Glaciaire. By any name, at home or abroad, it did not overwhelm his colleagues. He was attacked far more than defended. Von Buch literally threw up his hands, and not without the perspectives of the future partly on his side, for Agassiz—like the "plate-tectonics boys," as seen by Anita Harris—had not known where to stop. His remarks had gone beyond his reconstruction from observable phenomena of a cover of ice across the whole of northern Europe: he had concluded that the newborn Alps, rising under the ice, had caused it to break up.

Agassiz's friend and mentor Alexander von Humboldt, whose name reposes in the western Americas in the Humboldt Current, the Humboldt River, and the Humboldt Range, strongly urged Agassiz to go back to cataloguing fossil fishes, the work for which Agassiz was internationally known and for which the Geological Society of London had awarded him the Wollaston Medal. "You spread your intellect over too many subjects at once," he wrote to Agassiz. "I think that you should concentrate . . . on fossil fishes. In so doing, you will render a greater service to positive geology than by these general considerations (a little icy withal) on the revolutions of the primitive world. . . . You will say that this is making you the slave of others; perfectly true, but such is the pleasing position of affairs here below. Have I not been driven for thirty-three years to busy myself with that tiresome America . . . ? Your ice frightens me."

Agassiz's response was to address himself more intensively than ever to glaciers—glaciers of the present and the past. "Since I saw the glaciers I am quite of a snowy humor, and will have the whole surface of the earth covered with ice, and the whole prior creation dead by cold," he wrote in English to an English geologist. "In fact I am quite satisfied that ice must be taken in every complete explanation of the last changes which occurred at the surface of Europe." He found moraines on the plains of France. He found Swedish boulders in Germany. In Grindelwald, a stranger heard his name and, seeing his boyish appearance, asked if he was the son of the great and famous professor.

In 1839, Agassiz went to the glaciers on the apron of the Matterhorn, the glaciers under the Eiger and the Jungfrau. He walked up the Aar Glacier to the base of the Finsteraarhorn, the highest peak in the Bernese Oberland. "There I ascertained the most important fact that I now know concerning the advance of glaciers," he wrote later. From a message in a bottle in a cabin on the ice, he had learned that the monk who built the place in 1827 had returned nine years later to find it more than two thousand feet down the mountain. Agassiz established his own shelter on the Aar Glacier. He and his colleagues drove stakes into the ice—a row of them straight across the glacier—and before long discovered that glacier ice, like a river, flows more rapidly in the center and also tends to speed up toward the outsides of bends. Diverting a meltwater stream that was pouring into a deep hole in the ice, he set up a sturdy tripod at the surface and had himself lowered into the glacier. He was twenty fathoms down in a banded sapphire world when his feet touched water and he shouted instructions that his descent be stopped. His colleagues on the glacier misinterpreted his cry and lowered him into the water. The next shout was different and was clearly understood. The dripping Agassiz was raised toward the surface among stalactites of Damoclean ice, so big that had they broken they would have killed him. Concluding the experiment, he said, "Unless induced by some powerful scientific motive, I should not advise anyone to follow my example." The better to see the alpine-valley ice in its regional perspective, Agassiz and his team climbed mountains—they climbed the Jungfrau, the Schreckhorn, the Finsteraarhorn—and made their observations from the summits, completely unmindful atop a number of the mountains that no one had been there before.

Agassiz went to England, Scotland, Ireland, and Wales, looking for the tracks of glaciers. He found them in England, Scotland, Ireland, and Wales. As in Switzerland, he saw *roches moutonnées*—humps of exposed bedrock that were characteristically smooth on the side from which the ice had arrived, and plucked and shattered on the other. "The surface of Europe, adorned before by a tropical vegetation and inhabited by troops of large elephants, enormous hippopotami, and gigantic carnivora, was suddenly buried under a vast mantle of ice, covering alike plains, lakes, seas, and plateaus," he wrote in his *Études sur les Glaciers* (1840). "Upon the life and movement of a powerful creation fell the silence of death. Springs paused, rivers ceased to flow, the rays of the sun, rising upon this frozen shore (if, indeed, it was reached by them), were met only by the breath of the winter from the north and the thunders of the crevasses as they opened across the surface of this icy sea."

The reception all this got continued to be colder than the ice. Von Buch, author of the first geological map of Germany and already celebrated for his studies of volcanism, did not conceal his indignation. In fact, he had apparently removed Agassiz's name from consideration for a professorial chair at the University of Berlin. Sir Roderick Murchison, the Scottish geologist who had identified and named the Silurian system, warned that he was prepared to "make fight." Addressing the Geological Society of London, he said, "Once grant to Agassiz that his deepest valleys of Switzerland, such as the enormous Lake of Geneva, were formerly filled with snow and ice, and I see no stopping place. From that hypothesis you may proceed to fill the Baltic and the northern seas, cover southern England and half of Germany and Russia with similar icy sheets, on the surfaces of which all the northern boulders might have been shot off. So long as the greater number of the practical geologists of Europe are opposed to the wide extension of a terrestrial glacial theory, there can be little risk that such a doctrine should take too deep a hold of the mind."

Whatever the cause, the effects Agassiz was studying impressed von Humboldt as purely local phenomena. Agassiz's "*descente aux enfers*"—into the innards of the glacier—alarmed his friend as a physical risk commensurate with the risk Agassiz was taking with his paleontological reputation. Von Humboldt wrote to say that he had now "read and compared all that has been written for and against the ice-period" and that he was no closer to accepting the theory. He quoted Mme de Sévigné's saying that "grace from on high comes slowly." And added, "I especially desire it for the glacial period."

The turnabout was at hand, however. Charles Lyell, the most outstanding British geologist of the nineteenth century, closely read the *Études sur les Glaciers* and found himself enlightened. "Lyell has adopted your theory in toto!!!" a friend wrote to Agassiz. "On my showing him a beautiful cluster of moraines, within two miles of his father's house, he instantly accepted it, as solving a host of difficulties that have all his life embarrassed him." Charles Darwin hurried out into the countryside to see for himself if there were "marks left by extinct glaciers." He wrote to a friend, "I assure you, an extinct volcano could hardly leave more evident traces of its activity and vast powers. . . . The valley about here and the site of the inn at which I am now writing must once have been covered by at least eight hundred or a thousand feet in thickness of solid ice! Eleven years ago I spent a whole day in the valley where yesterday everything but the ice of the glaciers was palpably clear to me, and I then saw nothing but plain water and bare rock."

The scientific dons of Cambridge continued stubborn, but—as would happen with the theory of plate tectonics in the years following the revelations of the nineteen-sixties—geologists in expanding numbers accepted the glacial picture, and before long there was a low percentage that did not enthusiastically subscribe. Delivering an address in 1862 to the Geological Society of London, Sir Roderick Murchison declared without shame that he, too, now saw the picture. He sent a copy of his address to Agassiz with a note that said, "I have had the sincerest pleasure in avowing that I was wrong in opposing as I did your grand and original idea of my native mountains. Yes! I am now convinced that glaciers did descend from the mountains to the plains as they do now in Greenland."

Greenland is eighty-five per cent capped with ice. Anyone who doubts that we live in a glacial epoch need only note the great whiteness that Greenland contributes to a map. "The ice melted here eighteen thousand years ago," Anita said, with a nod toward the roadside in Ohio. "It melted twelve thousand years ago in Wisconsin and Maine. If you ask a penguin in the Antarctic, the Ice Age hasn't stopped yet."

The ice on Antarctica, six million square miles, is also (generally) two miles thick. "You get ice caps when you have landmasses in the polar positions," Anita went on. "The only thing worse would be if the Siberian landmass were sitting over the North Pole. Then, God help us, things would be really bad. As it is, the sea ice at the North Pole is only six feet thick. It takes a continent to support a really heavy sheet of ice. If the ice of Greenland and Antarctica were to melt now, sea level would go up at least a hundred feet. Think what the water would cover. Half the cities in the world. In the South, you can be three hundred miles from the coast and only fifty feet above sea level. Through most of time, the earth has been without ice caps. Twenty thousand years ago, when there was much more ice than there is now, the sea was three hundred feet lower. The coast was more than a hundred miles east of New York. You could have walked to the edge of the continental shelf. Baltimore Canyon, Hudson Canyon were exposed in the open air."

Outside the automobile window were three landscapes, trifocal, occupying separate levels in time and mind. Latently pictured in the rock beside the road was the epicratonic sea of three hundred and twenty million years ago, with the Cincinnatia Islands off to the west somewhere, in what is known to geologists as Ohio Bay. There was also evidence of the deep ice of twenty thousand years ago, with its lobate front some distance to the south, near Canton, Massillon, and Wooster. And there was, of course, the slightly rumpled surface of the modern state of Ohio, looking like a bedspread on which someone had taken a nap. Not nearly as flat as the rock below was the undulating interstate, where diesel exhausts were pluming and Winnebagos were yawing in the wind.

"The goal of many geologists is to make time-lapse maps of earth history," Anita remarked. "Look at topographic maps from just a hundred years ago for coastal areas of low relief, and the changes are tremendous."

We went through a ten-metre roadcut of massive sandstone so rich in iron it had rusted the road. Being tough by comparison with its neighboring rock, it stood high and formed a hill, and hence it had been blasted to convenience the interstate. "That is one hell of a sandstone," Anita said with enthusiasm, seeing in it something I could not discern.

We crossed a river. "That was the well-known Cuyahoga," she said. "If you swim in it, you dissolve."

The Cuyahoga was flowing south. It rises in northeasternmost Ohio, runs south into Akron, then reverses its direction, swinging north through Cleveland and into Lake Erie.

More warning signs flashed by. "STAY AWAKE! STAY ALIVE!"

Anita said, "I'm trying. I'm trying."

Now spanning the road was an Italianate steel-arch bridge, standing on Berea sandstone, a fragment of the Berea Delta, of early Mississippian age, which had extended its bird-foot shape far into Ohio Bay. We stopped, and picked quartz pebbles the size of golf balls out of a conglomerate there. "These would have been just offshore," she said. "You can take the pebbles out of the rock with your hands because it was never heated up like the conglomerate at the Delaware Water Gap. This was never buried much. It is not well lithified. It hasn't experienced enough heat to get tough."

A few miles west, we crossed the Cuyahoga River again, and looked down some distance from the interstate bridge into the Cuyahoga's extensively reamed-out valley, with its modest, meandering stream.

"It's an underfit stream," said Anita. "A little half-ass stream in a valley made wide by glacier ice. The Cuyahoga's valley was steepened and entrenched, like Yosemite."

"You are comparing the Cuyahoga Valley with Yosemite?"

"Technically."

We left the interstate and followed the valley into Cleveland.

The Cuyahoga River had suffered a bad press. When it caught fire some years before, it attracted national attention. Its percentage of water had become low relative to its content of hydrogen in various combinations with carbon. The river burned so fiercely that two railroad bridges were nearly destroyed. There was no mention in the papers of the good things the river had done. It had made parks. It had been there before the glacier ice and had cut down five hundred feet through Mississippian formations into Senecan and Chautauquan time—stages of the late Devonian. It cut deep ravines, which the ice later broadened into canyons. The ice augered through the V-shaped valley and turned it into a U. Which is what ice did at Yosemite—with the difference that the walls of Yosemite are speckled white granite, while the canyon walls of Cleveland are flaky black gasiferous anoxic shale. As mud, the shale was deposited in quiet water in a late Devonian sea. The rock contains the unoxidized remains of so many living things that it is by volume as much as twenty per cent organic. In thin laminations, it grew layer upon layer—paper shale. "The water was so quiet you can trace the same little lens forever," Anita said. "The formation produces gas like crazy. The gas migrates up into the sandstone above, which holds it. Berea sandstone. People drill their own wells to the Berea and heat their homes." Much of Cleveland's metropolitan-park system is in the deep Yosemite of the Cuyahoga River, under paper-flake carbon cliffs—a natural world of natural gas.

Like the Cuyahoga today, most rivers in Ohio before the recent ice sheets looked for outlets to the north and northwest. Nearly all were wiped away by the planing drive of ice. Water pooled against the glacial front and spilled away to the south and west. It skirted the ice, roughly tracing its southernmost outline, forming a new river system and a "periglacial valley"—the Ohio River, the Ohio Valley.

When Darwin published *The Origin of Species*, its affront to organized religion did not altogether exceed the dismay that was felt in science. Even Sir Charles Lyell said, "Darwin goes too far." Thomas Henry Huxley and a few others were supportive, but almost every paleontologist in the British Isles was flat negative, and the geologist Adam Sedgwick, of Cambridge University, who, with Murchison, had discerned and established the Devonian system, described himself reading Darwin "with more pain than pleasure." He said, "Parts of it I admired greatly, parts I laughed at till my sides were almost sore; other parts I read with absolute sorrow, because I think them utterly false and grievously mischievous. Many . . . wide conclusions are based upon assumptions which can neither be proved nor disproved. . . . Darwin has deserted utterly the inductive track and taken the broadway of hypothesis." Applause for Darwin was even sparer from scientists across the Channel, with the notable exception of the Belgian geologist J. J. d'Omalius d'Halloy, who, as it happened, had subscribed from the beginning to Louis Agassiz's glacial theory as well, and whose Terrain Crétacé was the discovery ground for the worldwide Cretaceous system.

In the United States, by contrast with Europe, geologists, biologists—the scientific community at large—were for the most part quick supporters and early participants in the sweep of evolution. In the United States, also, there was a notable exception. He was Professor Louis Agassiz, of Harvard University. He had crossed the Atlantic and given a few lectures. He stayed for the rest of his life. He became, as he has remained, one of the most celebrated professors in the history of American education. It was a renown that rested largely on his amazing and infectious capacity for talking about ice. Never mind that he could not speak schoolroom English. His words drew pictures of glaciers in motion, many thousands of feet thick and larger by far than the Sahara. His words drew pictures of glacier ice over Boston, in the act of depositing Cape Cod; of glacier ice over Bridgeport, in the act of depositing Long Island; of ice retreating from Concord, leaving Walden Pond. Harvard was, at core, a drumlin, a glacial coprolite, packed in recessional outwash. America excited Agassiz, as well it might, for it had held the greater part of the ice he had dreamed of, covering the world. He went to Lake Superior and paddled its shoreline in a bark canoe. The features he saw there he had known in Neuchâtel. He went to the Hudson Highlands and remembered the highlands of the Rhine. "The erratic phenomena and the traces of glaciers . . . everywhere cover the surface of the country," he wrote. "Polished rocks, as distinct as possible; moraines continuous over large spaces; stratified drift, as on the borders of the glacier of Grindelwald." He went to the Connecticut Valley: "The erratic phenomena are also very marked in this region; polished rocks everywhere, magnificent furrows on the sandstone and on the basalt, and parallel moraines defining themselves like ramparts upon the plain. . . . What a country is this! All along the road between Boston and Springfield are ancient moraines and polished rocks. No one who had seen them upon the track of our present glaciers could hesitate as to the real agency by which all these erratic masses, literally covering the country, have been transported. I have had the pleasure of converting already several of the most distinguished American geologists to my way of thinking."

Henry David Thoreau took Agassiz's book out of the Harvard library and returned it a few weeks later—perhaps unread. Apparently, Thoreau never knew that Walden Pond was a glacial kettle, had no idea that he lived among moraines and drumlins, ice-transported hills. Although he and Agassiz were acquainted and shared the same part of Massachusetts for sixteen years, there is in Thoreau's work no discussion of glaciation. Thoreau evidently never suspected that all his Nausets and Chesuncooks, Merrimacks and Middlesex ponds had been made and shaped by ice.

Agassiz was so caught up in glacial and general geology that he would try to teach it to stagecoach drivers. He believed that anyone, given a little help, could understand the nature of the earth. In Boston, in order to make his case perfectly and avoid the rockslides of his Franco-Germanic accent and syntax, he announced that he would give a series of lectures in French on the Époque Glaciaire. People paid to hear it, and he preserved their admiration in recrystallized *mots justes*. When he spoke of the Jura, the Pennine Alps, and the boulders in the valleys between, no one was as moved as Agassiz. His great range of expression did not exclude tears. With his large forehead, full lips, aquiline nose, and shoulder-flowing hair, he all but held a baton in his hand with which to conduct the movements of the ice. One Saturday a month, he met with his friends for a late, seven-course lunch from which no one was in a hurry to go home. They would meet, like as not, at "Parker's" in Boston, in a room looking out on City Hall. "Agassiz always sat at the head of the table by native right of his large good-fellowship and intense enjoyment of the scene," his friend Sam Ward eventually recalled. Henry Wadsworth Longfellow generally sat at the other end, with Oliver Wendell Holmes on his right. Holmes preferred his window-light over the shoulder. On around the table were James Russell Lowell, John Greenleaf Whittier, Nathaniel Hawthorne, Ralph Waldo Emerson, Richard Henry Dana, Jr., Ebenezer Hoar, Benjamin Peirce, Charles Eliot Norton, and James Elliot Cabot, among others. Agassiz, with a glass of wine at his elbow, would sometimes conduct the conversation with two lighted cigars, one in each hand. Holmes said of him that he had "the laugh of a big giant." Longfellow was relieved and pleased when Agassiz told him he liked the description of the glacier in "Hyperion." Emerson in his journal described Agassiz as "a broad-featured unctuous man, fat and plenteous." Sir Charles Lyell was invited, on his visits to America. United States Senator Charles Sumner was occasionally present as well. Agassiz was indifferent to him, because Sumner showed too much interest in politics.

The group was known as Agassiz's Club, more officially as the Saturday Club. One summer, when the club went off to the Adirondacks on a camping trip, Longfellow refused to go, because Emerson was taking a gun. "Somebody will be shot," said Longfellow, explaining that Emerson was too vague to be trusted with a gun. Longfellow's works of poetry include a birthday ballad in praise of Agassiz, which Longfellow read aloud at the Saturday Club, and in which Nature addressed the Professor:

"Come wander with me," she said,
"Into regions yet untrod;
And read what is still unread
In the manuscripts of God."

John Greenleaf Whittier also wrote a poem about Agassiz, more than a hundred lines in length, ten of which are these:

Said the Master to the youth:
"We have come in search of truth,
Trying with uncertain key

Door by door of mystery;
We are reaching, through His laws,
To the garment-hem of Cause,
Him, the endless, unbegun,
The unnameable, the One
Light of all our light the Source,
Life of life, and Force of force."

Longfellow, travelling in Europe in 1868, called on Charles Darwin. "What a set of men you have in Cambridge," Darwin said to him. "Both our universities put together cannot furnish the like. Why, there is Agassiz—he counts for three."

Darwin's generosity was remarkable in the light of Agassiz's reaction to *The Origin of Species*. As Agassiz summarized it: "The world has arisen in some way or other. How it originated is the great question, and Darwin's theory, like all other attempts to explain the origin of life, is thus far merely conjectural. I believe he has not even made the best conjecture possible in the present state of our knowledge." Agassiz never accepted Darwinian evolution. Many years earlier, as a young man, and as a result of his paleontological researches, he wrote the following:

More than fifteen hundred species of fossil fishes with which I have become acquainted say to me that the species do not pass gradually from one to the other, but appear and disappear suddenly without direct relations with their predecessors; for I do not think that it can be seriously maintained that the numerous types of Cycloids and Ctenoids, which are nearly all contemporaneous with each other, descend from the Placoids and Ganoids. It would be as well to affirm that the mammals, and man with them, descend directly from fishes. All these species have a fixed time for coming and going; their existence is even limited to a determined period. And still they present, as a whole, numerous, and more or less close affinities, a determined coördination in a system of organization which has an intimate relation with the mode of existence of each type, and even of each species. More still: there is an invisible thread which is unwinding itself, through all the ages, in this immense diversity, and offers as a final result a continuous progress in this development of which man is the termination, of which the four classes of vertebrates are the intermediate steps, and the invertebrates the constant accessory. Are not these facts manifestations of a thought as rich as it is powerful, acts of an intelligence as sublime as provident? . . . This is, at least, what my feeble intellect reads in the works of creation. . . . Such facts loudly proclaim principles which science has not yet discussed, but which paleontological researches place before the eyes of the observer with increasing persistency; I mean the relation of the Creation to the Creator.

Nothing that occurred during the rest of Agassiz's life caused him to revise what he had said. He died in 1873. Harvard appointed three professors to replace him. Nine years later, in a scientific journal Agassiz had founded, his successor in the chair of geology published a paper describing the Ice Age as a myth. "The so-called glacial epoch . . . so popular a few years ago among glacial geologists may now be rejected without hesitation," the article concluded. "The glacial epoch was a local phenomenon."

West of Cleveland, the terrain became increasingly flat. High outcrops disappeared, but now and again a blocky strip of rock would run along the road like a retaining wall—a glimpse of what underlay the surrounding fields. Berea sandstone. Bedford shale. Columbus limestone. "You could map this state at sixty miles an hour," Anita said. For some distance, the soil over the rock was fine glacial till—ground rock flour and sand—and then among white farms we moved out upon a black-earth plain where drainage ditches did the work of streams: a world of absolute level, until recently the bottom of a great lake. The limestone had formed in the clear salt sea of middle Devonian tropical Ohio. Eventually, the sea disappeared. Two eons later, ice slid over the limestone and, retreating, left a body of fresh water that included all of what is now Lake Erie and was twice as large.

"We wouldn't be able to feed this country the way we do if much of it had not been glaciated," Anita said. "South of the glaciers, ancient weathering removed soluble minerals and left a rather inert soil behind. After a couple of decades of planting, you need tremendous fertilizer additions there. This glacial stuff is full of unweathered mineral material—fresh-ground rock. And under it is limestone, which is what they *put* on fields. When early settlers came through here and saw no trees, they moved on to places like Missouri, beyond the glacial limit, and they missed some great farmland. In Egypt, they used to get fresh minerals with every flood, but those morons built the Aswan High Dam and stopped the floods. They're starving themselves out and making a salt pan of the delta."

We were crossing the Findlay Arch and had reached the edge of the Michigan Basin, features of the subsurface structure, invisible and unexpressed in the black level surface of silts and clays. In tropical Ohio, the arch had at one time held back a large piece of the retreating sea. As the isolated water slowly concentrated and eventually disappeared, it left Morton's salt and U.S. gypsum. It left even more limestone. It left dolomite, anhydrite—components of what is known as the evaporite sequence. North off the interstate, we went through Gypsum, Ohio, on Sandusky Bay, and on to the lake port Marblehead, where we boarded the Kelleys Island ferry. "VISIT HISTORIC GLACIAL GROOVES," said a sign beside the ticket booth, and soon, for a stiff toll, we were beating into an even stiffer wind, which was tearing the caps off the waves of Lake Erie. Kelleys Island is about four miles offshore, and other cars on the ferry were stuffed with a month's worth of groceries. A hundred and twenty people live there, year around, on four and a half square miles, and as we drove across the island we passed stone houses with red and black boulder walls—jaspers and amphibolites plucked up by the ice and brought south from the Canadian Shield.

Kelleys Island stands high because it is a piece of the structural arch. While the Wisconsinan ice sheet was excavating the Great Lakes, reaming out whole networks of streams and carrying away the prominent features of their valleys, it bevelled but could not destroy the resistant structural arch. An engulfed ridge stood up from the bottom of the primal lake. With the weight of the ice gone, all of northern America slowly rebounded. A large part of the water gradually drained away, leaving Kelleys Island dry in the air, sixty feet above the level of Lake Erie.

We passed the island cemetery, its names recorded in limestone. We came to the north shore, where the beginnings of a quarrying operation had revealed how the ice had cut its tracks into the rock. "GLACIAL GROOVES STATE MEMORIAL." It was as if a giant had drawn his fingers through an acre of soft butter. The grooves were parallel. They were larger than the gutters of bowling lanes. Aggregately, they suggested the fluted shafts of Greek columns. Their compass orientation was northeast-southwest —the established glide path of the moving ice. Nowhere had we seen or would we see more emphatic evidence of continental glaciation, with the obvious exception of the Great Lakes themselves. "If you were to hydraulically flush northern Ohio—wash off the soil from the bedrock—you'd see a hell of a lot of these grooves," Anita said. "In several hundred years, these won't be here. Limestone is soft enough to be grooved and hard enough to resist weather for a few hundred years. In shale, grooves like these would go quickly. The ice, carrying boulders in its underside—carrying those amphibolites and red jaspers in the people's houses—tore the hell out of this island. When Agassiz saw things like this, he went bananas."

There have been glacial geologists, even in the late twentieth century, who have believed that such impressive grooves were gouged by boulders rolling in the Flood. Exceptions notwithstanding, Louis Agassiz's theory of continental glaciation, like the theory of plate tectonics, achieved with extraordinary swiftness its general acceptance in the world. As Thomas Kuhn has demonstrated in *The Structure of Scientific Revolutions*, when a novel theory becomes relatively established it defines the patterns of amplifying research for many years and even centuries—until a new theory comes along to overturn the old, until an Einstein appears, outreaching the principles of Newton. Conceivably, the theory of plate tectonics will one day experience a general reformation. The theory of continental glaciation seems less prone to grand revision. The sun itself seems as likely to be banished from the center of the solar system as the ice from the Pleistocene continents. The ice made Lake Seneca, Lake Cayuga—all the so-called Finger Lakes, of western New York—cutting them into stream valleys in exactly the manner in which it cut the fjords of Patagonia, the fjords of Norway, Alaska, and Maine.

After the ice quarried the huge quantities of Canadian rock that it dumped in the United States, it melted back and filled the quarries with new Canadian lakes—hundreds of thousands of Canadian lakes. A sixth of all the fresh water on earth is in Canadian ponds, Canadian streams, Canadian rivers, Canadian lakes. In Greenland, Antarctica, and elsewhere, a much greater quantity of fresh water—four times as much—is still imprisoned as ice, leaving precious little fresh water for the rest of the world.

Our Époque Glaciaire has by now been illuminated by a century and a half of expanded research. Glacial outwash has been identified at the mouth of the Mississippi, six hundred miles from the terminal moraine—a suggestion of the power and the volume of the rivers that melted from the ice. Where the land tilted north and the meltwaters pooled against the glacial front—and where waters were trapped between moraines and retreating ice—gargantuan lakes formed, such as Glacial Lake Maumee, the one of which Lake Erie is all that remains. Lake Michigan is all that remains of Glacial Lake Chicago. Lake Ontario is all that remains of Glacial Lake Iroquois. Lake Winnipeg, Lake Manitoba, the Lake of the Woods are among the remains of a glacial lake whose bed and terraces, stream deltas and wave-cut shores reach seven hundred miles across Saskatchewan, Manitoba, and Ontario, and down into the United States as far as Milbank, South Dakota. With the exception of the Caspian Sea, this one was larger than any lake of the modern world. It was the supreme lake of the American Pleistocene—Glacial Lake Agassiz.

Cold air flowing off the ice sheets caused such heavy precipitation when it encountered warm and humid air to the south that whole regions there filled with water, too. The basins of Nevada became lakes and the ranges among them were islands. Lake Bonneville filled a third of Utah. Huge lakes grew in the Gobi Desert, in Australia's Great Artesian Basin, in various lowlands across North Africa. There were forests in the Sahara, as fossil pollen shows, and networks of flowing streams. Their dry channels remain.

In North America, where the ice started to go back about twenty thousand years ago, the first vegetation to spring up behind it was tundra. Carbon 14 can date the fossil tundra. The dates, particularly in the East, show a slow, and then accelerating, retreat.

After five thousand years, the front was still in Connecticut. In another twenty-five hundred years, it crossed the line to Canada. Human beings, living on the tundra near the ice, perforce were inventive and tough. Culture, in part, was a glacial effect. In response to the ice had come controlled fire, weapons, tools, and fur as clothing. Creativity is thought to have flourished in direct proportion to proximity to the glacier—an idea that must infuriate the equatorial mind. The ice drew back from Britain a geologic instant before the birth of Shakespeare. The fossils of *Homo sapiens* have never been found in sediments older than the Ice Age.

In the way that scenes of vanished mountains can be inferred from their debris, the vision of continents covered with ice came straightforwardly to Agassiz as the product of reasoning carried backward from evidence through time. It is one thing to say that the ice was there, quite another to say how it got there. If the origin of mountains is sublimely moot, so is the origin of the ice. Characteristically, the prime mover is not well understood. The ice did not come over the world like a can of paint poured out on the North Pole. It formed in places well below the Arctic Circle, and moved out in every direction—including north—until cut off by the unsupportive sea. Geologists call these places spreading centers, the same term they use for the rifted boundaries where plates tectonically divide. To the question "Why did the ice form?" they can answer only with speculation. The phenomenon is obviously rare. A pulsating series of ice sheets seems to have been set up in the discernible history of the world roughly once every three hundred million years. It happens so infrequently that it must be the result of coinciding circumstances that could not stand alone as explanations. There are components fast and slow. The atmosphere has been gradually cooling for sixty million years. Possibly this is explained by the great orogenies that have occurred during that time—the creation of the Rockies, the Andes, the Alps, the Himalaya—and the volcanism that is associated with mountain building. Volcanic ash in the stratosphere reflects sunlight back into space. Also, the weathering of mountains, particularly their granites, brings on a chemical reaction that removes carbon dioxide from the atmosphere, diminishing the greenhouse effect and chilling the earth. In any case, the essential requirement is a cool summer. A little snow from one winter must last into the next. Every forty thousand years, the earth's axis swings back and forth through three degrees. Summers are cooler when the earth is less tilted toward the sun. The sun, for that matter, is not consistent in the energy it produces. Moreover, the relative positions of the sun and the earth, in their lariat voyage through time, vary, too—enough for subtle influence on climate. Carbon dioxide also affects climate, and the amount of carbon dioxide in the atmosphere is not constant. Somewhere in such a list, which runs to many items, lie the simultaneous events that set the ice to growing. The change they bring is not at first dramatic. So critical is the earth's temperature that a drop of just a few degrees will cause ice to form and spread. A cool summer. Unmelted snow. An early fall in some penarctic valley. An overlap of snow. A long winter. A new cool summer. An enlarged residue of snow. It compacts and recrystallizes into granules, into ice. Because it is white, it repels the sun's heat and helps cool the air on its own. The process is self-enlarging, unstoppable, and once the ice is really growing it moves. Clear bands form near the base, along which the ice shears and slides upon itself in horizontal layers like the overthrust Appalachians. The thermal output of the earth melts a thin film of water on the glacier bottom, and the ice slides on that, too. Thrust sheets made of rock also slide on water. The lower part of a glacier is plastic, the upper part brittle—like the earth's moving plates and the plastic mantle beneath them. Where the brittle glacier surface bends, it cracks into crevasses, into fracture zones, as does the brittle ocean crust (the Clarion Fracture Zone, the Mendocino Fracture Zone). Such fractures are everywhere in the rock of continents, too. In fact, the ridged-and-valleyed surface of almost any flowing glacier is remarkably similar to the sinuous topography of the deformed Appalachian mountains. The continental ice sheet moves toward the equator and keeps on going until it cannot stand the heat. At the latitude of New York City, generally speaking, the ice melts as fast as it advances, and thus it goes no farther, and leaves on Staten Island its terminal moraine. Ocean temperatures will have dropped because of the cold, and therefore the oceans are providing less snow to feed the ice. On all fronts, the ice retreats—not necessarily to disappear. The climate warms. The oceans warm. The snow pack thickens in the Great North Woods. A glacier spreads again. Once the pattern is set, the rhythm is relatively steady. For us, the ice is due again in ninety thousand years.

We ran on through Ohio on the bed of the great former lake, Kelleys Island far behind us. Where there had been sand spits reaching into the water, with sandy hooks at their tips, there were farm buildings standing on the dry spits—the high prime ground, a few feet higher than the surrounding fields. Now, at spring plowing time, these things were visible as they would not be for a year again.

And then we went off the lakebed and up into roadcuts of vetch-covered till among the kettles, kames, and drumlins, the Wabash Moraine, the New England landscape of glacial Indiana. "This would be a good place for a golf course," Anita remarked. "If you want a golf course, go to a glacier." We left the interstate for a time, the better to inspect the rough country. "I grew up in topography like this—in Brooklyn," Anita said. "I didn't know what bedrock meant. You could plot the limit of glaciation in New York City by the subway system. Where it's underground, it's behind the glaciation. Where it's in the moraine and the outwash plain, it's either elevated or in cuts in the ground."

Back on I-80—and running now on a pitted outwash plain, now on a moraine—we crossed the St. Joseph River. Anita's thoughts were still in Brooklyn. "My father died twenty years ago," she said. "When he was a little boy, his mother told him that if he ever ate food with his yarmulke off he would be struck dead. When she wasn't looking, he lifted his yarmulke and ate a spoonful of cereal. He didn't die. He quit believing. His faith was shaken."

There was a gold dome on our left—like an egg resting in a bed of new green canopy leaves. It was the supreme roof of the University of Notre Dame. "They're on outwash," Anita said in passing, and returned to her reminiscences. "Religious prejudice in any form is despicable," she went on. "In Brooklyn, when the Jehovah's Witnesses tried to sell me *The Watchtower* I'd say, 'I'm illiterate.' If they persisted, I'd say, 'Let me tell you about *my* God.' "

Again the St. Joseph River intersected the highway, and we ran on through grass-covered roadcuts of a kame complex, and soon through others in a recessional moraine, locally called the Valparaiso Moraine. A road sign suggested the proximity of Valparaiso. "Where do they get a name like that in a lacklustre place like Indiana?" Anita said, and we swung out an exit for the Indiana Dunes.

They are higher than the highest dunes of Cape Cod, and they are lined up in rows four deep along the shore of Lake Michigan—longitudinal dunes, transverse dunes, parabolic dunes. Glacial effects. On our map of Indiana, three of them were called mountains. They were covered with sand cherries, marram grass, cottonwoods, jack pines, junipers, and bluebells, except where the wind that made them had returned to them later to tear great blowouts in their sides. On foot, we approached the base of Mt. Tom. Staring upward, Anita said, "Look at the size of *that* son of a bitch." We climbed to the summit—to a view that might have pleased Balboa, had he been fond of power plants. There was a power plant to our left, a power plant to our right—Gary, Michigan City. Chicago was a shimmer of structures up the lake. Chicago was underwater until two thousand years ago. The southern rim of Glacial Lake Chicago was the Valparaiso Moraine. As the lake level dropped, it left the makings of the Indiana Dunes. They are wind-built sands picked up from glacial till, so fresh that under a glass they are seen to be jagged. "It's been ground up so recently it's like the sand of Coney Island," said Anita, looking at it through her hand lens. She held the lens to her eye and a palmful of sand close to the lens. "I see angular grains of red chert," she said. "I see little fragments of igneous rock. I see amphibolite and red jasper—like the stuff that cut the grooves on Kelleys Island. I see red iron-oxide-coated quartz grains. You can see right through it to the quartz. I see little pieces of carbon. I see green chert. I see a bug crawling through the sand."

We sat in the lee of the top of Mt. Tom and watched whitecaps running on the lake. "As everybody knows, there are sand dunes in the Sahara," Anita said. "As everybody does not know, there are also grooves in the Sahara like the ones on Kelleys Island. They were cut in the bedrock in the Ordovician, when the Sahara was in a polar position and the equator was in Montreal."

I asked how the Sahara had accomplished such a journey.

"It is possible that the entire crust and some of the mantle can move around the interior earth—a sphere within a sphere," she said. "You've got to change the position of the continental landmasses with respect to latitude through time. That I can't deny."

"Do you believe that India smacked Asia?"

"I don't know. I know very little about the geology there, so how could I believe it? To many problems, plate tectonics is not the only solution. Often, it's a lazy man's out. It's a way of saying, 'I don't have to think any further.' It's a way of getting out of a problem. The geology has refuted plate-tectonic interpretations time and again in the Appalachians. Geology often refutes plate tectonics. So the plate-tectonics boys tend to ignore data. The horror is the ignoring of basic facts, not bothering to be constrained by data. It's like some modern art—done by people who throw paint at canvas and have never learned the fundamentals. Amateurs. Jumping into professionalism. Thinking they can get away with it. There are a lot of people out there in the profession like me who don't believe much of it. But we can't altogether complain. Plate tectonics has turned people on. It has brought a lot of new people into geology. You've always got to have devil's advocates, and with respect to plate tectonics I am a devil's advocate."

Book 3

Rising from the Plains

BEARTOOTH MOUNTAINS
MON
Yellowstone National Park
ABSAROKA RANGE
BIGHORN BASIN
Bighorn River
WASHAKIE RANGE
TETON RANGE
IDAHO
JACKSON HOLE
MT. LEIDY HIGHLANDS
Snake River
Thermop
OWL CREEK
Wind River
GROS VENTRE MTNS
WIND RIVER RANGE
WIND RIVER BA
Muskra
SALT RIVER RANGE
WYOMING RANGE
Cora
Pinedale
Lander
Afton
GREEN RIVER BASIN
GAS HILLS
Red Bluff Ranch
Sweetwater River
CROOKS GAP
OVERTHRUST BELT
STEAMBOAT MTN
LEUCITE HILLS
GREAT DIVIDE B
Green River
ROCK SPRINGS UPLIFT
Blacks Fork River
Rock Springs
Green River
80
UTAH
Evanston
Flaming Gorge Reservoir
WASHAKIE BASIN
COLOR
UINTA MOUNTAINS

Continental Divide
POWDER
RIVER
BASIN
Powder River
PUMPKIN
BUTTES
BLACK HILLS
SOUTH DAKOTA
North Platte River
Casper
LARAMIE
RANGE
HANNA
BASIN
SEMINOE MTNS
Laramie River
NEBRASKA
Rawlins
MEDICINE BOW
MOUNTAINS
SNOWY RANGE
Arlington
N. Platte River
LARAMIE
PLAINS
Laramie
SIERRA
MADRE
Cheyenne
Pine
Bluffs
80

This is about high-country geology and a Rocky Mountain regional geologist. I raise that semaphore here at the start so no one will feel misled by an opening passage in which a slim young woman who is not in any sense a geologist steps down from a train in Rawlins, Wyoming, in order to go north by stagecoach into country that was still very much the Old West. She arrived in the autumn of 1905, when she was twenty-three. Her hair was so blond it looked white. In Massachusetts, a few months before, she had graduated from Wellesley College and had been awarded a Phi Beta Kappa key, which now hung from a chain around her neck. Her field was classical studies. In addition to her skills in Latin and Greek, she could handle a horse expertly, but never had she made a journey into a region as remote as the one that lay before her.

Meanwhile, Rawlins surprised her: Rawlins, where shootings had once been so frequent that there seemed to be—as citizens put it—"a man for breakfast every morning"; Rawlins, halfway across a state that was spending per annum far more to kill wolves and coyotes than to support its nineteen-year-old university. She had expected a "backward" town, a "frontier" town, a street full of badmen like Big Nose George, the road agent, the plunderer of stagecoaches, who signed his hidden-treasure maps "B. N. George." Instead, this October evening, she was met at the station by a lackey with a hand-cart, who wheeled her luggage to the Ferris Hotel. A bellboy took over, his chest a constellation of buttons. The place was three stories high, and cozy with steam heat. The lights were electric. There were lace curtains. What does it matter, she reflected, if the pitchers lack spouts?

One spring day about three-quarters of a century later, a four-wheel-drive Bronco approached Rawlins from the east on Interstate 80. At the wheel was David Love, of the United States Geological Survey, supervisor of the Survey's environmental branch in Laramie, and—to an extent unusual at the highest levels of the science—an autochthonous geologist. The term refers to rock that has not moved. Love was born in the center of Wyoming in 1913, and grew up on an isolated ranch, where he was educated mainly by his mother. To be sure, experience had come to him beyond the borders—a Yale Ph.D., explorations for oil in the southern Appalachians and the midcontinent—but his career had been accomplished almost wholly in his home terrain. For several decades now, he had been regarded by colleagues as one of the two or three most influential field geologists in the Survey, and, in recent time, inevitably, as "the grand old man of Rocky Mountain geology." The grand old man had a full thatch of white hair, and crow's feet around pale-blue eyes. He wore old gray boots with broken laces, brown canvas trousers, and a jacket made of horsehide. Between his hips was a brass belt buckle of the sort that suggests a conveyor. Ambiguously, it was scrolled with the word "LOVE." On his head was a two-gallon Stetson, with a braided-horsehair band. He wore trifocals. There was stratigraphy even in his glasses.

A remarkably broad geologist, he had worked on everything from geochemistry to structural geology, environmental geology to Pleistocene geology, stratigraphy to areal geology and mapping—and he had published extensively in all these fields. In the Bronco, he seemed confined—a restlessness that derived from a lifetime of travel on foot or horseback. He was taking me across Wyoming, at my request, looking at the rock in roadcuts of the interstate, which in seasons that followed would serve as portals for long digressions elsewhere in Wyoming in pursuit of the geologies the roadcuts represented. Once, in the Bighorn Basin, as we were rolling out our sleeping bags, I asked him what portion of the nights of his life he had spent out under the stars, and he answered, "One-third." A few minutes later, half asleep, he added a correction: "Let's say one-quarter. I want to be careful not to exaggerate." He rolled over and was gone for the night. I passed out more slowly, while my brain tumbled heavily with calculation. Love was about seventy, and this, I figured, was something like his six-thousandth night on the ground. Well, not precisely on the ground. One must be careful not to exaggerate. He'd had the same old U.S.G.S. air mattress for forty years. When it was quite new, it sprang a leak. He poured evaporated milk in through the valve and stopped the leak.

Now, as we crossed the North Platte River and ran on toward Rawlins in May, over the road were veils of blowing snow. This was Wyoming, not some nice mild place like Baffin Island—Wyoming, a landlocked Spitsbergen—and gently, almost imperceptibly, we were climbing. The snow did not obscure the structure. We were running above—and, in the roadcuts, among—strata that were leaning toward us, strata that were influenced by the Rawlins Uplift, which could be regarded as a failed mountain range. The Medicine Bow Mountains and the Sierra Madre stood off to the south, and while they and other ranges were rising this one had tried, too, but had succeeded merely in warping the flat land. The tilt of the strata was steeper than the road. Therefore, as we moved from cut to cut we were descending in time, downsection, each successive layer stratigraphically lower and older than the one before. Had this been a May morning a hundred million years ago, in Cretaceous time, we would have been many fathoms underwater, in a broad arm of the sea, which covered the continental platform—reached across the North American craton, the Stable Interior Craton—from the Gulf of Mexico to the Arctic Ocean. The North Platte, scratching out the present landscape, had worked itself down into some dark shales that had been black muds in the organic richness of that epicratonic sea. The salt water rose and fell, spread and receded through time

—in Love's words, "advanced westward and then retreated, then advanced and retreated over and over again, leaving thick sequences of intertonguing sandstone and shale"—repeatedly exposing fresh coastal plains, and as surely flooding them once more. In what has become dry mountain country, vegetation flourished in coastal swamps. They would have been like the Florida Everglades, the peat fens of East Anglia, or borders of the Java Sea, which stand just as temporarily, and after they are flooded by a rising ocean may be buried under sand and mud, and reported to the future as coal. There were seams of coal in the roadcuts, under the layers of sandstone and shale. The Cretaceous swamps were particularly abundant in this part of Wyoming. A hundred million years later, the Union Pacific Railroad would choose this right-of-way so it could fuel itself with the coal.

In cyclic rhythm with the other rock was limestone. Here and again, the highway was running on this soft impure limestone. It was sea-bottom lime, from dissolved or fragmented shells, which had lithified at least ten thousand feet lower than it is now. Woody asters were in bloom in the median, and blooming, too, by the side of the road, prospering on the lime. Love pointed them out with an edge in his voice. He said they were not Wyoming plants. They had come into Wyoming with trail herds of cattle and sheep, and later in trucks and railroad cars bringing hay from hundreds of miles to the south; and disastrously they had the ability—actually, a need—to draw selenium from the rock below. Selenium, which in concentration is toxic to people and animals, is given to the wind in some volcanic ash. A hundred million years ago, stratovolcanoes stood in Idaho, and they sent up ash that fell out eastward in the sea. The selenium went into the lime muds, and now these alien asters were drawing it out of the limestone and spreading the poison across the surface world, as few other plants can do. Most plants ignore selenium. Woody asters and a few others require selenium in order to germinate. After they take it up from the rock, they convert it into a form that nearly all plants will, in turn, take up, too. Selenium-contaminated plants are eaten by sheep and cattle, which are served to people as chops and burgers. Concentrated selenium destroys an enzyme that transmits messages from brain to muscles. "Cattle and sheep get the blind staggers," Love went on. "People are also affected. They get dishrag heart. The liver is damaged, and the kidneys. Selenium causes sterility. Worse, it causes birth defects. It's a cumulative poison, like lead or arsenic. It's one of the ingredients of nerve gas."

He gestured left and right. "These were prize salt-sage flats for sheep-grazing once. They're now poisoned and dangerous. A bad selenium area stinks like rotten garlic. On a warmer day, you could smell it. Fifty years ago, one of my first jobs was to look for selenium-converting plants up the Gros Ventre River. We camped there for a week, hunted for them day after day, and found a handful. Now, in the same place, they're thicker than fleas on a dog. They can't cross non-seleniferous barriers, except with the help of human beings. In the Rocky Mountains generally, millions of acres have been converted. People sometimes think neighbors have poisoned their pastures."

Ten miles beyond the North Platte, a flat-topped ridge formed the horizon before us—a tough sandstone, disintegrating at a lower rate than surrounding shale. The interstate, encountering this obstacle, had dealt with it with dynamite, opening up what highway engineers called a benched throughcut and geologists finding such a thing in nature call a wind gap. When we reached it, we stopped, got out, and put our noses on the outcrop, for this high multitiered exposure was Frontier sandstone, and Love referred to it as "a published roadcut," studied to the last grain by paleontologists and stratigraphers. The reason for so much attention was not readily apparent in the gray and somewhat gloomy, sooty-looking rock, antiqued with fossil burrows. Nonetheless, it seemed to excite Love—as he picked at it with his hammer—at least as much as the woody asters had repelled him. The rock had been submarine sand, not far offshore. "Frontier is one of the great oil sands in the Rocky Mountain region," he said. At five, ten, twenty thousand feet, wildcat after wildcat had found handsome pay in this celebrated host formation, and here it was at the surface, fresh, unweathered, presenting clues to its wealth. Because oil is vulnerable to destruction by increased heat—in the earth as in the engine of a car—the oldest oil that has been recovered in large quantity is probably Cretaceous in age (loosely speaking, about a hundred million years). For about one human life span, geologists have had the ability to discern where, in the subsurface, oil should be. A large percentage of all the oil on earth has been burned up in fifty years. Around 1975, the quantity being discovered had diminished to the level of the quantity being burned. Love remarked that half a billion barrels of oil had been found in the Frontier sandstone in one field alone. With reverence, I collected a wormy chunk.

Less than a mile up the road, we stopped again—at a low, flaky roadcut of Mowry shale. Progressing thus across Wyoming with David Love struck me as being analogous to walking up and down outside a theatre in the company of David Garrick. The classic plays—Teton, Beartooth, Wind River—were not out here on the street, but meanwhile these roadcuts were like posters, advertising the dramatic events, suggesting their narratives, fabrics, and structures. This Mowry shale had been organic mud of the Cretaceous seafloor, wherein the oil of the Frontier could have formed. It was a shale so black it all but smelled of low tide. In it, like mica, were millions of fish scales. It was interlayered with bentonite, which is a rock so soft it is actually plastic—pliable and porous, color of cream, sometimes the color of chocolate. Bentonite is volcanic tuff—decomposed, devitrified. So much volcanic debris has settled on Wyoming that bentonite is widespread and, in many places, more than ten feet thick. To some extent, it covers every basin. Also known as mineral soap, it has a magical ability to adsorb water up to fifteen times its own volume, and when this happens it offers to a tire about as much resistance as soft butter. Wet, swollen bentonite soil is known as gumbo. We were crossing badlands of the Bighorn Basin one time when a light shower fell, and the surface of the road changed in moments from dust to colloidal suspension. The wheels began to skid as if they were climbing ice. Four-wheel drive was no help. Many a geologist has walked out forty miles from a vehicle shipwrecked in gumbo. Bentonite is mined in Wyoming and sold to the rest of the world. Blessed is the land that can sell its mud. Bentonite is used in adhesives, automobile polish, detergent, and paint. It is in the drilling "mud" of oil rigs, sent down the pipe and through apertures in the bit to carry rock chips to the surface. It sticks to the walls of the drill hole and keeps out unwanted water. It is used to line irrigation ditches and reservoirs, and in facial makeup. Indians drove buffalo into swamps full of bentonite. It is an ingredient of insecticides, insect repellents, and toothpaste. It is used to clarify beer.

If Love had ever tried bentonite to repair his air mattress, he did not mention it. He did remark that when there was rain in the Wind River Basin on the ranch where he grew up—an event that happened about as often as a birthday—wagons were stopped in their tracks. Much of Wyoming's bentonite is Cretaceous in age and consistent in composition. Since it lies on every side of the mountain ranges, it seems not so much to imply as to certify that when it was so broadly deposited the mountains were not there. The Cretaceous is not far back in the history of the world. It's in the last three per cent of time.

Love walked back to the Bronco with a look on his face that suggested a man who had long since had his last beer. He said he was hungry. He said, "My belly thinks my throat's been cut." Over the next rise was Rawlins, spread across the Union Pacific.

On October 20, 1905, the two-horse stage left Rawlins soon after dawn—not a lot of time for stretching out the comforts of the wonderful Ferris Hotel. Eggs were packed under the seats, also grapes and oysters. There were so many boxes and mailbags that they were piled up beside the driver. On the waybill, the passengers were given exactly the same status as the oysters and the grapes. The young woman from Wellesley, running her eye down the list of merchandise, encountered her own name: Miss Ethel Waxham.

The passenger compartment had a canvas roof, and canvas curtains at the front and sides.

> The driver, Bill Collins, a young fellow with a four days beard, untied the bow-knot of the reins around the wheel, and swung up on the seat, where he ensconced himself with one leg over the mail bags as high as his head and one arm over the back of his seat, putting up the curtain between. "Kind o' lonesome out here," he gave as his excuse.

There were two passengers. The other's name was Alice Amoss Welty, and she was the postmistress of Dubois, two hundred miles northwest. Her post office was unique, in that it was farther from a railroad than any other in the United States; but this did not inconvenience the style of Mrs. Welty. Not for her some false-fronted dress shop with a name like Tinnie Mercantile. She bought her clothes by mail from B. Altman & Co., Manhattan. Mrs. Welty was of upper middle age, and—"bless her white hairs"—her gossip range appeared to cover every living soul within thirty thousand square miles, an interesting handful of people. The remark about the white hairs—like the description of Bill Collins and the estimated radius of Mrs. Welty's gossip—is from a journal that Ethel Waxham had begun writing the day before.

The stage moved through town past houses built of railroad ties, past sheepfolds, past the cemetery and the state penitentiary, and was soon in the dust of open country, rounding a couple of hills before assuming a northwesterly course. There were limestone outcrops in the sides of the hills, and small ancient quarries at the base of the limestone. Indians had begun the quarries, removing an iron oxide—three hundred and fifty million years old—that made fierce and lasting warpaint. More recently, it had been used on Union Pacific railroad cars and, around 1880, on the Brooklyn Bridge. The hills above were the modest high points in a landscape that lacked exceptional relief. Here in the middle of the Rocky Mountains were no mountains worthy of the name.

Mountains were far away ahead of us, a range rising from the plains and sinking down again into them. Almost all the first day they were in sight.

As Wyoming ranges go, these distant summits were unprepossessing ridges, with altitudes of nine and ten thousand feet. In one sentence, though, Miss Waxham had intuitively written their geologic history, for they had indeed come out of the plains, and into the plains had in various ways returned.

Among rolling sweeps of prairie . . . we met two sheep herders with thousands of sheep each. "See them talking to their dogs," said the driver.

They raised their arms and made strange gestures, while the dogs, at the opposite sides of the flock, stood on their hind legs to watch for orders.

In Wyoming in 1905, three million sheep competed for range grass with eight hundred thousand cattle. Big winter winds, squeezed and therefore racing fast between the high ground and the stratosphere, blew the snow off the grass and favored the sheep. They were hardier, and their wool contended with the temperature and the velocity of the wind. Winter wind. There was a saying among homesteaders in Wyoming: "If summer falls on a weekend, let's have a picnic."

Twelve miles from Rawlins, the horses were changed at Bell Spring, where, in a kind of topographical staircase—consisting of the protruding edges of sediments that dipped away to the east—the Mesozoic era rose to view: the top step Cretaceous, the next Jurassic, at the bottom a low red Triassic bluff, against which was clustered a compound of buildings roofed with cool red mud. Miss Waxham had no idea then that she was looking at a hundred and seventy-five million years, let alone *which* hundred and seventy-five million years. She had no idea that those sediments had broken off just here, and that the other side of the break, two and three thousand feet below, contained prolific traps of gas and oil. Actually, no one knew that. Discovery was twenty years away.

The stage rolled onto Separation Flats—altitude seven thousand feet—still pursuing the chimeric mountains. One of them, she learned, was called Whiskey Peak. Collins looked around from the driver's seat and said a passenger had once asked him the name of the mountain, "and I told him that it was in this coach where I could put my hand on it—but he could not guess." In the far distance also appeared a "white speck"—a roadhouse—which they watched impatiently for hours.

It did not look larger when we reached it. . . . Mrs. Welty and I hurried in to get warm, for we were chilled through. Outside, hung from the roof, was half a carcass of a steer. . . . In a cluttered kitchen, a fat forlorn silent woman served us wearily with a plentiful but plain meal, and sat with her arms folded watching us eat. . . . We ate our baked potatoes and giant biscuits, onions and carrots and canned-apple pie in half silence, glad to be through. The stage horses were changed and we started on toward Lost Soldier.

Lost Soldier was another sixteen miles and thus would take three hours. Already, Mrs. Welty was talking about the Hog Back, more than twenty hours up the road—a steep descent from a high divide, where Wyoming's storied winds had helped many a stage-coach get to the bottom in seconds. Wreckage was strewn all over the ground there, among the bones of horses. A driver had been known to chain a coach to a tree to keep the coach from blowing away. Like the sails of boats and ships, the canvas sides of stage-coaches were often furled as they approached the Hog Back, to let the wind blow through. No one relied on brakes.

Always, going down, the wheels are rough-locked by a chain so that they slip along instead of turning. . . . A freight team went over the side a little while ago about Thanksgiving time. The load was partly supplies for Thanksgiving dinner, turkeys, oysters, fruit, etc. The driver called to the team behind for help. When it came, he was calmly seated on a stump peeling an orange while the wagon and debris were scattered below.

Oil would be discovered under Lost Soldier in 1916. It would yield the highest recovery per acre of any oil field that has ever been discovered in the Rocky Mountains. From level to level in a drill hole there—a hole about a mile deep—oil could be found in an amazing spectrum of host rocks: in the Cambrian Flathead sand-stone, in the Mississippian Madison limestone, in the Tensleep sands of Pennsylvanian time. Oil was in the Chugwater (red sands of the Triassic), and in the Morrison, Sundance, Nugget (celebrated for-mations of the Jurassic), and, of course, in the Cretaceous Frontier. A well at Lost Soldier was like grafted ornamental citrus—oranges, lemons, tangerines, grapefruit, all on a single tree. The discoverer of the oil-bearing structure was a young geologist from Princeton University, who not only found the structure but also helped to place the term "sheepherder anticline" in the geologic lexicon. A sheep-herder anticline is one that is particularly obvious, one that could be mapped by a Princeton geologist dressed as a shepherd and moving around with a flock of sheep—which is how he avoided attention as he studied the rock of Lost Soldier.

We rattled into the place at last, and were glad to get in to the fire to warm ourselves while the driver changed the load from one coach to another. With every change of drivers the coach is changed, making each man responsible for repairs on his own coach. The Kirks keep Lost Soldier. Mrs. Kirk is a short stocky figureless woman with untidy hair. She furnished me with an old soldier's overcoat to wear during the night to come. . . . Before long, we were started again, with Peggy Dougherty for driver. He is tall and grizzled. They say that when he goes to dances they make him take the spike out of the bottom of his wooden leg.

There were four horses now—"a wicked little team"—and immediately they kicked over the traces, tried to run away, became tangled like sled dogs twisting in harness, and set Peggy Dougherty to swearing.

Ye gods, how he could swear.

Mrs. Welty diverted him with questions about travellers marooned in snowdrifts. Mrs. Welty was aware that Mr. Dougherty—who was missing six fingers, one leg, and half of his remaining foot—was an authority on this topic. In 1883, a blizzard had overtaken him and his one passenger, a young woman comparable in age to Miss Waxham. When the snow became so deep that the coach ceased to move, he unhitched a horse. Already stiff with frostbite, he hung on to the harness while the animal hauled him through drifts. The horse dragged Dougherty for hours, until he finally lost his grip and let go, having nearly reached a stage-line station. Into the wind, he shouted successfully for help. When rescuers reached the stagecoach, the passenger was dead.

Dougherty remarked to Mrs. Welty that winters lately had not been so severe.

"No," she agreed. "And we haven't had a blizzard this summer."

The sun set, and the stars rose, and the cold grew more intense. . . . About half past nine, we reached the supper station, stiff with cold.

This was Rongis, a community of a few dozen people just south of Crooks Gap. "Rongis" was an ananym—so named by an employee of the stagecoach company whose own name was Eli Signor. Lost Soldier, Rongis—such names are absent now among the Zip Codes of Wyoming, but the ruins of the stations remain.

Supper was soon ready, a canned supper, with the usual dried-apple pie and monstrous biscuits and black coffee. About ten we started out again, with a new relay of horses. More wrapped up than ever, we sat close to each other to keep warm, and leaned against the sacks of mail behind us.

The night before, at Rongis, the temperature had gone to zero. As the stage moved into Crooks Gap, the bright starlight fell on fields of giant boulders black-and-silver in relief. Some were as large as houses. In time, it would be determined that they had come down off high mountains farther north that were no longer high—mountains that had somehow sunk into the plain. Meanwhile, anyone connecting the boulders to their source bedrock might wonder how they had made their way uphill. The big boulders were granite, and smaller ones among them—recognized by no one then—were jade: float boulders of gem jade, nephrite jade (green as emerald), rounded in streambeds and polished by weather. As she watched them in the moonlight from the stage, they must have seemed just rubble on the ground. There was uranium in Crooks Gap in great quantity—in pods and lenses for a thousand feet up either side. It would be discovered in 1955. There was petroleum under Crooks Gap, too. The year of discovery would be 1925. Crooks Creek flowed through Crooks Gap—straight through the highlands, from one side to the other. Above the gap was Crooks Mountain. Miss Waxham might well have wondered who the eponymous crook was. The possibilities in that country were bewilderingly numerous, but the honor belonged to Brigadier General George Crook, West Point '52, known among the Indians as the Gray Fox. General Crook, commander of the Department of the Platte, was at least a century ahead of his time in the integrity with which he dealt with aboriginal people, and deserved having his name writ in land if for no other reason than his reply when someone asked him if the campaigns of the Indian wars were difficult work. He said, "Yes, they are hard. But the hardest thing is to go out and fight against those who you know are in the right."

I watched for hours the shadow of the suitcase handle against the canvas to see the moon's change of position. The hours dragged by, and the cold grew worse. . . . Between three and four we reached Myersville.

They had come to the Sweetwater River, which they forded, with still another driver, who had a remarkably delicate cast of tongue. "Oh, good gracious!" he shouted at the team.

The driver had been on the road only once, did not know his horses, and had no whip. The Hog Back was ahead. . . . There was no more sleep for us then, not an eye wink.

The Hog Back was a knife-edged spur plunging off the Beaver Divide, which separates waters that flow east into the Platte from waters that flow north into the Wind, Bighorn, and Yellowstone rivers. The Hog Back was Frontier sandstone and Mowry shale, which had accumulated flat in the Cretaceous sea, and here, in subsequent time, had been bent upward sharply to make the jagged edge the travellers descended. Its shales were slick with bentonitic gumbo.

At the top of Beaver Mountain we saw the Wind River Range stretching white in the distance. The driver rough-locked the back wheels and we started down. It was a scramble for the horses to keep out of the way. There were sudden turns in the road and furrows cut by the freight wagons that almost threw the careening stage on its side. One of the horses fell, but was dragged along by the others until it finally regained its feet. We finally reached a place where the slope was less steep, the rough lock was taken off, and the driver began again to try to make time down the hills. The little leaders ran like rats and the heavier wheelers were carried along while the coach swung from side to side in the gullies.

Twenty-six hours out of Rawlins, the stage reached Hailey. Breakfast was waiting, and in Miss Waxham's opinion could have gone on waiting—"the same monstrous biscuits and black coffee." A rancher named Gardiner Mills arrived—"short, dark, of caustic speech"—and handed her a big fur overcoat to top her own and keep her warm in his springy buckboard. He had come to take the new schoolmarm the remaining ten miles to his Red Bluff Ranch, and into the afternoon they travelled northwest under six-hundred-foot walls of rose, vermilion, brick, and carmine—red Triassic rock. Near a big spring under the red bluff were the low buildings of the ranch.

The corral and bunkhouse, grain and milk house were log structures off to one side. When we drove up to the gate, and two little narrow-chested large-pompadoured girls came out the walk to meet us, all my fears as to obstreperous pupils were at an end.

The "chiffonier" in her room was a stack of boxes covered with muslin curtains. There was "a washstand for private individual use." There was a mirror a foot square. On her walls were Sargent and Gainsborough prints, and pictures of Ethel Barrymore and Psyche.

In the western outskirts of Rawlins, David Love pulled over onto the shoulder of the interstate, the better to fix the scene, although his purpose in doing so was not at all apparent. Rawlins reposed among low hills and prairie flats, and nothing in its setting would ever lift the stock of Eastman Kodak. In those western outskirts, we may have been scarcely a mile from the county courthouse, but we were very much back on the range—a dispassionate world of bare rock, brown grass, drab green patches of greasewood, and scattered colonies of sage. The interstate had lithified in 1965 as white concrete but was now dark with the remains of ocean algae, cremated and sprayed on the road. To the south were badlands—gullies and gulches, erosional debris. To the north were some ridgelines that ended sharply, like breaking waves, but the Rawlins Uplift had miserably fallen short in its bid to be counted among the Rocky

Mountains. So why was David Love, who had the geologic map of Wyoming in his head, stopping here?

The rock that outcropped around Rawlins, he said, contained a greater spread of time than any other suite of exposed rocks along Interstate 80 between New York and San Francisco. We were looking at many moments in well over half the existence of the earth, and we were seeing—as it happened—a good deal more time than one sees in the walls of the Grand Canyon, where the clock stops at the rimrock, aged two hundred and fifty million years. The rock before us here at Rawlins reached back into the Archean Eon and up to the Miocene epoch. Any spendthrift with a camera could aim it into that scene and—in a two-hundred-and-fiftieth of a second at f/16—capture twenty-six hundred million years. The most arresting thing in the picture, however, would be Rawlins' municipal standpipe—that white, squat water-storage tank over there on the hill.

The hill, though, was Archean granite and Cambrian sandstone and Mississippian limestone. If you could have taken pictures when *they* were forming, the collection would be something to see. There would be a deep and uncontinented ocean sluggish with amorphous scums (above cooling invisible magmas). There would be a risen continent reaching its coast, with rivers running over bare rock past not so much as a lichen. There would be rich-red soil on a broad lowland plain resembling Alabama (but near the equator). There would be clear, warm shelf seas.

There would also be a picture of dry hot dunes, all of them facing the morning sun—the rising Miocene sun. Other—and much older—dunes would settle a great question, for it is impossible to tell now whether they were just under or just out of water. They covered all of Wyoming and a great deal more, and may have been very much like the Libyan Desert: the Tensleep-Casper-Fountain Pennsylvanian sands. There would be a picture, too, of a meandering stream, with overbank deposits, natural levees, cycads growing by the stream. Footprints the size of washtubs. A head above the trees. In the background, swamp tussocks by the shore of an oxbow lake. What was left of that picture was the Morrison formation—the Jurassic landscape of particularly dramatic dinosaurs—outcropping just up the road. There would be various views of the great Cretaceous seaway, with its plesiosaurs, its giant turtles, its crocodiles. There would be a picture from the Paleocene of a humid subtropical swamp, and a picture from the Eocene of gravel bars in a fast river running off a mountain onto lush subtropical plains, where puppy-size horses were hiding for their lives.

Such pictures, made in this place, could form a tall stack—scene after scene, no two of them alike. Taken together, of course—set one above another, in order—they would be the rock column for this part of Wyoming. They would correlate with what one would see in the well log of a deep-drilling rig. There would be hiatuses, to be sure. In the rock column, anywhere, more time is missing than is there; so much has been eroded away. Besides, the rock in the column is more apt to commemorate a moment—an eruption, a flood, a fallen drop of rain—than it is to report a millennium. Like a news broadcast, it is more often a montage of disasters than a cumulative record of time.

I asked Love why so much of the earth's history happened to be here on the surface in this nondescript part of the state.

He said, "It just came up in the soup. Why it is out here all by itself is a matter of fierce debate." The Rawlins Uplift had not accomplished nothing.

The Precambrian granite on the ridge was from the late Archean Eon and was 2.6 billion years old. It dipped below us. Close to the interstate, the Union Pacific had been blasted through some sandstone that rested on, and was derived from, the granite—littoral sands of Cambrian time, when the American west coast was at Rawlins. Between this Flathead sand and the Madison limestone above it, lying here and there in pockets in an unconformity of a hundred and seventy million years, was the rich-red soil of the Paleozoic plain. A streak of it showed in a low hillside even closer to us than the railroad cut, so we walked over to collect some and put it in a bag. As rock it was so incompetent that it could easily be crushed to powder—a beautiful rose-brick powder with the texture of cocoa. It had been known in the paint business as Rawlins Red, and in the warpaint business as effective medicine, this paleosol (fossil soil) three hundred and fifty million years old. As we returned to the road, a couple of Consolidated Freightways three-unit twenty-six-wheel tractor-trailers went by, imitating thunder. Love said, "First we had the Conestoga, then the big freight wagons with twelve to sixteen oxen. Now we have those things."

The spread of time at Rawlins, like the rock column in a great many places in Wyoming, was so impressively detailed that it seemed to suggest that Wyoming, in its one-thirty-seventh of the United States, contains a disproportionate percentage of American geology. Geologists tend to have been strongly influenced by the rocks among which they grew up. The branch of the science called structural geology, for example, has traditionally been dominated by Swiss, who spend their youth hiking and schussing in a national textbook of structure. When a multinational oil company held a conference in Houston that brought together structural geologists from posts all over the world, the coffee breaks were in Schweizerdeutsch. The wizards of sedimentology tend to be Dutch, as one would expect of a people who have figured out a way to borrow against unrecorded deposits. Cincinnati has produced an amazingly long list of American paleontologists—Cincinnati, with profuse exceptional fossils in its Ordovician hills. Houston—the capital city of the oil geologist—is a hundred and fifty miles from the first place where you can hit a hammer on a rock. Houston geologists come from somewhere else.

Geologists who have grown up on shield rock—Precambrian craton—tend to be interested in copper, diamonds, iron, and gold. Most of the world's large metal deposits are Precambrian. Diamonds, after starting upward from the mantle, seem to need the thickness of a craton to survive their journey to the surface world.

Geologists who grow up in California start out with strange complex structures, highly deformed rock—mélanges and turbidites that seem less in need of a G. K. Gilbert than of an Alfred Adler or a Carl Jung. Shell, in its rosters, used to put an asterisk beside the names of geologists from California. The asterisk meant that while they were in, say, Texas they might be quite useful among the Gulf Coast turbidites of the Hackberry Embayment, but assign them with caution almost anywhere else. A former Shell geologist (not David Love) once said to me, "The asterisk also meant 'Ship them back to California when they're done.' Shell considered them a separate race."

A geologist who grew up in Wyoming would have something of everything above—with the probable exception of the asterisk. A geologist who grew up in Wyoming could not ignore economic geology, could not ignore vertebrate paleontology, could not ignore the narrative details in any chapter of time (every period in the history of the world was represented in Wyoming). Wyoming geology would above all tend to produce a generalist, with an eye that had seen a lot of rocks, and a four-dimensional gift for fitting them together and arriving at the substance of their story—a scenarist and lithographer of what geologists like to call the Big Picture.

Wyoming, at first glance, would appear to be an arbitrary segment of the country. Wyoming and Colorado are the only states whose borders consist of four straight lines. That could be looked upon as an affront to nature, an utterly political conception, an ignoring of the outlines of physiographic worlds, in disregard of rivers and divides. Rivers and divides, however, are in some ways unworthy as boundaries, which are meant to imply a durability that is belied by the function of rivers and divides. They move, they change, and they go away. Rivers, almost by definition, are young. The oldest river in the United States is called the New River. It has existed (in North Carolina, Virginia, and West Virginia) for a little more than one and a half per cent of the history of the world. In epochs and eras before there ever was a Colorado River, the formations of the Grand Canyon were crossed and crisscrossed, scoured and dissolved, deposited and moved by innumerable rivers. The Colorado River, which has only recently appeared on earth, has excavated the Grand Canyon in very little time. From its beginning, human beings could have watched the Grand Canyon being made. The Green River has cut down through the Uinta Mountains in the last few million years, the Wind River through the Owl Creek Mountains, the Laramie River through the Laramie Range. The mountains themselves came up and moved. Several thousand feet of basin fill has recently disappeared. As the rock around Rawlins amply shows, the face of the country has frequently changed. Wyoming suggests with emphasis the page-one principle of reading in rock the record of the earth: Surface appearances are only that; topography grows, shrinks, compresses, spreads, disintegrates, and disappears; every scene is temporary, and is composed of fragments from other scenes. Four straight lines—like a plug cut in the side of a watermelon—should do as well as any to frame Wyoming and its former worlds.

A geologist who grew up in Wyoming has grown up among

mountains that in terms of plate-tectonic theory are the least explainable in the world. A geologist who grew up in Wyoming—with its volcanic activity, its mountains eroding, and its basins receiving sediment—would inherently comprehend the cycles of the earth: geology repeating itself as people watch. G. K. Gilbert, the first Chief Geologist of the United States Geological Survey, once remarked that it is "the natural and legitimate ambition of a properly constituted geologist to see a glacier, witness an eruption and feel an earthquake." A geologist could do all that as a child in Wyoming, and not have to look far for more.

Miss Waxham's school was a log cabin on Twin Creek near the mouth of Skull Gulch, a mile from the Mills ranch. Students came from much greater distances, even through deep snow. Many mornings, ink was frozen in the inkwells, and the day began with ink-thawing, followed by reading, spelling, chemistry, and civil government. Sometimes snow blew through the walls, forming drifts in the schoolroom. Water was carried from the creek—drawn from a hole that was chopped in the ice. If the creek was frozen to the bottom, the students melted snow. Their school was fourteen by sixteen feet—smaller than a bathroom at Wellesley. The door was perforated with bullet holes from "some passerby's six-shooter." Over the ceiling poles were old gunnysacks and overalls, to prevent the sod roof from shedding sediment on the students. Often, however, the air sparkled with descending dust, struck by sunlight coming in through the windows, which were all in the south wall. There was a table and chair for Miss Waxham, and eight desks for her pupils. Miss Waxham's job was to deliver a hundred per cent of the formal education available in District Eleven, Fremont County, Wyoming.

The first fifteen minutes or half hour are given to reading "Uncle Tom's Cabin" or "Kidnapped," while we all sit about the stove to keep warm. Usually in the middle of a reading the sound of a horse galloping down the frozen road distracts the attention of the boys, until a few moments later six-foot George opens the door, a sack of oats in one hand, his lunch tied up in a dish rag in the other. Cold from his five-mile ride, he sits down on the floor by the stove, unbuckles his spurs, pulls off his leather chaps, drops his hat, unwinds two or three red handkerchiefs from about his neck and ears, takes off one or two coats, according to the temperature, unbuttons his vest and straightens his leather cuffs. At last he is ready for business.

Sandford is the largest scholar, six feet, big, slow in the school room, careful of every move of his big hands and feet. His voice is subdued and full of awe as he calls me "ma'am." Outside while we play chickens he is another person—there is room for his bigness. Next largest of the boys is Otto Schlicting, thin and dark, a strange combination of shrewdness and stupidity. His problems always prove, whether they are right or not! He is a boaster, too, tries to make a big impression. But there is something very attractive about him. I was showing his little sister how to add and subtract by making little lines and adding or crossing off others. Later I found on the back of Otto's papers hundreds and hundreds of little lines—trying to add that way as far as a hundred evidently. He is nearly fifteen and studying division. . . . Arithmetic is the family failing. "How many eights in ninety-six?" I ask him. He thinks for a long time. Finally he says—with such a winsome smile that I wish with all my heart it were true—"Two." "What feeds the cells in your body?" I ask him. He thinks. He says, "I guess it's vinegar." He has no idea of form. His maps of North America on the board are all like turnips.

Students' ages ranged through one and two digits, and their intelligence even more widely. When Miss Waxham called upon Emmons Schlicting, asking, "Where does digestion take place?," Emmons answered, "In the Erie Canal." She developed a special interest in George Ehler, whose life at home was troubled.

He is only thirteen, but taller than Sandford, and fair and handsome. I should like to get him away from his family—kidnap him. To think that it was he who tried to kill his father! His face is good as can be.

At lunchtime, over beans, everyone traded the news of the country, news of whatever might have stirred in seven thousand square miles: a buffalo wolf trapped by Old Hanley; missing horses and cattle, brand by brand; the sheepherder most recently lost in a storm. If you went up Skull Gulch, behind the school, and climbed to the high ground beyond, you could see seventy, eighty, a hundred miles. You "could see the faint outlines of Crowheart Butte, against the Wind River Range." There was a Wyoming-history lesson in the naming of Crowheart Butte, which rises a thousand feet above the surrounding landscape and is capped with flat sandstone. To this day, there are tepee rings on Crowheart Butte. One of the more arresting sights in remote parts of the West are rings of stones that once resisted the wind and now recall what blew away. The Crows liked the hunting country in the area of the butte, and so did the Shoshonis. The two tribes fought, and lost a lot of blood, over this ground. Eventually, the chief of the Shoshonis said, in effect, to the chief of the Crows: this is pointless; I will fight you, one against one; the hunting ground goes to the winner. The chief of the Shoshonis was the great Washakie, whose name rests in six places on the map of Wyoming, including a mountain range and a county. Washakie was at least fifty, but fit. The Crow would have been wise to demur. Washakie destroyed him in the hand-to-hand combat, then cut out his heart and ate it.

Despite her relative disadvantages as a newcomer, an outlander, and an educational ingénue, Miss Waxham was a quick study. Insight was her long suit, and in no time she understood Wyoming. For example, an entry in her journal says of George Ehler's father, "He came to the country with one mare. The first summer, she had six colts! She must have had calves, too, by the way the Ehlers' cattle increased." These remarks were dated October 22, 1905—the day after her stagecoach arrived. In months that followed, she sketched her neighbors (the word applied over many tens of miles). "By the door was Mrs. Frink, about 18, with Frink junior, a large husky baby. Ida Franklin, Mrs. Frink's sister and almost her double, was beside her, frivolous even in her silence." There was the story of Dirty Bill Collins, who had died as a result of taking a bath. And she fondly recorded Mrs. Mills' description of the libertine Guy Signor: "He has a cabbage heart with a leaf for every girl." She noted that the nearest barber had learned his trade shearing sheep, and a blacksmith doubled as dentist. Old Pelon, a French Canadian, impressed her, because he had refused to ask for money from the government after Indians killed his brother. "Him better dead," said Old Pelon. Old Pelon was fond of the masculine objective pronoun. Miss Waxham wrote, "Pelon used to have a wife, whom he spoke of always as 'him.' " Miss Waxham herself became a character in this tableau. People sometimes called her the White-Haired Kid.

"There's many a person I should be glad to meet," read an early entry in her journal. She wanted to meet Indian Dick, who had been raised by Indians and had no idea who he was—probably the orphan of emigrants the Indians killed. She wanted to meet "the woman called Sour Dough; Three Fingered Bill, or Suffering Jim; Sam Omera, Reub Roe. . . ." (Reub Roe held up wagons and stagecoaches looking for members of the Royal Family.) Meanwhile, there was one flockmaster and itinerant cowboy who seemed more than pleased to meet her.

In the first reference to him in her journal she calls him "Mr. Love—Johnny Love." His place was sixty miles away, and he had a good many sheep and cattle to look after, but somehow he managed to be right there when the new young schoolmarm arrived. In the days, weeks, and months that followed, he showed a pronounced tendency to reappear. He came, generally, in the dead of night, unexpected. Quietly he slipped into the corral, fed and watered his horse, slept in the bunkhouse, and was there at the table for breakfast in the morning—this dark-haired, blue-eyed, handsome man with a woolly Midlothian accent.

> Mr. Love is a Scotchman about thirty-five years old. At first sight he made me think of a hired man, as he lounged stiffly on the couch, in overalls, his feet covered with enormous red and black striped stockings that reached to his knees, and were edged with blue around the top. He seemed to wear them instead of house shoes. His face was kindly, with shrewd blue twinkling eyes. A moustache grew over his mouth, like willows bending over a brook. But his voice was most peculiar and characteristic. . . . A little Scotch dialect, a little slow drawl, a little nasal quality, a bit of falsetto once in a while, and a tone as if he were speaking out of doors. There is a kind of twinkle in his voice as well as his eyes, and he is full of quaint turns of speech, and unusual expressions.

Mr. Love travelled eleven hours on these journeys, each way. He did not suffer from the tedium, in part because he frequently rode in a little buggy and, after telling his horses his destination, would lie on the seat and sleep. He may have been from Edinburgh, but he had adapted to the range as much as anyone from anywhere. He had slept out, in one stretch, under no shelter for seven years. On horseback, he was fit for his best horses: he had stamina for long distances at sustained high speed. When he used a gun, he hit what he was shooting at. In 1897, he had begun homesteading on Muskrat Creek, quite near the geographical center of Wyoming, and he had since proved up. One way and another, he had acquired a number of thousands of acres, but acreage was not what mattered most in a country of dry and open range. Water rights mattered most, and the area over which John Love controlled the water amounted to a thousand square miles—about one per cent of Wyoming. He had come into the country walking, in 1891, and now, in 1905, he had many horses, a couple of hundred cattle, and several thousand sheep. Miss Waxham, in her journal, called him a "muttonaire."

He was a mirthful Scot—in abiding contrast to the more prevalent kind. He was a wicked mimic, a connoisseur of the absurd. If he seemed to know everyone in the high country, he knew even better the conditions it imposed. After one of her conversations with him, Miss Waxham wrote in her journal:

> It is a cruel country as well as beautiful. Men seem here only on sufferance. After every severe storm we hear of people's being lost. Yesterday it was a sheep camp mover who was lost in the Red Desert. People had hunted for him for a week, and found no trace. Mr. Love—Johnny Love—told of a man who had just been lost up in his country, around the Muskrat. "Stranger?" asked Mr. Mills. "No; born and brought up here." "Old man?" "No; in the prime of life. Left Lost Cabin sober, too."

Mr. Love had been born near Portage, Wisconsin, on the farm of his uncle the environmentalist John Muir. The baby's mother died that day. His father, a Scottish physician who was also a professional photographer and lecturer on world travel, ended his travels and took his family home. The infant had three older sisters to look after him in Scotland. The doctor died when John was twelve. The sisters emigrated to Broken Bow, Nebraska, where in the eighteen-seventies and eighties they all proved up on homesteads. When John was in his middle teens, he joined them there, in time to experience the Blizzard of '88—a full week of blowing snow, with visibility so short that guide ropes led from house to barn.

He was expelled from the University of Nebraska for erecting a sign in a dean's flower bed, so he went to work as a cowboy, and soon began to think about moving farther west. When he had saved enough money, he bought matching black horses and a buggy, and set out for Wyoming. On his first night there, scarcely over the border, his horses drank from a poison spring and died. What he did next is probably the most encapsulating moment in his story. In Nebraska were three homes he could return to. He left the buggy beside the dead horses, abandoned almost every possession he had in the world, and walked on into Wyoming. He walked about two hundred miles. At Split Rock, on the Oregon Trail—near Crooks Gap, near Independence Rock—he signed on as a cowboy with the 71 Ranch. The year was 1891, and the State of Wyoming was ten months old.

Through the eighteen-nineties, there are various hiatuses in the résumé of John Love, but as cowboy and homesteader he very evidently prospered, and he also formed durable friendships—with Chief Washakie, for example, and with the stagecoach driver Peggy Dougherty, and with Robert LeRoy Parker and Harry Longabaugh (Butch Cassidy and the Sundance Kid). There came a day when Love could not contain his developed curiosity in the presence of the aging chief. He asked him what truth there was in the story of Crowheart Butte. Had Washakie really eaten his enemy's heart? The chief said, "Well, Johnny, when you're young and full of life you do strange things."

Robert LeRoy Parker was an occasional visitor at Love's homestead on Muskrat Creek, which was halfway between Hole-in-the-Wall and the Sweetwater River—that is, between Parker's hideout and his woman. Love's descendants sometimes stare bemusedly at a photograph discovered a few years ago in a cabin in Jackson Hole that had belonged to a member of the Wild Bunch. The photograph, made in the middle eighteen-nineties, shows eighteen men with Parker, who is wearing a dark business suit, a tie and a starchy white collar, a bowler hat. Two of the bunch are identified only by question marks. One of these is a jaunty man of middle height and strong frame, his hat at a rakish angle—a man with a kindly face, twinkling shrewd eyes, and a mustache growing over his mouth like willows bending over a brook. It may be doubtful whether John Love would have joined such a group, but when you are young and full of life you do strange things.

At Red Bluff Ranch, Mrs. Mills once twitted Mr. Love for being Scottish when other Scots were around and American in the presence of Americans. For a split second, Mr. Love thought this over before he said, "That leaves me eligible for the Presidency." Out of Mr. Love's buggy came a constant supply of delicacies and exotic gifts—including candy, nuts, apples—which he came by who knows where and liberally distributed to all. Miss Waxham began to look upon him as "a veritable Santa Claus"; and, predictably, at Christmastime Santa appeared.

> And the next day was Christmas. . . . Just before supper the joyful cry went up that Mr. Love was coming, and actually in time for dinner. He had broken his record and arrived by day!

A pitch pine had been set up indoors and its boughs painted with dissolved alum to simulate frost. Hanging from the branches were wooden balls covered with tobacco tinfoil. Flakes of mica were glued to paper stars. On Christmas, Mr. Mills and Mr. Love dressed in linen collars and what Miss Waxham called "fried shirts." When Miss Waxham turned to a package from home that she knew contained pajamas, she went into her bedroom to open it.

The following day, Miss Waxham was meant to go to something called Institute, in Lander—a convocation of Fremont County schoolteachers for lectures, instruction, and professional review. By phenomenal coincidence, Mr. Love announced that he had business in Lander, too.

> It was decided that I should go with him. I rather dreaded it. . . . I confess I was somewhat afraid of him. . . . I was wrapped up in a coat of my own with Mrs. Mills' sealskin over it, muffler, fur hat, fur gloves, leggings, and overshoes. Then truly I was so bundled up that it was next to impossible to move. "Absolutely helpless," laughed Mr. Love.

Whatever business Mr. Love had in Lander did not in any way seem to press him. Miss Waxham stayed with Miss Davis, the county superintendent, and while other people came and went from the premises Mr. Love was inclined to remain.

Supper time came and Mr. Love remained. We had a miserable canned goods cold supper. Miss MacBride left, Mr. Love remained.

In the afternoon, Mr. Love called. It certainly was a surprise. I explained why Miss Davis was out, but he didn't seem to mind. I said that she would be back soon. He asked if I should not like to take a drive and see the suburbs. Of course I would. . . . We went for a long drive in the reservation, with a box of chocolates between us, and a merry gossip we had. . . . He was bemoaning the fact that there is no place for a man to spend the evening in Lander except in a saloon. "Come and toast marshmallows," I said, and he took it as a good suggestion.

When she went to church on Sunday, Love was there—John Santa Love, who had not been to church in ten years. After the service, it was time to leave Lander.

There had been snow falling since morning, and the road was barely visible. The light faded to a soft whiteness that hardly grew darker when the sun set and the pale outline of the moon showed through the snow. Everywhere was the soft enveloping snow shutting out all sounds and sights. The horses knew the way and travelled on steadily. Fortunately it was not cold, and the multitudinous rugs and robes with the new footwarmer beneath kept us warm and comfortable. More pleasant it was travelling through the storm than sitting at home by the fire and watching it outside. When the conversation ran low and we travelled on quietly, Mr. Love discovered bags of candy under the robes . . . and he fed us both, for I was worse than entangled in wraps and the long sleeves of Mrs. Mills' sealskin. The miles fell away behind us easily and quietly.

Even as those words were written, the editor and publishers of the *Shoshone Pathfinder*, in Lander, were completing a special issue urging young people to make their lives in central Wyoming. "We beg leave to extend to each and every one of you a most cordial welcome to come, remain, and help develop a country so rich in natural resources as to be beyond the computation of mortal man," wrote the publishers. It was a country "clothed in a mantle of the most nutritious grasses and sage brush browse." In its Wind River Mountains were "thousands of square miles of dense forests, which the foot of man has never invaded, and . . . as to the supply and quality of timber in this county it will meet the requirements of all demands for all time to come." Moreover, there was coal: "It has been said of our coal fields that the entire United States would be unable to exhaust them in a century. . . . It is in excess of the imagination to contemplate the vastness of this tremendous supply of fuel or what would ever transpire to exhaust it." And there was oil: "It is a recognized fact of long standing that the quantity of oil stored in the natural reservoirs of this county is so great that no estimate can be made." And there was gold. At the south end of the Wind Rivers, nearly five million dollars had come out of small mines with names like Hard Scrabble, Ground Hog, Hidden Hand, Mormon Crevice, Iron Duke, Midget, Rustler, Cariboo, and Irish Jew. "None of the mines have been exhausted, but merely sunk to a depth where more and better machinery is required." There was uranium, too, but as yet no compelling need to find it, and as yet no geologist equal to the task.

As the winter continued, with its apparently inexhaustible resources of biting wind and blinding snow, temperatures now and again approached fifty below zero. Miss Waxham developed such an advanced case of cabin fever that she wrote in her journal, "My spirit has a chair sore." Even when drifts were at their deepest, though, Mr. Love somehow managed to get through. "Much wrapped up" on one occasion, he rode "all the way from Alkali Butte." On another, he spent an entire day advancing his education at the Twin Creek school.

These attentions went on in much the same way for five years. He pursued her to Colorado, and even to Wisconsin. They were married on the twentieth of June, 1910, and drove in a sheep wagon to his ranch, in the Wind River Basin. It was plain country with gently swelling hills. Looking around from almost any one of them, you could see eighty miles to the Wind River Range, thirty to the Owl Creeks, twenty to the Rattlesnake Hills, fifteen to the Beaver

Divide, and a hundred into the Bighorns. No buildings were visible in any direction. In this place, they would flourish. Here, too, they would suffer calamitous loss. Here they would raise three children —a pair of sons close in age, and, a dozen years after them, a daughter. The county from time to time would supply a schoolmarm, but basically the children would be educated by their mother. One would become a petroleum chemist, another a design engineer for the New Jersey Turnpike and the New York State Thruway, another the preeminent geologist of the Rocky Mountains.

Along the Nebraska-Wyoming line, in the region of the forty-first parallel, is a long lumpy break in the plains, called Pine Bluffs. It is rock of about the same age and story as Scotts Bluff, which is not far away. David Love—standing on top of Pine Bluffs—remarked that for a great many emigrants with their wagons and carts these had been the first breaks in the horizon west of Missouri. From the top of the bluffs, the emigrants had their first view of the front ranges of the Rockies, and the mountains gave them hope and courage. For our part, looking west from the same place, we could not see very far across the spring wildflowers into the swirling snow. The Laramie Range was directly ahead and the Never Summer Mountains off somewhere to the southwest—at ten and nearly thirteen thousand feet indeed a stirring sight, but not today. Love said a spring snowstorm was "sort of like a kiss—it's temporary, and it will go away." (That one stopped us for three days.) Meanwhile, there were large roadcuts to examine where Interstate 80 sliced through the bluffs, and scenes to envision that were veiled by more than snow.

The bluffs stand above the surrounding country because—like other mesas and buttes—they are all that is left of what was the surrounding country. The rock of Pine Bluffs is sedimentary (limestone, sandstone) and seems to lie flat, but in fact it tilts very slightly, and if its bedding planes are projected westward a hundred miles they describe the former landscape, rising about sixty feet a mile to touch the summits of mountains. "There was a continuous surface," Love said. "It came over the top of the Laramie Range and out here onto the High Plains. Pine Bluffs was part of that surface."

In earth history, that was not long ago. He said the best general date for it was ten million years—when the central Rocky Mountains, which had long since taken form as we know them, were buried up to their chins. Only the highest peaks remained uncovered, like nunataks protruding through continental ice, or scattered islands in a sea. The summits of the Wind River Range were hills above that Miocene plain. The highest of the Bighorns stuck out, too—as did the crests of other ranges. Forty million years before that, in Eocene time, most parts of the Rocky Mountains seem to have looked much as they do today, and so did the broad basins among them. The region as a whole was closer to sea level, but its relief was essentially the same.

It was a bizarre story, full of odd detail. Limestone, for example, is ordinarily a marine rock, derived from corals and shells. What sort of limestone would form on a surface that came sloping down like a tent roof from the ridgelines of buried mountains? The surface had been laced with streams, Love said, and in the rumpled topography east of the buried ranges the streams filled countless lakes. Old carbonates dissolved and were carried by the streams to the lakes. Lime also leached out of the granite in the mountain cores. Freshwater limestones formed in the lakes, self-certified by the fossils of freshwater snails. There were other fossils as well—discovered in dense compilations in confined areas that have been described as concentration camps and fossil graveyards. They suggest a modern plain in south-central Brazil where heavy seasonal rains so elevate the waters of innumerable lakes that animals crowd up on small islands and perish. Limestone, being soluble in water, forms fertile valleys in eastern North America but in the dry West remains largely undissolved. Its inherent ruggedness holds it high, while weaker rock around it falls away. Limestone is the protective caprock of Pine Bluffs. Junipers were flourishing in it, as were ponderosa pines and Spanish bayonets.

In the Bronco, we moved through the snow toward the mountains, crossing the last of the Great Plains, which had been shaped like ocean swells by eastbound streams. Now and again, a pump jack was visible near the road, sucking up oil from deep Cretaceous sand, bobbing solemnly at its task—a giant grasshopper absorbed in its devotions. As we passed Cheyenne, absolutely all we could see in the whiteout was a raging, wind-whipped flame, two hundred feet in the air, at the top of a refinery tower. "Such a waste," said Love.

Had we been moving west across Wyoming about seventy-five million years ago, in the Campanian age of late Cretaceous time, we would, of course, have been at sea level in the most literal sense. The Laramie Range did not exist, nor did the Bighorns, the Beartooths, the Wind Rivers. There is no evidence of mountains at that time anywhere in Wyoming. In an oceangoing boat (which the Bronco in some ways resembled), we would have raised the coastline not far east of Rawlins. Beyond the beach and at least as far as Utah was flat marshy terrain.

Earlier, there had been mountains—a few ranges that were largely in Colorado and poked some miles into Wyoming. They have been called the Ancestral Rockies; but they and the Rockies are scarcely more related than two families who happen at different times to live in the same house. Those Pennsylvanian mountains had worn down flat two hundred and thirty million years before. There had been other mountains as well—in the same region—some hundreds of millions of years before that, in various periods of Precambrian time. The Precambrian evidence, in fact, suggests numerous episodes, across two thousand million years, of the rise of big mountains and their subsequent wearing away—any of them as deserving as others to be called ancestral Rockies.

In the middle Precambrian, not long after the end of the Archean Eon, lava ran down the sides of big volcanoes and far out onto a seafloor that is now a part of Wyoming. It is impossible to say where the volcanoes stood, but the fact that they existed is stated by the lava. Somewhat later, the lava was folded and faulted, apparently in the making of mountains. Still more Precambrian ranges, of vast dimension, came up in the region, and shed twenty-five thousand feet of sediment into seas that covered parts of Wyoming. After the sediment formed into rock, even more episodes of mountain building heated and changed the rock: the limestone to marble, the sandstone to quartzite, the shale to slate. Meanwhile, coming into the crust at depths on the order of six miles were vast bodies of fiery-hot magma, much of which happened to have the chemistry of granite. Under eastern Wyoming, where Interstate 80 now crosses the Laramie Range, the magma contained enough iron to tint the feldspar and make the granite pink. It is axiomatic that big crystals grow slowly. Slowly, the magma cooled, forming quartz and feldspar crystals of exceptional size.

All of that occurred in Precambrian time—during the first eighty-eight per cent of the history of the world. Often, Precambrian rock is collectively mentioned as "the basement"—the basement of continents—as if that is all there is to say about it before setting up on top of it the wonders of the world. This scientific metaphor is at best ambiguous—connoting, as it does, in one sense a firm foundation, in another an obscure cellar. In either case, it dismisses four billion years. It attempts to compress the uncompressible. It foreshortens a regional history wherein numerous ancestral mountain ranges developed and were annihilated—where a minor string of Pennsylvanian ridges could scarcely be said to represent the incunabular genealogy of the Rockies.

In late Cretaceous and early Tertiary time, mountains began to rise beneath the wide seas and marsh flats of Wyoming. The seawater drained away to the Gulf of Mexico, to the Arctic Ocean. And, in David Love's summary description, "all hell broke loose." In westernmost Wyoming, detached crustal sheets came planing eastward —rode fifty, sixty, and seventy-five miles over younger rock—and piled up like shingles, one overlapping another. In the four hundred miles east of these overthrust mountains, other mountains began to appear, and in a very different way. They came right up out of the earth. In Love's phrase, they simply "pooched out." Basins flexed between them, filling as they downwarped—folding, too, especially at their edges. These mountains moved, but not much—five miles here, eight miles there. They moved in highly miscellaneous and ultimately perplexing directions. The Wind River Range crept southwest, about five inches every ten years for a million years. The Bighorns split. One part went south, the other east. Similarly, the Beartooths went east and southwest. The Medicine Bows moved east. The Washakies west. The Uintas north. All distances were short, because the mountains were essentially rooted. The Sierra

Madre did not move at all. The spines of the ranges trended in as many directions as a weathervane. The Laramie Range trended north-south. The Wind Rivers and Bighorns northwest-southeast. The spectacularly anomalous Uintas, lining themselves up at right angles to the axis of the Western cordillera, ran east-west, and so did the Owl Creeks. All these mountain ranges were coming up out of the craton—heartland of the continent, the Stable Interior Craton. It was as if mountains had appeared in Ohio, inboard of the Appalachian thrust sheets, like a family of hogs waking up beneath a large blanket. An authentic enigma on a grand scale, this was one of the oddest occurrences in the tectonic history of the world. It would probe anybody's theories. It happened rapidly. As David Love at one point remarked about the Medicine Bow Mountains, "It didn't take very long for those mountains to come up, to be deroofed, and to be thrust eastward. Then the motion stopped. That happened in maybe ten million years, and to a geologist that's really fast."

Twenty thousand feet of rock was deroofed from the rising mountains. The entire stratigraphy from the Cretaceous down to the Precambrian was broken to bits and sent off to Natchez, as the mountains were denuded to their crystalline and metamorphic cores. In half a billion years of history, this was the great event. In the words of the *Geologic Atlas of the Rocky Mountain Region*, it was "tectonically unique in the Western Hemisphere and, therefore, it seems to require a somewhat unusual if not unique tectonic interpretation." The foreland ranges, as the mountains east of the overthrust are called (the Wind Rivers, Uintas, Bighorns, Medicine Bows, Laramie Range, and so forth), came into the world with their own odd syncopation, albeit the general chronology went from west to east and the Laramie Range was among the last to rise. "The mountains were restless," Love was saying now. "They didn't all pooch out at once. They moved in fits and starts over a span of time. The Owl Creeks rose in the early Eocene, as did the Uintas. The Medicine Bows, which are farther east, came up before the Uintas. They are all separate mountains with the same general type of origin. They are cohesive in the way that a family is cohesive. They are part of the same event."

The event is known in geology as the Laramide Orogeny. Alternatively it is called the Laramide Revolution.

Mountains always come down, of course, as they are coming up. In the contest between erosion and orogeny, erosion never loses. For a relatively short time, though, the mountains prevail by rising faster than they are destroyed. In what Love has called "some of the greatest localized vertical displacement known anywhere in the world," the Wind Rivers rose sixty thousand feet with respect to the rock around them, the Uintas fifty thousand, others as much. Frequent rains and many streams helped melt them away. West of Wyoming, in the Eocene, there were no Coast Ranges, no Sierra Nevada. Warm winds off the Pacific brought rains to the Rockies, and a climate similar to the present climate of Florida. In the early Eocene, when the ranges in general looked much as they do today, the mountain building ceased. In the tectonic quiet, erosion of course continued, and the broad downwarps among the ranges continued to fill.

Then came a footnote to the revolution. "In latest early Eocene, fifty-two million years ago, all hell broke loose again," Love said. From thousands of fissures in northwest Wyoming, lava poured forth by the cubic mile. Torn apart by weather and rearranged by streams, it has since been etched out as the Absaroka Range. "After that, everything went blah," he went on. "In the Oligocene, the tectonic activity was totally dead, and it stayed dead at least until the early Miocene. Thirty million years. Then, in the late Miocene, all hell broke loose again. And all hell has been breaking loose time and again for the last ten million years. This is not a static science."

During those thirty million years after things went blah, the Rockies were quietly buried ever deeper in their own debris—and, not so peacefully, in materials oozing overland or falling from the sky. Much came in on the wind from remote explosive volcanoes—stratovolcanoes of huge size in Idaho, Oregon, Nevada. "And maybe Arizona and California, for all we know," Love said. "Clinical details are still inadequate. By the end of the Eocene, the Washakie and Owl Creek Mountains were so deeply buried that the Wind River and Bighorn Basins had coalesced above them. At the end of the Oligocene, only a thousand to four thousand feet of the highest mountains protruded above the aggradational plain. Streams were slow and sluggish and so choked with ash they were unable to erode."

Rhinoceroses lived through those changes, and ancestral deer and antelope, and little horses with three toes. As altitude and aridity increased, a subtropical world of figs, magnolias, and breadfruit cooled into forests of maple, oak, and beech. Altitude alone could not account for the increasing coolness. It foreshadowed the coming ice.

The burial of the mountains continued far into the Miocene, with—as Love described it—"surprising thicknesses of sandstone and tuffaceous debris." Volcanic sands, from Yellowstone and from elsewhere to the west, were spread by the wind, and in places formed giant dunes. Two thousand feet of sand accumulated in central Wyoming. Nineteen thousand—the thickest Miocene deposit in America—went into the sinking Jackson Hole. From the Wind River Mountains southward to Colorado and eastward to Nebraska, the plain was unbroken except for the tops of the highest peaks. Rivers were several thousand feet higher than they are now. The ranges, buried almost to their summits, were separated by hundreds of miles of essentially flat terrain. Mountains that were completely covered —lost to view somewhere below the water-laid sediments and deep volcanic sand—outnumbered the mountains that barely showed through. At its maximum, the broad planar surface occupied nearly all of Wyoming—upward of ninety per cent—and on it meandered slow streams, making huge bends and oxbows. As events were about to prove, the deposition would rise no higher. This—in the late Miocene—was the level of maximum fill.

For something began to elevate the region—the whole terrane, the complete interred family of underthrust, upthrust, overthrust mountains—to lift them swiftly about a mile. "The uplift was not absolutely uniform everywhere," Love said. "But nothing ever is." What produced this so-called epeirogeny is a subject of vigorous and sometimes virulent argument, but the result, continuing to this day, is as indisputable as it has been dramatic. It is known in geology as the Exhumation of the Rockies.

From around and over the Wyoming ranges alone, about fifty thousand cubic miles have been dug out and taken away, not to mention comparable excavations in the neighboring cordillera. Though the process has been going on for ten million years, it is believed to have been particularly energetic in the past million and a half, in part because of the amount of rain that fell on the peripheries of continental ice. In response to the uplift, the easygoing streams that had aimlessly wandered the Miocene plain began to straighten, rush, and cut, moving their boulders and gravels in the way that chain saws move their teeth. The streams lay in patterns that had no relationship to the Eocene topography buried far below. Some of them, rushing along through what is now the Wyoming sky, happened to cross the crests of buried ranges. After they worked their way down to the ranges, they sawed through them. Some effects were even odder than that. If a river happened to be lying above a spur of a buried range, it would cut down through the spur, and seem, eventually—without logic, with considerable magic—to flow into a mountain range, change its mind, and come back out another way. "Eventually," of course, is now. The North Platte River now flows into the rocks of the Medicine Bow Mountains, comes out again to cross the Hanna Basin, and then runs through the Seminoe Mountains and the Granite Mountains. It is joined by the Sweetwater River on the crest of the Granite Mountains. Irrespective of modern topography, the pattern of the rivers is Miocene. On the Laramie Plains, the Laramie River behaves for a while in a deceptively conventional manner. It establishes itself as the centerpiece of the basin, pretending to be the original architect of the circumvallate scene, but then takes a sharp right and, like a bull with lowered head, charges the Laramie Range. The canyon it has made is deep and wild. Water roars through it. When, in the exhumation, the river got down to the mountains, it packed the abrasive power to cut them in half.

In fact, there is no obvious relationship between most of the major rivers in Wyoming and the landscapes they traverse. While rivers elsewhere, running in their dendritic patterns like the veins in a leaf, shape in harmony the landscapes they dominate, almost all the rivers of the Rockies seem to argue with nature as well as with common sense. At Devil's Gate, on the Oregon Trail, the Sweetwater River flows into a hill of granite and out the other side. The Wind River addresses itself to the Owl Creek Mountains and flows right at them. It, too, breaks through and comes out the other side. It, too, flowed across the totally buried mountains in the Miocene, and descended upon them during the exhumation. The anomaly is so startling that early explorers, and even aborigines, did not put one and one together. To the waters on the south side of the mountains they gave the name Wind River. The waters on the north side they called the Bighorn. Eventually, they discovered Wind River Canyon.

On the east flank of the Laramie Range is a piece of ground that somehow escaped exhumation. Actually contiguous with Miocene remains that extend far into Nebraska, it is the only place between Mexico and Canada where the surface that covered the mountains still reaches up to a summit. To the north and south of it, excavation has been deep and wide, and the mountain front is of formidable demeanor. Yet this one piece of the Great Plains—extremely narrow but still intact—extends like a finger and, as ever, touches the mountain core: the pink deroofed Precambrian granite, the top of the range. At this place, as nowhere else, you can step off the Great Plains directly onto a Rocky Mountain summit. It is known to geologists as the gangplank.

Now the Bronco began to rise through the snow, and Love remarked that we were on the gangplank. The land fell away on either side, and in the low visibility we seemed indeed to be on a plank going up into the sky. As we continued to climb, the strip of earth became narrower and narrower. We pulled over onto the shoulder, shut off the motor, and squinted. We appeared to be on a bridge—built of disassembled Rockies and travelled ash—crossing a great excavation through flapping veils of snow. "There are twelve inches of precipitation per annum here, and it's mostly snow," he said. "The mean temperature is thirty-eight degrees. The growing season is less than ninety days. Conditions are about the same in this part of Wyoming as at the Arctic Circle." With that, we gave up the geology and crawled off to his home in Laramie, defeated by the snow. We went back to the gangplank in clear weather.

It was half a mile long. To the north and south, the land fell away along the mountain front in profound excavation of the sediments that once had been there. The excavation had exposed the broken, upturned ends of Pennsylvanian sandstones, dipping steeply eastward and leaning on the mountains. They rested there like lumber stood against a barn. These red sandstones lean against the Laramie Range on both sides. By themselves, they tell the story of the Laramide Orogeny, for they are a part of what was deroofed. They are a part of the Paleozoic package that once rested flat on the deep

Precambrian granite. They are thought by some to have been Pennsylvanian beach sands. Whatever they may have been, they were indubitably horizontal, and for roughly two hundred and fifty million years remained horizontal while layer after layer of sediment accumulated above them, finally including the floors of the Cretaceous seas. Then all hell broke loose, and the granite rose beneath them. The granite core came up like a basement elevator that rises through a city sidewalk, pushing to either side a pair of hinged doors. That was the chronology of numerous ranges—the old hard stuff from far below breaking upward through roofrock and ultimately standing highest, while the ends of the roofrock lean on the flanks in gradations of age that are younger with distance from the core. The broken ends of that Pennsylvanian sand—the outcropping edges of the tilted strata—had weathered out as a rough, serrate ridge along the border of the range. Rocky Mountain ranges are typically flanked by such hogbacks. Boulder, Colorado, is backdropped with hogbacks (the Flatirons), which are more of the same Pennsylvanian strata leaning against the Front Range. Now, on the gangplank, Love said parenthetically, "You are seeing Paleozoic rocks for the first time since the Mississippi River. They go all the way through—under Iowa and Nebraska—but they're buried."

In the fall of 1865, Major General Grenville Dodge and his pack trains and cavalry and other troops were coming south along the St. Vrain Trail, under the front of the Laramie Range. The Powder River campaign, behind them, had been, if not a military defeat, a signal failure in its purpose: to cow the North Cheyennes and the Ogallala Sioux. General Dodge, though, was preoccupied with something else. President Lincoln, not long before he died, had instructed Dodge to choose a route for the Union Pacific Railroad. Dodge, like others before him, had sought the counsel of Jim Bridger, the much celebrated trapper, explorer, fur trader, commercial entrepreneur, and all-around mountain man. Bridger, who was sixty by then, had preceded almost everybody else into the West by two or three decades and knew the country as few other whites ever would. It was he who discovered the Great Salt Lake, reporting his find as the Pacific Ocean. It was he whose descriptions of Jackson Hole, Yellowstone Lake, Yellowstone Falls, the Fire Hole geysers, and the Madison River had once been known as "Jim Bridger's lies." His father-in-law was Chief Washakie. And now this bluecoat general wanted to know where to put a railroad. The Oregon Trail went around the north end of the Laramie Range and up the Sweetwater to South Pass—to say the least, an easy grade. But for a competitive transcontinental railroad the Sweetwater was a route of wide digression and no coal. Bridger mentioned Lodgepole Creek and said the high ground above it was the low point on the crest of the Laramie Range (a fact that theodolites would in time confirm). The route could go there.

So Dodge, in 1865, coming south from the Powder River, left his pack trains and cavalry on the St. Vrain Trail and led a small patrol up Lodgepole Creek. At the top, he turned south and did reconnaissance of the summit terrain. In the small valley of a high tributary of Crow Creek—five or ten miles south of Bridger's recommendation—he surprised a band of Indians. His report does not say of what tribe. They were hostiles—or at least became so after Dodge started firing at them. At the moment of mutual surprise, they were between him and his main column, and that made him tactically nervous. The patrol dismounted and walked due east —"holding the Indians at bay, when they came too near, with our Winchesters." In this manner, the gangplank was discovered. As Dodge kept going east, expecting to reach the escarpment from which he would signal with smoke, he reached no escarpment. Instead, he reached the remnant of the high ancient surface—this interfluvial isthmus between Crow Creek and Lone Tree Creek—touching the mountain summit.

It led down to the plains without a break. I then said to my guide that if we saved our scalps I believed we had found the crossing.

General Dodge went back east, and in the spring of 1867 returned with his route approved. The Union Pacific at that time ended in the middle of Nebraska. He got off the train, went up the North Platte, up the Lodgepole, and, as he approached the mountains, went directly overland to Crow Creek, where he staked out the western end of the railroad's next division. Without much pleasing anybody, he named the place Cheyenne. In no time, he was defending himself against furious Cheyennes. They killed soldiers and laborers, pulled up survey stakes, stole animals, and destroyed equipment. When some politicians, bureaucrats, and financiers arrived on a see-it-yourself junket west, the Cheyennes attacked them. With drawn revolver, General Dodge told his visitors, "We've got to clean these damn Indians out or give up building the Union Pacific Railroad. The government may take its choice."

The narrowest point on the gangplank is wide enough for the Union Pacific and nothing else. The interstate highway clings to one side. The tracks and lanes are so close that the gangplank resembles the neck of a guitar. A long coal freight slid by us. "The coal isn't piled higher than the tops of the gondolas," Love commented. "It's an environmental move—to keep the dust from blowing downwind." He said it was a good idea, no doubt, but he had experienced so many cinder showers earlier in his life that he could not help thinking that this latter-day assault on dust was "like bringing a fire under control at timberline." A cinder shower was what happened when an old-time locomotive pulled into a town and blew its stack. He also said that this could not have been an important emigrant route, because there was a lack of grass and water—absolute necessities for animal-powered travel. To the Union Pacific, however, the gangplank offered speed, efficiency, and hence predominance with respect to the competition. When the Denver & Rio Grande was laboring up switchbacks in a hampering expenditure of money and time—and the Santa Fe was struggling not only with mountains but also with desert terrain—the Union Pacific had already run up the gangplank, opened the West, and become everybody's Uncle Pete. Love said, "Out here, Uncle Sam is a gnat under a blanket compared to Uncle Pete. The Union Pacific had the best of it. This Miocene Ogallala formation was the youngest of the high-plains deposits that lapped onto the mountain front. It's subtle and seems academic until you try to build a railroad. This is the only place in the whole Rocky Mountain front where you can go from the Great Plains to the summit of the mountains without snaking your way up a mountain face or going through a tunnel. This one feature had more to do with the building of the West than any other factor. I don't diminish the importance of the Oregon Trail, but here you had everything going for you. This point hasn't been made before."

When the railroad was built, it was given (by the federal government) fifty per cent of the land in a forty-mile swath along its route—in checkerboard fashion, one square mile in every two. Uncle Pete is so big that he has spun off, among many things, the Rocky Mountain Energy Company, the Upland Industries Corporation, the Champlin Petroleum Company, and enough unmined uranium to send Wyoming to the moon. In Cheyenne, the Union Pacific station and the state capitol face each other at opposite ends of Capitol Avenue. The Union Pacific station came out of the Laramie Range, forty miles west, and, like the range itself, is sheathed in the russet Pennsylvanian sandstone and has a foundation of Precambrian granite. At least as imposing as the capitol, it is a baronially escutcheoned mountain of grandeur.

Indians, of course, had used the gangplank for who knows how long before General Dodge surprised them on the Laramie summit. They had crossed it on their journeys from the Great Plains to the Laramie Basin and on up to hunting grounds in the Medicine Bow Mountains. And the Indians, from the beginning, were themselves following a trail. Buffalo discovered the gangplank. "It was a buffalo trail," Love said. "Buffalo were the real trailmakers—trails you wouldn't believe. They were as good as the best civil engineers. It remains true today. If you're in Yellowstone, in the backcountry, and you have trouble finding your way across swamps, mountains, and thermal areas, you look for a buffalo trail and you'll get through." Beside Interstate 80 on the gangplank, a sign said, "GAME CROSSING."

We moved off the gangplank and into a highway throughcut of pink granite. Love said, "Now we are on the mountain, on the Precambrian core. You have to watch closely. This fantastic geology is subtle. I-80 was not built to show it off but to take advantage of its beneficences." There were more pink granite cuts and also some dark, shattered amphibolite that had been the country rock into which the granite intruded 1.4 billion years ago. The interstate had sliced through a section where the bright-pink granite and the charcoal-gray amphibolite met. It was as if a wall painter had changed colors there. The dark rock was full of fracture planes and cleavage planes. "That rock probably had been messed around for a long time before the granite came," Love said. "It could be two, three billion years old. We don't know."

As our altitude increased, the granite roadcuts became deeper and higher and seemingly more rutilant. The rock was competent. There were no benches, and the cuts were as much as fifty metres high. Resembling marbled steak, they were shot through with veins of quartz, where, long after the granite formed, it cracked and quartz filled it in. The walls were indented with vertical parallel grooves, like giant wormtrails in some exotic sediment. These were actually fossil shot holes and unloaded guide holes from the process of pre-splitting. The highway builders drilled the holes and then dynamited one of three. In this manner, they—and we—reached eight thousand six hundred and forty feet, the highest point on Interstate 80 between the Atlantic and the Pacific. What appeared to be the head of a chicken sat at the top of a big granite block, as if it had been chopped off there. Only when we drew close did I glance up and see that it was Abraham Lincoln. It was, in fact, an artful likeness, resting on an outsized plinth. Years ago, this had been the summit of the Lincoln Highway, which was now incorporated in its substance, if not in its novel spirit, into the innards of the interstate.

We left I-80 there and bucked the southwest wind, crossing the surprisingly flat mountain-crest terrain on a pair of ruts in the pink granite, which had crystals the size of silver dollars. The view from that high wide surface took in a large piece of the front of the Rockies, with the Never Summer Mountains standing out clearly in Colorado, to the south, and, to the west, the bright peaks of the Snowy Range. The Snowy Range—rising white above a dark high forest—appeared to be on top of the Medicine Bow Mountains. Remarkable as it seemed, that was the case. At the ten-thousand-foot level, the bottom of the Snowy Range rests on the broad flat top of the Medicine Bows like a sloop on water, its sails flying upward another two thousand feet. In the Miocene, the high flat Medicine Bow surface at the base of the Snowy Range was the level of maximum fill. In the fifty miles between the Snowy Range and our position on top of the Laramie Range lay the gulf of the excavated Laramie Plains. Our line of sight to the tree line of the Medicine Bows had been landscape in the Miocene. From twelve thousand feet it had gently sloped to about nine thousand where we were, and as we turned and faced east and gazed on down that mostly vanished plane we could all but see the Miocene surface continuing—as Love expressed it—"on out to East G-string."

Everywhere in the central Rockies, that highest level of basin fill touched the eminent ranges at altitudes that are now between ten and twelve thousand feet with results that are as beautiful as they are anomalous in the morphology of the world's mountains. In the Beartooths, for example, you can ascend a glacial valley that—in its U shape and high cirques—closely resembles any hanging valley in the Pennine Alps; but after you climb from ten to eleven to twelve thousand feet you do not find a Weisshorn fingering the sky. Instead, you move into an unexpectable physiographic setting, which, after steep slopes above a dry Wyoming basin, is lush and paradisal to the point of detachment from the world. Alpine meadows with meandering brooks are spread across a rolling but essentially horizontal scene, in part forested, in part punctuated with discrete stands of conifers and small cool lakes. The Medicine Bows are also like that—and the Uintas, the Bighorns. Their high flat surfaces, with peaks that seem to rest on them like crowns on tables, make no sense unless—as you look a hundred miles from one such surface to another across a deep dividing basin—you imagine earth instead of air: the Miocene fill, the continuous terrain. The high plateaus on the shoulders of the ranges, remaining from that broad erosional plane, have been given various names in the science, of which the most prominent at the moment is subsummit surface. "There's a plateau above Union Pass in the Wind River Range that's twelve thousand feet and flatter than a turd on a hot day," Love recalled, and went on to say that at such an altitude in flat country he sometimes becomes panicky—which does not happen if he is among craggy peaks, and seems to be a form of acrophobia directly related to the oddity of being in southern Iowa at twelve thousand feet.

With those big crystals, the granite under our feet was about as coarse as granite ever gets, and, as a result, was particularly vulnerable to weather. Its pink feldspar, black mica, and clear glass quartz had been so exposed there for millions of years that gravels could be scraped off without the help of dynamite. The Union Pacific took advantage of this, ballasting its roadbed with pink granite for eight hundred miles. There was almost no soil in that part of the range—just twelve miles' breadth of rough pink rock. "As you go from Chicago west, soil diminishes in thickness and fertility, and when you get to the gangplank and up here on top of the Laramie Range there is virtually none," Love said. "It's had ten million years to develop, and there's none. Why? Wind—that's why. The wind blows away everything smaller than gravel."

Standing in that wind was like standing in river rapids. It was a wind embellished with gusts, but, over all, it was primordially steady: a consistent southwest wind, which had been blowing that way not just through human history but in every age since the creation of the mountains—a record written clearly in wind-scored rock. Trees were widely scattered up there and, where they existed, appeared to be rooted in the rock itself. Their crowns looked like umbrellas that had been turned inside out and were streaming off the trunks downwind. "Wind erosion has tremendous significance in this part of the Rocky Mountain region," Love said. "Even down in Laramie, the trees are tilted. Old-timers used to say that a Wyoming wind gauge was an anvil on a length of chain. When the land was surveyed, the surveyors couldn't keep their tripods steady. They had to work by night or near sunrise. People went insane because of the wind." His mother, in her 1905 journal, said that Old Hanley, passing by the Twin Creek school, would disrupt lessons by making some excuse to step inside and light his pipe. She also described a man who was evidently losing to the wind his struggle to build a cabin:

> He was putting up a ridgepole when the wind was blowing. He looked up and saw the chipmunks blowing over his head. By and by, along came some sheep, dead. At last one was flying over who was not quite gone. He turned around and said, "Baa"—and then he was in Montana.

Erosion, giving the landscape its appearance, is said to be the work of water, ice, and wind; but wind is, almost everywhere, a minimal or negligible factor, with exceptional exceptions like Wyoming. Looking back across the interstate—north up the crest of the range—among ponderosas, aspens, and limber pines we could see the granites of Vedauwoo Glen, which had weathered out in large blocks, as granite does, along intersecting planes of weakness, while wind-borne grit had rounded off the corners of the blocks. Where some had tumbled and become freestanding, grit flying close above the ground had abraded them so rigorously that the sub-summit surface was, in that place, a flat of giant mushrooms. The cliffs behind them also looked organic—high piles of rounded blocks, topped in many places by narrowly balanced boulders that were undercut almost to the point of falling. Love, contemplative, appeared to be puzzling out some deep question in geomorphology. At length, he said, "When wild horses defecate, they back up to a place where other wild horses have defecated, and so on, until they build turd towers, like those, in the air. Domestic horses do not do this."

At the Wyoming Information Center, beside Interstate 80 just south of Cheyenne, eleven picnic tables are enclosed in brick silos, and each silo has a picture window, so that visitors to Wyoming can picnic more or less al fresco and not be blown home. On the range, virtually every house has a shelter belt of trees—and for the most part the houses are of one story. Used tires cover the tops of mobile homes. Otherwise, wind tears off the roofs. Mary Kraus, a sedimentologist from the University of Colorado, got out of her car one day in north-central Wyoming and went to work on an outcrop. The wind blew the car off a cliff. A propeller-drawn airplane that serves Wyoming is known as the Vomit Comet. When people step off it, they look like spotted slate.

"Most people today don't realize the power of wind and sand," Love said. "Roads are paved. But in the first fifty years of the Lincoln Highway you didn't like to travel west in the afternoon. You'd lose the finish on your car. Your windshield became so pitted you could hardly see out." The Highway Department has not yet paved the wind. On I-80, wind will capsize tractor-trailers. When snow falls on Wyoming, its travels are only beginning. Snow snows again, from the ground up, moves along the surface in ground blizzards that can blind whole counties. Ground blizzards bury houses. In roadcuts, they make drifts fifty feet deep. The wind may return ahead of the plows and take the snow away. The old-timers used to say, "Snow doesn't melt here; it just wears out." Interstate 80 has been closed by snow in Wyoming in every month but August—sometimes closed for days. Before Amtrak dropped its Wyoming passenger service, people stranded on I-80 used to abandon their cars and make their escape by train. The most inclement stretch of 80 is east of Rawlins where it skirts the tip of the Medicine Bows, where anemometers set on guardrails beside the highway frequently catch the wind exceeding the speed limit.

Now, looking from mountains to mountains west over the Laramie Plains—his gaze bridging fifty miles of what had fairly recently been solid ground—Love said he thought the role of the wind had been much greater than hitherto suspected in the Exhumation of the Rockies. Water, of course, was the obvious agent for the digging and removal of the basin fill, as a look at the Mississippi Delta would tend to confirm. Many miles off the coast there, you could drill down into the muck and after fifteen thousand feet the bit would still be in the Miocene. He continued, "We know, however, the approximate volume of sediment from the Powder River Basin, the Bighorn Basin, the Wind River Basin, the Laramie Basin, and so forth. We can say it all went downhill to the Mississippi Delta. But go to the delta. Look at the volumes. There's an enormous discrepancy. You add up what's down there in the Gulf and what was removed here, and they don't square. A great deal more has been removed from here than is down there. Streams only account for about half the material that was taken up and out of here. Since it is not all in the delta, where did it go? So much has been taken away that it's got to be explained in some other manner. I think the wind took it. My personal feeling is that a lot of it blew eastward to the Atlantic. Possibly some went to Hudson Bay. We don't know. These are problems we are trying to grapple with at the present time. How much did the wind take? Again, we don't know, but in one dust storm several years ago a great deal of debris from Kansas and Nebraska and Colorado went into the Atlantic—a storm that lasted only a couple of days."

Such storms are frequent, and this one was not unusual in size or duration. It is noteworthy because its effects were studied and published, in the *Journal of Sedimentary Petrology*. When the dust appeared above the coast of Georgia—as thick haze—it attracted the attention of researchers at the Skidaway Institute of Oceanography, near Savannah. The cloud of particles was two miles in height, and satellite photographs showed its other dimensions: four hundred thousand square miles. With air-sampling-and-measuring equipment, the Skidaway people collected particles. They reported in that one storm enough dust to account for twenty-five per cent of the annual rate of sedimentation—from all rivers as well as the air—in the proximate North Atlantic. Moreover, about eighty-five per cent of it was a clay mineral called illite. Silts coming out of east-coast rivers include very little illite, and yet illite is predominant among the sediments of the ocean floor. By Skidaway's calculations, that one storm's deposits in the ocean amounted to a million tons.

Moving even farther from the interstate on the subsummit surface, we came upon a granite pyramid, sixty feet high, sixty feet wide at the base. It had been designed by the architect H. H. Richardson and weighed six thousand tons—enough to prevent its blowing over. We stood in its lee. The wind was coming in pulses that made percussions in the ears. The incongruity of this monument was in direct proportion to its stark isolation. It was Uncle Pete's version of Interstate 80's Abraham Lincoln. It commemorated the brothers Oakes and Oliver Ames—Massachusetts shovel-makers, railroad financiers—whose Crédit Mobilier of America made construction contracts with itself in enjoying the fruits of subsidy of the Union Pacific Railroad. If you belonged to the United States Congress, you could buy shares of Crédit Mobilier stock for fifty per cent of their value. Near the apex of the east side of the pyramid was Oliver's face in a portrait plaque, sculptured in 1881 by Augustus Saint-Gaudens, whose William Tecumseh Sherman stands in Manhattan's Grand Army Plaza and Robert Gould Shaw in Boston Common. Saint-Gaudens' plaque of Oakes Ames was on the west side of the pyramid, facing the wind. The monument had been built beside the Union Pacific at the railroad's highest point, but the railroad's highest point was somewhere else now; the alignment had been changed in 1901, and the track was three or four miles away. The original roadbed had become so indistinct that a geologist was required to point out where it had been, which he did. Oakes' nose had been shot off with a high-powered rifle. Oliver's nose had been shot away, too, and a large part of his face. Love remarked that Greek, Roman, and Saracen vandals broke off the noses from pieces of sculpture. Probably the Vandals did, too.

Back on the interstate and just west of Abraham Lincoln, the rock became younger again, as we left the Precambrian range core and encountered the same Pennsylvanian red sandstone that had leaned on the mountain on the other side. It was rich red, and the cuts were very big as the road plunged through them in christie turns, running down the mountains through Telephone Canyon. Somewhere overhead had been the first telephone wire ever strung across the Rockies. The President of the United States, with a dozen horses and companions, rode up Telephone Canyon on his way to Cheyenne in 1903. His mustache was an airfoil with a fineness ratio that must have impressed the Wright brothers. He wore a three-gallon hat. His paunch at the time was under control. The interstate trail was more than a little wild then, but manifestly so was he. The red rock is of so much beauty there, and competence, that people collect it for building material, banging it free from the shattered roadcuts and loading it into pickups, much as ranchers did when they first came to the Laramie Plains and ascended the mountains in wagons and collected the rock to build their homes. It is a porous and permeable, fine-grained, hard, brittle sandstone; and because it rests on impermeable granite water moves through it downhill. Released in a fault zone at the bottom, the water leaps to the surface in artesian fountains—the springs that established Laramie. The bright-red roadcuts, ten and twenty metres high, were capped with a buff-colored limestone, which had been deposited in tropical waters on top of the Pennsylvanian sand. After a mountain range rises under layers of flat-lying rock and bends them upward until they all but stand on end, the slopes of the eroding mountains will descend more gently than the dip of the molested strata. And so, as we plunged down Telephone Canyon, the interstate was tilting less than the rock of the roadcuts, and the red sandstone yielded gradually, interstitially, to the younger limestones, until the sandstone was gone altogether and we were moving through the floor of an ocean. It was full of crinoids, brachiopods, and algal buttons, which had lived near the equator in a place like the Bismarck Archipelago or an arm of the Celebes Sea.

The canyon opened to the plains—a broad dry sea of the interior Rockies—and soon we were on Grand Avenue, Laramie, passing the University of Wyoming, whose buff buildings on wide soft lawns could never be said to resemble roadcuts, notwithstanding the crinoids in their walls, the brachiopods and algal buttons. We passed Love's home, on Eleventh Street, and his office on the campus, adjacent to a life-size two-story sculpture of *Tyrannosaurus rex*, the toughest-looking critter in the history of the earth, a native, needless to say, of Wyoming. We passed St. Matthew's Episcopal Cathedral, which also—as Love had reason to regret—contained in its walls brachiopods, crinoids, and algal buttons. He once taught Sunday school there. He took the kids outside and showed them the fossils in the church walls. He described the environment in which the creatures had lived. He mentioned the age of the rock. He explained how things evolve and the fit prosper. Here endeth his career in sedimentary theology.

A few miles north of town, we passed the quarry out of which had emerged not only the university and St. Matthew's Cathedral but also the Ivinson Home for Aged Ladies and the Albany County Courthouse. "It's a limey sandstone, slightly fossiliferous," he said. "It holds up pretty well." We continued north along the foot of the Laramie Range and then turned east into the mountains, climbing a canyon downsection until we had returned to Precambrian time. The rock in this place was even older than the neighboring subsummit granite, and some of it was chatoyant: flashing like a cat's eye. It flashed every color in the spectrum. The rock was anorthosite, nearly fifteen per cent aluminum, Love said. When the bauxites of the Caribbean run out, anorthosite will be a source of aluminum. "Anorthosite is tough, has a high melting point, and doesn't fracture easily," he continued. "Hence it might be useful for containing atomic waste." Anorthosite is rare on earth. It began forming during the Archean Eon and predominantly dates from an age of the later Precambrian known as Helikian time. Yet the high Adirondacks are largely anorthosite. The choice they present is to seal up our spent nuclear fuel inside them or dismantle them one at a time to make beer cans. Anorthosite is more plentiful elsewhere. It is most of what you are looking at when you are looking at the moon.

Moving on west, another day, we crossed the Laramie Plains on I-80 through a world of what to me were surprising lakes. They were not glacial lakes or man-made lakes or—as in Florida—sink-hole lakes filling bowls of dissolved limestone. For the most part, they had no outlets, and were therefore bitter lakes—some alkaline, some saline, some altogether dry. Of Knadler Lake, about a mile long, Love said, "That's bitter water—sodium sulphate. It would physic you something awful." A herd of twenty antelopes galloped up the shore of Knadler Lake. Most of the lakes of the world are the resting places of rivers, where rivers seek their way through landscapes that have been roughed up and otherwise left chaotic by moving ice. Ice had never covered the Laramie Plains. What, then, had dug out these lakes?

Love's response to that question was "What do you suppose?"

We had seen—a mile or two away—a hole in the ground eleven miles long, four miles wide, and deeper than the Yellow Sea. There were some puddles in it, but it did not happen to intersect any kind of aquifer, and basically it was dry. With a talent for understatement, the people of Laramie call it the Big Hollow. Geologists call it a deflation basin, a wind-scoured basin, or—more succinctly—a blow-out. The wind at the Big Hollow, after finding its way into some weak Cretaceous shales, had in short order dug out four million acre-feet and blown it all away. Wind not only makes such basins but maintains them—usually within frameworks of resistant rock. On the Laramie Plains, the resistant rock is heavy quartzite gravel—Precambrian pieces of the Snowy Range which were brought to the plains as the beds of Pleistocene rivers. Wet or dry, all the lakes we passed had been excavated by the wind. It was a bright cloudless morning with a spring breeze. Spheres of tumbleweed, tumbling east, came at us on the interstate at high speed, like gymfuls of bouncing basketballs dribbled by the dexterous wind. "It's a Russian thistle," Love said. "It's one of nature's marvels. As it tumbles, seeds are exploded out."

Across the green plains, the Medicine Bow Mountains and the Snowy Range stood high, sharp, and clear, each so unlike the other that they gave the impression of actually being two ranges: in the middle distance, the flat-crested Medicine Bows, dark with balsam, spruce, and pine; and, in the far high background, the white and treeless Snowy Range. That the one was in fact directly on top of the other was a nomenclatural Tower of Babel that contained in its central paradox the narrative of the Rockies: the burial of the ranges, the subsequent uplifting of the entire region, the exhumation of the mountains. As if to emphasize all that, people had not only named this single mountain range as if it were two but also bestowed upon the highest summit of the Snowy Range the name Medicine Bow Peak. It was up there making its point, at twelve thousand thirteen feet.

We passed a stone ranch house a century old, and a set of faded ruts in the rangeland that were older than the house. This was the Overland Trail, abandoned in 1868 after seven dismal years. "A nasty route," Love remarked. "Steep grades. Many rocks. Poor water. Poor grass. It was three days across the Laramie Plains at ten miles a day. It was often muddy and boggy. A disaster."

When, in the orogeny, the Medicine Bow Mountains were shoved a few miles east, the rock in front of them folded. The anticlines among the folds formed traps for migrating fluids. All about us were pump jacks bobbing for oil.

Boulder beds in the roadcuts represented, as Love put it, "the deroofing of the Medicine Bow Mountains in the first pulsation of the Laramide Revolution." The beds were of Paleocene age. In a knife-edge ridge a few miles farther on, the interstate had exposed the same conglomerates tilted forty-five degrees as mountain building continued. And soon after that came a flat-lying Eocene deposit. "So you have a time frame for the orogeny," Love said, and this was when he added, "It didn't take very long for those mountains to come up, to be deroofed, and to be thrust eastward. Then the motion stopped. That happened in maybe ten million years, and to a geologist that's really fast."

Near Arlington, an anomalous piece of landscape reached straight out from the mountains like a causeway heading north. It was capped with stream gravel, brought off the mountains by furious rivers rushing through the tundras of Pleistocene time. The gravel had resisted subsequent erosion, while lighter stuff was washed away on either side. Geologists call such things pediments, and Love remarked that the one before us was "the most striking pediment in this region." In my mind's eye I could see the braided rivers coming off the Alaska Range, thickly spreading gravels, perhaps to preserve beneath them the scenes of former worlds. Where I-80 cut through the Arlington pediment, the Pleistocene gravel rested on Eocene sandstones, on red and green claystones; and they in turn covered conglomerates that came from the mountains when the mountains were new. One could read upward from one world to another: the boulders falling from rising mountains, the quiet landscapes after the violence stopped—all preserved in a perplexing memento from the climate of an age of ice.

In a cut eight miles farther on, that early conglomerate was in contact with Cretaceous rock bent upward even more steeply as the Laramide Orogeny lifted the mountains. Picking through the evidence in the conglomerate was like sorting out debris from an explosion. One after another, I chose a cobble from the roadcut, handed it to Love, and asked him what it was. A Paleozoic quartzitic sandstone, for example—probably Mississippian. Grains rounded. No biotite. In fact, no mica of any kind. A Cretaceous sandstone. That would be from nearby, not from the mountains. A Paleozoic or Precambrian chert. Some Hanna formation sandstone, Paleocene in age—the matrix of the conglomerate. Some Precambrian quartzite from the Snowy Range, two billion years old. Some bull quartz from a vein in the Precambrian. And one he didn't know.

While the orogeny was making mountains, it was also making basins, for which it is less noted, even where the basins are a good deal deeper than Mt. Everest is high. As we crossed the Medicine Bow River and approached the North Platte and Rawlins, we moved out upon the surface of the Hanna Basin. It was choppy but essentially level nondescript ground, like all the rest of the rangeland on the apron of the mountains. It was not water, and we were not in a boat, but in some ways it seemed so as we crossed a basin forty-two thousand feet deep. It is the deepest structural basin in North America. It is Cretaceous, Paleocene, and Eocene rock, bent in U's, with seams of coal as much as fifty feet thick in the arms of the U's. Union Pacific.

We crossed the North Platte, climbed some long grades, examined a few roadcuts, and pulled off on the shoulder at Rawlins to absorb, in the multiple exposures of the Rawlins Uplift, its comprehensive spread of time—Rawlins, where his mother had boarded the stage north, three-quarters of a century before.

In the United States Geological Survey's seven-and-a-half-minute series of topographic maps is a quadrangle named Love Ranch. The landscape it depicts lies just under the forty-third parallel and west of the hundred-and-seventh meridian—coordinates that place it twelve miles from the geographic center of Wyoming. The names of its natural features are names that more or less materialized around the kitchen table when David Love was young: Corral Draw, Castle Gardens, Buffalo Wallows, Jumping-Off Draw. To the fact that he grew up there his vernacular, his outlook, his pragmatic skills, and his professional absorptions about equally attest. The term "store-bought" once brightened his eyes. When one or another of the cowpunchers used a revolver, the man did not so much fire a shot as "slam a bullet." If a ranch hand was tough enough, he would "ride anything with hair on it." Coffee had been brewed properly if it would "float a horseshoe." Blankets were "sougans." A tarpaulin was a "henskin." To be off in the distant ranges was to be "gouging around the mountains." In Love's stories of the ranch, horses come and go by the "cavvy." If they are unowned and untamed, they are a "wild bunch"—led to capture by a rider "riding point." In the flavor of his speech the word "ornery" endures.

He describes his father as a "rough, kindly, strong-willed man" who would put a small son on each knee and—reciting "Ride a cockhorse to Banbury Cross to see a fine lady upon a white horse" —give the children bronco rides after dinner, explaining that his purpose was "to settle their stomachs." Their mother's complaints went straight up the stovepipe and away with the wind. When their father was not reciting such Sassenach doggerel, he could draw Scottish poems out of the air like bolts of silk. He had the right voice, the Midlothian timbre. He knew every syllable of "The Lady of the

Lake." Putting his arms around the shoulders of his wee lads, he would roll it to them by the canto, and when they tired of Scott there were in his memory more than enough ballads to sketch the whole of Scotland, from the Caithness headlands to the Lammermuir Hills.

David was fifteen months younger than his brother, Allan. Their sister, Phoebe, was born so many years later that she does not figure in most of these scenes. They were the only children in a thousand square miles, where children outnumbered the indigenous trees. From the ranch buildings, by Muskrat Creek, the Wind River Basin reached out in buffalo grass, grama grass, and edible salt sage across the cambered erosional swells of the vast dry range. When the wind dropped, this whole wide world was silent, and they could hear from a great distance the squeak of a horned lark. The nearest neighbor was thirteen miles away. On the clearest night, they saw no light but their own.

Old buffalo trails followed the creek and branched from the creek: old but not ancient—there were buffalo skulls beside them, and some were attached to hide. The boys used the buffalo trails when they rode off on ranch chores for their father. They rode young and rode long, and often went without water. Even now, so many decades later, David will pass up a cool spring, saying, "If I drink now, I'll be thirsty all day." To cut cedar fence posts, they went with a wagon to Green Mountain, near Crooks Gap—a round trip of two weeks. In early fall, each year, they spent ten days going back and forth to the Rattlesnake Hills for stove wood. They took two wagons—four horses pulling each wagon—and they filled them with limber pine. They used axes, a two-handled saw. Near home, they mined coal with their father—from the erosional wonderland they called Castle Gardens, where a horse-drawn scraper stripped the overburden and exposed the seams of coal. Their father was adept at corralling wild horses, a skill that called for a horse and rider who could outrun these closest rivals to the wind. He caught more than he kept, put his Flatiron brand on the best ones and sold the others. Some of them escaped. David remembers seeing one clear a seven-foot bar in the wild-horse corral and not so much as touch it. When he and Allan were in their early teens, his father sent them repping—representing Love Ranch in the general roundup—and they stayed in cow camp with other cowboys, and often enough their sougans included snow. When they were out on the range, they slept out on the range, never a night in a tent. This was not a choice. It was a family custom.

In the earlier stretch of his life when John Love had slept out for seven years, he would wrap himself in his sougans and finish the package with the spring hooks and D-rings that closed his henskin. During big gales and exceptional blizzards, he looked around for a dry wash and the crease of an overhanging cutbank. He gathered sage and built a long fire—a campfire with the dimensions of a cot. He cooked his beans and bacon, his mutton, his sourdough, his whatever. After dinner, he kicked the fire aside and spread out his bedroll. He opened his waterproof packet of books and read by kerosene lamp. Then he blew out the light and went to sleep on warm sand. His annual expenditures were seventy-five dollars. This was a man who wore a long bearskin coat fastened with bone pegs in loops of rope. This was a man who, oddly enough, carried with him on the range a huge black umbrella—his summer parasol. This was a man whose Uncle John Muir had invented a device that started a fire in the morning while the great outdoorsman stayed in bed. And now this wee bairn with the light-gold hair was, in effect, questioning Love Ranch policy by asking his father what he had against tents. "Laddie, you don't always have one available," his father said patiently. "You want to get used to living without it." Tents, he made clear, were for a class of people he referred to as "pilgrims."

When David was nine, he set up a trap line between the Hay Meadow and the Pinnacles (small sandstone buttes in Castle Gardens). He trapped coyotes, bobcats, badgers. He shot rabbits. He ran the line on foot, through late-autumn and early-winter snow. His father was with him one cold and blizzarding January day when David's rifle and the rabbits he was carrying slipped from his hands and fell to the snow. David picked up the gun and soon dropped it again. "It was a cardinal sin to drop a rifle," he says. "Snow and ice in the gun barrel could cause the gun to blow up when it was fired." Like holding on to a saddle horn, it was something you just did not do. It would not have crossed his father's mind that David was being careless. In sharp tones, his father said, "Laddie, leave the rabbits and rifle and run for home. Run!" He knew hypothermia when he saw it, no matter that it lacked a name.

Even in October, a blizzard could cover the house and make a tunnel of the front veranda. As winter progressed, rime grew on the nailheads of interior walls until white spikes projected some inches into the rooms. There were eleven rooms. His mother could tell the outside temperature by the movement of the frost. It climbed the nails about an inch for each degree below zero. Sometimes there was frost on nailheads fifty-five inches up the walls. The house was chinked with slaked lime, wood shavings, and cow manure. In the wild wind, snow came through the slightest crack, and the nickel disks on the dampers of the heat stove were constantly jingling. There came a sound of hooves in cold dry snow, of heavy bodies slamming against the walls, seeking heat. John Love insulated his boots with newspapers—as like as not the *New York Times*. To warm the boys in their beds on cold nights, their mother wrapped heated flatirons in copies of the *New York Times*. The family were subscribers. Sundays only. The *Times*, David Love recalls, was "precious." They used it to insulate the house: pasted it against the walls beside the *Des Moines Register*, the *Tacoma News Tribune*—any paper from anywhere, without fine distinction. With the same indiscriminate voracity, any paper from anywhere was first read and reread by every literate eye in every cow camp and sheep camp within tens of miles, read to shreds and passed along, in tattered circulation on the range. There was, as Love expresses it, "a starvation of print." Almost anybody's first question on encountering a neighbor was "Have you got any newspapers?"

The ranch steadings were more than a dozen buildings facing south, and most of them were secondhand. When a stage route that ran through the ranch was abandoned, in 1905, John Love went down the line shopping for moribund towns. He bought Old Muskrat—including the hotel, the post office, Joe Lacey's Muskrat Saloon—and moved the whole of it eighteen miles. He bought Golden Lake and moved it thirty-three. He arranged the buildings in a rough semicircle that embraced a corral so large and solidly constructed that other ranchers travelled long distances to use it. Joe Lacey's place became the hay house, the hotel became in part a saddlery and cookhouse, and the other buildings, many of them connected, became all or parts of the blacksmith shop, the chicken hatchery, the ice shed, the buggy shed, the sod cellar, and the bunkhouse—social center for all the workingmen from a great many miles around. There was a granary made of gigantic cottonwood logs from the banks of the Wind River, thirty miles away. There were wool-sack towers, and a wooden windmill over a hand-dug well. The big house itself was a widespread log collage of old town parts and original construction. It had wings attached to wings. In the windows were air bubbles in distorted glass. For its twenty tiers of logs, John had journeyed a hundred miles to the lodgepole-pine groves of the Wind River Range, returning with ten logs at a time, each round trip requiring two weeks. He collected a hundred and fifty logs. There were no toilets, of course, and the family had to walk a hundred feet on a sometimes gumbo-slick path to a four-hole structure built by a ranch hand, with decorative panelling that matched the bookcases in the house. The cabinetmaker was Peggy Dougherty, the stagecoach driver who had first brought Miss Waxham through Crooks Gap and into the Wind River country.

The family grew weary of carrying water into the house from the well under the windmill. And so, as she would write in later years:

> After experiments using an earth auger and sand point, John triumphantly installed a pitcher pump in the kitchen, a sink, and drain pipe to a barrel, buried in the ground at some distance from the house. This was the best, the first, and at that time the only water system in an area the size of Rhode Island.

In the evenings, kerosene lamps threw subdued yellow light. Framed needlework on a wall said "WASH & BE CLEAN." Everyone bathed in the portable galvanized tub, children last. The more expensive galvanized tubs of that era had built-in seats, but the Loves could not afford the top of the line. On the plank floor were horsehide rugs—a gray, a pinto—and the pelt of a large wolf, and two soft bobcat rugs. Chairs were woven with rawhide or cane. John recorded the boys' height on a board nailed to the inside of the kitchen doorframe. A brass knocker on the front door was a replica of a gargoyle at Notre-Dame de Paris.

The family's main sitting and dining room was a restaurant from Old Muskrat. On the walls were polished buffalo horns mounted on shields. The central piece of furniture was a gambling table from Joe Lacey's Muskrat Saloon. It was a poker-and-roulette table—round, covered with felt. Still intact were the subtle flanges that had caused the roulette wheel to stop just where the operator wished it to. And if you reached in under the table in the right place you could feel the brass slots where the dealer kept wild cards that he could call upon when the fiscal integrity of the house was threatened. If you put your nose down on the felt, you could almost smell the gunsmoke. At this table David Love received his basic education—his schoolroom a restaurant, his desk a gaming table from a saloon. His mother may have been trying to academize the table when she covered it with a red-and-white India print.

When other schoolmarms were provided by the district, they came for three months in summer. One came for the better part of a year. By and large, though, the boys were taught by their mother. She had a rolltop desk, and Peggy Dougherty's glassed-in bookcases. She had the 1911 Encyclopædia Britannica, the Redpath Library, a hundred volumes of Greek and Roman literature, Shakespeare, Dickens, Emerson, Thoreau, Longfellow, Kipling, Twain. She taught her sons French, Latin, and a bit of Greek. She read to them from books in German, translating as she went along. They read the *Iliad* and the *Odyssey*. The room was at the west end of the ranch house and was brightly illuminated by the setting sun. When David as a child saw sunbeams leaping off the books, he thought the contents were escaping.

In some ways, there was more chaos in this remote academic setting than there could ever be in a grade school in the heart of a city.

The house might be full of men, waiting out a storm, or riding on a round-up. I was baking, canning, washing clothes, making soap. Allan and David stood by the gasoline washing machine reading history or geography while I put sheets through the wringer. I ironed. They did spelling beside the ironing board, or while I kneaded bread; they gave the tables up to 15 times 15 to the treadle of the sewing machine. Mental problems, printed in figures on large cards, they solved while they raced across the . . . room to write the answers . . . and learned to think on their feet. Nine written problems done correctly, without help, meant no tenth problem. . . . It was surprising in how little time they finished their work—to watch the butchering, to help drive the bawling calves into the weaning pen, or to get to the corral, when they heard the hoofbeats of running horses and the cries of cowboys crossing the creek.

No amount of intellectual curiosity or academic discipline was ever going to hold a boy's attention if someone came in saying that the milk cow was mired in a bog hole or that old George was out by the wild-horse corral with the biggest coyote ever killed in the region, or if the door opened and, as David recalls an all too typical event, "they were carrying in a cowboy with guts ripped out by a saddle horn." The lessons stopped, the treadle stopped, and she sewed up the cowboy.

Across a short span of time, she had come a long way with these bunkhouse buckaroos. In her early years on the ranch, she had a lesser sense of fitting in than she would have had had she been a mare, a cow, or a ewe. She did not see another woman for as much as six months at a stretch, and if she happened to approach a group of working ranch hands they would loudly call out, "Church time!" She found "the sudden silence . . . appalling." Women were so rare in the country that when she lost a glove on the open range, at least twenty miles from home, a stranger who found it learned easily whose it must be and rode to the ranch to return it. Men did the housekeeping and the cooking, and went off to buy provisions at distant markets. Meals prepared in the bunkhouse were carried to a sheep wagon, where she and John lived while the big house was being built and otherwise assembled. The Wyoming sheep wagon was the ancestral Winnebago. It had a spring bed and a kitchenette.

After her two sons were born and became old enough to coin phrases, they called her Dainty Dish and sometimes Hooty the Owl. They renamed their food, calling it, for example, dog. They called other entrées caterpillar and coyote. The kitchen stool was Sam. They named a Christmas-tree ornament Hopping John. It had a talent for remaining unbroken. They assured each other that the cotton on the branches would not melt. David decided that he was a camel, but later changed his mind and insisted that he was "Mr. and Mrs. Booth." His mother described him as "a light-footed little elf." She noted his developing sense of scale when he said to her, "A coyote is the whole world to a flea."

One day, he asked her, "How long does a germ live?"

She answered, "A germ may become a grandfather in twenty minutes."

He said, "That's a long time to a germ, isn't it?"

She also made note that while David was the youngest person on the ranch he was nonetheless the most adroit at spotting arrowheads and chippings.

> When David was five or six we began hunting arrowheads and chippings. While the rest of us labored along scanning gulches and anthills, David rushed by chattering and picking up arrowheads right and left. He told me once, "There's a god of chippings that sends us anthills. He lives in the sky and tinkers with the clouds."

The cowboys competed with Homer in the entertainment of Allan and David. There was one who—as David remembers him—"could do magic tricks with a lariat rope, making it come alive all around his horse, over our heads, under our feet, zipping it back and forth around us as we jumped up and down and squealed with delight." Somber tableaux, such as butcherings, were played out before them as well. Years later, David would write in a letter:

> We always watched the killing with horror and curiosity, although we were never permitted to participate at that age. It seemed so sad and so irrevocable to see the gushing blood when throats were cut, the desperate gasps for breath through severed windpipes, the struggle for and the rapid ebbing of life, the dimming and glazing of wide terrified eyes. We realized and accepted the fact that this was one of the procedures that were a part of our life on the range and that other lives had to be sacrificed to feed us. Throat-cutting, however, became a symbol of immediate death in our young minds, the ultimate horror, so dreadful that we tried not to use the word "throat."

He has written a recollection of the cowboys, no less frank in its bequested fact, and quite evidently the work of the son of his mother.

> The cowboys and horse runners who drifted in to the ranch in ever-increasing numbers as the spring advanced were lean, very strong, hard-

muscled, taciturn bachelors, nearly all in their twenties and early thirties. They had been born poor, had only rudimentary education, and accepted their lot without resentment. They worked days that knew no hour limitations but only daylight and dark, and weeks that had no holidays. . . . Most were homely, with prematurely lined faces but with lively eyes that missed little. None wore glasses; people with glasses went into other kinds of work. Many were already stooped from chronic saddle-weariness, bow-legged, hip-sprung, with unrepaired hernias that required trusses, and spinal injuries that required a "hanging pole" in the bunkhouse. This was a horizontal bar from which the cowboys would hang by their hands for 5-10 minutes to relieve pressure on ruptured spinal disks that came from too much bronc-fighting. Some wore eight-inch-wide heavy leather belts to keep their kidneys in place during prolonged hard rides.

When in a sense it was truly church time—when cowboys were badly injured and in need of help—they had long since learned where to go. David vividly remembers a moment in his education which was truncated when a cowboy rode up holding a bleeding hand. He had been roping a wild horse, and one of his fingers had become caught between the lariat and the saddle horn. The finger was still a part of his hand but was hanging by two tendons. His mother boiled water, sterilized a pair of surgical scissors, and scrubbed her hands and arms. With magisterial nonchalance, she "snipped the tendons, dropped the finger into the hot coals of the fire box, sewed a flap of skin over the stump, smiled sweetly, and said: 'Joe, in a month you'll never know the difference.' "

There was a pack of ferocious wolfhounds in the country, kept by another flockmaster for the purpose of killing coyotes. The dogs seemed to relish killing rattlesnakes as well, shaking the life out of them until the festive serpents hung from the hounds' jaws like fettuccine. The ranch hand in charge of them said, "They ain't happy in the spring till they've been bit. They're used to it now, and their heads don't swell up no more." Human beings (on foot) who happened to encounter these dogs might have preferred to encounter the rattlesnakes instead. One summer afternoon, John Love was working on a woodpile when he saw two of the wolfhounds streaking down the creek in the direction of his sons, whose ages were maybe three and four. "Laddies! Run! Run to the house!" he shouted.

"Here come the hounds!" The boys ran, reached the door just ahead of the dogs, and slammed it in their faces. Their mother was in the kitchen:

> The hounds, not to be thwarted so easily, leaped together furiously at the kitchen windows, high above the ground. They shattered the glass of the small panes, and tried to struggle through, their front feet catching over the inside ledge of the window frame, and their heads, with slavering mouths, reaching through the broken glass. I had only time to snatch a heavy iron frying pan from the stove and face them, beating at those clutching feet and snarling heads. The terrified boys cowered behind me. The window sashes held against the onslaught of the hounds, and my blows must have daunted them. They dropped back to the ground and raced away.

In the boys' vocabulary, the word "hound" joined the word "throat" in the deep shadows, and to this day when David sees a wolfhound there is a drop in the temperature of the center of his spine.

The milieu of Love Ranch was not all wind, snow, freezing cattle, and killer dogs. There were quiet, lyrical days on end under blue, unthreatening skies. There were the redwing blackbirds on the corral fence, and the scent of moss flowers in spring. In a light breeze, the windmill turned slowly beside the wide log house, which was edged with flowers in bloom. Sometimes there were teal on the creek—and goldeneyes, pintails, mallards. When the wild hay was ready for cutting, the harvest lasted a week.

> John liked to have me ride with them for the last load. Sometimes I held the reins and called "Whoa, Dan!" while the men pitched up the hay. Then while the wagon swayed slowly back over the uneven road, I lay nestled deeply beside Allan and David in the fragrant hay. The billowy white clouds moving across the wide blue sky were close, so close, it seemed there was nothing else in the universe but clouds and hay.

When the hay house was not absolutely full, the boys cleared off the dance floor of Joe Lacey's Muskrat Saloon and strapped on their roller skates. Improbable as it may seem, there was also a Love

Ranch croquet ground. And in winter the boys clamped ice skates to their shoes and flew with the wind up the creek. Alternatively, they lay down on their sleds and propelled themselves swiftly over wind-cleared, wind-polished black ice, with an anchor pin from a coyote trap in each hand. Almost every evening, with their parents, they played mah-jongg.

One fall, their mother went to Riverton, sixty-five miles away, to await the birth of Phoebe. For her sons, eleven and twelve, she left behind a carefully prepared program of study. In the weeks that followed, they were in effect enrolled in a correspondence school run by their mother. They did their French, their spelling, their arithmetic lessons, put them in envelopes, rode fifteen miles to the post office and mailed them to her. She graded the lessons and sent them back—before and after the birth of the baby.

Her hair was the color of my wedding ring. On her cheek the fingers of one hand were outspread like a small, pink starfish.

From time to time, dust would appear on the horizon, behind a figure coming toward the ranch. The boys, in their curiosity, would climb a rooftop to watch and wait as the rider covered the intervening miles. Almost everyone who went through the region stopped at Love Ranch. It had not only the sizable bunkhouse and the most capacious horse corrals in a thousand square miles but also a spring of good water. Moreover, it had Scottish hospitality—not to mention the forbidding distance to the nearest alternative cup of coffee. Soon after Mr. Love and Miss Waxham were married, Nathaniel Thomas, the Episcopal Bishop of Wyoming, came through in his Gospel Wagon, accompanied by his colleague the Reverend Theodore Sedgwick. Sedgwick later reported (in a publication called *The Spirit of Missions*):

We saw a distant building. It meant water. At this lonely ranch, in the midst of a sandy desert, we found a young woman. Her husband had gone for the day over the range. Around her neck hung a gold chain with a Phi Beta Kappa key. She was a graduate of Wellesley College, and was now a Wyoming bride. She knew her Greek and Latin, and loved her horse on the care-free prairie.

The bishop said he was searching for "heathen," and he did not linger.

Fugitive criminals stopped at the ranch fairly often. They had to—in much the way that fugitive criminals in lonely country today will sooner or later have to stop at a filling station. A lone rider arrived at the ranch one day with a big cloud of dust on the horizon behind him. The dust might as well have formed in the air the letters of the word "posse." John Love knew the rider, knew that he was wanted for murder, and knew that throughout the country the consensus was that the victim had "needed killing." The murderer asked John Love to give him five dollars, and said he would leave his pocket watch as collateral. If his offer was refused, the man said, he would find a way to take the money. The watch was as honest as the day is long. When David does his field geology, he has it in his pocket.

People like that came along with such frequency that David's mother eventually assembled a chronicle called "Murderers I Have Known." She did not publish the manuscript, or even give it much private circulation, in her regard for the sensitivities of some of the first families of Wyoming. As David would one day comment, "they were nice men, family friends, who had put away people who needed killing, and she did not wish to offend them—so many of them were such decent people."

One of these was Bill Grace. Homesteader and cowboy, he was one of the most celebrated murderers in central Wyoming, and he had served time, but people generally disagreed with the judiciary and felt that Bill, in the acts for which he was convicted, had only been "doing his civic duty." At the height of his fame, he stopped at the ranch one afternoon and stayed for dinner. Although David and Allan were young boys, they knew exactly who he was, and in his presence were struck dumb with awe. As it happened, they had come upon and dispatched a rattlesnake that day—a big one, over five feet long. Their mother decided to serve it creamed on toast for dinner. She and their father sternly instructed David and Allan not to use the word "rattlesnake" at the table. They were to refer to it as chicken, since a possibility existed that Bill Grace might not be an eater of adequate sophistication to enjoy the truth. The excitement was too much for the boys. Despite the parental injunction, gradually their conversation at the table fished its way toward the snake. Casually—while the meal was going down—the boys raised the subject of poisonous vipers, gave their estimates of the contents of local dens, told stories of snake encounters, and so forth. Finally, one of them remarked on how very good rattlers were to eat.

Bill Grace said, "By God, if anybody ever gave me rattlesnake meat I'd kill them."

The boys went into a state of catatonic paralysis. In the pure silence, their mother said, "More chicken, Bill?"

"Don't mind if I do," said Bill Grace.

Muskrat Creek was the second homestead on which John Love had filed in Wyoming. The first—thirty miles away—was in the Big Sand Draw, where the grass was inadequate, the snows were exceptionally deep, and the water was marginally potable. In 1897, he collapsed his umbrella and moved. At Muskrat Creek, long before he bought the stagecoach towns, he lived in an earth dugout roofed with pine poles and clay. It was warm in winter, cool in summer, and danker than Scotland all year round. He was prepared to run risks. In Lander, sixty miles west, he made an extraordinary bet with a bank, whose assets included a number of thousands of sheep. John Love bet that he could take them for a summer and return them in the fall, fatter on the average by at least ten pounds. If he succeeded, he would be paid handsomely. If he failed, he would receive a scant wage. He was taking a chance on the weather, because a bad storm could wipe out the flock. By November, the sheep were as round as poker chips, ready to be cashed in. Leaving them in the care of a herder, he rode to Thermopolis, where he made a down payment on a flock of his own. The conditions of the deal were rigid: the rest of the money was to be paid in seven days or the deposit was forfeit and the animals, too. Within the week, he would have to return to his fattened sheep, move them to Lander, collect his money, and return to Thermopolis—a round trip of two hundred and fifty miles.

The sky over Thermopolis was dark with snowcloud. In his bearskin cap, his bearskin coat, his fleece-lined leather chaps, he saddled up Big Red—Big Red, whose life had begun somewhere in the Red Desert in 1888, a wild horse. The blizzard began as horse and rider were climbing the Owl Creek Mountains. Through steep terrain that would have been hazardous in warm clear weather, they felt their way in whiteouts and darkness, in wind-chill factors greater than fifty below zero. Covering about six miles an hour, they reached the herd in twenty-one hours, and almost immediately began the gingerly walk to Lander, conserving the animals' weight. John won his bet, got back on Big Red, and flew across the mountains with the money. He and the horse beat the deadline. He collected his ewes, took them home, and bred them. In seven days, he had, among other things, set himself forward one year. By 1910, when he married Miss Waxham, he owned more than eleven thousand sheep and hundreds of cattle and horses—a fortune in livestock which today would be valued at roughly five million dollars.

In the early days of his marriage, John Love used to ride around his place reciting the verse of William Cowper:

I am monarch of all I survey,
My right there is none to dispute.

As he built up his new home, he did not seem worried that in recent years herders had been killed, wagons had been burned, and sheep had been clubbed to death or driven over cliffs by the thousand. As anyone who has seen three Western movies cannot help but know, there was bloody warfare between cattlemen and sheepmen; and well into the new century the strife continued. According to David, his father stocked the ranch with both cattle and sheep specifically as a way of getting along with both sides. His monarchy would be disputed only by nature and bankers.

Cowboys, meanwhile, made unlikely paperhangers.

Rolls of green figured wallpaper had arrived from a mail-order company. What to do with them, no one quite knew, but there were directions. I made dishpans full of paste. In the evening John called in the half dozen cowboys from the bunkhouse. They carried planks and benches. They put all the leaves in the wobbly dinner table. I measured and cut, pasted and trimmed lengths of wallpaper. Then in chaps and jingling spurs the cowpunchers strode along the benches, slapping paste brushes and dangling strips of torn wallpaper over the dining room ceiling. We were all surprised and tremendously pleased with the results and celebrated over a ten-gallon keg of cider.

John put a roof on the ranch house that was half clay and a foot thick. It consisted of hundreds of two-inch poles covered with burlap covered with canvas covered with rafters embedded in the clay, with corrugated iron above that, coated with black asphaltum. It helped the house be cool in summer, warm in winter—and in the Wind River Basin was unique. But while this durable roof could defend against Wyoming weather the rest of the ranch could not. In the winter of 1912, winds with velocities up to a hundred miles an hour caused sheep to seek haven in dry gulches, where snows soon buried them as if in avalanche. Going without sleep for forty and fifty hours, John Love and his ranch hands struggled to rescue them. They dug some out, but many thousands died. Even on the milder days, when the temperature came up near zero, sheep could not penetrate the wind-crusted drifts and get at the grass below. The crust cut into their legs. Their tracks were reddened with blood. Cattle, lacking the brains even to imagine buried grass, ate their own value in cottonseed cake. John Love had to borrow from his bankers in Lander to pay his ranch hands and buy supplies.

That spring, a flood such as no one remembered all but destroyed the ranch. The Loves fled into the night, carrying their baby, Allan.

At daylight we returned to the house. Stench, wreckage and debris met us. The flood had gone. Its force had burst open the front door and swept a tub full of rainwater into the dining room. Chairs and other furniture were overturned in deep mud. Mattresses had floated. Doors and drawers were already too much swollen for us to open or shut. The large wardrobe trunk of baby clothes was upset. Everything in it was soaked and stained. Around all the rooms at the height of the tabletops was a water mark, fringed with dirt, on the new wallpaper.

Almost immediately, the bankers arrived from Lander. They stayed for several amiable days, looked over the herd tallies, counted surviving animals, checked John Love's accounts. Then, at dinner one evening, the bank's vice-president rubbed his hands together and said to his valued customer, his trusted borrower, his first-name-basis longtime friend, "Mr. Love, we need more collateral." The banker also said that while John Love was a reliable debtor, other ranchers were not, and others' losses were even greater than Love's. The bank, to protect its depositors, had to use Love Ranch to cover itself generally. "We are obliged to cash in on your sheep," the man went on. "We will let you keep your cattle—on one condition." The condition was a mortgage on the ranch. They were asking for an interest in the land of a homesteader who had proved up.

John Love shouted, "I'll have that land when your bones are rotting in the grave!" And he asked the man to step outside, where he could curse him. To the banker's credit, he got up and went out to be cursed. Buyers came over the hill as if on cue. All surviving sheep were taken, all surviving cattle, all horses—even dogs. The sheep wagons went, and a large amount of equipment and supplies. John Love paid the men in the bunkhouse, and they left. As his wife watched the finish of this scene, standing silent with Allan in her arms, the banker turned to her kindly and said, "What will you do with the baby?"

She said, "I think I'll keep him."

It was into this situation that John David Love was born—a family that had lost almost everything but itself, yet was not about to lose that. Slowly, his father assembled more modest cavvies and herds, beginning with the capture of wild horses in flat-out all-day rides, maneuvering them in ever tighter circles until they were beguiled into entering the wild-horse corral or—a few miles away—the natural cul-de-sac (a small box canyon) known to the family as the Corral Draw. Watching one day from the granary roof, the boys—four and five—in one moment saw their father on horseback crossing the terrain like the shadow of a cloud and in the next saw his body smash the ground. The horse had stepped in a badger hole. The rider—limp and full of greasewood punctures, covered with blood and grit—was unconscious and appeared to be dead. He was carried into the house. After some hours, he began to stir, and through his pain mumbled, "That damned horse. That damned horse—I never did trust him." It was the only time in their lives that his sons would hear him swear.

There were periods of drought, and more floods, and long, killing winters, but John Love never sold out. He contracted and survived Rocky Mountain spotted fever. One year, after he shipped cattle to Omaha he got back a bill for twenty-seven dollars, the amount by which the cost of shipment exceeded the sale price of the cattle. One spring, after a winter that killed many sheep, the boys and their father plucked good wool off the bloated and stinking corpses, sold the wool, and deposited the money in a bank in Shoshoni, where the words "STRENGTH," "SAFETY," and "SECURITY" made an arc above the door. The bank failed, and they lost the money. Of many bad winters, the worst began in 1919. Both David and his father nearly died of Spanish influenza, and were slow to recuperate, spending months in bed. There were no ranch hands. At the point when the patients seemed most in danger, his mother in her desperation decided to try to have them moved to a hospital (a hundred miles away), and prepared to ride for help. She had the Hobson's choice of a large, rebellious horse. She stood on a bench and tried to harness him. He kicked the bench from under her, and stepped on her feet. She gave up her plan.

The bull broke into the high granary. Our only, and small, supply of horse and chicken feed was there. Foolishly, I went in after him and drove him out down the step. Cows began to die, one here, one there. Every morning some were unable to rise. By day, one walking would fall suddenly, as if it had no more life than a paper animal, blown over by a gust of wind.

The bull actually charged her in the granary and came close to crushing her against the back wall. She confused it, sweeping its eyes with a broom. It would probably have killed her, though, had it not stepped on a weak plank, which snapped. The animal panicked and turned for the door. (In decades to follow, John Love never fixed the plank.)

Snow hissed around the buildings, wind blew some snow into every room of the closed house, down the chimney, between window sashes, even in a straight shaft through a keyhole. The wood pile was buried in snow. The small heap of coal was frozen into an almost solid chunk of coal and ice. In the numbing cold, it took me five hours a day to bring in fuel, to carry water and feed to the chickens, to put out hay and cottonseed cake for the cattle and horses.

John began to complain, a favorable sign. Why was I outside so much? Why didn't I stay with him? To try to make up to him for being gone so long, I sat on the bed at night, wrapped in a blanket, reading to him by lamplight.

Somewhere among her possessions was a letter written to her by a Wellesley friend asking, "What do you do with your spare time?"

Where the stage route from Casper to Fort Washakie had crossed a tributary of Muskrat Creek, the banks were so high and the drop to the creekbed so precipitous that the site was littered with split wagon reaches and broken wheels. Allan and David called it Jumping-Off Draw, its name on the map today. Finding numerous large bones in a meadowy bog, they named the place Buffalo Wallows. Indians had apparently driven the bison into the swamp to kill them. One could infer that. One could also see that the swamp was there because water was bleeding from rock outcrops above the meadows. In a youth spent on horseback, there was not a lot to do but look at the landscape. The rock that was bleeding water was not just porous but permeable. It was also strong. It was the same red rock that the granary stood on, and the bunkhouse. Very evidently, it was made of naturally cemented sand. The water could not have come from the creek. The Buffalo Wallows were sixty feet higher than the creek. The sandstone layers tilted north. They therefore reached out to the east and west. There was high ground to the east. The water must be coming down from there. One did not need a Ph.D. from Yale to figure that out—especially if one was growing up in a place where so much rock was exposed. Pending further study, his interpretation of the Buffalo Wallows was just a horseback guess. All through his life, when he would make a shrewd surmise he would call it a horseback guess.

The water in the sandstone produced not only the bogs but the adjacent meadows as well—in this otherwise desiccated terrain. From the meadows came hay. There was an obvious and close relationship between bedrock geology and ranching. David would not have articulated that in just those words, of course, but he thought about the subject much of the time, and he was drawn to be a geologist in much the way that someone growing up in Gloucester, Massachusetts, would be drawn to be a fisherman. "It was something to think about on long rides day after day when everything was so monotonous," he once remarked. "Monotony was what we fought out there. Day after day, you had nothing but the terrain around you—you had nothing to think about but why the shale had stripes on it, why the boggy places were boggy, why the vegetation grew where it did, why trees grew only on certain types of rock, why water was good in some places and bad in others, why the meadows were where they were, why some creek crossings were so sandy they were all but impassable. These things were very real, very practical. If you're in bedrock, caliche, or gumbo, the going is hard. Caliche is lime precipitate at the water table—you learn some geology the hard way. There was nothing else to be interested in. Everything depended on geology. Any damn fool could see that the vegetation was directly responsive to the bedrock. Hence birds and wildlife were responsive to it. We were responsive to it. In winter, our life was governed by where the wind blew, where snow accumulated. We could see that these natural phenomena were not random—that they were controlled, that there was a system. The processes of erosion and deposition were things we grew up with. An insulated society does not see how important terrain is to someone who has to understand it in order to live with it. Much of it meant life or death for the animals, and therefore survival for us. If there was one thing we learned, it was that you don't fight nature. You live with it. And you make the accommodations—because nature does not accommodate."

In the driest months, he saw mud cracks so firm a horse could step on them without breaking their polygonal form. When he saw the same patterns in rock, he had no difficulty discerning that the rock had once been mud and that the cracks within it were the preserved summer of a former world. In the Chalk Hills (multicolored badlands), getting down from his mount, he found the tiny jaws and small black teeth of what he eventually learned were Eocene horses—the first horses on earth, three hands tall.

Among the figures that appeared on the horizon and slowly approached the ranch—and sometimes stayed indefinitely—were geologists. The first he met were from the United States Geological Survey. Others worked for oil companies. The oilmen were well dressed and had shiny boots that caught his eye. Some of these people were famous in the science—for example, Charles T. Lupton, a structural geologist who had located the wildcats of the Cat Creek Anticline and discovered the oil of Montana. He did something like it in the Bighorn Basin. David particularly remembers him on two counts: first, that he "talked about the outside world," and, second, that he came in off the range with fragments of huge ammonites—index fossils of the late Cretaceous—and demonstrated by extrapolation that these spiral cephalopods had approached the size of wagon wheels. Lupton's obituary in the *Bulletin of the American Association of Petroleum Geologists* says, "Always he had a word to say to the children of his friends."

That was written in 1936, by Charles J. Hares, who had also made frequent stops at Love Ranch. Hares (1881–1970), in the course of a career in the Geological Survey and private business, became "the dean of Rocky Mountain petroleum geologists" and was one of the founders of the Wyoming Geological Association. His work on the anticlines of central Wyoming set up most of the major oil discoveries in the region. He was a celebrated teacher as well, and his roster of youthful field assistants in time became a list of some of the most accomplished geologists in America. Geologists who came to the ranch were reconnaissance geologists of the first rank, who went into unknown country and mapped it with an accuracy that is remarkable to this day. In David's words: "They raised a magic curtain. They showed us things we'd never seen. There was mother-of-pearl on some of the ammonites. There were Mesozoic oyster shells with both valves intact. You could open them up and see inside. All these things were marine—known only from ocean floors. They also brought in beautiful leaves, fifty million years old, from non-marine rocks of the Eocene. The seas were gone. The mountains had come up. Day after day, we could look around us and see, in the mind's eye, those things happening."

David's mother owned Joseph LeConte's *Elements of Geology*. He read it when he was nine years old. Did he grasp structure and stratigraphy then? Could he have begun to understand faulting? "To some extent, yes," he says. "After all, we could see it out in front of us."

On the southern horizon were the Gas Hills—a line of blue-banded ridges formed in a wedge like the prow of a ship (actually, an arch of shale). David would find uranium there in 1953. Riding over those ridges as a boy, he smelled gas. There were oil seeps as well. ("It was something you could relate to. The Gas Hills weren't called that for nothing.")

Oil and gas had entered the conversation at the ranch when David was four years old. In that summer (1917), derricks suddenly appeared in six different places within twenty miles; and, like other ranchers, the Loves began to muse upon a solvency giddily transcending the wool of frozen sheep. David's mother referred to all this as the family mirage.

Oil to us was once just a word recurring through the story of Wyoming. Indians and trappers told of curious oil seeps. Captain Bonneville in 1832 wrote about finding the "great tar springs" near what is now Lander. His party used the oil as a remedy for the cracked hoofs and harness sores of their horses, and as a "balsam" for their own aches and pains. Jim Bridger, scout, Indian fighter, and fort builder, mixed tar with flour and sold it along the Oregon Trail to emigrants, who needed axle grease for their wagons. They found, too, that buffalo chips made a hotter fire when a little tar was added to them.

And now she told the visiting geologists that if oil was what they were looking for they would surely find it under the ranch, because her younger son's initials were J.D.

Such excitement was contagious. Into our repetitive talk of sheep, cattle, horses, weather, and markets, new words appeared: anticline, syncline, red beds, sump, casing, drill stem, bits, crow's nest, cat walk, headache beam. Almost every herder had his own oil dome. We took up oil claims.

A range ne'er-do-well, grizzled and tattered, caught a ride to our house. He inquired importantly whether he might stay with us a few days while he did some validating work on his oil claims. Then he asked John if he might borrow a shovel. But to get to his claims, he said, he needed a team and a wagon. Having succeeded so far, he demanded, "Now, where's your oil?"

The boys might be far from sidewalks, but they would not grow up naïve.

A man named Jim Roush had a way of finding oil without a drill bit. Arriving at the ranch, he offered his services. Jim Roush was sort of a Music Man—an itinerant alchemist of structure, a hydrocarbonic dowser. He had a bottle that was wrapped in black friction tape. It dangled from a cord, and contained a secret fluid tomographically syndetic with oil. While David Love looked on—with his brother, his mother, and his father—Roush stood a few feet from their house and suspended the bottle, which began to spin. Light flashed on one hand—from a large apparent diamond. In silent concentration, he counted. Ypresian, Albian, Hauterivian, Valanginian—there was a geologic age in every spin. When the bottle stopped, its aggregate revolutions could be factored as depth to oil. David never saw Jim Roush again, except to the extent that his ghost might haunt the Geological Survey.

In 1918, a hundred-million-dollar oil-and-gas field was discovered at Big Sand Draw, where John Love gave up his first homestead, in 1897. After the Mineral Lands Leasing Act of 1920, oil companies could obtain leases directly from the government. A rancher's claims no longer intervened. A rancher needed fifty thousand dollars to drill on his own.

It was the general opinion on the range that if a man had that kind of money he did not need an oil well. Our mirage disappeared completely.

Emblematically, fire broke out in the oil fields of Lost Soldier, fifty miles southwest, and for weeks lighted up the night sky. In Horseshoe Gulch, six miles from the ranch, Sinclair Wyoming drilled forty-three hundred feet and found gas, which came out with such force that it destroyed the drill stem and blew the wooden derrick to pieces.

When the blast came, the driller was carrying a hundred-pound anvil across the rig floor. He told us that he raced half a mile over the sagebrush before he realized that he still held the anvil.

The Loves hitched up a wagon and went after the wood. They would burn the entire derrick in their kitchen stove.

David picked up a small, rough chunk of soft gray shale, blasted out of the depths of the well. He saw in it tiny marine fossils and fragile, lustrous pieces of mother-of-pearl, the size of his fingernail, that had once held the bodies of living clams; they came from more than half a mile underground; they lived before the time of men on the earth; they had been buried, how many ages, since they moved about that unseen shore. The driller told him that those shells predicted the presence of oil. . . . He brought home the rock with the delicate shells embedded in it and has kept it ever since in an Indian bowl.

When the boys were teen-aged, they occasionally saddled up and rode twenty-six miles to dances in Shoshoni. (I once asked David if they were square dances, and he said, "No. It was contact sport.") After dancing half the night, they rode twenty-six miles home. Their mother rented a house in Lander and stayed there with them while they attended Fremont County Vocational High School. One of their classmates was William Shakespeare, whose other name was War Bonnet. Lander at that time was the remotest town in Wyoming. It advertised itself as "the end of the rails and the start of the trails." Now and again, when the boys and their mother went visiting, they went through Red Canyon. A scene of great beauty, long sinuous Red Canyon was a presentation of the Mesozoic, framed in wide margins of time. In the eastern escarpment, the rock, tilting upward, protruded from the earth in cliffs six hundred feet high, and these were Eocene over Paleocene over Cretaceous and Jurassic benches above the red Triassic wall. Upward to the west ran a sage-covered Permian slope, on a line of sight that led higher in altitude and lower in time to the Precambrian roof of the Wind River Range—peaks on the western horizon.

Within the Triassic red was a distinctive white line that ran on as far as the eye could see. It was amazingly consistent, five feet thick, a limestone. In time, David would learn that this uniform bed covered fifty thousand square miles and was one of the most unusual rock units anywhere in the world, with such an absence of diagnostic fossils that no one could tell if the water it formed in was fresh or salt. ("It is a major marker bed of the Rockies," he once said, pointing it out to me. "Fifty thousand square miles—try to imagine any place in the world today where you can find that kind of stability. I can't. It's unique geologically.") As a youth, though, he was less fascinated by this Alcova limestone than by other aspects of Red Canyon. A woman who lived there was known as Red Canyon Red, for her striking Triassic hair. She was —as he would in later years describe her—"a whore *par excellence*." This may have been one reason that the boys' mother routinely accompanied her sons when they went through Red Canyon Red's canyon.

To shorten the trip to Lander, and make his own visits more frequent, John Love bought a used Buick.

Under severe nervous and vocal strain, he taught himself to drive, alone, on a wide expanse of level ground. Automatically he called "Whoa, there!" when he wanted to stop.

He knew a stockman who, in a similar effort, had failed, and had destroyed his car with an axe. John was resolved not to let that happen to him. He triumphed, of course, and the family was soon cruising to the Sweetwater Divide, with a picnic lunch and a jug of lemonade. Their horizons, already wide, before long rapidly expanded as John decided to take the first vacation of his life in order that the boys might see the Pacific Ocean before they went to college. The Loves headed west in the Buick. They had a breadbox, a camp stove, a nest of aluminum pots. The back seat was stacked high with blankets. Suitcases rode on the running boards, and on top of the luggage was a tepee.

The boys went to the University of Wyoming, where Allan majored in civil engineering. David majored in geology, and was elected to Phi Beta Kappa. Words carved into the university sandstone said:

STRIVE ON—THE
CONTROL OF NATURE
IS WON NOT GIVEN

David stayed in Laramie to earn a master's degree, and later, on a scholarship, went east to seek a Ph.D. at Yale. Arriving with some bewilderment in that awesome human topography, he noticed a line from Rafael Sabatini carved in stone in a courtyard of the Hall of Graduate Studies: "He was born with a gift for laughter and a sense that the world is mad." Those words steadied him at Yale, and helped prepare him for a lifetime in government and science. As a graduate student, he had to advance his reading knowledge of German, which he did over campfires on summer field work in the mountains of Wyoming. One book mentioned an inscription above a doorway at the German Naval Officers School, in Kiel—an unlikely place for a Rocky Mountain geologist to discover what became for him a lifelong professional axiom. As he renders it in English: "Say not 'This is the truth' but 'So it seems to me to be as I now see the things I think I see.' "

Yale had one of the better geology departments in the world, and its interests were commensurately global. It was syllogistic, encyclopedic, and stirred its students to extended effort—causing him to disappear into the library for months on end in what he calls his golden years. It was a department preoccupied with the Big Picture, and as a result it was not overcrowded with people who had seen a lot of outcrops. That, at any rate, is how it seemed to a student who had seen almost nothing but outcrops—close at hand, or slowly turning from the perspective of a saddle. In no way did this distinction diminish the reverence he felt for these eastern petrologues. "Their field geology was, let's say, incomplete," he will remark tenderly.

He did his field work in exceptionally rugged country—in the Tetons for a time, during those grad-school summers, but mainly along the southern margin of the Absaroka Range, roughly a hundred miles northwest of the ranch. He chose an area of about three hundred thousand acres (five hundred square miles) with intent to develop an understanding of it sufficient for the completion of a doctoral thesis. Geologically, it was a blank piece of the earth. Virtually nothing was known. The area had been mapped topographically. He took the map with him. Some of its streams ran uphill.

The Absarokas, it seemed, were a multilayered pile of pyroclastic debris—sedimentary rock whose components had once been volcanic outpourings. It was material that—after hardening—had been crumbled by weather and collected and moved by streams. The Absaroka volcanic sediments were a local part of the vast fill that had buried the central Rocky Mountains—a hard and therefore durable part. Their huge boulders indicated close proximity to the vents from which the rock had poured. (On a relief map, the Absarokas seem to spill out of Yellowstone Park.) During the Exhumation of the Rockies, the durability of these formations had left them in place. They resemble a battlement, standing seven thousand feet above the adjacent plains.

When Love chose this thesis topic, he was not choosing a journey from A to B. He did not mark off a little basin somewhere and essay to describe the porridge it contained. He picked a spinning pinwheel of geology, with highlights flashing from every vane. For example, the western end of the Owl Creek Mountains is still buried—under the Absarokas in the area of Love's thesis. Another chain of mountains is largely buried there, too. He discovered it and named it the Washakie Range. That the Absarokas were structurally separate from the Owl Creeks was obvious. It was not so readily apparent that the Owl Creeks were younger than the Washakies, or that the Washakies in their early days had been thrust southwestward over undistorted shales on the floor of the Wind River Basin. Gradually, he figured these things out, alone in the country, on foot. His thesis area reached a short distance into the Wind River Basin, and thus completed in its varied elements the panoply of the Rockies. It included folded mountains and dissected plateaus. It included basin sediments and alpine peaks, dry gulches and superposed master streams, desert sageland and evergreen forest rising to a timberline at ten thousand four hundred feet. It contained the story unabridged—from the preserved subsummit surface to the fossil topographies exhumed far below. At any given place in the area, temperatures could change eighty degrees in less than a day. The territory was roadless, and after a couple of years on foot he was ready to bring in a horse. Methodically cross-referencing the lithology, the paleontology, the stratigraphy, the structure, he mapped the region geologically—discovering and naming seven formations.

In one of those Yale summers, while taking some time away from the rock, he badly cut his foot in a lake near Lander. He made a tourniquet with his bandanna, and limped into town to see Doc Smith. This was Francis Smith, M.D., who had coaxed David's father past the tick fever, had seen David's mother through a strep infection that nearly killed her, and, over the years, had put enough stitches in David to complete a baseball. Now, as he worked on the foot, he told David that one of his recent office visitors had been Robert LeRoy Parker himself (Butch Cassidy).

David said politely that Cassidy was dead in Bolivia, and everybody knew that.

Smith said everybody was wrong. The patient had appeared in the doorway, and had stood there long and thoughtfully, searching the face of the doctor. Pleased by what he did not find, he said. "You don't know who I am, do you?"

The doctor said, "You look familiar, but I can't quite say."

The patient remarked that his face had been altered by a surgeon in Paris. Then he lifted his shirt, exposing the deep crease of a repaired bullet wound—craftsmanship that Doc Smith recognized precisely as his own.

The work that David Love was doing in Wyoming attracted the attention of the Geological Society of America. He was invited to speak at the society's national meeting, which, in the course of its migrations, happened to be scheduled for Washington, D.C. He suffered great anticipatory fright. It was most unusual for a graduate student to be asked to speak to the G.S.A. He was intimidated—by the East in general, by the capital city, by the fact that the foremost geologists in America would be there. Bailey Willis, of Stanford, would be there; Andrew Lawson, of Berkeley; Walter Granger, of the American Museum of Natural History; Taylor Thom, of Princeton. The paper that Love presented was on folding and faulting in Tertiary rocks. The Tertiary period runs from sixty-five million to something under two million years before the present. When Love went to Yale, the conventional wisdom in geology held that all folded and faulted rock was older than the Tertiary—that all Tertiary rock was undeformed. For his thesis in the Absarokas, he had mapped many areas of folded and faulted Tertiary rock. He knew the fossils, the stratigraphy. This was in no sense a horseback guess. He practiced carefully what he would say, and, when his moment came, there he was on a platform in the ballroom of the Hotel Washington struggling to control his voice, unaware that he had forgotten to button his suspenders. They were hanging down in back, exposed and flapping. His embarrassment had scarcely begun. At the climax of his presentation—as he described the deformation that had made clear to him that fifteen million years into Tertiary time the Laramide Revolution had not quite ended—he heard what he describes as hoots of derision, and when he finished there was no applause. The big room was silent. A moment passed, and then the structural geologist Taylor Thom, some of whose work was challenged by Love's paper, stood up and said, "This paper is a milestone in Rocky Mountain geology."

At his first G.S.A. meeting, a couple of years earlier in New York, about the only person he knew was Samuel H. Knight, his distinguished tutor from the University of Wyoming. Gradually, the faces and forms of strangers attached themselves to names long familiar to him on scientific papers and the spines of books. His personal pantheon came alive around him, and he was pleased to discover how easily approachable they were. When he tells of the experience—now that he is the Grand Old Man of Rocky Mountain Geology—he could be describing himself: "They put their pants on one leg at a time. They were very human individuals. They encouraged young people to speak out." As they discussed one another's papers, he relished their candor, their style of disagreement. A paleontologist named Asa Mathews got up and presented what he believed was the discovery that the world's first bird had come into existence in Permian time—roughly a hundred million years earlier than *Archaeopteryx*, the incumbent first bird. Mathews detailed some remarkable trace fossils in Permian rock in Utah, which sequentially recorded, he said, a bird as it ran along the ground, its wingtips awkwardly scraping until, finally, it took off. Walter Granger arose before the assembly to greet this unusual news. He said, "Professor Mathews has undoubtedly demonstrated that the first bird flew over that part of Utah, but he has not demonstrated that it landed."

David Griggs, who was not much older than Love, gave a superb demonstration of some fresh ideas about mountain building. Afterward, Bailey Willis, whom Love describes as "one of the grandfathers of structural geology in the world," anointed Griggs with praise. The great Andrew Lawson, who named the San Andreas Fault, was on his feet next, and virtually conferred upon Griggs an honorary degree by saying, "For the first and only time in my life, I agree with Bailey Willis."

In another context, a young geologist challenged Walter Granger, saying, "Dr. Granger, are you sure you're right?" Granger answered, without a flicker of hesitation, "Young man, I will consider myself a great success in life if I prove to be right fifty per cent of the time."

After Yale, Love worked a season for the U.S. Geological Survey in the Wasatch Mountains of Utah, scratching his way a step at a time through the dense, stiff branches of scrub oak that—ninety years before—had held up the Donner party enough to set the schedule for its eventual rendezvous with snow. Employed by the Shell Oil Company, he spent five years looking for areas of possible oil accumulation in the structures of Illinois, Indiana, Missouri, Kentucky, Georgia, Arkansas, Michigan, Alabama, and Tennessee. This increased his experience in ways that included a good deal more than rocks—especially in the southern Appalachians. Looking for outcrops, he walked many hundreds of miles of streambeds, "brushing water moccasins to one side." He studied roadcuts and railroad cuts. He slept where the day ended. Sometimes he stayed, for weeks at a time, with farmers. A farmer in Tennessee took him aside one day and offered him his unusually beautiful teen-age daughter in marriage if David would buy her a pair of shoes—her first. Apparently, the girl had developed such a longing for the young geologist that her father wanted her to have him. David felt that he was receiving "the ultimate compliment," for the farmer wanted to give him what the farmer valued most in the world. Earnestly, he wished not to offend, lest, among other things, he lose his advantage in a complicated country dissected by entrenched streams, a prime piece of the Mississippi Embayment, as geologists call the great bulb of sediments that reaches from the mouth of the Mississippi River as far north as Paducah, Kentucky. The farmer was in a strategic position to permit the young geologist to find certain permeable sandstones wedging out between layers of shale in updips where migrating petroleum might have become trapped. So he was anxious not to ruffle the feelings of his host. Besides, he was engaged.

In Laramie in 1934, David had met a geology student from Bryn Mawr College who had come west to spend a couple of semesters at the University of Wyoming under an arrangement that he would ever after refer to as her junior year abroad. Her father, in granting her permission to go to Wyoming, had commented that everyone has a right—at least once in a lifetime—to run away to sea. Her name was Jane Matteson, and she had grown up in the quiet streets and private educational enclaves of Providence, Rhode Island. She had twice the sophistication of this ranch hand, notwithstanding the fact that she wrote "crick" in her early field notes, believing that western geologists had taught her a new term. Moreover, she considered him "too good-looking." (Had he ever been inclined to, he could have answered the complaint in kind.) He appealed to her, though—in part because he kept his distance. She liked her cowboys unaggressive, and this one (at the time) was so shy—so reserved and respectful—that he stayed on his own side of his little Ford coupé. Her philosophy of conjugal evolution was contained in the phrase "A kiss is a promise," and for the time being there were no promises. He took her to the top of the Laramie Range, and up there on the pink granite under the luminous constellations gave her some gallant, if cryptic, advice. He said, "Whatever you do, don't come up here to look at the stars with a geologist."

When a letter arrived containing his acceptance at Yale, they went over the mountains to celebrate in Cheyenne. On the return trip, on old U.S. 30, they were met at the gangplank by a spring storm, and they worried that they would be snowed in for the night, with resulting damage to their reputations. Forging on through the blizzard, they made it to Laramie. After Jane finished Bryn Mawr, she did graduate work at Smith, returning to Wyoming for the summer field work in the Black Hills and the Bighorn Mountains which led to her master's thesis on Pennsylvanian-Permian rock. More apart than together while he completed his work at Yale, they exchanged long and frequent letters, which was her idea of a way for two people to get to know each other well and review their approaches to life. (When she was telling me this, not long ago, she added, with a bit of gemflash in her dark eyes, "I suppose that's why young people live together now—instead of writing letters.") In the nineteen-seventies, she edited a book on Mesozoic mammals for the University of California Press, but her own work in geology has been at best sporadic. "In our generation," she once remarked to me, "a woman's place was in the home raising a family if that was what she chose to do. I brooded a little bit about this but not much. You get to be twenty-five. You kick around the options. You decide you would rather be a wife and mother than a geologist. The fact that I can talk geology with him is just gravy." She offers her thoughts as an advocate of the geological devil to assist in the refinement of David's ideas. "What makes you think you know that? How could you possibly infer that?" she will say to him, not always stopping short of his Celtic irascibilities.

At Love Ranch, in 1910 or so, and apropos of who knows what, David's mother asked his father, "Is it true that it is necessary to kill a Scot or agree with him?"

John Love took the question seriously. After thinking it over, he answered, without elaboration, "No."

"David is not afraid of a new idea," Jane continues. "He's a pragmatist. He never looks back. He is both ingenious and practical. On the ranch when he grew up there was no plumbing, no electricity, no automobile—and the equipment they had they repaired themselves. If he has a piece of baling wire, he can fix anything. He fixes everything from plumbing to cars. He applies the same practicality to geology. If a slide block suggests that it might go downhill, he has the physics and he knows if it will work. His talent lies particularly in his sense of cause and effect. His knowledge, experience, and curiosity extend far beyond the mere presence of a rock. He is the most creative geologist I know."

They were married in 1940. Two of their four children were conceived after a summer field season and born during the next one, with him off in the wild, hundreds of miles away. That, says Jane, is just one more facet of geology. On a field trip somewhere in the late nineteen-seventies, David said to me, "I've been hearing about it for thirty-four years." The summer field season begins in June and ends about four months later, during which time she seldom saw much of him in their early years, a condition she regards as follows: "My father was a lawyer in Providence. After that junior year in Wyoming, the thought of being married to a lawyer in Providence gave me claustrophobia."

Their first child was a little white-haired kid named Frances, who arrived in Centralia, Illinois, where her father was based while he worked for Shell. Centralia was as rough as a frontier town, but "gone sour." Much of what happened there offended David's sense of fair play. In his words: "It was a boomtown, full of run-out seed. There hadn't been a fair killing there since 1823." There was union trouble. The hod carriers were trying to organize the oil drillers. The drillers resisted, and had no intention of paying what they regarded as tribute. The hod carriers attacked. They scalped a driller with a hunting knife and then broke his bones with hammers. "That was macho stuff, to them," David comments. "They played rough. They were a mean bunch of bastards." In some of the towns he visited were signs that said "NO DOGS OR OILMEN." He posed as a travelling salesman.

In Tennessee, he was sometimes mistaken for a revenue agent, which could have led to an unpleasant fate. And on one occasion he was taken for a railroad detective by some fugitives from the law. Jane happened to be with him, and as they made their way along the tracks, pausing like detectives to examine the rock, the fugitives—who had dumped into a railroad cut a corpse that had needed killing—were watching from the woods. They drew beads with their rifles but held their fire. They didn't want to include Jane. Eventually, David learned all this from the fugitives themselves, and he asked them if they were not made uneasy by the discovery that they might have killed an innocent man. They gave him a jug of sorghum.

In the evening in Centralia, David could read his newspaper by the light of gas flares over the oil fields. The company was burning off the gas because, at the time, it lacked economic value. This impinged his Scottish temper. "I don't consider that good stewardship," he explains. "We're stewards here—of land and resources. If you gut the irreplaceable resources, you're not doing your job. There were thousands of flares in Centralia. You could see them for a hundred miles." He was troubled as well by the secrecy of the oil company, which was otherwise an agreeable employer. As a scientist, he believed in the open publication of research, and meanwhile his work was being locked in a safe for the benefit of one commercial interest. Moreover, he moved around so much that in the first two years of his daughter's life she had been in thirteen states, while he was "looking for oil for some damn fool to burn up on the road." With Jane, he reached an obvious conclusion: "There has to be more to life than this."

He resigned to return to Wyoming with the U.S. Geological Survey, at first to pursue an assignment critical to the Second World War. The year was 1942, and the United States was desperately short of vanadium, an alloy that enables steel to be effective as armor plate. Working out of Afton, in the Overthrust Belt, he looked for the metal in Permian rock. He first identified vanadium habitat (where it was in beds of black shale), and then—in winter—built a sawmill and cabins and made his own timbers for eight new mines. Afton was a Mormon redoubt. The municipal patriarch had thirty-four children. The Loves and the haberdasher Isidor Schuster were the only Gentiles in town. Love enlisted some Mormon farmers and taught them how to mine. They worked in narrow canyons, where avalanches occurred many times a day, coming two thousand feet down the canyon walls. David was caught in one, swept away as if he were in a cold tornado, and badly injured. He lay in a hospital many weeks. Not much of him was not damaged. His complete statement of the diagnosis is "It stretched me out."

After the mines were well established, the Loves moved to Laramie, where he set up the field office that he would stay in for the rest of his U.S.G.S. career. His children grew up in Laramie. Frances now teaches French in public schools in Oklahoma. The Loves' two sons—Charles and David—are both geologists. Barbara is director of academic programs at Mukogawa Fort Wright Institute, in Spokane, a branch campus of Mukogawa Women's University, in Nishinomiya, Japan. Gradually—as regional field offices were closed and geologists were being consolidated in large federal centers in Menlo Park (California), Reston (Virginia), and Denver—Love became vestigial in the structure of the Survey. He resisted these bureaucratic winds even when they were stiffer than the winds that come over the Medicine Bows. "The tendency has been to have all the geologists play in one sand pile," he once explained diplomatically. And here his friend Malcolm McKenna, curator of vertebrate paleontology at the American Museum of Natural History, took up the theme: "Dave chose to stay where the geology is, and not to go up the ladder. He was so competent the Survey tried to get him to go out of Wyoming, but he wouldn't go. His is one of the few field offices left. The Survey gets information from people in addition to providing it. People stop in to see Dave. When he goes, the office in Laramie will close—and that will be a loss to Wyoming. While a whole bunch of people sit in little cubicles in Denver, Dave is close to the subject. He can walk out the door in the morning and do important stuff."

Love said that a part of his job was to find anything from oil to agates, and then, in effect, say, "Fly at it, folks," to the people of the United States. Within the law, he was always free to resign and then fly at something himself, but—whether by oil, uranium, gemstones, or gold—time after time he was not so much as tempted. Very evidently, he is not interested in money, and would not have joined the Survey in the first place if its services had been limited to commerce. The Survey evaluates the nation's terrain for academic purposes as well, there being no good way to comprehend any one aspect of geology without studying the wider matrix in which it rests. Within the geologic profession, the Survey has particular prestige—as much as, or even more than, the geology faculties of major universities, where chair professors have been known to mutter about the U.S.G.S., "They think they are God's helpers." Academic geologists tend to look upon the Survey as "stuffy." And, as Love discovered long ago, there is such an authoritarian atmosphere in the Survey—so much review of anything to be published, and so much hierarchical attention to a given piece of work—that sometimes when it is all done you cannot see the science for the initials that cover the paper. Established in 1879, the Survey had become so august that McKenna referred to it as "an inertial organization, a remnant of medieval scholasticism," but went on to say, "University people have two months a year; company people are restricted. The Survey can do things no one else can do." Many people in the profession tend to think that a geologist who has not at some point worked for the Survey has not been rigorously trained.

Love also established a base in Jackson Hole—a small house, eventually a couple of cabins. This would be the point of orientation for much of his summer field work. His absorptions over the years would take him to every sector of Wyoming, to other parts of the Rockies, and elsewhere in the world. Always, though, from his earliest days in geology, he would be drawn and drawn back to the Teton landscape—to the completeness of its history, the enigmas of its valley. To come to an understanding of one such scene is to understand a great deal about the geologic province of which it is a part, and more than any other segment of the Rockies he assigned himself to investigate the story of Jackson Hole.

"**H**ole" was a term used by the earliest whites to describe any valley that was closely framed by very high mountains. It was used by David Jackson, who essentially had his valley to himself, running his trap lines in the eighteen-twenties in the afternoon shadows of the Teton Range. Over time, bands of outlaws followed him, then cattlemen, and eventually homesteading farmers, whose fences invaded the rangeland, creating incendiary tension and setting the scene for the arrival of Shane, who came into the valley wearing no gun and "riding a lone trail out of a closed and guarded past." A farmer offered him employment, and he accepted—earnest in his quest for a peaceful life. The farmer asked Shane almost nothing of his history but felt he could trust him and imagined a number of ways in which the man might be needed on the farm. Deep in the stranger's saddle roll was an ivory-handled Colt revolver that came out of its holster with no apparent friction, had a filed-down hammer and no front sight, and would balance firmly on one extended finger. The farmer's young son quite innocently discovered the gun one day, and hurried to his father.

"Father, do you know what Shane has rolled up in his blankets?"

"Probably a gun."

"But—how did you know? Have you seen it?"

"No. That's what he would have."

"Well, why doesn't he ever carry it? Do you suppose maybe it's because he doesn't know how to use it very well?"

"Son, I wouldn't be surprised if he could take that gun and shoot the buttons off your shirt with you a-wearing it and all you'd feel would be a breeze."

Shane, of course, was a fictional character, but the era he represented was a stratum of the region. In the opening words of the novel, by Jack Schaefer, "He rode into our valley in the summer of '89." He also glanced "over the valley to the mountains marching along the horizon." The geography is vague, but Schaefer evidently had in mind a place beside the Bighorn Mountains. When Hollywood took up the story, though, and prepared to spread it from Cheyenne to Bombay, the valley that Shane would ride into seemed an almost automatic choice. Its floor, as he slowly moved across it, was generally as flat as the bottom of a lake. Incongruous in its center were forested buttes, with clear cold streams running past them. In many places, the flatness was illusory, for there was random undulation and, for no apparent reason, a lyrical quilting of stands of dark pine and broad open stretches of pale-green sage. There were ponds, some of them warm enough to hold trumpeter swans for the winter; and lying against the higher mountains were considerable lakes. Mountains were everywhere. On three sides of the valley, they went up in fairly stiff gradients—the Mt. Leidy Highlands, the Gros Ventre Mountains, the Snake River Range. On the western side—without preamble, without foothills, with a sharp conjunctive line at the meeting of flat and sheer—were the Tetons, which seemed to have lifted themselves rapidly past timberline in kinetic penetration of the sky. The Tetons resemble breasts, as will any ice-sculpted horn—Weisshorn, Matterhorn, Zinalrothorn—at some phase in the progress of its making. Hollywood cannot resist the Tetons. If you have seen Western movies, you have seen the Tetons. They have appeared in the background of countless pictures, and must surely be the most tectonically active mountains on film, drifting about, as they will, from Canada to Mexico, and from Kansas nearly to the coast. After the wagon trains leave Independence and begin to move westward, the Tetons soon appear on the distant horizon, predicting the beauty, threat, and promise of the quested land. After the wagons have been moving for a month, the Tetons are still out there ahead. Another fortnight and the Tetons are a little closer. The Teton Range is forty miles long and less than ten across—a surface area inverse in proportion not only to its extraordinary ubiquity but also to its grandeur. The Tetons—with Jackson Hole beneath them—are in a category with Mt. McKinley, Monument Valley, and the Grand Canyon of the Colorado River as what conservation organizations and the Washington bureaucracy like to call a scenic climax.

In the Teton landscape are forms of motion that would not be apparent in a motion picture. Features of the valley are cryptic, paradoxical, and bizarre. In 1983, divers went down into Jenny Lake, at the base of the Grand Teton, and reported a pair of Engelmann spruce, rooted in the lake bottom, standing upright, enclosed in eighty feet of water. Spread Creek, emerging from the Mt. Leidy Highlands, is called Spread Creek because it has two mouths, which is about as common among creeks as it is among human beings. They are three miles apart. Another tributary stream is lower than the master river. Called Fish Creek, it steals along the mountain base. Meanwhile, at elevations as much as fifteen feet higher—and with flood-control levees to keep the water from spilling sideways—down the middle of the valley flows the Snake.

One year, with David Love, I made a field trip that included the Beartooth Mountains, the Yellowstone Plateau, the Hebgen earthquake zone of the Madison River, the Island Park Caldera, and parts of the Snake River Plain. Near the end of the journey, we

MONTANA
WYOMING
ABSAROKA RANGE
Hebgen Lake
adison R
Yellowstone Lake
IDAHO
WYOMING
WASHAKIE RANGE
Jackson Lake
TETON RANGE
Signal Mtn
Spread Creek
Jenny Lake
MT. LEIDY HIGHLANDS
Moose
JACKSON HOLE
Fish Creek
Gros Ventre River
SNAKE RIVER RANGE
Snake River
Jackson
GROS VENTRE MTNS
Green River

came over Teton Pass and looked down into Jackson Hole. In a tone of sudden refreshment, he said, "Now, there is a place for a kid to cut his eyeteeth on dynamic geology."

Among others, he was referring to himself. He rode into the valley in the summer of '34. Aged twenty-one, he set up a base camp, and went off to work in the mountains. There were a number of small lakes among the Tetons at altitudes up to ten thousand five hundred feet—Cirque Lake, Mink Lake, Grizzly Bear Lake, Icefloe Lake, Snowdrift Lake, Lake Solitude—and no one knew how deep they were or how much water they might contain. The Wyoming Geological Survey wanted to know, and had offered him a summer job and a collapsible boat. He climbed the Tetons, and rowed the lakes, like Thoreau sounding depths on Walden Pond. He likes to say that the first time he was ever seasick was above timberline. If the Teton peaks were like the Alps—a transplanted segment of the Pennine Alps—there was the huge difference that just up the road from the Pennine Alps there are no geyser basins, boiling springs, bubbling muds, or lavas that froze in human time. His base camp was on Signal Mountain—by Teton standards, a hill—rising from the valley floor a thousand feet above Jackson Lake. More than fifty summers later, one day on Signal Mountain he said, "When I was a pup, I used to come up here to get away from it all."

I said, "By yourself?"

And he answered, "Oh, yes. Always. No concubines. I've always been pretty solitary. I still am."

Gouging around the mountains in his free time—and traversing the valley—he would get off his horse here and again, sit down, and think. ("You can't do geology in a hurry.") On horseback or on foot—from that summer forward, whenever he was there—he gathered with his eyes and his hammer details of the landscape. If he happened to come to a summit or an overlook with a wide view, he would try to spend as much of a day as possible there, gradually absorbing the country, sensing the control from its concealed and evident structure, wondering—as if it were a formal composition—how it had been done. ("It doesn't matter that I don't know what I'm looking at. Later on, it becomes clear—maybe. And maybe not. You try to put the petals back on the flower.") Some of those summits had not been visited before, but almost without exception he did not make a cairn or leave his name. ("I left my name on two peaks. When you're young and full of life, you do strange things.") Having no way to know what would or would not yield insight, he noticed almost anything. The mountain asters always faced east. Boulders were far from the bedrock from which they derived. There was no quartzite in any of the surrounding mountains, but the valley was deeply filled with gold-bearing quartzite boulders. He discovered many faults in the valley floor, and failed for years to discern among them anything close to a logical sequence. There were different episodes of volcanism in two adjacent buttes. From high lookoffs he saw the barbed headwaters of streams that started flowing in one direction and then looped about and went the other way—the sort of action that might be noticed by a person carrying water on a tray. Something must have tilted this tray. From Signal Mountain he looked down at the Snake River close below, locally sluggish and ponded, with elaborate meanders that had turned into oxbows—the classic appearance of an old river moving through low country. This was scarcely low country, and the Snake was anything but old. Several miles downstream, it took a sharp right, straightened itself out, picked up speed, and turned white. Looking down from Signal Mountain, he also noticed that moose, elk, and deer all drank from one spring just before their time of rut, crowding in, pushing and shoving to get at it ("They honk and holler and carry on"), ignoring the nearby waters of river, swamp, and lake. He named the place Aphrodisiac Spring. Over the decades, a stretch at a time, he completely circumambulated the skyline of Jackson Hole, camping where darkness came upon him, casting grasshoppers or Mormon crickets to catch his dinner. There were trout in the streams as big as Virginia hams. Sometimes he preferred grouse. ("I could throw a geology hammer through the air and easily knock off a blue grouse or a sage chicken. In season, of course. Hammer-throwing season. In the Absarokas, I threw at rattlesnakes, too. I don't kill rattlesnakes anymore. I've come to realize they're a part of the natural scene, and I don't want to upset it.") He carried no gun. He carried a bear bell instead. One day, when he forgot the bell, a sow grizzly stood up out of nowhere—six feet tall—and squinted at him. Suddenly, his skin felt dry and tight. ("Guess who went away.") A number of times, he was charged by moose. He climbed a tree. On one occasion, there was no tree. He and the moose were above timberline. He happened to be on the higher ground, so he rolled boulders at the moose. One of them shattered, and sprayed the moose with shrapnel. ("The moose thought it over, and left.")

The Gros Ventre River entered the valley almost opposite the high Teton peaks. A short way up the Gros Ventre was a denuded mountainside, where seventy-five million tons of rock had recently avalanched and dammed the river. He saw glacial grooves running north-south, and remembered the levees that kept the Snake from spilling west. This suggested to him that the valley floor had tilted westward since the glacier went by. Curvilinear pine-covered mounds cupped the valley's various lakes and held them close to the Tetons. Each lake was at the foot of a canyon. Evidently, alpine glaciers had come down the canyons to drop their moraines in the valley and, melting backward, fill the lakes. Some of the effects of ice were as fresh as that; others were less and less discernible, dating back from one episode of glaciation to another, separated by tens of thousands of years. Love's son Charlie, who teaches geology and anthropology at Western Wyoming Community College, was hiking one day in 1967 along the ridgelines of the Gros Ventre Mountains when he discovered boulders whose source bedrock was fifty miles away in the Absaroka Range. If they were glacial, as they seemed to be, they recorded an episode until then unknown, and of greater magnitude than any other. The evidence remains scant, but what else could have carried those boulders fifty miles and set them down on mountain summits at ten thousand feet? David answered the question by coining the term "ghost glaciation."

From lookoffs in and around Jackson Hole, the view to the north concluded with a high and essentially level tree-covered terrain that seemed to be advancing from the direction of Yellowstone, as indeed it had done, spreading southward, concealing the earlier topography, filling every creek bed, pond, and gulch. When he first rode in that terrain, he saw with no surprise that the rock was rhyolite, which has the same chemistry as granite but not its crystalline texture, because rhyolite cools quickly as a result of coming out upon the surface of the earth. This rhyolite, in a fiery cloud rolling down from Yellowstone, had buried the north end of the Teton Range, where it split and flowed along both sides.

From one end to the other of the valley were outcrops that from a distance looked like snow. Close up, they were white limestone, white shale, and white ash. After noting strikes and dips, and compiling the data, he calculated the thickness of the deposit as approximating six thousand feet. Top to bottom, it was full of freshwater clams and snails, and some beavers, aquatic mice, and other creatures that live in shallows. So the valley had been filled with a lake. The lake was always shallow. Yet its accumulated sediments were more than a mile thick. There was no rational explanation—unless the floor of the valley was steadily sinking throughout the life of the lake.

Volcanic rocks around the valley were white, brown, red, purple, and numerous hues of yellow and green. Quartzite boulders—stream-rounded, and scattered far and wide—had come from a source far to the northwest, in Idaho, and could not have been transported by ice. In the Mt. Leidy Highlands and along the eastern edge of Jackson Hole he saw other boulders, larger than human heads. Like the quartzites, they asked questions that, for the time being, he could not answer. He found black and gray sediments of the Cretaceous seas. He measured them, and they were two miles thick. Just above them in time, he found coal. In red and salmon rock nearby were the small tracks and tiny bones of dinosaurs. Larger ones, too. There were beds of marine phosphate. He collected cherty black shales, pure dolomites, dark dolomites, the massive sandstones of an ocean beach. He went into blue-gray caves in beautiful marine limestone. He found mud-crack-bearing shales. He saw mounds resembling anthills, which had been built by blue-green algae.

Chipping with his hammer, he bagged folded and fractured schist, amphibolite, and banded gneiss—and granite that had come welling up as magma, intruding these older rocks at a time when they were far below the earth's surface, a time that was eventually determined by potassium-argon dating. The time was 2.5 billion years before the present. Therefore, the rock that the granite invaded was a good deal older, but it had been metamorphosed, and there was no telling how long it had existed before it was changed—how far it reached back toward the age of the oldest dated rock on earth: a number approaching four billion years.

In these lithic archives—randomly assembled, subsequently arranged and filed—was a completeness in every way proportionate to the valley's unexceedable beauty. From three thousand million years ago to the tectonically restless present, a very high percentage of the epochs in the history of the earth were represented. It was no wonder that a geologist would especially be drawn to this valley. As he moved from panorama to panorama and outcrop to outcrop —relating this rock type to that mountain, this formation to that river—David gradually began to form a tentative regional picture, and after thirty years or so had placed in sequential narrative the history of the valley. When new evidence and insight came along, what had once seemed logical sometimes fell into discard. When plate tectonics arrived, he embraced its revelations, or accommodated them, but by no means readily accepted them. He wrote more than twenty professional papers on subjects researched in the vicinity, and, with his colleague John Reed, published a summary volume for the general public called *Creation of the Teton Landscape*. When the Department of the Interior honored him with a Citation for Meritorious Service, it said, in part, that he had "established the fundamental stratigraphic and structural framework for a region." In short, he had put the petals back on the flower.

And it was some flower. The Teton landscape contained not only the most complete geologic history in North America but also the most complex. ("One reason I've put in a part of my life here is that we have so much coming together. I don't want to waste my time. I can make more of a contribution by concentrating here than on any other place.") After more than half a century with the story assembling in his mind, he can roll it like a Roman scroll. From the Precambrian beginnings, he can watch the landscape change, see it move, grow, collapse, and shuffle itself in an intricate, imbricate manner, not in spatial chaos but by cause and effect through time. He can see it in motion now, in several ways responsively moving in the present—its appearance indebted to the paradox that while the region generally appears to have been rising the valley has collapsed.

Splitting the wall of the Tetons is a diabase dike a hundred and fifty feet wide, running like a dark streak of warpaint straight up the face of the mountains. Diabase: a brother of gabbro, a distant relative of granite. Four miles below the surface of the earth, the space occupied by this now solid dike was once a fissure through which the dark rock flowed upward as magma. At the same point in the narrative—1.3 billion years before the present, in the age of the Precambrian called Helikian time—marine beaches are not far to the west, and beyond them is a modest continental shelf. There is no Oregon, no Washington, not much Idaho—instead, blue ocean over ocean crust. Down toward the beaches flow sluggish rivers across a featureless plain. Folded and faulted schists and gneisses are bevelled under the plain, preserving in their deformation compressive crustal movements that have long since driven skyward uncounted ranges that have worn away. The Helikian beaches in their turn disappear, in burial becoming sandstone, which in the heat and pressure of more folding mountains is altered to quartzite. The mountains dissolve, and still another quiet plain vanishes below waves. The water advances into this piece of the world that will one day form as Jackson Hole. It lies close to the latitude of Holocene Sri Lanka—or Malaya, or Panama—and is moving toward the equator. The water is warm but not always quiet or clear. Blue-green algae build mounds in the shallows. There is a drop in sea level. Polygonal mud cracks become ceramic in the tropical sun. The sea returns. The water is virtually transparent, and the skeletons of billions of creatures form a pure blue-gray limestone. Like Debussy's engulfed cathedral, the site comes up now and again into the light and air, but for the most part seas stay over it. Sands accumulate—broad, deep sands—but they preserve almost no fossil record, so not even David Love will ever say with certainty whether they are underwater or out in the air. (What he cannot say with certainty he will readily say without certainty, provided the difference is clear. He prefers not to be, as he likes to put it, "a man walking with one foot on each side of a fence." He thinks that some of those sands were terrestrial dunes and coastlines, reddened as oxides in the air.) Jackson Hole is close to the equator, and phosphates form in the shallow evaporating sea. Tidal flats appear—wide red flats, thickened by slow rivers coming from an uplift far to the east. In the muds are small tracks and tiny bones of dinosaurs. Rapidly—and possibly as a result of the breaking up of the earth's only continent—the region travels north, moving about a thousand miles in thirty million years. Big dunes form upon the flats: dry, windblown dunes—a Sahara in salmon and red, at the precise latitudes of the modern Sahara. The red sands in turn are covered by the Sundance Sea. Coming from the north, it not only buries the big dunes under mud and sand but covers them with galaxies of clams. When the water drops, floodplains emerge, and flooding rivers band the country—

pink, purple, red, and green. Dinosaurs wander this chromatic landscape—a dinosaur as large as a corgi, a dinosaur as large as a bear, a dinosaur larger than a Trailways bus. Seas return, filled with a viciousness of life. Black and gray sediments pour into them from stratovolcanoes off to the west. In these times, the piece of sea bottom which is the future site of Jackson Hole overshoots the latitude of modern Wyoming and continues north to a kind of apogee near modern Saskatoon.

The land arches. Deep miles of sediments lying over schists and granites rise and bend. The seas drain eastward. The dinosaurs fade. Mountains rise northwest, rooted firmly to their Precambrian cores. Braided rivers descend from them, lugging quartzite boulders, and spreading fields of gold-bearing gravel tens of miles wide. Other mountains—as rootless sheets of rock—appear in the west, sliding like floorboards, overlapping, stacking up, covering younger rock, colliding with the rooted mountains, while to the east more big ranges and huge downflexing basins appear in the random geometries of the Laramide Revolution.

For all that is going on around it, the amount of activity at the site of Jackson Hole is relatively low. Across the future valley runs a northwest-trending hump that might be the beginnings of a big range but is destined not to become one. Miles below, however, a great fault develops among the Precambrian granites, amphibolites, gneisses, and schists—and a crustal block moves upward at least two thousand feet, stopping, for the time being, far below the surface.

New volcanoes rise to the north and east. Fissures spread open. Materials ranging from viscous lavas to flying ash obliterate the existing topography. Streams disintegrate these materials and rearrange them in layers a few miles away. So far, these scenes—each one of which is preserved in the rock of Jackson Hole—have advanced to a point that is 99.8 per cent of the way through the history of the earth, yet nothing is in sight that even vaguely resembles the Tetons. The Precambrian rock remains buried under younger sediments. At the surface is a country of undramatic hills. The movements that brought the Overthrust Belt to western Wyoming—and caused the more easterly ranges to leap up out of the ground—have all been compressional: crust driven against crust, folded, faulted, and otherwise deformed. Now the crust extends, the earth stretches, the land pulls apart—and one result is a north-south-trending normal fault, fifty miles long. On the two sides of this fault, blocks of country swing on distant hinges like a facing pair of trapdoors—one rising, one sagging. The rising side is the rock of the nascent Tetons, carrying upward on its back the stratified deposits of half a billion years. One after another, erosion shucks them off. Even more rapidly, the east side falls—into a growing void. Magma, in motion below, is continually being drawn toward volcanoes, vents, and fissures to the north. Just as magma moving under Idaho is causing land to collapse and form the Snake River Plain, magma drawn north from this place is increasing the vacuity of Jackson Hole. As the magma reaches Yellowstone, it rises to the surface, spreads out in all directions, and in a fiery cloud rolls down from Yellowstone to bury the north end of the Tetons, where it splits and flows along both sides. The descending valley floor breaks into blocks, like ice cubes in a bucket of water. Some of them stick up as buttes. A lake now fills the valley—shallow, forty miles long—and in it forms a limestone so white it looks like snow. There are white shales as well, and water-laid strata of white volcanic ash. As these sediments thicken to a depth approaching six thousand feet, the lake that rests upon them is always shallow, and full of freshwater clams and snails, and some beavers and aquatic mice. While the lake is accepting sediment, the bottom of its bottom is sinking at the same rate. With a loud terminal hissing, lavas flowing down from Yellowstone cool in the lake as obsidian. Fiery billows of sticky fog come down the valley as well. It cools as tuff. The big lake vanishes. In successive earthquakes, there is more valley faulting, damming the valley streams to form deep narrow lakes, which appear suddenly and as quickly go. Off the fast-rising block of mountains, erosion has by now removed fifteen thousand feet of layered sediments, and the Precambrian granites—with their attendant amphibolites, schists, and gneisses, and a vertical streak of diabase—are the highest rock below the sky. Bent upward against the flanks of the Precambrian are the broken-off strata of the Paleozoic era, and the broken-off strata of the Mesozoic era—serrated, ragged hogbacks, continually pushed aside. Perched on the granite at the skyline is a bit of Cambrian sandstone that the weather has yet to take away. On the opposite side of the Teton Fault, the same sandstone lies beneath the valley. The vertical distance between the two sides of this once contiguous formation is thirty thousand feet.

That brings the chronicle essentially to the present, but still the blockish mountains look more like hips than breasts. Now off the Absarokas, off the Wind Rivers, off the central Yellowstone Plateau—and, to a lesser extent, down the canyons of the Tetons—comes a thousand cubic miles of ice. A coalesced glacier more than half a mile thick enters and plows the valley. The west side of this glacier scrapes along the Tetons above the level of the modern timberline. Melting away, the glacier leaves a barren ground of boulders. More ice comes—a lesser but not insignificant volume—and a third episode, which is smaller still. The ice cuts headward up canyons into the mountains, making cirques. As rings of cirques further erode, they form the spires known as horns. The ice signs the valley with lakes, and as it shrinks back into the mountains human beings have come to watch it go. Long after it is gone, the valley floor, continuing to be unstable as magmas are drawn north from below, drops even more. Big spruce go down with it—trees with diameters of five feet—and are enveloped by the water of Jenny Lake. The mountains jump upward at the same time, many feet in a few seconds near the end of the fourteenth century, emphasizing the fact that they are active in our time. In 1925, seventy-five million tons of rock fall into the Gros Ventre River. In 1983, the year that the trees are discovered at the bottom of Jenny Lake, an earthquake halfway up the Richter scale rumbles through Jackson Hole.

A geologic map is a textbook on one sheet of paper. In its cryptic manner—its codes of color and sign—it reflects (or should reflect) all the important research that has been done on any geologic topic within its boundaries. From broad formational measurements down to patterns in the fabric of the rock, a map should serve as an epitome of what is known and not known about a region, up to date. Regional maps have traditionally been presented state by state, and the dates they are up to vary: Nevada 1978, New York 1970, New

Jersey 1910. On a geologic map, as on any scientific publication, the name of the person primarily responsible appears first. The job involves so many years and such a prodigious bibliography that the completion of a state geologic map can be regarded as the work of a lifetime, and David Love is only the second person in the history of American geology who has served as senior author of a state map twice (Wyoming 1955, Wyoming 1985). Geology is a descriptive, interpretive science, and conflict is commonplace among its practitioners. Where two or more geologists have come to divergent conclusions, Love has had to go out and rehearse their field work, in order to decide what to show on the map. People tend to become ornery when the validity of their assertions is challenged, and figuratively some of his colleagues have reached for their holsters, which may have been a mistake, as the buttons fell off their shirts and they felt a little breeze.

The 1955 Wyoming map set a standard for state geologic maps in the detail of its coverage, in its fossil dating, in its delivery of the essence of the region—a standard set anew in the 1985 edition. In the words of Malcolm McKenna, of the American Museum of Natural History, "Most maps are patched together from various papers and reports. Dave has looked at all the rock. It's all in one mind. Most geologic maps are maps of time, not rocks. They will say something like 'undifferentiated Jurassic' and omit saying what the rocks are. There is little of that on Dave's map. Mapping is below the salt now. Yet you can't look at satellite photos for everything. You've got to have high-resolution basic mapping. You have to keep your hand in with the real stuff. When the solid foundations aren't there, geologists are talking complete mush. Dave is making sure the foundation is there. He does not write about geology from a distance. He does not sit in high councils figuring out how the earth works. He is field-oriented. Some geologists think field work is wheeling their machines out into the yard. Dave has his hand on the pulse. He knows geology from having found it out himself. He has set an example of the way geology is done—one hell of an example. To compete with Dave, you'd have to do a lot of walking."

Love once picked up a mail-order catalogue and saw an item described as "Thousand Mile Socks." He sent for them skeptically but later discovered that there was truth in the catalogue's claim. They were indeed thousand-mile socks. He had rapidly worn them out, but that was beside the point.

Years ago, almost anybody going into geology could look forward to walking some tens of thousands of miles and seriously studying a comparable number of outcrops. Geology, by definition, was something you did in the field. You sifted fine dirt for fossils the eye could barely see. You chiselled into lithified mud to remove the legs of dinosaurs. You established time-stratigraphic relationships as you moved from rock to rock. You developed a sense of structure from, among other things, your own mapping of strikes and dips. In the vernacular of geology, your nose was on the outcrop. Through experience with structure, you reached for the implied tectonics. Gradually, as you gathered a piece here, a piece there, the pieces framed a story. Feeling a segment of the earth, you were touching a body so great in its dimensions that you were something less than humble if you did not look upon your conclusions as tentative. Like many geologists, Love became fond of the Hindu fable of the blind men and the elephant, because the poem in a few short verses allegorized for him the history and the practice of his science. "We are blind men feeling the elephant," he would say, almost ritually, as a way of reminding anyone that the crust is so extensive and complicated—and contains so little evidence of most events in earth history—that every relevant outcrop must be experienced before a regional outline can so much as be suggested, let alone a global picture.

In recent years, the number of ways to feel the elephant has importantly increased. While the science has assimilated such instruments as the scanning transmission electron microscope, the inductively coupled plasma spectrophotometer, and the $^{39}Ar/^{40}Ar$ laser microprobe—not to mention devices like Vibroseis that thump the earth to reflect deep structures through data reported by seismic waves—the percentage of geologists has steadily diminished who go out in the summer and deal with rock, and the number of people has commensurately risen who work the year around in fluorescent light with their noses on printouts. This is the age of the analog geologist, who, like a watch with a pair of hands, now requires a defining word. For David Love, the defining word is "field." Whereas all geologists were once like him, they are no longer, and his division of the science is field geology. He is the quintessential field geologist—the person with the rock hammer and the Brunton compass to whom weather is just one more garment to wear with his thousand-mile socks, the geologist who carries his two-hundred-gigabyte hard disk between his ears. There are young people following in his steps, people who still go out to scuff their boots and fray their jeans, but they have become greatly outnumbered by their contemporaries who feed facts and fragments of the earth into laboratory machines—activity that field people describe as black-box geology. Inevitably, some touches of tension have appeared between these worlds:

"Who is the new structural geologist?"

"Dorkney."

"Is he a field-oriented person?"

"He's a geophysicist, but he's a good guy."

"That would be difficult."

Black-box geologists—also referred to as office geologists and laboratory geologists—have been known to say that field work is an escape mechanism by which their colleagues avoid serious scholarship. Their remarks may rarely be that overt, but the continuing relevance of field geology is not—to say the least—universally acknowledged. Some laboratory geologists, on the other hand, are nothing less than eloquent in expressing their symbiosis with people of wide experience out in the terrain. "I spend most of my time working on computers and waving my arms," the geophysicist Robert Phinney once said to me, adding that he required the help of someone's field knowledge as a check, and without it would be in difficulty. "Without such people there would be no such thing as a geological enterprise," he went on. "Every box of samples that comes into the lab should include a worn-out pair of field boots. There's a group of senior geologists who have met on the outcrops and share a large body of knowledge. They paste together different perceptions of the world by visiting each other's areas. When I meet them, I chat them up like the guys at the corner store, because what I do is conceptual and idealized, and I'd like to know that it relates to what they have seen. These people are generally above fifty. Their kind is being diminished, which is a major intellectual crime. It has to do with the nature of science and what we're doing. Reality is not something you capture on a blackboard."

Such sentiments notwithstanding, within university geology departments black-box people tend to outvote field people on questions of curriculum and directions of research, and to outperform them in pursuit of funds. "The black-box era has been caused by the availability of money for esoteric types of work," Love remarked one day. "The Department of Defense, the National Science Foundation, and so forth have had money to spend on—let's say—unusual quests. The experience you get from collecting rocks in the field is lost to the lab geologist. For example, there's a boom in remote-sensing techniques—in satellite imagery. From that, you get a megapicture without going into the field. But it's two-dimensional. To get the third dimension—to study what's underground—they consult another sacred cow, which is geophysics. They can make a lot of these interpretations in the office. They can go off the mark easily, because for field relationships they often rely on data collected years ago. They use samples from museums, or samples collected by somebody else—perhaps out of context. I'm afraid I'm rather harsh about it, but we see misinterpretations, because of lack of knowledge of field relationships. Many of the megathinkers are doing their interpretations on the basis of second- and third-hand information. The name of the game now is 'modelling.' A lot of it I can't see for sour owl shit. How can you write or talk authoritatively about something if you haven't seen it? It isn't adequate to trust that the other guy is correct. You should be able to evaluate things in your own right. Laboratory geology is where the money is, though. The money is in the black box. I think eventually it will get out. You can't blame the kids for doing this kind of office research when they're financed. I don't want to do it myself. Putting the geologic scene into a broad perspective is for me more satisfying. I want to know what's over the next hill."

He was saying some of this in the Mt. Leidy Highlands one day when we were sitting on an outcrop at ninety-two hundred feet and looking at a two-hundred-and-seventy-degree view that ran across the pinnacled Absarokas to a mountain of lava of Pleistocene age and then on up the ridgeline of the Continental Divide to the glaciers and summits of the Wind River Range, thirty-eight feet higher than the Tetons. The skyline sloped gently thereafter, flattened, and became the subsummit surface of Miocene age, the level of maximum burial. There followed, across the southern horizon, the whole breadth of the Gros Ventre Mountains, with afternoon light on bright salmon cliffs of Nugget sandstone, at least four hundred feet high. The eye moved west over other summits and ultimately came to rest on the full front of the Tetons. We looked at it all for a considerable time in silence. Love said he liked this place because he could see so much from it, and had stopped here many times across the decades, to lean against a piñon pine and sort through the country, like an astronomer with the whole sky above him sorting through the stars. He also said, reflectively, "I guess I've been on every summit I can see from here."

Below us was Dry Cottonwood Creek. It ran southeast several miles, and then turned through a tight bend to head west toward the Tetons. We could see other streams almost identical in configuration, like a collection of shepherd's crooks. "The land tilted east, and then south, before it tilted west," Love said. "This is the tilting block that stops at the foot of the Tetons. The barbed streams are evidence that the hinge is east of us here. The hinge is probably the Continental Divide. We can learn a lot from streams. They're so sensitive. They respond to the slightest amount of tilting. I think this is underestimated." Pointing down to some sandstone ledges along the bank of Dry Cottonwood Creek, he said that Indians had frequently camped there because long ago the stream was so full of trout you could reach in under the ledges and catch them with your hand. He asked if I knew why the water was so clear. "There's no shale upstream," he said. "No fines to contaminate it. If you look at a stream, you can see in the sediments the whole history of a watershed. It's as plain as the lines on the palm of your hand."

On the way up to the lookoff, we had stopped at a spring, where I buried my face in watercress and simultaneously drank and ate. Love said that F. V. Hayden, the first reconnaissance geologist in Wyoming Territory, also happened to be a medical doctor, and he went around dropping watercress in springs and streams to prevent scurvy from becoming the manifest destiny of emigrants. Hayden, who taught at the University of Pennsylvania, led one of the several groups that in 1879 combined to become the United States Geological Survey. When he came into the country in the late eighteen-fifties, he was so galvanized by seeing the composition of the earth in clear unvegetated view that he regularly went off on his own, moved hurriedly from outcrop to outcrop, and filled canvas bags with samples. This puzzled the Sioux. Wondering what he could be collecting, they watched him, discussed him, and finally attacked him. Seizing his canvas bags, they shook out the contents. Rocks fell on the ground. In that instant, Professor Hayden was accorded the special status that all benevolent people reserve for the mentally disadvantaged. In their own words, the Sioux named him He Who Picks Up Rocks Running, and to all hostilities thereafter Hayden remained immune.

I remarked at the spring that Love was having nothing to drink.

He said, "If I drink, I'll be thirsty all afternoon."

And now, on the high outcrop, turning again from the Eocene volcanic Absarokas to the Wind River Range (the supreme expression in Wyoming of the Laramide Orogeny) and on to the newly risen Tetons (by far the youngest range in the Rockies), I mentioned the belief of some geologists that of all places in the world the Rocky Mountains will be the last to be deciphered in terms of the theory of plate tectonics.

"I don't think I would necessarily agree," Love said. "I think it is one of the more difficult ones, yes. I've thought a lot about it. At this stage, I'm uncomfortable with a direct tie-in. Until we have a detailed chronology of all the mountains, how can we plug them into a megapicture of plate tectonics? I don't want to give a premature birth to anything."

Plate-tectonic theorists pondering the Rockies have been more than a little inconvenienced by the great distances that separate the mountains from the nearest plate boundaries, where mountains theoretically are built. The question to which all other questions lead is, What could have hit the continent with force enough to drive the overthrust and cause the foreland mountains to rise? In the absence of a colliding continent—playing the role that Europe and Africa are said to have played in the making of the Appalachians—theorists have lately turned to the concept of exotic terranes: island arcs like Japan slamming up against the North American mainland one after another, accreting what are now the far western states, and erecting in the course of these collisions the evidential mountains. Whatever the truth in that may be, a tectonic coincidence very much worth noting is that the development of the western mountain ranges begins at the same time as the opening of the Atlantic Ocean. In the middle Mesozoic, as the Atlantic opens, the North American lithosphere, like a great rug, begins to slide west, abutting, for the most part, the Pacific Plate. A rug sliding across a room will crumple up against the far wall.

"We're about a thousand miles from the nearest plate boundary," Love was saying. "We should not tie in the landscape here with events that have taken place along the coast. This doesn't neutralize or dispose of the theory of plate tectonics, but applied here it's incongruous—it's kind of like a rabbit screwing a horse. There is no evidence of plates grinding against each other here. The thrust sheets are probably symptoms of plate-tectonic activity fifty million years ago, but the chief problem is that tectonism is not adequately placed in a time framework here. Almost everybody now agrees that there is tremendous significance to plate tectonics—also that the concept is valid. Most people don't argue about that anymore. Our arguments come in the details. We should dissect all these mountain ranges before we get diarrhea of the pen trying to clue them in to plate theory. There's nothing wrong with ideas, with working hypotheses, but unsubstantiated glittering generalities are a waste of time. Most of the megathinkers are basing sweeping interpretations on pretty inadequate data. There are swarms of papers being written by people who have been looking at state and federal and worldwide geologic maps and coming to sweeping conclusions on how mountains were formed and what the forces involved were. Until we know the anatomy of each mountain range, how are we going to say what came up when—or if they all came up in one great spasm? You can't assume they're all the same. In order to know the anatomy of each mountain range, you have to know details of sedimentary history. To know the details of sedimentary history, you have to know stratigraphy. I didn't know until recently that stratigraphy is dead. Many schools don't teach it anymore. To me, that's writing the story without knowing the alphabet. The geologic literature is a graveyard of skeletons who worked the structure of mountain ranges without knowing the stratigraphy. In Jackson Hole in the late Miocene, you had a lake that collected six thousand feet of sediment, half of which was limestone that was chemically precipitated. There had to be a source. It came from broad exposures of Madison limestone in the ancestral Teton–Gros Ventre uplift, chemically dissolved and then precipitated with cool-climate fossils. Therefore, that lake lay under a cool, humid climate. First, a basin had to be created in which the material was deposited—a basin ultimately thirteen thousand feet deep to accommodate all the lake and river sediments we find there, which puts it two miles below sea level at a time when the region is supposedly uplifting. All this is basic to structure, and the structure is basic to tectonics. The Owl Creek Mountains and the Uinta Mountains trend east-west. Why? Why are their axes ninety degrees from what you would expect if the tectonic force came from the west? You can do a torsion experiment with a rubber sheet and get folds in various directions—you can get east-west uplifts in the rubber sheet—but I would not say that is conclusive. You have mountains foundering. You have thrusting in the Laramide and sinking forty to fifty million years later, causing parts of basins to tilt this way and that like broken pieces of piecrust. The Granite Mountains were once as high as the Wind Rivers. Why did they go down? How did they go down? I don't think we're ready yet to put together a real megapicture. The 1985 geologic map of Wyoming consists of eighty-five-per-cent new mapping since 1955. The amount remapped shows how much new information was acquired in thirty years. The Big Picture is not static. It will always include new ideas, new tectonics, new stratigraphy. This information is an essential part of the mega-thinking of the plate-tectonics people, and twenty-five, fifty, a hundred years from now it will be very different."

West of Rawlins on Interstate 80, Love and I in the Bronco came into a region a good deal flatter than most of Iowa, with so little relief that there were no roadcuts for more than fifty miles. Among dry lakebeds dimpling the Separation Flats, our altitude was seven thousand feet, yet the distant horizon was close to the curve of the earth. In this unroughened milieu, we passed a sign informing us that we were crossing the Continental Divide. So level is the land there that the divide is somewhat moot. Cartographers seem to have difficulty determining where it is. Its location will vary from map to map. Moreover, it frays, separates, and, like an eye in old rope, surrounds a couple of million acres that do not drain either to the Atlantic or the Pacific—adding ambiguity to the word "divide."

With respect to underlying strata, we were running along the crest of an arch between two sedimentary basins, although nothing on the surface suggested that this was so, for the basins were completely filled. The flats to our left were the Washakie Basin, to our right the Great Divide Basin—each like a bowl brimming over with

Eocene alluvial soup. Younger deposits—maybe a mile's thickness —had long since been washed or blown away, leaving a fifty-million-year-old surface on which anything modern might fall.

A shepherd on horseback stood out against the sky, more so than his sheep. Even from a distance, he looked cold and uncomfortable. On this robust May afternoon, gray clouds, moving fast, were beginning to throw down hail. Love turned off the interstate, and the vehicle bucked south for a couple of miles in drab brown ruts that suddenly turned bright, almost white, as the ground jumped forward in time roughly fifty million years. This patch of thirty or forty acres was all that remained locally of a volcanic-ash fall that had covered large parts of Wyoming, Colorado, Kansas, and Nebraska and had reached as far as Texas. Radiometric dating has established the event at six hundred thousand years before the present—far along in Pleistocene time, and an extremely recent date in the history of the world when you reflect that the age of the earth is more than seven thousand times the age of that ash fall, just as the United States of America is more than seven thousand times as old as something that happened in the middle of last week. The ash—consisting of very small shards of glass—had travelled about two hundred miles downwind from its volcanic source. Two hundred miles downwind from Mt. St. Helens, in the state of Washington, the amount of ash that has accumulated as a result of Mt. St. Helens' recent eruptions is three inches. The ash here at the Continental Divide was sixty feet thick. A hundred and more miles northwest are remnants of the same fallout, suggesting the dimensions of the great regional blanket of six hundred thousand years ago, now almost wholly lost to erosion. Love said cryptically, "We have to assume it fell on saint and sinner alike." It had not been milled around by streams. It was a pure ash, distinctly wind-borne, containing no sand, no clay. He said that some woolly-mammoth bones had been found not far away, and with them as a minor exception this ash marked the only firm Pleistocene date in an area of twenty thousand square miles. After settling, it had not consolidated—as volcanic ash sometimes will, forming welded tuff. (The Vesuvian air-fall ash that settled on Pompeii also flew too high to weld. Rising rapidly like smoke, it actually pooled up against the stratosphere. Pliny said it looked like a flat-topped Italian pine. The geologic term for such an event is "Plinian eruption.") A couple of hundred miles northwest of us were the paint pots and fumaroles, the geysers and calderas of Yellowstone. Love said that this Lava Creek ash represented one of the great outpourings in Yellowstone history. The hail now was pelting us. It collected like roe on the brim of his Stetson. Love seemed to regard it as a form of light rain, as something that would not last even for six hundred thousand nanoseconds and was therefore beneath notice.

Most volcanoes and related phenomena—most manifestations of the sort represented by the surface history of Yellowstone—are lined up along boundaries of the twenty-odd plates that collectively compose the earth's outer shell. The plates, which are something like a sixtieth of the earth's radius, slide around on a layer of the mantle hot enough to be lubricious. Where plates spread apart (the Red Sea, the mid-Atlantic), fresh magma wells up to fill the gap. Where plates slide by one another (San Francisco, Jericho), the ground is torn and walls collapse. Where plates collide (Denali, Aconcagua, Kanchenjunga), impressive mountains form. In collision, one plate usually slides beneath the other, plunging—in the so-called subduction zone—as much as four hundred miles. The material carried down there tends to melt, and to rise as magma, reaching the surface in volcanic form, as in the Cascade Range, the Andes, the Aleutians, and Japan. Yellowstone, with all its magmatic products and bubbling sulphurs, attracts special attention in the light of this story, because Yellowstone is eight hundred miles from the nearest plate boundary.

When the theory of plate tectonics developed, it asked as well as answered questions—and not a few of the questions were inconvenient to the theory. Many had to do with volcanism. For example, why was the island of Hawaii pouring out lava in the dead center of the Pacific Plate? Similarly, if volcanoes were the products of subduction zones, where was the nearest subduction zone to the Tibesti Mountains of Saharan Chad? The Tibesti massif—a couple of thousand kilometres from the leading edge of the African Plate—consists of shield volcanoes like Mauna Kea and Mauna Loa. Where was the closest subduction zone to the chain of peaks that culminates in Mt. Cameroon, a stratovolcano fifteen hundred miles from the nearest plate boundary of any kind? Moreover, some of the fine old conundrums of geology—problems that antedated the plate-tectonics revolution—remained standing in its aftermath. What could explain the Canadian Shield? The South American Shield? The South African

Shield? How could so much Precambrian rock lie close to sea level and not have been buried in a thousand million years? What, in recent time, had lifted the platform of the Rockies, causing their exhumation? Why were Love and I, there on the platform, not at sea level? What had lifted the Colorado Plateau, subjecting it to incision by canyon-cutting rivers? What explained flood basalts? Plate tectonics seemed to have no relevance to them. With plate theory, you would think you could predict the sedimentary history of continents, but you couldn't. Why were continental basins—the Michigan Basin, the Illinois Basin, the Williston Basin—several kilometres deep? If you expect a shieldlike situation as the ultimate scene, what could explain these anomalous deep basins? Oil people wanted to know most of all. They asked plate theorists, "What does plate tectonics tell us about these basins?" The answer was "Nothing." Why were the granites of New Hampshire relatively young, and therefore anachronistic in the Appalachian story? What explained great crustal swells, like random blisters on the ocean floor, rising high above the abyssal plains? What could explain Bermuda—a mountain summit seventeen thousand six hundred and fifty-nine feet above the Hatteras Abyssal Plain? What created the Marshall Islands, the Gilbert Islands, the Line Islands, the Tuamotu Archipelago—where corals veneer the peaks of twenty-thousand-foot mountains that tend to run in chains? Like Yellowstone, like Bermuda, like Hawaii, like Mt. Cameroon, they lie great distances from the nearest intersections of plates.

Yellowstone draws its name from rich golden splashes of chemically altered volcanic rock. The place smokes and spits—the effects of proximate magma. On a geologic map of North America, Yellowstone appears at the eastern end of a bright streak of volcanic debris, coming off it like a contrail, extending across Idaho. With distance from Yellowstone, rock on that track is progressively older, descending age by age to the Columbia River flood basalts, which emerged from the ground like melted iron in early Miocene time, spread out across three hundred thousand square kilometres (in some places two and three miles deep), filled the Columbia Valley, and pooled against the North Cascades. By comparing the dates of the rock, one could be led to conclude that the geologic phenomenon now called Yellowstone has somehow been moving east at a rate of two and a half centimetres a year. As it happens, that is the rate at which, according to plate-tectonic theorists, North America is moving in exactly the opposite direction. In increasing numbers, geologists have come to believe that in a deep geophysical sense Yellowstone is not what is moving. They believe that the great heat that has expressed itself in so many ways on the topographic surface of the modern park derives from a source in the mantle far below the hull of North America. They believe that as North America slides over this fixed locus of thermal energy the rising heat is so intense that it penetrates the plate.

The geologic term for such a place is "hot spot." The earth seems to have about sixty of them—most older, and many less productive, than Yellowstone. Despite its position under thick continental crust, the Yellowstone hot spot has driven to the surface an amount of magma that is about equal to the over-all production of Hawaii, which has written a clear signature on the Pacific floor. Hawaii is the world's most preserved and trackable hot spot. You can see its geologic history on an ordinary map if the map shows even the rudiments of what lies below the sea. The Pacific Plate is moving northwest. It dives into the Japan Trench, the Aleutian Trench, and regurgitates the volcanic islands that lie on the far side. The plate used to move in a direction closer to true north, but forty-three million years ago it shifted course. Any hot spot now active under the Pacific Plate will produce islands or other crustal effects that appear to be moving in the opposite direction—southeast. Mauna Kea and Mauna Loa—the shield volcanoes that from seafloor to summit are the highest mountains on earth—stand close to the southeasterly tip of the Hawaiian Islands. The extremely eruptive Kilauea is making the tip. The islands become lower, quieter, older—the farther they lie northwest. Islands older still—defeated by erosion—now stand below the waves. These engulfed ancestors of Hawaii form a clear track in the Pacific crust for more than five thousand miles. When their age reaches forty-three million years, their direction bends about sixty degrees to the north. Above the bend, they are known as the Emperor Seamounts. Ever older, they continue to the juncture of the Kuril and Aleutian Trenches, into which they disappear. The oldest of the Emperor Seamounts is Cretaceous in age. Mauna Loa, of course, is modern. Under the ocean forty miles southeast of Mauna Loa is Loihi, a mountain of new basalt, which has already risen about twelve thousand feet and should make it to the surface in Holocene time.

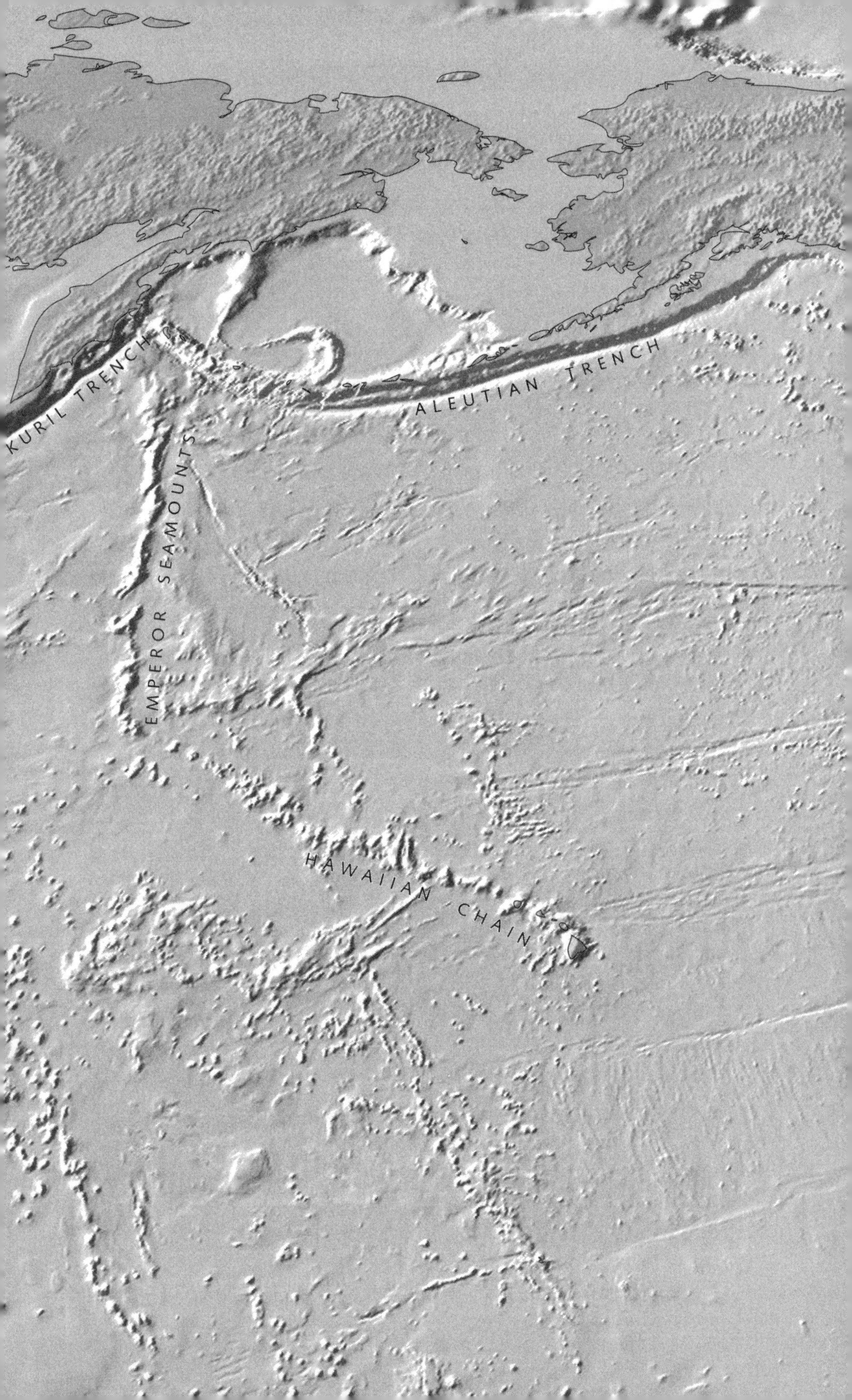
KURIL TRENCH
ALEUTIAN TRENCH
EMPEROR SEAMOUNTS
HAWAIIAN CHAIN

The ages of the Emperor Seamounts and the familial Hawaiian islands create the illusion that Hawaii is propagating southeast at a rate of nine centimetres a year, while the message from plate tectonics is, of course, that the Pacific Plate is what is moving. The speeds and directions of the plates have been established by a number of corroborant observations. Offsets in faults like the San Andreas have been measured as expressions of time. Places in California that were once side by side and are now six hundred kilometres apart are also separated by eleven million years. A great deal of ocean-crustal rock has been dredged up and radioactively dated. The ages have been divided by distance from the spreading center to determine the rate at which the rock has moved. More recently, methods have been refined for making annual measurements of plate motions by satellite triangulation. Hot spots provide one more way of calculating plate velocities, for hot spots are to the drift of plates as stars to navigation.

Conversely, it is possible to use established tectonic velocities to chart the tracks of hot spots with respect to the overriding plates. Given just one position and one date (the present will do), it is possible to say where, under the world, a hot spot would have been at any time across a dozen epochs. W. Jason Morgan, a geophysicist at Princeton, has sketched out many such tracks and reported them in various publications. Morgan can fairly be described as an office geologist who spends his working year indoors, and he is a figure of first importance in the history of the science. In 1968, at the age of thirty-two, he published one of the last of the primal papers that, taken together, constituted the plate-tectonics revolution. Morgan had been trained as a physicist, and his Ph.D. thesis was an application of celestial mechanics in a search for fluctuations in the gravitational constant. Only as a postdoctoral fellow was he drawn into geology, and assigned to deal with data on gravity anomalies in the Puerto Rico Trench. Fortuitously, he was assigned as well an office that he shared for two years with Fred Vine, the young English geologist who, with his Cambridge colleague Drummond Matthews, had discovered the bilateral symmetry of the spreading ocean floor. This insight was fundamental to the revolutionary theory then developing, and sharing that office with Fred Vine drew Morgan into the subject—as he puts it—"with a bang." A paper written by

H. W. Menard caused him to begin musing on his own about great faults and fracture zones, and how they might relate to theorems on the geometry of spheres. No one had any idea how the world's great faults—like, say, the San Andreas and Queen Charlotte faults—might relate to one another in a system, let alone how the system might figure in a much larger story. Morgan looked up the work of field geologists to learn the orientations of great faults, and found remarkable consistencies across thousands of miles. He tested them—and ocean rises and trenches as well—against the laws of geometry for the motions of rigid segments of a sphere. At the 1967 meeting of the American Geophysical Union, he was scheduled to deliver a paper on the Puerto Rico Trench. When the day came, he got up and said he was not going to deal with that topic. Instead, reading the paper he called "Rises, Trenches, Great Faults, and Crustal Blocks," he revealed to the geological profession the existence of plate tectonics. What he was saying was compressed in his title. He was saying that the plates are rigid—that they do not internally deform—and he was identifying rises, trenches, and great faults as the three kinds of plate boundaries. Subsequently, he worked out plate motions: the variations of direction and speed that have resulted in exceptional scenery. It was about a decade later when Morgan's Princeton colleague Ken Deffeyes asked him what he could possibly do as an encore, and Morgan—who is shy and speaks softly in accents that faintly echo his youth in Savannah, Georgia—answered with a shrug and a smile, "I don't know. Prove it wrong, I guess."

Instead, he developed an interest in hot spots and the thermal plumes that are thought to connect their obscure roots in the mantle with their surface manifestations—a theory that would harvest many of the questions raised or bypassed by plate tectonics, and similarly collect in one story numerous disparate phenomena.

In 1937, an oceanographic vessel called Great Meteor, using a newly invented depth finder, discovered under the North Atlantic a massif that stood seventeen thousand feet above the neighboring abyssal plains. It was fifteen hundred miles west of Casablanca. No one in those days could begin to guess at the origins of such a thing. They could only describe it, and name it Great Meteor Seamount. Today, Jason Morgan, with other hot-spot theorists, is prepared not only to suggest its general origin but to indicate what part of the world has lain above it at any point in time across two hundred million years. Roughly that long ago, they place Great Meteor under the district of Keewaytin, in the Northwest Territories of Canada, about halfway between Port Radium and Repulse Bay. That the present Great Meteor Seamount was created by a hot spot seems evident from the size and configuration of its base, which is about eight hundred kilometres wide and closely matches the domal base of Hawaii and numerous other hot spots. If a submarine swell is of that size, there is not much else it can be. That it was once, theoretically, somewhere between Port Radium and Repulse Bay is a matter of tracing and dating small circles on the sphere traversed by moving plates.

Keewaytin is in the center of the Canadian Shield. If the shield once had younger sediments on it, a hot spot underneath it would have lifted it up and cleaned it off, creating the enigma of the Canadian Shield. Morgan believes that various hot spots positioned in various eras under shield rock are what have kept it generally free of latter-day deposits. Stubborn fragments of the Paleozoic here and there on the shield suggest that this is so, as does the relatively modest number of meteorite craters. If the shield rock had been sitting there uncovered since Precambrian time, its surface could be expected to be more widely pockmarked, not unlike the plains of the moon.

Later in the Jurassic, the Great Meteor Hot Spot was under the west side of Hudson Bay, and in the early Cretaceous under Moose Factory, Ontario. All this is postulated not on any direct field evidence but simply on a charted extrapolation from an ocean dome nearly four thousand miles away. As time comes forward, however, the calculations place the hot spot—with its huge volumes of magma—under New Hampshire a hundred and twenty million years ago. The radioactively derived age of a good deal of granite in the White Mountains, so puzzlingly "anachronistic" in Appalachian history, is a hundred and twenty million years.

East of the North American continental shelf, lined up like bell jars on the Sohm Abyssal Plain, are the New England Seamounts. Their average height is eleven thousand feet. They are very well dated, and their ages decrease with distance east. Their positions and their ages—ninety-five million years, ninety million years, eighty-five million years—coincide with Morgan's mathematical biography of the Great Meteor Hot Spot.

A development that has greatly improved the precision of these measurements is argon-argon dating. A stream of neutrons in a nuclear reactor bombards a rock sample and causes a known fraction of its atoms of potassium to change into argon-39. Also in the sample are atoms of the isotope argon-40, which are unaffected by the bombardment and are the result of the natural decay of potassium through geologic time. The rate of decay is known and constant. The higher the proportion of argon-40, the older the rock. A mass spectrometer measures these ratios to establish a date. The older procedure known as potassium-argon dating—hitherto the best way of determining the age of something more than a few tens of thousands of years old—is done in two steps, requiring two samples. First, a chemical process determines how much potassium is present. Then a mass spectrometer looks at the second sample to see how much potassium has altered radioactively to become its daughter argon. The procedure suffers from the effects of weathering, which occur not only on the surface of rock but from grain to grain within. Argon slips away from weathered material, thus changing its overall ratio to potassium and making any date determined by this method all the more approximate. Argon-argon dating is accomplished in the microscopic core of a single grain, beyond even the faintest disturbances of weather. The newer method is significantly more consistent and accurate than the older one. Results have shown—notably among the New England Seamounts—that where many potassium-argon dates fall into general approximation with Morgan's calculations, the dates derived by argon-argon follow the track exactly.

Eighty million years ago, in the Campanian age of late Cretaceous time, Great Meteor would have underlain the American-African plate boundary, the Mid-Atlantic Ridge. Since then, Great Meteor has cut a gentle curve southward through the African Plate. From late Cretaceous, Paleocene, and Eocene time, the path is as well defined as it is on the American side. After the Eocene, the hot spot made the big seamount that bears its name. Then it began to go cold, to evanesce, to fade like a shooting star.

Shooting star. Almost everyone who describes hot spots is tempted to reverse reality and go for illusion at the expense of fact—that is, to narrate the apparent travels of hot spots as if they were in motion leaving trails like shooting stars, instead of telling the actual story of slow crustal drift over the fixed positions of thermal plumes. Myself included. With words, it is much easier to move a hot spot than it is to move a continent. Here, for example, is the story of another of the world's hot spots told in terms of its illusory motion. With the flood basalts of Serra Geral, in southern Brazil, a hot spot is said to have begun in late Jurassic time. It moved east under Brazil for several million years and then crossed over to Africa, which at that time was not much separated from South America. It lifted mountains in Angola, and then, doubling back, headed southwest under the ocean to form the Walvis Ridge, a line of seamounts leading to the hot spot's present position—Tristan da Cunha.

From the Serra Geral to the present island, the Tristan da Cunha hot-spot track is so well defined and dated that, as Morgan says, "it really ties down Africa." Not to mention South America.

An automatic inference from the theory is that hot spots perforating the same plates at the same times must make parallel tracks. On the floor of the Pacific, the tracks of the Line Islands, the Tuamotu Archipelago, the Marshalls, and the Gilberts parallel the track of Hawaii and the Emperor Seamounts. In the Atlantic, the Canary Islands have traced a curve parallel to Madeira's. Both are hot spots, and have left tracks that conform to Great Meteor. The Cape Verde Islands are a hot spot. A hundred and seventy million years ago, it was under New Hampshire, on a track nearly coincident with the later track of Great Meteor. The most voluminous intrusions of granite in the White Mountains are dated around a hundred and seventy million years. Cape Verde was where Charles Darwin first got off the Beagle with Charles Lyell's *Principles of Geology* in his hand, and quickly developed such admiration for Lyell's presentation of the science. Had Lyell told him that the Cape Verde Islands had also been on a voyage—that in a deep geophysical sense they had come from New England—Darwin might have thrown the book overboard.

Even on the best-defined tracks, not everything falls patly into place. There is some granite in New Hampshire that is two hundred million years old—still too young to be part of the Appalachian orogenic story but too old to be explained in terms of the two passing hot spots that left other granites. Possibly the two-hundred-million-year-old rock has something to do with magmas that came up at that time as the crust tore apart to admit the Atlantic. When Great Meteor arrived at the edge of the Canadian Shield, under the present site of Montreal, it presumably made the Monteregian hills, for one of which the city is named. The Monteregian hills are volcanic, but their potassium-argon age disagrees by twenty million years with the date when, by all other calculations, Montreal was over the hot spot—an exception that probes the theory. Morgan attributes the inconsistency to "random things you can't explain" and mentions the possibility of faulty dating. He also says, quite equably, "If the Monteregian hills really don't fit the model, you have to come up with another model."

The hot-spot hypothesis was put forward in the early nineteen-sixties by J. Tuzo Wilson, of the University of Toronto, as a consequence of a stopover in Hawaii and one look at the islands. The situation seemed obvious. James Hutton, on whose eighteenth-century *Theory of the Earth* the science of geology has been built, understood in a general way that great heat from deep sources stirs the actions of the earth ("There has been exerted an extreme degree of heat below the strata formed at the bottom of the sea"), but no one to this day knows exactly how it works. Heat rising from hot spots apparently lubricates the asthenosphere—the layer on which the plates slide. According to theory, the plates would stop moving if the hot spots were not there. Why the hot spots are there in the first place is a question that seeks its own Hutton. For the moment, all Jason Morgan can offer is another shrug and smile. "I don't know," he says. "It must have something to do with the way heat gets out of the lower mantle."

From very deep in the mantle (and perhaps all the way from the core) the heat is thought to rise in a concentrated column, and for this reason is alternatively called a plume. Its surface features are not proof in themselves that they are the product of some plant-stem phenomenon that is (or was) standing in the mantle far below. The chemistry of hot-spot lavas suggests that the rock is coming from below the asthenosphere, but there is no direct evidence of fixed hot spots in the mantle. They exist on inference alone. There is no way to sample the mantle. It can only be sensed—with vibrational waves, with viscosity computations, with thermodynamic calculations of what minerals do at different temperatures and pressures. Sound waves move slowly in soft rock, and some modes of the sound can be stopped completely where the rock is molten. The speed and patterns of seismic waves tell the story of the rock. Seismology is not quite sophisticated enough to look through the earth and count hot spots, but it approaches that capability, and when it gets there hot spots should appear on the screen like downspouts in a summer storm. If they don't, that may be the end of the second-greatest story in the youthful explorations of geological geophysics.

Hot spots seem to be active for roughly a hundred million years. Some of their effects on overriding plates last, of course, longer than they do. If they begin under continents, their initial manifestations at the surface are likely to be flood basalts. Hot-spot tracks have gone forth not only from the flood basalts of the Columbia River and the Serra Geral but in India from the flood basalts of the Deccan Plateau, in South Africa from the flood basalts of the Great Karroo, in East Africa from the flood basalts of the Ethiopian Plateau, in Russia from the flood basalts of the Siberian platform. Flood basalts are what the term implies—geologically fast, and voluminous in their declaration of the presence of a hot spot. In Oregon and Washington, in the middle Miocene, two hundred and fifty thousand cubic kilometres flowed out within three million years. Having achieved the surface in this form, the plume begins to make its track as the plate above slides by, just as Yellowstone, starting off from the flood basalts of Oregon and Washington, stretched out the pathway that has become the Snake River Plain.

An event of the brevity and magnitude of a great basalt flood is an obvious shock to the surface world. "We don't know what flood basalts do to the atmosphere," Morgan remarked one day in 1985, showing me a chronology he had been making of the great flood basalts that not only filled every valley "like water" and killed every creature in areas as large as a million square kilometres but also may have spread around the world lethal effects through the sky. Morgan's time chart of flood basalts matched almost exactly the cycles of death that are currently prominent in the dialogue of mass-extinction theorists, including the flood basalts of the Deccan Plateau, which are contemporaneous with the death of the dinosaurs—the event that is known as the Cretaceous Extinction.

The perforations made by hot spots may be analogous to the perforations in sheets of postage stamps. Plume tracks might weaken the plates through which they pass, so that tens of millions of years later the plates would break apart along those lines. Madeira, for example, first drew the line where Greenland broke away from Canada. The Kerguelen Hot Spot, in the Indian Ocean, may have helped India break away from Antarctica. The Crozet Hot Spot, also in the Indian Ocean, seems to have helped Madagascar get away from Africa. In the interior of the southern supercontinent of three hundred million years ago, a hot spot punched out the line that is now the north coast of Brazil. The same line is the Gold Coast and Ivory Coast of Africa. The hot spot now stands in the Atlantic as the island St. Helena.

The oldest rocks in Iceland are at the eastern and western extremes of the island, because Iceland is a hot spot whose track comes down from the northwest and at present intersects the Mid-Atlantic Ridge where Europe and America diverge. Iceland, for the time being, is spreading with the Atlantic. A hundred million years ago, the Mt. Etna Hot Spot was under the Ukraine, and seems to have cleaned off the Ukrainian Shield. A hot spot has made Ascension Island, on the South American Plate beside the Mid-Atlantic Ridge, fourteen hundred miles east of Brazil. It spent a hundred and ten million years under Africa after starting off from the Bahamas in early Jurassic time, when the transatlantic crossing was instantaneous, because there was no Atlantic. The high-standing Bahamas—eighteen thousand feet above the Hatteras Abyssal Plain—are defined as a carbonate platform, its wide shallow seas underlain by limestones and corals. Morgan says, "I would hope that if you drilled through them you would end up with basalt." The Labrador Hot Spot is thought to be "blind"—a hot spot that has not found a way to drive a plume to the surface but has nonetheless raised the terrain. This would account for the otherwise unaccountable altitudes of Labrador, not to mention more cleansing of the Canadian Shield. The Guiana Shield is also thought to lie above a blind hot spot, which has lifted the country and produced, among other things, the world's highest falls—a plume of water twenty times the height of Niagara.

Bermuda is the last edifice of a faint but evident hot spot, which underlies the ocean crust east of the present islands. The domal swell of the seafloor is classic—like Hawaii's, a thousand kilometres wide. (Under continents, upwelled masses analogous to the Bermudian and Hawaiian swells can be shown by satellite measurements of gravity anomalies.) Bermuda has not been active for thirty million years, but its track can be extrapolated westward in conformity with the track of Great Meteor and the well-established motions of the North American Plate. Seen in its former contexts, Bermuda proves to be a good bit less interesting for where it is now than for where it has been. If you could somehow look into the side of the American continent from Georgia to Virginia, you would see a great suite of Cretaceous strata dipping north and south, descending like a rooftop from an apex at Cape Fear. Something lifted up that arch, and, as one can readily discern from the stratigraphy and structure, whatever did the lifting did it in Paleocene time. Since the Paleocene, the North American Plate has moved the exact distance from Bermuda to Cape Fear.

Bermuda came through there like a train coming out of a tunnel. Or so it would appear. In the Campanian age of late Cretaceous time, when Great Meteor was in mid-Atlantic, Bermuda was under the Great Smoky Mountains. The Appalachian system consists of parallel bands of kindred geology sinuously winding from Newfoundland to Alabama, where they disappear under the sediments of the Gulf Coastal Plain. Why this long ropy package would stand high in two places and sink low in others is not explained by plate tectonics. It can be explained by hot spots. Great Meteor and Cape Verde seem to have lifted New England's high mountains, Bermuda the Smokies. Uplift accelerates erosion. The rock of the Permian period—the last chapter in the Appalachian mountain-building story—has been removed everywhere in eastern America except in West Virginia and nearby parts of Ohio and Pennsylvania, halfway between the hot-spot tracks, halfway between New Hampshire and North Carolina. Because plate motions have shifted over time, the tracks of all hot spots, ancient and modern, form a plexus on the face of the earth. Untouched areas between lines often prove to be continental basins—the Michigan Basin, the Illinois Basin, the Mississippi Embayment, the Williston Basin—while the rims of the basins are structural arches lined up on the tracks of the hot spots. Morgan thinks the large continental basins may have been created when hot spots elevated the edges. The Great Meteor track runs between the Hudson Bay Basin and the Michigan Basin. A Paleozoic hot spot seems to have made the Kankakee Arch, which separates the Michigan and Illinois basins. The Bermuda track runs between the Illinois Basin and the Mississippi Embayment. "Every basin gets missed," comments Morgan, with his hand on a map. "I don't think that's a coincidence."

Bermuda made the Nashville Dome. It lifted the Ozark Plateau, in middle Cretaceous time. "How much erodes off the top when a hot spot lifts something up depends on the durability of what's there," Morgan goes on. "If it's coastal mush, or Mississippi River mush, it goes quickly and in great volume. If it's quartzite, it resists. The resistant stuff stands up higher." Much of a hot spot's energy is expended in thinning the plate above it. Where the plate is already thin, most of the energy will appear at the surface in outpourings such as lava flows. When a plume has to come up through thick old craton, it makes kimberlites, carbonatites, gas-rich blowouts. The plume is expressing itself as a diatreme, the extremely focussed volcanic event that brings diamonds out of the mantle and explodes them into the air at Mach 2. The conduit is called a pipe because it is so narrow. The rock left inside it after the explosion is kimberlite. When Bermuda was under Kansas, it sent up the Riley County kimberlites. For many years, these diamond pipes were described as cryptovolcanic structures, meaning that nobody knew what they were. Later, they were thought to be meteorite strikes. In 1975, in Riley County, a hole was drilled with a tungsten-carbide bit that could smoothly cut its way through anything but diamonds. It went down sixty feet, where all penetration ceased. The bit was pulled. It was grooved and scarred. There are "meteor impacts" along the Bermuda track in Tennessee, southern Kentucky, and Missouri. Morgan thinks they are diatremes, or, as he puts it, "hot-spot blasts." They lie in a matrix of Paleozoic rock. If in fact they are meteor impacts, the hot spot would have lifted up the country and caused the erosion that exposed them to view. In Morgan's summation, "the thing works for me either way."

When Bermuda was under Wyoming, in Neocomian time, the

Rockies did not exist, but the magmas of the Idaho batholith had recently come in, a short distance up the track. When Bermuda was under the State of Washington, the State of Washington was blue ocean. If the track is followed back to two hundred million years, Bermuda seems to have been under Yakutat, Alaska. Hearing most of this for the first time at a colloquium in Princeton, a graduate student said, "This is like playing chess without the rules."

During the past twenty million years, the region that we like to call the Old West is thought to have been passing over not one but two hot spots, which have done much to affect the appearance of the whole terrain. The other one is less intense than Yellowstone, and is at present centered under Raton, New Mexico. Volcanoes are at the surface there. The Raton plume has lifted the Texas panhandle, the southern Colorado high plains. Its easternmost lava flow is in western Oklahoma. Its track, parallel to Yellowstone's, includes the Jemez Caldera, above Los Alamos, and may have begun in the Pacific. To the question "What lifted the Colorado Plateau, the Great Plains, and the Rocky Mountain platform?" the answer given by this theory is "The plumes of Raton and Yellowstone." As Utah and Nevada crossed the hot spots, the plumes are thought to have initiated the extensional faulting that has separated the sites of Reno and Salt Lake City by sixty miles in eight million years, breaking the earth into fault blocks and creating the physiographic province of the Basin and Range. Work done in the rock-dating laboratory of Richard Armstrong, a geochemist and geochronologist at the University of British Columbia, showed that Basin and Range faulting began at the western extreme of the region and moved eastward at a general rate of twenty-eight miles per million years—a frame commensurate in time and space with the continent's progress over the hot spots now positioned under Yellowstone and Raton. The Tetons began to rise eight million years ago and are clearly not products of the Laramide Orogeny. They are a result of extensional faulting, and conform to hot-spot theory as the easternmost expression of the Basin and Range. The Colorado Plateau lies between the two hot-spot tracks, and Morgan believes that their combined influence is what lifted it, setting up the hydraulic energy that has etched out the canyonlands. How the plateau avoided the rifting and extension that went on all around it—why it, too, did not break into blocks—is a question that leaves him baffled. That the two hot spots, at any rate, are progressively lifting the country is a point reinforced by a remarkable observation: a line drawn between them is the Continental Divide.

Inevitably, it has been suggested that someday North America may split apart along the Yellowstone perforations of the Snake River Plain. "That gives me a caution," says David Love. "I think there are some problems there. I have a feeling that the hot-spot ideas have been somewhat enlarged beyond the facts. The term itself probably means different things to different people. To me, a hot spot is an area of abnormally high temperature gradients, so high that it can be interpreted as having an igneous mush down below. In the Snake River Plain, the volcanics do get older east to west—in a broad sense, yes. But when you get down to details you get down to discrepancies. We don't know all the ages we should, on the various sets of volcanics. We need to learn them, and plot them up in geographic and time perspective. We will—but to my satisfaction we have not, yet. I would like to see a lot more regional information. In northwest Wyoming, volcanism began in the early Eocene, fifty-two million years ago. You got the Absaroka volcanic centers. Volcanic debris from them was spread by water and wind across the Wind River Basin, the Green River Basin. Then what happened? Everything went blah. The Yellowstone-Absaroka hot spot abruptly terminated at the end of Eocene time. Where the hell did that hot spot go? Twenty-five to thirty million years later, it was reactivated in the same place. What was that plume doing for all those millions of years? How do you reactivate a plume? We need answers to this sort of thing, and we don't have them. If the plume theory is correct, you've *got* to answer those questions."

The hail over the interstate turned to snow, and we passed a Consolidated Freightways tandem trailer lying off the shoulder with twenty-six wheels in the air—apparently overturned (a day or two before) by the wind. Abruptly, the weather changed, and we climbed the Rock Springs Uplift under blue-and-white marble skies. As we moved on to Green River and Evanston—across lake deposits and badlands, and up the western overthrust—the sun was with us to the end of Wyoming. On the state line was a flock of seagulls, in the slow lane, unperturbed, emblematically announcing Utah—these birds that saved the Mormons. Mormon traffic, heading home, did not seem intent on returning the favor.

If Wyoming can be said to have been acupunctured for energy, nowhere was this so variously evident as in the southwestern quadrant of the state, from the new coalfields near Rock Springs to the new oil fields of the Overthrust Belt, not to mention experimental attempts to extract petroleum from Eocene lacustrine shale, which —in that corner of Wyoming and adjacent parts of Colorado and Utah—contains more oil than all the rock of Saudi Arabia. More than the Union Pacific was after such provender now. "We are at the mercy of the east-coast and west-coast establishments," Love said. "It's been called energy colonization." And while we traversed the region, with scene after scene returning us to this theme, his reactions were not always predictable. There were moments that emphasized the scientist in him, others that brought out the fly-at-it-folks discoverer of resources, and others that brought forth a vigorous environmentalist, conserving his native ground, fulminating in the face of effronteries to humanity and the earth. Love is a prospector in the name of the people, who looks for the wealth in exploitable rock. He is also a pure scientist, who will follow his instincts wherever they lead. And he is a frequent public lecturer who turns over every honorarium he receives to organizations like the Teton Science School and *High Country News*, whose charter is to understand the environment in order to defend it. Thus, he carries within himself the whole spectrum of tensions that have accompanied the rise of the environmental movement. He carries within himself some of the central paradoxes of his time. Among environmentalists, he seems to me to be a good deal less lopsided than many, although beset by contradictory interests, like the society he serves. He cares passionately about Wyoming. It may be acupunctured for energy, but it is still Wyoming, and only words and images, in their inevitable concentration, can effectively clutter its space: a space so great that you can stand on a hilltop and see not only what Jim Bridger saw but also—through dimming tracts of time—what no one saw.

The Rock Springs Uplift, like the Rawlins Uplift, is a minor product of the Laramide Revolution, a hump in the terrain which did not keep rising as mountains. There was "red dog"—red clinker beds—in low cuts beside the road. When a patch of coal is ignited by lightning or by spontaneous combustion, it will oxidize the rock above it, turning it red. The sight of clinker is a sign of coal. Love said that this clinker was radioactive. Like coal, it was adept at picking up leached uranium. As the cuts became higher, we could see in the way they had been blasted the types of rock they contained. Where the cuts were nearly vertical, the rock was competent sandstone. Where the backslope angle was low, you knew you were looking at shale. Cuts that went up from the road through sandstone, then shale, then more sandstone, had the profiles of flying buttresses, firmly rising to their catch points, where they came to the natural ground. The shallower the slope, the softer the rock. The shallowest were streaked with coal.

At Point of Rocks, a hamlet from the stagecoach era, was a long roadcut forty metres high, exposing the massive sands of a big-river delta, built out from rising Rockies at the start of the Laramide Orogeny into the retreating sea. We left the interstate there and went north on a five-mile road with no outlet, which followed the flank of the Rock Springs Uplift and soon curved into a sweeping view: east over pastel buttes into the sheep country of the Great Divide Basin, and north to the white Wind Rivers over Steamboat Mountain and the Leucite Hills (magmatic flows and intrusions, of Pleistocene time), across sixty miles of barchan dunes, and, in the foreground—in isolation in the desert—the tallest building in Wyoming. This was Jim Bridger, a coal-fired steam electric plant, built in the middle nineteen-seventies, with a generating capacity of two million kilowatts—four times what is needed to meet the demands of Wyoming. Twenty-four stories high, the big building was more than twice as tall as the Federal Center in Cheyenne, which is higher than Wyoming's capitol dome. Rising beside the generating plant were four freestanding columnar chimneys so tall that they were obscured in cumulus from the cooling towers, which swirled and billowed and from time to time parted to reveal the summits of the chimneys, five hundred feet in the air. "This place is smoking the hell out of the country," Love said. "The wind blows a plume of corruption. In cold weather, sulphuric acid precipitates as a yellow cloud. It's not so good for people, or for vegetation. Whenever I think of this plant, I feel sadness and frustration. We could have got baseline data on air and water quality before the plant was built, and we muffed it." He blames himself, although at that time he had arsenic poisoning from springwater in the backcountry and was sick for many months.

The idea behind Jim Bridger was to ship energy out of Wyoming in wires instead of railway gondolas. Ballerina towers, with electric drapery on their outstretched arms, ran from point to point to the end of perspective, relieving pressure on the Oregon-Idaho grid. The coal was in the Fort Union formation—in a sense, the bottom layer of modern time. Locally, it was the basal rock of the Cenozoic, the first formation after the Cretaceous Extinction—when the big animals were gone, but not their woods and vegetal swamps. Wyoming had drifted a few hundred miles farther north than it is now, and around the low swamplands were rising forests of oak, elm, and pine. The terrain was near sea level. Mountains had begun to stir—Uintas, Wind Rivers, Owl Creeks, Medicine Bows—and off their young slopes they shed the Fort Union, its muds burying the compiled vegetation, cutting off oxygen, preserving the carbon. As the mountains themselves became buried, the fallen vegetation in the thickening basins was ever more covered as well, to depths and pressures that caused it to become a soft and flaky sub-bituminous low-rent grade of coal, a nonetheless combustible low-sulphur coal. With the Exhumation of the Rockies, nature, in the form of wind and water, worked its way down toward this coal. By the middle nineteen-seventies, nature had removed a mile of overburden, and had only sixty feet to go. At that point, something called the Marion 8200, an eight-million-pound landship also known as a walking dragline, took over the job.

The machine was so big it had to be assembled on the site—a procedure that required fourteen months. Now working within a mile or two of the generating plant, it could swing its four-chord deep-section boom and touch any spot in six acres, its bucket biting, typically, a hundred tons of rock, and dumping it to one side. The 8200 had dug a box canyon, its walls of solid coal about thirty feet thick. The inside of the machine was painted Navy gray, and had non-skid deck surfaces, thick steel bulkheads, handrails, and oval doors that looked watertight. They led from compartment to compartment, and eventually into the air-conditioned sanctum of Centralized Power Control, where, lined up in ranks, were electric motors. The foremost irony of this machine was that it was far too large and powerful to operate on diesel engines. Although the chassis was nine stories high, it could not begin to contain enough diesels to make the machine work. Only electric motors are compact enough. Out the back of the machine, like the tail of a four-thousand-ton rat, ran a huge black cable, through gully and gulch, over hill and draw, to the generating plant—whose No. 1 customer was the big machine.

Once every couple of hours, the 8200 walked—raised itself up on its pontoonlike shoes and awkwardly lurched backward seven feet, so traumatically compressing the dirt it landed on that smoke squirted out the sides and the ground became instant slate. This machine—with its crowned splines, its precise driveline mating, its shop-lapped helical gears, its ball-swivel mounting of the boom-point sheaves, its anti-tightline devices and walking-shoe position indicators—had unsurprisingly attracted the attention of Russian engineers, who came in a large committee to see Jim Bridger, because they were about to build twenty-five similar generating stations in one relatively concentrated area of Siberia, which, they confided, closely resembled Sweetwater County, Wyoming.

This strip mine, no less than an erupting volcano, was a point in the world where geologic time and human time overtly commingled. Ordinarily, the close relationship between the two is masked: human time, full of beepers and board meetings, sirens and Senate caucuses, all happening in microtemporal units that physicists call picoseconds; geologic time, with its forty-six hundred million years, delivering a message that living creatures prefer to return unopened to the sender. In this place, though, geology had come up out of its depths to join the present world, and, as Love would put it, all hell had broken loose. "How people look at it depends on whose ox is being gored," he said. "If you're in a brownout, you think it's great. If you're downwind, you don't. Wyoming's ox is being gored."

When the Bridger operation was under construction, hundreds of tents and trailers lined most of the five miles of the spur road to the site—an "impact" that ultimately shifted to Rock Springs, thirty miles away, and Superior, and other small towns in the region. Populations doubled during the coal rush, which was close in time to the booms in trona mining and oil. Even after the booms had settled down, twenty-eight per cent of the people of Wyoming were living in mobile homes. During the construction of Jim Bridger, Rock Springs, especially, became a heavy-duty town, attracting people with no strong attachments elsewhere who came into the country in pickups painted with flames. With its bar fights and prostitutes, it was wild frontier territory, or seemed so to almost everyone but David Love. "Fights were once fights," he commented. "Now the fight starts and your friends hold you back while you throw insults." Cars were stripped of anything that would come off. Pushers arrived with every kind of substance that could stun the human brain. A McDonald's sprang up, of course, decorated with archaic rifles, with plastic cattle brands lighted from the inside, with romantic paintings of Western gunfights—horses rearing under blazing pistols on dusty streets lined with false-fronted stores. A Rock Springs policeman shot another Rock Springs policeman at point-blank range and later explained in court that he had sensed that his colleague was about to kill him. How was that again? The defendant said, "When a man has the urge to kill, you can see it in his eyes." The jury saw it that way, too. Not guilty. Some people in Sweetwater County seemed to be of the opinion that the dead policeman needed killing.

Love's son Charlie, who lives and teaches in Rock Springs, once told us that the community's underworld connections were "only at the hoodlum level." He explained, "The petty gangsters here aren't intelligent enough for the Mafia to want to contact. You can't make silk purses out of sows' ears."

The number of cowboys in Wyoming dropped from six thousand to four thousand as they rushed into town to join the boom, disregarding the needed ratio of one man per thousand head of cattle. In desperation for help at branding time, calving time, and haying time, ranchers had to go to the nearest oil rig and beg the roughnecks to moonlight.

For a steam-driven water-cooled power plant, this one seemed to have a remarkably absent feature. It seemed to be missing a river. The brown surrounding landscape was a craquelure of dry gulches. In one of them, though—a desiccated arroyo called Dead Man Draw—was a seventy-five-acre lake, fringed with life rings, boats, and barbecue grills. At the rate of twenty-one thousand gallons a minute, Jim Bridger was sucking water from the Green River, forty miles to the west. To cool an even drier power station, some hundreds of miles away in northeast Wyoming, a proposal had been made to pump Green River water over the Continental Divide to the Sweetwater River, which runs into the North Platte, from which the water would be pumped over a lesser divide and into the Powder River Basin. Love said, "That would destroy the whole Sweetwater regimen, destroy the Platte, and destroy the Powder River, all for coal in the Powder River Basin—a slurry pipeline or something of the sort. It's very much on the books. If they go in for the gasification of coal, they're going to need it. It's known as the transbasin diversion of the Green River. The water has fluorine in it. Wherever it gets into the ground, it can pollute the water table in ten to fifteen years. The river also picks up sodium from trona. In the town of Green River, the sodium in the drinking water greatly exceeds E.P.A. standards. If they decide to pipe the water over the Continental Divide, water quality could be lowered in the Powder River Basin to the point of needing a desalinization plant."

We moved on toward Green River, where the most spectacular suite of roadcuts and rock exposures anywhere on Interstate 80 contained in its sediments the history of these evils. Dark mountains, spread low across the horizon, might have been a storm coming on—and in a sense they were, or had been. They were the Overthrust Belt, cumulate from the west. Looking north to the even more distant Gros Ventres and Wind Rivers, and south to the high cirques of the Uintas, we were encompassing in a wide glance about sixteen thousand square miles of land, much of it so dry, stacked flat like crumbling hardtack, that only a geologist could absorb such a scene and see in it a lake that would rank seventh in the world.

In the Eocene, when the lake existed, the appearance of North America approached its present form. The journey from New York to Paris may have been eight hundred miles shorter than it is now, but the North Atlantic was a maturing ocean. The Appalachians were much higher. There were no Great Lakes. There was no westerly rise to the Great Plains. The foreland ranges of the Rockies had pooched out from their sea-level platform, and west-running rivers were flowing around them to pool against the overthrust mountains. In California there was no Sierra, in Nevada and Utah no mountains of the Basin and Range—only moist gentle country coming in from the Pacific Coast. This Eocene time line, drawn from either end of the continent, would have converged in western Wyoming in something comparable to the Sea of Azov. A hundred and fifty miles long, a hundred miles wide, it was larger by far than Erie, larger than Lake Tanganyika, larger than Great Bear. It was two hundred times Lake Maggiore. It had no name until a century ago, when a geologist called it Lake Gosiute.

Lakes are so ephemeral that they are seldom developed in the geologic record. They are places where rivers bulge, as a temporary consequence of topography. Lakes fill in, drain themselves, or just evaporate and disappear. They don't last. The Great Lakes are less than twenty thousand years old. The Great Salt Lake is less than twenty thousand years old. When Lake Gosiute took in the finishing touch of sediment that ended its life, it was eight million years old.

West of Rock Springs, we came to an escarpment known as White Mountain, standing a thousand feet above the valley of Killpecker Creek. In no tectonic sense was this a true mountain—a folded-and-faulted, volcanic, or overthrust mountain. This was just a Catskill, a Pocono, a water-sliced segment of layered flat rock, a geological piece of cake. In fact, it was the bed of Lake Gosiute, and contained almost all of the eight million years. Apparently, the initial freshwater lake eventually shrank, became bitter and saline, and intermittently may have gone dry. Later, as the climate remoistened, water again filled the basin, and the lake reached its greatest size. As we looked at White Mountain, we could see these phases. It was the dry, salt-lake interval in the middle—straw and hay pastels so pale they were nearly white—that had given the bluff its name. The streams that had opened it to view were lying at its base. Killpecker Creek (full of saltpeter) flowed into Bitter Creek, and that soon joined the Green River.

Down the road a couple of miles was a pair of tunnels—snake eyes in the lakebed. They were one of the three sets of tunnels on

Interstate 80 between New York and San Francisco, and they had to be there in the nose of White Mountain, or the interstate, flexing left, would destroy the town of Green River. Tower sandstone stood on the ridgeline in castellated buttes. With each mile, they increased in number, like buildings on the outskirts of a city. Off to the left was the island from which the geologist John Wesley Powell—seven years before the battle of the Little Bighorn—set off in a flotilla of dinghies to follow the Green River into its master stream, and to survive the preeminent rapids of North America on the first known voyage through the Grand Canyon. A huge sandstone broch stood in brown shale above the tunnels, which penetrated the lakebed's saline phase.

We burst into the light at the western end among concentrated stands of lofted redoubts, a garden of buttes, and huge walls of flat strata in roadcuts and rivercuts extending a full mile. William Henry Jackson photographed this scene, in 1870, for the Hayden Survey. The buttes have been given names like Tollgate Rock, Teakettle Rock, Sugar Bowl Rock, Giant's Thumb. Love said there were Indian petroglyphs on Tollgate Rock but they were far too high to see. "You'd have to be a mountain goat to get up there," he went on, and scarce had he uttered the words when a figure white as gypsum appeared on a cone of talus at the base of the Tower sandstone, close by the petroglyphs, its head in motionless silhouette. "You can tell people just to look for that goat if they want to see where the petroglyphs are," he advised me. "They can always find the petroglyphs by looking for the goat."

About halfway up White Mountain was a layer of sandstone that happened to be phosphatic and contained uranium. Love said he knew this because he had discovered the uranium. Non-marine phosphate, largely unknown elsewhere in the world, was one of the many legacies of this strange vanished lake. A few miles back, the uraniferous phosphatic sandstone had formed a low ridge in the path of the interstate, which cut straight through it, dosing all drivers with a few milliroentgens to keep them awake.

He also remarked that the sedimentary story was reflecting a lot of tectonic history. You could see the orchestration of the mountain ranges by reading backward through the layers of sediment. For example, from the age and position of the sedimentary rock derived from the Uinta Mountains and the Wind River Range you could see the Wind Rivers developing first.

At all moments in the history of Lake Gosiute, it was replete with organic life, from the foul clouds of brine flies that obscured its salty flats to the twelve-foot crocodiles and forty-pound gars in the waters at their widest reach. For this was Wyoming in the Eocene, and in the lake at varying times were ictalurid catfish, bowfins, dogfish, bony tongues, donkey faces, stingrays, herring. The American Museum of Natural History has a whole Gosiute trout perch in the act of swallowing a herring, recording in its violence two or three seconds from forty-six million years ago. In the museum's worldwide vertebrate collection, roughly one fossil in five comes from Wyoming, and a high percentage of those are from Gosiute and neighboring lakes. Around the shores were red roses and climbing ferns, hibiscus and soapberries, balloon vines, goldenrain. The trees would generally have been recognizable as well: pines, palms, redwoods, poplars, sycamores, cypresses, maples, willows, oaks. There were water striders, plant hoppers, snout beetles, crickets. The air was full of frigate birds. Dense beds of algae matted the shallows. In all phases through the eight million years, quantities of organic material mixed with the accumulating sediments and are preserved with them today in the form of oil shale. On the far side of the Uinta Mountains was another great lake, reaching from western Colorado well into Utah. Lake Uinta, as it has come to be called, and Lake Gosiute and several smaller lakes left in their shales a potential oil reserve estimated at about one and a half trillion barrels. This is the world's largest deposit of hydrocarbons. It is actually nine times the amount of crude oil under Saudi Arabia, and about ten times as much oil as has so far been pumped from American rock.

Distinct in the long suite of cuts at Green River were the so-called mahogany ledges, where oil shale is particularly rich. They looked less like wood than like bluish-white slabs of thinly bedded slate. Oil shale always weathers bluish white but is dark inside, and grainy like wood. The thinner the laminae, the higher the ratio of organic material. The richest of the dark oleaginous flakes—each representing the sedimentation of one year—were fifteen-thousandths of a millimetre thick. Love dropped some hydrochloric acid on the rock, and the acid beaded up like an arching cat. "It's actually kerogen," he said. "It converts to high-paraffin oil. It's not like Pennsylvania crude."

To mining engineers, oil shale had presented an as yet unsolved and completely unambiguous problem: how to remove the shale without destroying the face of the earth. So far, three principal methods had been considered. One was to strip-mine it, crush it, separate the oil, then smooth out the tailings—a process that could result in the absolute rearrangement of twenty-five thousand square miles. Another was to go underground, excavate a percentage of the rock, and refill the caverns with tailings. That was known as the "modified in situ" approach. And finally someone thought of drilling a hole, pumping in propane, and starting a fire. The heat would cause liquid oil to run out of the shale. The oil could be forced up through another well before the fire destroyed it. A burn would not, like a clinker fire, continue indefinitely. If oxygen was not fed to the flames, they would die. This was known as "true in situ" mining; and there in White Mountain, a few miles away, the federal government had been perfecting the technique. The experiments thus far had brought down the recovery cost to a million dollars a barrel. In Cheyenne one time, I saw a Peter Pan Crunchy Peanut Butter jar filled with such oil. It looked and smelled like the contents of a long unemptied spittoon.

The one-and-a-half-trillion-barrel estimate was somewhat extravagant, because it included every last drop—referring, as it did, to all shale with any content of kerogen. In the richer rock—in the shales that contained from twenty-five to sixty-five gallons of oil per ton—were no more than six hundred billion barrels. That would do. That was more petroleum in place than all the petroleum produced in the world to date. Love remarked that oil shale had been "trumpeted to the skies" but, with the energy crisis in perigee, both government and industry were losing interest and pulling out. Temporarily pulling out. Sooner or later, people were going to want that shale.

For Lake Gosiute to have lasted so long in a mountain setting, Love said, an amazing delicacy of crustal balance was required. As the lakebed thickened, it had to subside at an appropriate rate if it was to continue to hold water and accept sediment. Gosiute sediments average about half a mile, top to bottom. The oil is at all levels. The evaporite phase, in the middle, reports a Gosiute of dense and complex brine that was surrounded by mud flats sickeningly attended by the hum of flies. Trona—sodium sesquicarbonate—precipitated out of the brine in concentrations rare in the world. It was discovered in 1938, but the boom did not begin until the sixties. We tasted some salty crystals in the rock at Green River, in beds that dipped west and pointed into the ground toward mines. Trona is an important component of ceramics and textiles, pulp and paper, iron and steel, and, most especially, glass. Love commented that more than two tons of trona had been going into the Green River every day merely from the washing of freight cars—and that was a lot of sodium. The Wyoming Department of Environmental Quality had put a stop to the practice. He said there had been a brewery in Green River that drew its water from a well drilled to trona. The beer had a head like a stomach tablet. A few miles south of us were the headwaters of the reservoir that covered Flaming Gorge. Before the federal Bureau of Reclamation built a dam there, Flaming Gorge was one of the scenic climaxes of the American West—a seven-hundred-foot canyon in arching Triassic red beds so bright they did indeed suggest flame. Afterward, not much was left but the hiss, and an eyebrow of rock above the water. The reservoir stilled fifty miles of river. Some of the high water penetrates beds of trona. When the reservoir drops, dissolved trona comes out of the rock and drips into the reservoir. When water rises again, it goes back into the rock for more trona. Love said that Lake Powell and Lake Mead—reservoirs downstream—were turning into chemical lakes as a result. "And a lot of it winds up with the poor farmers in Mexico," he said. "We are going to have to desalinate their water." Some miles along the interstate, when we crossed the Blacks Fork River, we would see alkali deposits lying in the floodplain like dried white scum. On both sides of the road were abandoned farmhouses, abandoned barns, their darkly weathered boards warping away from empty structures out of plumb. The river precipitates and the abandoned farms were not unrelated. This was the Lyman irrigation project, Love explained—a conception of the Bureau of Reclamation, an attempt to make southwestern Wyoming competitive with Wisconsin. The

Blacks Fork River was dammed in 1971, and its waters were used to soak the land. The land became whiter than a bleached femur. It still appeared to be covered with light snow. "Alkali sours the land," Love said. "The drainage here is just too poor to flush it out. Imagine the sodium those farmers drank in their water."

Meanwhile, west of Green River, a tall incongruous chimney seemed to rise up out of the range, streaming a white plume downwind. Below the chimney, but hidden by the roll of the land, was a trona refinery, and, below the refinery, a mine. I had gone down into it one winter day half a dozen months before, and I now remarked that the people there had told me that the white cloud issuing from the chimney was pure steam.

"It goes clear across the state," Love said. "That's pretty durable for steam."

He said that fluorine, among other things, was coming out of the refinery with the steam. Settling downwind, it could cause fluorosis. He thought it might be damaging forests in the Wind River Range. The afternoon sky was cloudless but not exactly clear. "The haze you see is the trona haze that goes across Wyoming," he continued. "We never used to have this. You could clearly see distant mountains on any average day."

Trona is about as hard as a fingernail, and much of it looks like maple sugar or honey-colored butter crunch. I remembered drinking coffee at a picnic table nine hundred feet below us, in a twilighted Kafkan dusty world where dynamite provoked reverberate thunders that moved from room to room and eventually clapped themselves. Chain saws with bars ten feet long sliced into the rock to define the next blast. Stickers on lunch pails said:

DON'T TEMPT FATE.

I HAVE MET THE A-O DUST DEMON.

WHEN ESCAPE IS CUT OFF: 1. BARRICADE 2. LISTEN FOR 3 SHOTS 3. SIGNAL BY POUNDING HARD 10 TIMES 4. REST 15 MINUTES THEN REPEAT SIGNAL UNTIL 5. YOU HEAR 5 SHOTS, WHICH MEANS YOU ARE LOCATED AND HELP IS ON THE WAY.

"The Southeast is the stroke and hypertension belt of the United States," Love was saying. "That is blamed on sodium, including sodium in the water. We're not far behind. Perhaps we can overtake them."

By the Gros Ventre River near Crystal Creek, some years ago, Love noticed horses eating the Cloverly formation—putting their noses right on the outcrop and slurping up nodules of soft Cretaceous lime. He could guess where the horses had come from. They were from Cora, near Pinedale, at the western base of the Wind River Mountains. When the Wind Rivers came rising up during the Laramide Revolution and moved a few miles west, they completely covered the only limestone in the region. As a result, he said, it is not unusual for a college freshman who has grown up in Pinedale to require false teeth. Pinedale has one of the two or three highest records of dental decay in Wyoming. Pinedale is to caries as Savannah is to coronary thrombosis, in each case for a geological reason.

He said that somewhere in limbo on the industrial drawing board was a geothermal project that would mine the hot groundwaters of the Island Park Caldera, southwest of Yellowstone. The question uppermost in many people's minds seemed to be: What would happen to Old Faithful and other Yellowstone geysers? In New Zealand, when the government tapped the fifth-largest geyser field in the world for geothermal energy the Karapiti Blowhole shut down as promptly as if a hand had turned a valve. A geyser field in Nevada once rivalled Yellowstone's—until 1961, when geothermal well drilling killed the Nevada geysers. Old Faithful was having trouble enough without help from the hand of man. For a century, and who knows how much longer, Old Faithful had erupted at intervals averaging seventy minutes, but in 1959 the earthquake at Hebgen Lake, nearby in Montana, slowed the geyser down. Additional earthquakes in 1975 and 1983 caused Old Faithful to become so erratic that visitors complained. Constructed around the geyser is something that resembles a stadium, where crowds collect in bleachers and expect Old Faithful to be faithful: "to play," as hydrologists put it—to burst in timely fashion from its fissures, like a cuckoo clock made of water and steam. Frustrated travellers, sometimes clapping their hands in unison, seemed to be calling on the National Park Service to repair the geyser. A scientist confronted with these facts could only shrug, make observations, and formulate a law: *The volume of the complaints varies inversely with the number of miles per gallon attained by the vehicles that bring people to the park.*

Love, who has made a subspecialty of the medical effects of geology, had other matters on his mind. In public lectures and in meetings with United States senators, he asked what consideration was being given to radioactive water from geothermal wells, which would be released into the Snake River through Henrys Fork and carried a thousand miles downstream. After all, radioactive water was known from Crawfish Creek, Polecat Creek, and Huckleberry Hot Springs, not to mention the Pitchstone Plateau. On the Pitchstone Plateau were colonies of radioactive plants, and radioactive animals that had eaten the plants: gophers, mice, and squirrels with so much radium in them that their bodies could be placed on photographic paper and they would take their own pictures. A senator answered the question, saying, "No one has brought that up."

At the Overthrust Belt—glamorized of late for its fresh new yields of regular and unleaded—we moved up in topography, down in time, because the great thrust sheets are older than the rock on which they came to rest. The first high ridge was Cretaceous in age, and we left the interstate to climb it, on an extremely steep double-rutted dirt road that led to a mountain valley—a so-called strike valley, of a type that will form where upturned strata angle into the sky and a section is softer than those that flank it. The high valley was fringed with junipers, and, from its eastern rim, presented a view that would impress an astronaut. To look from left to right was to see a hundred and fifty miles, from the Uinta Mountains to the Wind River Range, interspersed with badlands. The badlands were late-Eocene river muds and sands chaotically distributed on top of the filled-in lake, and now being further strewn about at the whim of cloudbursts.

In the center of the high swale were the silvery-gray remains of hundreds of cut trees, which had been dragged into the open and arranged as a fence in kidney-bean shape, all but enclosing about fifteen acres. Vaguely, they formed a double corral, with an aperture in one place only, and had apparently been used—for uncounted years—to trap antelopes. Antelopes don't climb fences, as people fond of roast pronghorn discovered centuries ago. Love's son Charlie, the professor of anthropology and geology at Western Wyoming Community College, knew of the trap and had thought out the strategies by which it was effective. His father expressed pride in Charlie for "thinking as intelligently as the aborigines." The high valley held fast an aesthetic silence—a silence reminiscent of the Basin and Range, a silence equal to the winter Yukon. About the only sign of humankind was the antelope trap. This was the Overthrust Belt as it had appeared before white people—thinking intelligently but not like the aborigines—mapped the terrain, modelled its structure, and went after what lay beneath it. There were mountain bluebells and salt sage in the valley, ground phlox and prickly pear. Love reached down and plucked up a plant and asked if I knew what it was. It looked familiar, and I said, "Wild onion."

He said, "It's death camas. It brings death quickly. It killed many pioneer children. They thought it looked like wild onion."

Suddenly, the great silence was smashed by running gunfire as two four-wheel-drive vehicles, each with a lone rider, appeared over the western rim and thundered up the valley, leaving behind them puffs of blue smoke. They disappeared to the north, still shooting. This was boom country now, however temporarily—another world of pickups painted with flames. It had been described in journals as "the hottest oil-and-gas province in North America"—a phrase in which Love found bemusement and irony, because for three-quarters of a century the hottest oil-and-gas province in North America had been lying there neglected.

"This region was written up in 1907 as containing possible oil fields," Love said. "They're 'finding' them now. That 1907 paper, by A. C. Veatch, of the U.S.G.S., was simply ignored. Until 1975, people said there was no oil in the thrust belt. Now it's the hot area. Veatch did his work in the part of the thrust belt that straddles I-80. He said oil should be there, and he said where. His paper is a classic. That it was ignored shows the myopia of oil companies, and of geologists in general. The La Barge oil field, in the Green River Basin off the edge of the thrust belt, was discovered in 1924. Twenty years later it became evident that the La Barge field was producing more oil than the structure could contain. The oil was migrating into it from the thrust belt. The evidence was there before us, and we didn't see it. We talked about it. We wondered why. Now the margins of basins have become new frontiers for oil. Anywhere that mountains have overridden a basin, there are likely to be Cretaceous and Paleocene rocks below, quite possibly with oil and gas. The Moncrief oil company drilled through nine thousand feet of granite at Arminto and into Cretaceous rocks and got the god-damnedest field you ever saw."

On I-80 to the end of Wyoming, we moved among the drilling rigs and pump jacks of some of the most productive fields you ever saw. Love said, "These rigs are not damaging the landscape very much. It isn't all or nothing. It doesn't have to be." I remembered a time when we had gazed down into the Precambrian metasediments of a taconite mine off the southern tip of the Wind River Range. It was an open pit, square, more than a mile on a side. I asked him how he felt about a thing like that, and he said, "They've only ruined one side of the mountain. Behind the pit, the range top is covered with snow. I can live with this. This is a part of the lifeblood of our nation." I recalled also that when the Beartooth Highway was built, ascending the wall of a Swiss-like valley to subsummit meadows of unique beauty, Love defended the project, saying that people who could not get around so well would be enabled to see those scenes.

Love is an unsalaried adjunct professor at the University of Wyoming, an adviser to graduate students in Laramie and in the field. The imaginations of graduate students have a tendency to go dark when the time arrives to choose a topic for a thesis. Typically, they say to him, "Everything has been done."

"Nonsense," says the adjunct professor. "I can blindfold you and have you throw a dart at the geologic map of Wyoming, and wherever it hits you'll find a subject for a thesis."

One day in Jackson Hole, in a small log cabin Love for many years used as a field office, I asked him if I could throw a dart at the geologic map of Wyoming. "Be my guest," he said, and I sent one flying three times, scoring my first, third, and only Ph.D.s. The second one, for me, struck closest to home. It landed by Sweetwater Creek under Nipple Mesa, a couple of miles from Sunlight Peak in the North Absaroka Wilderness—eight miles from Yellowstone Park. "You have hit the Sunlight intrusives," Love said, and somehow I expected the sound of falling coins. "The area has not been surveyed," he continued. "There's no grid. Along Sweetwater Creek are mineral springs and oil seeps. A consortium of major oil companies wants the region removed from wilderness designation." The oil fields of the Bighorn Basin march across the sageland right to the feet of the Absarokas, he said, and their presence asks a great structural question: How far does the basin reach under the mountains? Since the Absarokas are made of volcanic debris, the oil seeping out of the banks of Sweetwater Creek could not have originated in Absarokan rock. He said he thought that the oil-bearing rock of the Bighorn Basin might go under the mountains all the way to Mammoth.

I repeated the name Mammoth, trying to remember where it was, and then said, "That's on the Montana border. It's all the way across Yellowstone Park."

He said, "Yes."

In 1970, Love and his colleague J. M. Good had published a paper on this subject. After considering and rejecting a number of titles—seeking to fashion the flattest and drabbest appropriate phrase—they settled upon "Hydrocarbons in Thermal Areas, Northwestern Wyoming." Now, with regard to my dart in the map, he said, "If you are interested in geochemistry, the composition of the oil from those seeps has not been studied. Is it Paleozoic high-sulphur oil? Mesozoic low-sulphur oil? Tertiary low-sulphur oil? One needs to know the quality of the oil and the depth of the reservoir rock." His tone seemed to exclude both emotion and opinion. "If you're interested in geophysics, what kinds of seismic reflections do you get from rocks below the volcanics?" he went on. "Can they be interpreted in a way that works out the prevolcanic structures? In terms of volcanic chemistry, what kinds of alteration of these Eocene volcanic rocks have occurred because of thermal activity and migration of oil into these rocks? None of this has ever been explored. In the regional context, a geologist cannot ignore the possibilities where that dart hit. A scientist, as a scientist, does not determine what should be the public policy in terms of exploration for oil and gas."

No rock could be more volcanic than the rock of Yellowstone Gorge—rose-and-burgundy, burnt-sienna, yellowcake-yellow Yellowstone Gorge—where petroleum comes out of the walls with hot water and steam. In 1939, when the National Park Service was digging abutments for a bridge downriver from the gorge, the National Park Service struck oil. Several workers, overcome by fumes of sulphur, died. These nagging facts notwithstanding, it was conventional wisdom in geology that where you found volcanics you would not find oil. In the nineteen-sixties, Love went out for a wider look. For example, he went on horseback into the Yellowstone backcountry carrying a four-foot steel rod. Twenty miles from the trailhead, he found swamps that were something like tar pits. When he jammed the rod into a swamp, a cream-colored fluid welled up. He put it in a bottle. In a day's time, the mixture had separated, and much of it was clear amber oil. In pursuing this project, the environmentalist within him balked, the user of resources preferred the resources somewhere else, but the scientist rode on with the rod. He knew he would bring scorn upon himself, but he was not about to stifle his science for anybody's beliefs or opinions. He did lose friends, including some Friends of the Earth. He lost friends in the Wilderness Society and the Sierra Club as well. To them, the Yellowstone oil was only the beginning of the threat he might be raising. The Designated Wilderness Areas of the United States had been selected on the assumption that they were barren of anything as vital as petroleum. "I will admit that it bothers me that I have provoked the wrath of organizations like the Sierra Club," Love remarked that day in the cabin. "My great-uncle John Muir founded the Sierra Club, and here I am, being a traitor."

Passing through Yellowstone on one of our journeys, Love and I found ourselves in foggy mists beside a boiling spring, and on impulse he got out a scintillometer and held it over the water. The scintillometer clicked away at a hundred and fifty counts per second, indicating that the radioactivity in the spring was about three times background. Interesting—but not exactly adrenalizing to a man who had seen the thing going at five thousand and upward.

In the years that immediately followed the Second World War, the worldwide search for uranium was so feverish that geologists themselves seemed to be about three times background. Not only was the arms race getting under way—with the security of the United States thought to be enhanced by the fashioning of ever larger and ever smaller uranium bombs—but also there was promise of a panacean new deal in which this heaviest of all elements found in nature would cheaply heat homes and light cities. The rock that destroyed Hiroshima had come out of the Colorado Plateau, and it was to that region that prospectors were principally drawn.

As any geologist would tell you, metal deposits were the result of hydrothermal activity. Geochemists imagined that water circulating deep in the crust picked up whatever it encountered—gold, silver, uranium, tin, all of which would go into solution with enough heat and pressure. They imagined the metal rising with the water and precipitating near the surface. By definition, a vein of ore was the filling of a fissure near a hot spring. This theory was so correct that it tended to seal off the conversation from intrusion by other ideas.

Three geologists working in South Dakota in 1950 and 1951 found uranium in a deposit of coal. Locally, there was no hydrothermal history. Oligocene tuff—volcanic detritus blown east a great distance—overlay the coal. There were people who thought that ordinary groundwater had leached the uranium out of the tuff and carried it into the coal. Love was one of the people. If such a process—contravening all accepted theory—had in fact occurred, then uranium might be found not only in hydrothermal settings but also in sedimentary basins. When Love proposed a search of Wyoming basins, hydrothermalists in the United States Geological Survey not only mocked the project but attempted to block it. So goes, sometimes, the spirit of science. The tuffs of the Oligocene were a part of the burial of the Rockies, and most had been removed during the exhumation. Love looked around for sedimentary basins where there was evidence that potential host rocks had once been covered with tuff. He had a DC-3 do surveys with an airborne scintillometer over the Powder River Basin. Some of the readings were remarkably hot—notably in the vicinity of some high-standing erosional remnants called Pumpkin Buttes. He went there in a jeep, taking with him for confirming consultation the sedimentologist Franklyn B. Van Houten, who has described himself ever after as "Dave Love's human scintillometer." Love wanted to see if there had been enough fill by Oligocene time to allow the tuff to get over the buried Bighorn Mountains and be spread across the Powder River Basin. He and Van Houten climbed to the top of North Pumpkin Butte and found volcanic pebbles from west of the Bighorns in Oligocene tuff. Then Love went down among the sandstones of the formation lying below, where, at many sites, his Halross Gamma Scintillometer gave six thousand counts per second.

In time, he and others developed the concept of roll fronts to explain what he had found. In configuration, they were something like comets, or crescent moons with trailing horns—convex in the direction in which groundwater had flowed. As Love and his colleagues worked out the chemistry, they began with the fact that six-valent uranium is very soluble, and in oxidized water easily turns into uranyl ions. As the solution moves down the aquifer, a roll front will develop where the water finds an unusual concentration of organic matter. The organic matter goes after the oxygen. The uranium, dropping to a four-valent state, precipitates out as UO_2—the ore that is called uraninite.

One way to find deeply buried uraninite, therefore, would be to drill test holes in inclined aquifers. Wherever you found unusual concentrations of organic matter, you would move up the aquifer and drill again. If you found red oxidized sandstone, you would know that uraninite was somewhere between the two holes.

Drafting his report to the Geological Survey, Love described the "soft porous, pink or tan concretionary sandstone rolls in which the uranium was discovered," and added that "the commercial grade of some of the ore, the easy accessibility throughout the area, the soft character of the host rocks and associated strata, and the fact that strip-mining methods can be applied to all the deposits known at the present time, make the area attractive for exploitation." With those sentences he had become, in both a specific and a general sense, the discoverer of uranium in commercial quantity in Wyoming and the progenitor of the Wyoming uranium industry—facts that were not at once apparent. Within the Survey, the initial effect of Love's published report was to irritate many of his colleagues who were committed hydrothermalists and were prepared not to believe that uranium deposits could occur in any other way. They were joined in this opinion by the director of the Division of Raw Materials of the United States Atomic Energy Commission. A committee was convened in the Powder River Basin to confirm or deny the suspect discovery. All the members but one were hydrothermalists, and the committee report said, "It is true that high-grade mineral specimens of uranium ore were found, but there is nothing of any economic significance." Within weeks, mines began to open in that part of the Powder River Basin. Eventually, there were sixty-four, the largest of which was Exxon's Highland Mine. They operated for thirty-two years. They had removed fifteen million tons of uranium ore when Three Mile Island shut them down.

In 1952, after Love's report was published, the *Laramie Republican and Boomerang* proclaimed in a banner headline, "LARAMIE MAN DISCOVERS URANIUM ORE IN STATE." The announcement set off what Love described as "the first and wildest" of Wyoming's uranium booms. "Hundreds came to Laramie," he continued. "I was offered a million dollars cash and the presidency of a company to leave the U.S.G.S. At that time, my salary was $8,640.19 per year."

The discovery predicted uranium in other sedimentary basins, and Love went on to find it. In the autumn of 1953, he and two amateurs, all working independently, found uranium in the Gas Hills—in the Wind River Basin, twelve miles from Love Ranch. By his description: "Gas Hills attracted everybody and his dog. It was Mecca for weekend prospectors. They swarmed like maggots on a carcass. There was claim jumping. There were fistfights, shootouts. Mechanics and clothing salesmen were instant millionaires."

As it happened, he made those remarks one summery afternoon on the crest of the Gas Hills, where fifty open-pit uranium mines were round about us, and in the low middle ground of the view to the north were Muskrat Creek and Love Ranch. The pits were roughly circular, generally half a mile in diameter, and five hundred feet deep. Some four hundred feet of overburden had been stripped off to get down to the ore horizons. The place was an unearthly mess. War damage could not look worse, and in a sense that is what it was. "If you had to do this with a pick and shovel, it would take you quite a while," Love said. The pits were scattered across a hundred square miles.

We picked up some sooty black uraninite. It crumbled easily in the hand. I asked him if it was dangerously radioactive.

"What is 'dangerously radioactive'?" he said. "We have no real standards. We don't know. All I can say is the cancer rate here is very high. There are four synergistic elements in the Gas Hills: uranium, molybdenum, selenium, and arsenic. They are more toxic together than individually. You can't just cover the tailings and forget about it. Those things are bad for the environment. They get into groundwater, surface water. The mines are below the water table, so they're pumping water from the uranium horizon to the surface. There has been a seven-hundred-per-cent increase of uranium in Muskrat Creek at our ranch."

We could see in a sweeping glance—from the ranch southwest to Green Mountain—the whole of the route he had taken as a boy to cut pine and cedar for corral poles and fence posts. An hour before, we had looked in at the ranch, where most of those posts were still in use—gnarled and twisted, but standing and not rotted. From John Love's early years there, when he slept in a cutbank of the creek, the ranch had belonged only to him and his family. The land was leased now—as was most of the surrounding range—to cattle companies. In the last half mile before we reached the creek, David counted fifty Hereford bulls and remarked that the lessees seemed to be overgrazing. "The sons of bitches," he said. "That's way too many for this time of year." Noticing some uranium claim stakes, he said, "People stake illegally right over land that has been deeded nearly a century."

Over the low and widespread house, John Love's multilaminate roof was scarcely sagging. No one had lived there in nearly forty years. The bookcases and the rolltop desk had been removed by thieves, who had destroyed doorframes to get them out. The kitchen doorframe was intact, and nailed there still was the board that showed John Love's marks recording his children's height. The green-figured wallpaper that had been hung by the cowboys was long since totally gone, and much of what it had covered, but between the studs and against the pine siding were fragments of the newspapers pasted there as insulation.

POSSE AFTER FIVE BANDITS
BATTLE NEAR ROCK ISLAND TRAIN

Robbers Are Found in Haystack
and Chase Becomes Hot

BOTH SIDES ARE HEAVILY ARMED

Fugitives Are Desperate, but Running Fight
Is Expected to End in Their Capture

Spinach had run wild in the yard. In the blacksmith shop, the forge and the anvil were gone. Ducks flew up from the creek. There were dead English currant bushes. A Chinese elm was dead. A Russian olive was still alive. David had planted a number of these trees. There was a balm of Gilead broadleaf cottonwood he had planted when he was eleven years old. "It's going to make it for another year anyway," he said. "It's going to leaf out."

I said I wondered why the only trees anywhere were those that he and his father had planted.

"Not enough moisture," he replied. "Trees never have grown here."

"What does 'never' mean?" I asked him.

He said, "The last ten thousand years."

An antelope, barking at us, sounded like a bullfrog. Of the dozen or so ranch buildings, some were missing and some were breaking down. The corrals had collapsed. The bunkhouse was gone. The cottonwood-log granary was gone, but not Joe Lacey's Muskrat Saloon, which the Loves had used for storing hay. Its door was swinging in the wind. David found a plank and firmly propped the door shut. The freight wagon was there that he had used on trips for wood. It was missing its wheels, stolen as souvenirs of the Old West. We looked into a storage cellar that was covered with sod above hand-hewn eighteen-inch beams. He said that nothing ever froze in there and food stayed cold all summer. More recently, a mountain lion had lived there, but the cellar was vacant now.

In the house, while I became further absorbed by the insulation against the walls, Love walked silently from room to room.

Bizerta, Tunis, May 4—At a reception tendered him by the municipality, M. Pelletan, French Minister of Marine, in a brief speech, declared that France no longer dreamed of conquests, and that her resources would hereafter be employed to fortify her present possessions.

Cattle chips and coyote scat were everywhere on the floors. The clothes cupboards and toy cupboards in the bedroom he had shared with Allan were two feet deep in pack-rat debris.

Have you lost a friend or relative in the Klondike or Alaska? If so, write to us and we will find them, quietly and quickly. Private information on all subjects. All correspondence strictly confidential. Enclose $1.00. Address the Klondike Information Bureau, Box 727, Dawson, Y.T.

David came back into the space that had been his schoolroom, saying, "I can't stand this. Let's get out of here."

In the Gas Hills, as we traced with our eyes his journeys to Green Mountain, he said, "You can see it was quite a trek by wagon. Am I troubled? Yes. At places like this, we thought we were doing a great service to the nation. In hindsight, we do not know if we were performing a service or a disservice. Sometimes I think I might regret it. Yes. It's close to home."

Book 4

Assembling California

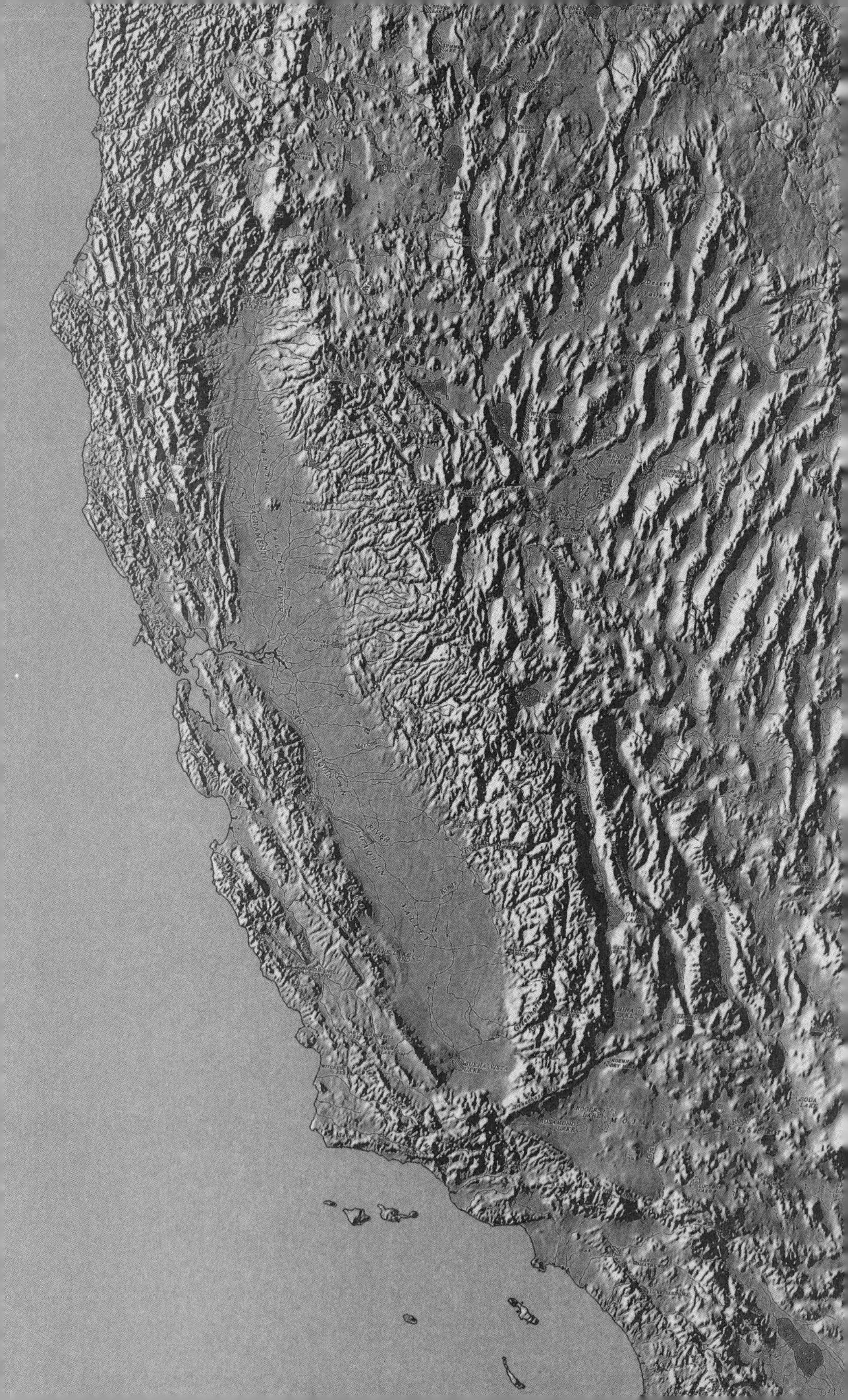

SACRAMENTO
VALLEY
RIVER
SAN JOAQUIN
Merced
RIVER
JOAQUIN
Kings
VALLEY
TULARE LAKE BED
BUENA VISTA LAKE
Greenhorn
ROSAMOND LAKE
MOJAVE
SODA LAKE
CARSON SINK
Desert
Valley

You go down through the Ocean View district of San Francisco to the first freeway exit after Daly City, where you describe, in effect, a hairpin turn to head north past a McDonald's to a dead end in a local dump. You leave your car and walk north on a high contour some hundreds of yards through deep grasses until a path to your left takes you down a steep slope a quarter of a mile to the ocean. You double back along the water, south to Mussel Rock.

Mussel Rock is a horse. As any geologist will tell you, a horse is a displaced rock mass that has been caught between the walls of a fault. This one appeared to have got away. It seemed to have strained successfully to jump out of the continent. Or so I thought the first time I was there. It loomed in fog. Green seas slammed against it and turned white. It was not a small rock. It was like a three-story building, standing in the Pacific, with brown pelicans on the roof. You could walk out on a ledge and look up through the fog at the pelicans. When you looked around and faced inland, you saw that you were at the base of a fifty-foot cliff, its lithology shattered beyond identification. A huge crack split the cliff from top to bottom and ran on out through the ledge and under the waves. After a five-hundred-mile northwesterly drift through southern and central California, this was where the San Andreas Fault intersected the sea.

I went to Mussel Rock that foggy afternoon in 1978 with the geologist Kenneth Deffeyes. I have returned a number of times since, alone or in the company of others. With regard to the lithosphere, it's a good place to sit and watch the plates move. It is a moment in geography that does your thinking for you. The San Andreas Fault, of course, is not a single strand. It is something like a wire rope, as much as half a mile wide, each strand the signature of one or many earthquakes. Mussel Rock is near the outboard edge of the zone. You cannot really say that on one side of the big crack is the North American Plate and on the other side is the Pacific Plate, but it's tempting to do so. Almost automatically, you stand with one foot on each side and imagine your stride lengthening—your right foot, say, riding backward toward Mexico, your left foot in motion toward Alaska. There's some truth in such a picture, but the actual plate boundary is not so sharply defined. Not only is the San Andreas of varying width in its complexity of strands, it is merely the senior fault in a large family of more or less parallel faults in an over-all swath at least fifty miles wide. Some of the faults are to the west and under the ocean; more are inland. Whether the plate boundary is five miles wide or fifty miles wide or extends all the way to central Utah is a matter that geologists currently debate. Nonetheless, there is granite under the sea off Mussel Rock that is evidently from the southern Sierra Nevada, has travelled three hundred miles along the San Andreas system, and continues to move northwest. As evidence of the motion of the plates, that granite will do.

For an extremely large percentage of the history of the world, there was no California. That is, according to present theory. I don't mean to suggest that California was underwater and has since come up. I mean to say that of the varied terranes and physiographic provinces that we now call California nothing whatever was there. The continent ended far to the east, the continental shelf as well. Where California has come to be, there was only blue sea reaching down some miles to ocean-crustal rock, which was moving, as it does, into subduction zones to be consumed. Ocean floors with an aggregate area many times the size of the present Pacific were made at spreading centers, moved around the curve of the earth, and melted in trenches before there ever was so much as a kilogram of California. Then, a piece at a time—according to present theory—parts began to assemble. An island arc here, a piece of a continent there—a Japan at a time, a New Zealand, a Madagascar—came crunching in upon the continent and have thus far adhered. Baja is about to detach. A great deal more may go with it. Some parts of California arrived head-on, and others came sliding in on transform faults, in the manner of that Sierra granite west of the San Andreas. In 1906, the jump of the great earthquake—the throw, the offset, the maximum amount of local displacement as one plate moved with respect to the other—was something like twenty feet. The dynamics that have pieced together the whole of California have consisted of tens of thousands of earthquakes as great as that—tens of thousands of examples of what people like to singularize as "the big one"—and many millions of earthquakes of lesser magnitude. In 1914, Andrew Lawson, writing the San Francisco Folio of the Geologic Atlas of the United States, wistfully said, "Most of the faults are the expression of energies that have been long spent and are not in any sense a menace. It is, moreover, barely possible that stresses in the San Andreas fault zone have been completely and permanently relieved by the fault movement of 1906." Andrew Lawson—who named the San Andreas Fault—was a structural geologist of the first order, whose theoretical conclusions were as revered in his time as others' are at present. For the next six decades in California, a growing population tended to imagine that the stresses were indeed gone—that the greatest of historic earthquakes (in this part of the fault) had relieved the pressure and settled the risk forever. In the nineteen-sixties, though, when the work of several scientists from various parts of the world coalesced to form the theory of plate tectonics, it became apparent—at least to geologists—that those twenty feet of 1906 were a minuscule part of a shifting global geometry. The twenty-odd lithospheric plates of which the rind of the earth consists are nearly all in continual motion; in these plate movements, earthquakes are the incremental steps. Fifty thousand major earthquakes will move something about a hundred miles. After there was nothing, earthquakes brought things from far parts of the world to fashion California.

Deffeyes and I had been working in Utah and Nevada, in the physiographic province of the Basin and Range. Now he was about to go east and home, and we wandered around San Francisco while waiting for his plane. Downtown, we walked by the Transamerica Building, with its wide base, its high sides narrowing to a point, and other buildings immensely tall and straight. Deffeyes said, "There are two earthquake-resistant structures—the pyramids and the redwoods. These guys are working both sides of the street." The skyscrapers were new, in 1978. In an earthquake, buildings of different height would have different sway periods, he noted. They would "creak and groan, skin to skin." The expansion joints in freeways attracted his eye. He said they might open up in an earthquake, causing roadways to fall. He called the freeways "disposable—Kleenexes good for one blow." He made these remarks in the shadowy space of Second Street and Stillman, under the elevated terminus of Interstate 80, the beginnings of the San Francisco Skyway, the two-level structure of the Embarcadero Freeway, and so many additional looping ramps and rights-of-way that Deffeyes referred to it all as the Spaghetti Bowl. He said it was resting on a bog that had once surrounded a tidal creek. The multiple roadways were held in the air by large steel Ts. Deffeyes said, "It's the engineer in a game against nature. In a great earthquake, the ground will turn to gray jello. Those Ts may uproot like tomato stakes. And that will seal everyone in town. Under the landfill, the preexisting mud in the old tidal channel will liquefy. You could wiggle your feet a bit and go up to your knees." In 1906, the shaking over the old tidal channel that is now under the freeways was second in intensity only to the San Andreas fault zone itself, seven miles away. "Los Angeles, someday, will be sealed in worse than this," he continued. "In the critical hours after a great earthquake, they will be cut off from help, food, water. Take one piece out of each freeway and they're through."

In a rented pickup, we had entered California the day before, climbing the staircase of fault blocks west of Reno that had led the Donner party to the crest of the mountains named for snow. In California was the prow of the North American Plate—in these latitudes, the sliding boundary. California was also among the freshest acquisitions of the continent. So radical and contemporary were the regional tectonics that the highest and the lowest points in the contiguous United States were within eighty miles of each other in California. As nowhere else along the fortieth parallel in North America, this was where the theory of plate tectonics was announcing its agenda.

Over the years, I would crisscross the country many times, revisiting people and places, yet the first morning with Deffeyes among the rocks of California retains a certain burnish, because it exemplified not only how abrupt the transition can be as you move from one physiographic province to another but also the jurisdictional differences in the world of the geologist. As we crossed the state line under a clear sky and ascended toward Truckee, we passed big masses of competent, blocky, beautiful rocks bright in their quartzes and feldspars and peppered with shining black mica. The ebullient Deffeyes said, "Come into the Sierra and commune with the granite."

A bend or two later, his mood extending even to the diamond-shaped warnings at the side of the road, he said, "Falling-rock signs are always good news to us."

Then a big pink-and-buff roadcut confused him. He said he thought it consisted of "young volcanics," but preferred to let it remain "mysterious for the moment." The moment stretched. Deffeyes is as eclectic as a geologist can become, a generalist of remarkable range, but his particular expertise—he wrote his dissertation in Nevada and has done much work there since—was fading in the distance behind him. Up the road was a metasediment in dark and narrow blocks going every which way, like jackstraws. Deffeyes got out of the pickup and put his nose on the outcrop, but he had an easier time identifying a bald eagle that watched him from an overhanging pine.

"You need a new geologist," he said to me.

We took a rock sample, washed our hands in melting snow, and ate a couple of sandwiches as we watched wet traffic with bright headlights come down from Donner Summit. Looking back to the cloudless Basin and Range and seeing what lay ahead for us, Deffeyes said, "Out of the rain shadow, into the rain."

After we got up into the high country ourselves, some additional metasediment left him colder than the rain. "The time has come to turn you over to Eldridge Moores," he said.

A few miles farther on, we came to a big, gravelly roadcut that looked like an ash fall, a mudflow, glacial till, and fresh oatmeal, imperfectly blended. "I don't know what this glop is," he said, in final capitulation. "You need a new geologist. You need a Californian."

Moores could be found on a one-acre farm in the Great Central Valley—in a tract surrounded on three sides by the vegetable-crop field labs of the University of California, Davis. Twenty years earlier, Davis had been an agricultural college, but it had since expanded in numerous directions to take its place beside Berkeley, attracting to the Geology Department, for example, such youthful figures of future reputation as the mantle petrologist Ian MacGregor and the paleobiologist Jere Lipps, not to mention the tectonicist Eldridge Moores.

At one time and another, over a span of fifteen years, Moores and I would not so much traverse California as go into it in both directions from the middle. We would hammer the outcrops of Interstate 80 from Nevada to San Francisco, reaching out to related rock even farther than Timbuctoo. Timbuctoo is in Yuba County. The better to understand California, I would follow him to analogous geological field areas in Macedonia and Cyprus—journeys much enhanced by his knowledge of modern Greek. He has read widely in Greek history as well as geologic history, and standing on the steps of the Parthenon he sounds like any other tour guide—recounting wars, explosions, orations, and stolen marbles—until he tells you where the hill itself arrived from, and when, and why the Greeks sited their temple on soluble rock that they knew to be riddled with caverns. Moores has been a counsellor through all my projects in geology, across which time our beards have turned gray. He and his wife, Judy, still live in their turn-of-the-century farmhouse, with its high ceilings, its old two-light windows, its pools of sun on cedar floors. Their children—who were five, eight, and eleven when I met them—are grown and gone. On each of two porches lie big chunks of serpentine—smooth as talc, mottled black and green. When you see rocks like that on a porch, a geologist is inside.

In the living room is a framed montage of nine covers from *Geology*, a magazine introduced in the nineteen-seventies by the Geological Society of America and raised during the editorship of Eldridge Moores (1981–88) to a level of world importance in the science. Moores is the sort of person who runs up flights of stairs circling elevator shafts, because elevators are so slow. He edited *Geology* while teaching full time and advancing his own widespread research. The montage was a gift to him from people at the G.S.A. It includes fumaroles in Iceland, dunes in southern Colorado, orange-hot lava on Kilauea, and a painting of a *Triceratops* being eaten alive by a *Tyrannosaurus rex*. In the heavens close above the struggling creatures is the Apollo Object—an asteroid, roughly six miles in diameter—that is believed to have collided with the earth and caused the extinction of the dinosaurs. In the editor's notes on the contents page, Moores referred to the painting as "the Last Supper." There were outraged complaints from geologists.

The centerpiece of the montage is a 1988 cover showing Moores on a coastal outcrop playing a cello. Moores grew up in Arizona's central highlands, in a community so remote and sparse that it was called a camp. A very great distance from pavement, it was far up the switchbacks of a mountain ridge and among the open mouths of small, hard-rock mines. At the age of thirteen, he learned to play the cello, and he practiced long in the afternoons. The miners, his father included, could not understand why he would want to do that. Moores has played with symphony orchestras in Davis and Sacramento. The coastal outcrop on the cover of *Geology* is the brecciated limestone of Petra tou Romiou, Cyprus. Moores in the field has long since overcome the most obvious drawback of a cello. He travels with an instrument handcrafted by Ernest Nussbaum in a workshop in Maryland. Essentially, it is just like any other cello but it has no belly. Neck, pegbox, fingerboard, bridge—everything from scroll to spike fits into a slim rectangular case wired to serve as an electronic belly. This is a Sherpa's cello, a Chomolungma cello, a base-camp viol. In Moores' living room is a grand piano. Still on a shelf behind it are the sheet-music boxes of his children, labelled "Brian Clarinet," "Brian Bassoon," "Kathryn Cello," "Geneva Piano," and "Geneva Violin," and three additional boxes labelled "Eldridge Cello," "Eldridge Cello and Piano," "Eldridge Cello Concertos and Trios."

Judy grew up in farming country in Orange County, New York. On her California acre of the Great Valley she grows vegetables twelve months a year, and has also raised bush strawberries, grapes, blackberries, goats, pigs, chickens, pears, nectarines, plums, cherries, peaches, apricots, asparagus, ziziphus, figs, apples, persimmons, and pineapple guavas—but not so prolifically in recent years, because she has been working with a group that provides food and emergency assistance for homeless people and for people who have run out of money and are about to be evicted. She has worked in regional science centers since she was a teen-ager, and, with others, she founded one in Davis. School buses bring children there from sixty miles around to get their hands on spotting scopes, microscopes, oscilloscopes, and living snakes, on u-build-it skeletons, on take-apart anatomies and disassembled brains. Judy, trim and teacherly, puts her hands palms down on a table to show the interaction of lithospheric plates. Lithosphere, she explains in simpler words, is crustal rock and mantle rock down to a zone in the mantle that is lubricious enough to allow the plates to move. Thumbs tucked, fingers flat, the hands side by side, she presses them hard together until they buckle upward. The hands are two continents, or other landmasses, converging, colliding—making mountains. The Himalaya was made that way. Placing the hands flat again, she slowly moves them apart. These are two plates separating, one on either side of a spreading center. The Atlantic Ocean was made that way. She begins to slide one hand under the other. This is subduction. Ocean floors are consumed that way. Thumbs tucked, fingers flat, palms again side by side, she slides one hand forward, one back, the index fingers rubbing. This is the motion of a transform fault, a strike-slip fault—the San Andreas Fault. Parts of California have slid into present place that way. Convergent margins, divergent margins, transform faults: she has outlined the boundaries of the earth's plates. There is enough complexity in tectonics to lithify the nimblest mind, but the basic model is that simple. Take your hands with you—she smiles—and you are ready for the mountains.

When I first went into the Sierra with Judy's husband, in 1978, he had an oyster-gray Volkswagen bus with a sticker on its bumper that said "Stop Continental Drift." I guess he thought that was funny. There were not a few geologists then who really would have stopped it in its tracks if they could have figured out a mechanism for doing so, but, since no one knew then (or knows for certain now) what drives the plates, no one knew how to stop them. Plate tectonics had arrived in geology just about when Moores did, and—in his metaphor—he hit the beach in the second wave. He has called it "the realization wave": when geologists began to see the full dimensions and implications of the new theory, and the research possibilities it afforded—a scientific revolution literally on a global scale.

Physiographic California, for much of its length, is divided into three parts. Where Interstate 80 crosses them, from Reno to San Francisco, they make a profile that is acutely defined: the Sierra Nevada, highest mountain range in the Lower Forty-eight; the Great Central Valley, essentially at sea level and very much flatter than Iowa or Kansas; and the Coast Ranges, a marine medley, still ascending from the adjacent sea.

VERTICAL EXAGGERATION 10X

In this cross section, the Coast Ranges occupy forty miles, the valley fifty miles, the mountains ninety. All of it added together is not a great distance. It is not as much as New York to Boston. It is Harrisburg to Pittsburgh. In breadth and in profile, a comparable country lies between Genoa and Zurich—the Apennines, the Po Plain, the Alps.

An old VW bus is best off climbing the Sierra from the west. Often likened to a raised trapdoor, the Sierra has a long and planar western slope and—near the state line—a plunging escarpment facing east. The shape of the Sierra is also like an airfoil, or a woodshed, with its long sloping back and its sheer front. The nineteenth-century geologist Clarence King compared it to "a sea-wave"—a crested ocean roller about to break upon Nevada. The image of the trapdoor best serves the tectonics. Hinged somewhere beneath the Great Valley, and sharply faulted on its eastern face, the range began to rise only a very short geologic time ago—perhaps three million years, or four million years—and it is still rising, still active, continually at play with the Richter scale and occasionally driven by great earthquakes (Owens Valley, 1872). In geologic ages just before the uplift, volcanic andesite flows spread themselves over the terrain like butterscotch syrup over ice cream. Successive andesite flows filled in local landscapes and hardened flat upon them. As the trapdoor rises—as this immense crustal block, the Sierra Nevada, tilts upward—the andesite flows tilt with it, and to see them now in the roadcuts of the interstate is to see the angle of the uplift.

Bear in mind how young all this is. Until the latter part of the present geologic era, there was no Sierra Nevada—no mountain range, no rain shadow, no ten-thousand-foot wall. Big rivers ran west through the space now filled by the mountains. They crossed a plain to the ocean.

Remember about mountains: what they are made of is not what made them. With the exception of volcanoes, when mountains rise, as a result of some tectonic force, they consist of what happened to be there. If bands of phyllites and folded metasediments happen to be there, up they go as part of the mountains. If serpentinized peridotites and gold-bearing gravels happen to be there, up they go as part of the mountains. If a great granite batholith happens to be there, up it goes as part of the mountains. And while everything is going up it is being eroded as well, by water and (sometimes) ice. Cirques are cut, and U-shaped valleys, ravines, minarets. Parts tumble on one another, increasing, with each confusion, the landscape's beauty.

On the first of our numerous trips to the Sierra, Moores pulled over to the shoulder of the interstate to have a look at the outcrop that had frustrated Ken Deffeyes—the one that Deffeyes had identified as glop. It was sixteen miles west of Donner Summit, beside a bridge over the road to Yuba Gap. Moores in the field looks something like what Sigmund Freud might have looked like had Freud gone into geology. Above Moores' round face and gray-rimmed glasses and diagnostic beard is a white, broad-brimmed, canvas fedora featuring a panama block. There are weather creases at the edges of his eyes. He typically wears plaid shirts, blue twill trousers, blue running shoes. On one hip is a notebook bag, on the other a Brunton compass in a cracked leather case. He is a chunky man with a long large chest and a short stretch between his hips and the terrain. From cords around his neck dangle two Hastings Triplets, the small and powerful lenses that geologists hold close to outcrops in order to study crystals. He did not need them to see what was incorporated in this massive paradox of glop. It contained jagged rock splinters and smoothly rounded pebbles as well. "It's hard at first blush to tell that it's mudflow and not wholly glacial," Moores said. "It is mostly andesite mudflow breccia with reworked stream gravel in it and glacial till on top, which appears to be moraine but is not."

In the early Pliocene, a volcano grew into the range there. It has long since eroded away. Andesite lavas poured from the volcano. Lighter eruptive material settled around the crater. In the moist atmosphere, the volcano's eruptions caused prolonged heavy rains. The water mobilized the unstable slopes. Volcanic muds—full of the sharp rock fragments that would cement together as breccia—slid into the country. In quiet periods between eruptions, streams flowing down the volcano tumbled some of the rock fragments, rounding pebbles. In recent time, alpine glaciers dug into the country and dozed away much of what was left of the volcano, and as the ice melted it left upon the brecciated mudflows heaps of lateral till. ("It is mostly andesite mudflow breccia with reworked stream gravel in it and glacial till on top, which appears to be moraine but is not.")

All this had happened in one areal spot. All this was represented in that one roadcut. Anyone could be pardoned if, at first glance, the complete narrative seemed less than apparent. The story had repeated itself through much of the Sierra during the same band of time: other volcanoes extruding andesite and shedding mud, their remains disturbed by ice. It was a surface story, a latter-day account. The brecciated mudflows and andesite lava flows had come to rest on rock that was older by as much as five hundred million years—rock with a deep and different story, rock that just happened to be there when the mountains rose. The layers of the Grand Canyon are full of the temporal stratigraphic gaps known as unconformities. In the Grand Canyon, much more time is absent than is represented. If a gap of five hundred million years were the right five hundred million years, it could erase the Grand Canyon. In eastern California, the infinitesimal space between the andesite flows and the rock on which they hardened is known as the Great Sierra Nevada Unconformity. To understand what that was and how it had come to be was to understand the relationship between just two of the parts in a millipartite structure.

Moores and I went on to California's eastern boundary, turned around, and recrossed the Sierra, as we would do repeatedly in the coming years. Climbing the steep east face of the mountains, you see granite and more granite and andesite capping the granite. So far so comprehensible. But before you have crossed the range you have seen rock of such varied type, age, and provenance that time itself becomes nervous—Pliocene, Miocene, Eocene non-marine, Jurassic here, Triassic there, Ypresian, Lutetian, Tithonian, Rhaetian, Messinian, Maastrichtian, Valanginian, Kimmeridgian, upper Paleozoic. The rocks seem to change as fast as the traffic. You see olivine-rich, badly deformed metamorphic rock. You see serpentine. Gabbro. One thing follows another in a manner that seems random—a collection of relics from varied ages and many ancestral landscapes, transported from far or near, set beside or upon one another, lifted en masse in fresh young mountains and exposed in roadcuts by the state. You cannot be expected, just by looking at it, to fit it all together in mobile space and sequential time, to see in the congestion within this lithic barn—this Sierra Nevada, this atticful of objects from around the Pacific world—the events and the vistas that each item represents.

Suppose you were to find in a spacious loft a whale-oil lamp of pressed lead glass. What would you think, know, guess, and wonder about the origin and the travels of that lamp? And suppose you were to find near it a Joseph Meeks laminated-rosewood chair, and an English silver porringer and stand, and an eight-lobed dish with birds in a flowering thicket. It is possible that you would not immediately think 1850, 1833, 1662, and 1620. It is possible that you would not envision the place in which each object was made or the milieu in which it was first used, and even more possible that you would not discern how or when any of these pieces moved through the world and came to be in this loft. You also see, lined up in close ranks, a Queen Anne maple side chair, a Federal mahogany shield-back side chair, a Chippendale shell-carved walnut side chair, and a William and Mary carved and caned American armchair. Stratigraphically, they are out of order. How did that happen? Why are they here? Only one thing is indisputable: this is some loft. Jammed to the trusses, it also contains a Queen Anne carved-mahogany block-front kneehole dressing table, a Hepplewhite mahogany-and-satin-wood breakfront bookcase, a rosewood Neo-Gothic chair, an Empire mahogany step-back cupboard, and a Regency mahogany metamorphic library bergère. It contains a classical brass-mounted mahogany gilt wood-and-gesso bed with pressed-brass repoussé. It contains a Federal cherry-wood-and-bird's-eye-maple bowfront chest of drawers, an early Victorian mahogany dining chair with a compressed balloon back, a Federal carved and inlaid curly-maple-and-walnut fall-front desk, a Windsor sackback writing armchair, and a Louis XV ormolu-mounted kingwood parquetry commode. There's a temple bell dating to Auspicion Day of the fifth month of the first year of Tembrun. There's a Federal carved-mahogany armchair with a cornucopian splat.

Sort that out. Complete a title search for each piece. Tell each story backward through shifting space to differing points in time. Imagine the palace, the pavilion, the house, the hall for which each piece was fashioned, the climate and location of the country outside.

Naturally, you can't do that—not in a single reconnaissance. Don't fret it. Don't fret that you can't see the story whole. You cannot tell whence each of these items has come, any more than its maker could have known where it would go.

"Nature is messy," Moores remarked. "Don't expect it to be uniform and consistent."

I remembered the sedimentologist Karen Kleinspehn saying to me in these same mountains, "You can't cope with this in an organized way, because the rocks aren't organized."

Gradually, though—outcrop to outcrop, roadcut to roadcut—Moores revived enough related scenes in the distinct origins of the random rock to frame a cohesive chronological story. That is what geologists do. "You spend a lot of time working over rocks and you have a lot of time to do nothing but think," he said. "These mountains, for example, are Tertiary normal faulted, confusing the topography with regard to structure. They show different levels of structure in different places. To see through the topography and see how the rocks lie in three dimensions beneath the topography is the hardest thing to get across to a student." After a mile of silence, he added cryptically, "Left-handed people do it better."

I said nothing for a while, and then asked him, "Are you left-handed?"

He said, "I'm ambidextrous."

As it happens, I am left-handed, but I kept it to myself.

From the east, the climb is rapid to Donner Summit—less than thirty miles, and the road is not straight. Yet elsewhere along the Sierra front the rise is so much shorter and steeper that nothing on wheels could ever climb it. From the basin below (altitude four thousand feet), you bend your neck and look ten thousand feet up a granite mass that was lifted intact, whereas here, on the route above Reno, the "Tertiary normal faulting" that Moores referred to has tiered the escarpment and lowered the crestline as well. The early trappers found a native trail here. In all likelihood, the natives who made the trail were animals, followed, in time, by people.

Under ponderosas and western cedars at the Nevada-California line, the granite reveals itself and then is quickly gone, as the roadside rock becomes something like dark cordwood, fallen in columnar blocks. This is the caprock andesite, which cracks into columns as it cools. Another five miles, and the interstate moves through a long cut that is buff, gray, buff, gray, and buff again as lava flows and mudflows intersperse. Perhaps a hundred thousand years separate the lava flows, while the laminating muds come ten times as often. The volcanic cap over the granite is still a kilometre thick here. Among the trees are erratic boulders—granite boulders out of place on the andesite, transported a few thousand years ago by a descending ribbon of ice.

Three miles before the summit, the granite reappears, not in ice-transported bits but in bedrock at the side of the road. And then more granite, under Jeffrey pines—weathered granite, light and sparkling sliced granite. It ends abruptly, at a contact with andesite. This particular granite had been sitting here eroding quietly for maybe ninety million years when the andesite lava flowed upon it, coating hills and filling valleys, plastering over the granitic terrain, concealing and preserving a Miocene landscape. Differentially, randomly, erosion has eaten through the caprock. So the road encounters both formations. Granite reappears at the summit.

Donner Summit, at seven thousand two hundred and thirty-nine feet, is half the height of the range. Locally, engineers found a way for the interstate which is considerably less precipitous than the trail used by the emigrants in the eighteen-forties. The place that came to be known as Donner Pass is a couple of miles south, on a relic stretch of U.S. 40. Moores and I once went over there and stood on a cliff edge, looking east. Tens of thousands of square miles of basin-and-range topography fanned out into Nevada, all of it aimed, within converging lines, at the pass. The drop to Donner Lake, more than a thousand feet below, was almost giddy. To get over the pass, everything on feet or wheels had to come up that grade. In a normal year, about seventy inches of water falls on the High Sierra, nearly all of it as snow. Seventy inches of water is roughly one and a half times what falls on New York City and twice what falls on Seattle. The snow on the Sierra Nevada can be forty feet deep. At the end of October, 1846, the Donner party came up to this pass and were forced to retreat by a mountain of snow. The winter camp where they starved and died was by the shore of Donner Lake, in the cirque below the pass.

In deep winter, I have stayed near Donner Lake in a ski condo where a previous guest left a peevish note: "The peace and beauty are marred by a noisy refrigerator and heating unit." Now, in midsummer, there were, around the pass, spreads of tenacious snow. A bicyclist, standing as he pumped but scarcely puffing, came up the route of the emigrants. Seating himself as he reached the zenith, he coasted on to the west. To the east, the deep gulf of scenery that he had come out of owed itself less to the finishing touches of ice than to large parallel north-south faults that had lowered a large piece of country—a crustal block, dropped between two other crustal blocks, and now a graben. Lake Tahoe, southeast across a partitioning ridge from Donner Lake, lies in the same graben. The small lake and the large one would be connected but for a recent pouring of andesite, which formed the ridge.

Moores said to notice how the mechanical lowering of a large piece of the mountains had caused varying levels of the original structure to turn up in unexpected places. To try to sense a structure, he repeated, one must develop a talent for "seeing through the topography" and into the rock on which the topography was carved.

When rocks in their variety arrive in a given place, like furniture going into storage, they hold within themselves their individual histories: their dates of solidification, their environments of deposition, or their metamorphic experience, as the case may be. Their unit-to-unit relationship—their stratigraphy and other juxtapositions—pondered as a whole is structure. Structure on the move is tectonics.

When topography is as beautiful as at Donner Pass, it is not an easy matter to see through it, but if you're looking for structure you might start with the granite. In all the country from Nevada to the pass, the volcanic cap makes its appearances, but always as veneer —eroding everywhere, opening windows, and ultimately suggesting the bewildering mass of the underlying granite. This is the Sierra batholith. Geologists reserve that term for the largest bodies of magmatic rock. A batholith, as defined in the science, has a surface of at least forty square miles and no known bottom. For the latter reason, it is also called an abyssolith. The one in California has a surface of about twenty-five thousand square miles. It lies inside the Sierra like a big zeppelin. Geologists in their field boots mapping outcrops may not have been able to find a bottom, but geophysicists can, or think they can, and they say it is six miles down. If so, the batholith weighs a quadrillion tons, and its volume is at least a hundred and fifty thousand cubic miles.

It reminds me of a big rigid airship because the rigids contained, within their metal frames, rows of giant bags that resembled aerial balloons. Batholiths develop not as single chambers of magma but as contiguous balloons of molten rock called plutons. As red-hot rising fluid, the great Sierra batholith came into the country in successive pulses during a hundred and thirty million years between early Jurassic and latest Cretaceous time. There were three peak periods—the first nearly two hundred million years before the present, the second at a hundred and forty million years before the present, and the third at eighty. The most extensive is the "80 pulse." All this went on some ten to thirty kilometres below the earth's surface, where continental crust and subducting ocean crust (coming under the continent) were melting. Through Maastrichtian time and nearly all the Cenozoic epochs, the cooled and cooling magma lay buried. The topography above changed and changed again, like a carrousel of slides. And eventually, recently, the batholith came up, to serve as the lithic medium for the erosive sculpting of Olancha Peak, of Wheeler Peak, of Mt. Whitney.

Dark cliffs above Donner Pass were Pliocene volcanics, but the rock beside the trail was granite—poetically weathered organic billows of granite. There were small black shapes within it. Thousands of them. Alien pebbles. These were bits of the country rock that the batholith intruded. They had fallen into the magma while it was still molten or, if cooler than that, sufficiently yielding to be receptive. They had been softened and rounded but not melted and destroyed. On Interstate 80 west of Donner Summit, we saw larger chips of such metasediment in the granite of the roadcuts. Another mile, and they were larger still. Moores referred to them as "abundant xenoliths—Jura-Triassic pieces of the wall or the roof." These were not the andesites and other outpourings that had been spread upon the granite in fairly recent times; these were parts of the intruding batholith's containing walls or roof. They had fallen into the soft granite eighty million years ago, and, before that, had been crustal rock for something like a hundred million years.

In the interstate median, under Jeffrey pines, were bedrock outcrops that had been scoured and polished by overriding ice eleven thousand years ago. There were plenty of erratic boulders. For eleven miles after Donner Summit, the xenoliths in the granite increased in volume until what we had first seen as pebbles were now the size of bears. To sense the implication of what was coming, a structural geologist would need no further sign. We were fast approaching the wall of the batholith—the magma's contact with the country rock. That highway engineers would blast out a roadcut at just such a place is fortuitous, a matter of random chance, but when Ken Deffeyes and I had come into this same right-hand bend he had shouted, "Whoa! Whoa! Pull over!" And a moment later he was saying, "This is the best outcrop on all of I-80. You can walk up and touch the wall of the great batholith."

Moores now called it "about as classic and neat a contact as you'll ever see." As cars shot past us like F-18s, he added, "Right here. Bang!" The contact was essentially vertical. It ran on up the mountainside and vanished under the trees. It could not have been more distinct had it been the line between a granite building and a brick building adjacent in a city. The granite of the batholith looked almost white beside the reddish country rock, which Moores described as the metamorphosed remains of what had once been an island arc. The granite was customary, competent—a lot of salt and less pepper. The arc rock was flaky, slaty—like aged iron in a state of ulcerated rust. In the first yards after the contact, tongues of granite reached into the country rock, preserved in the act of eating xenoliths. Within a short distance, they gave up.

As the rock ran on in the long continuous cut, it turned black, burgundy, buff, and green, in vertical stripes, in tight drapefolds with long limbs. Obviously, it had been caught up—before the arrival of the batholith—not in some minor local slumpage but in a regional and pervasive tectonic event. With two more miles, the story again made a radical change, as we came to a roadcut of gabbro. Charcoal-gray and sparkling, it was perfect gabbro. There are rich, handsome houses on the Upper East Side of Manhattan that are made of less perfect gabbro. Gabbro, too, is cooled magma. Lacking quartz, it is at the dark end of a spectrum the light end of which is, for the most part, granite. Peridotite—the rock of the earth's mantle—was in the roadcut as well, and Moores said that in his opinion these mafics and ultramafics (rocks low in silica and high in magnesium and iron) arrived after the event that had drapefolded the rock up the road and before the intrusion of the batholith. As we moved on, the gabbro-peridotite interdigitated with granite and then disappeared as the road once again descended into the Sierra batholith. After corridors of granite, there were more volcanics, in the topographic scramble of structure.

When panoramic views came along, they showed the uniformity of the sixty-mile slope—the low-angle plane of the western Sierra. The great surface (the top of the trapdoor) was completed in the eye rather than the rock. It was deeply eaten out by river gorges. To the north and the south, the vistas were wide over deep valleys to tilting planar skylines. We came to Emigrant Gap, where the erosional dissection was particularly deep. Nineteen miles from Donner Pass, the scene demonstrated with emphasis that once emigrants were across the summit they were scarcely free of trouble. From Emigrant Gap into Bear Valley they lowered their wagons on ropes. We looked into the valley, where an alpine meadow was flanked with incense cedars. Above it to the north, under the smoothly sloping skyline, were west-dipping sediments that Moores described as mudflow breccias over Paleozoic sandstones. A deep gorge cut through this ridge. It contained the Yuba River, where the Yuba, with the help of alpine ice, had captured the Bear. To the northeast, under high white peaks, was a lake gouged in granite by an alpine glacier, which had left its moraines on the volcanic muds among the sharp shards and round pebbles that had caused Deffeyes to throw in his towel. Rocks between us and that lake, Moores said, were "lower Paleozoic quartz-rich sediments metamorphosed and folded at least twice." And the rocks in the peaks above the lake were remains of a Jurassic island arc.

Moores spoke reflectively of "the joy of being alone with the geology," of spending enough time walking such a scene to learn how some of it fits together, and then adding what you can to the scientific literature, "which is not like a solo but like an orchestral piece." To Moores, what had happened to create California where no surviving rock had been before was much in evidence in the scene around us, as it had been in the rocks we saw along the road. As terrane—the homonym that refers not merely to surface configurations but to a full three-dimensional piece of the earth's crust—this region had become known in geology as Sonomia. It reached from the Sonoma Range in central Nevada to the Sierra foothills west of where we stood. As plate theorists reconstruct plate motions backward through time, they see landmasses converging to form superterranes and breaking apart to form new continents. Swept up in these great events are islands and island arcs—Newfoundlands, Madagascars, New Zealands, Sumatras, Japans—that slide in or collide in toward continental cores. They become the outermost laminations of new landscapes.

When terranes coming via the ocean attach themselves to a continent, they are said to have "docked." Never shy about metaphors, geologists are not encumbered by the fact that they also call the docking place a suture. In early Triassic time, in the narrative according to present theory, Sonomia docked against western North America. The suture is on the longitude where Golconda, Nevada, is now. For a century or so before plate tectonics, the obviously overriding rock was known as the Golconda Thrust. It was an event that happened about two hundred and fifty million years before the present. Sonomia was an island arc. North to south, Moores said, it might have stretched two thousand miles. It brought with it those Paleozoic sandstones above Bear Valley and the quartz-rich sediments we could also see to the northeast. Volcanoes grew in the newly docked terrane. Bits of them would become the xenoliths in the granites of the summit. Along the western margin of Sonomia, where ocean crust was subducting in a trench, more volcanoes developed. Their rock was in peaks above us. Roughly where we stood, a coastal region of exceptional beauty had lain at the base of the volcanoes. Stratovolcanoes. Kilimanjaros and Fujis.

Sonomia was actually the second terrane to attach itself to the western edge of ancestral North America. The first had arrived in Mississippian time. It had thrust itself almost to Utah. At this latitude, a third terrane would follow Sonomia in the Mesozoic, smashing into it with crumpling, mountain-building effects that would propagate eastward through the whole of Sonomia, metamorphosing its sediments—turning siltstones into slates, sandstones into quartzites—and folding them at least twice: the multicolored drapefolds we had seen beside the road. This was the country rock the batholith intruded.

A granite batholith will not appear just anywhere. You will wait eons for one to develop under Kansas. A great tectonic event must come first. Then granite—or, rather, the magma that will cool and produce granite—comes in beneath the mountains. Volcanoes appear at the surface. Lava flows.

To create the magma, you must in some way melt the bottom of the crust. Subduction—one plate sliding beneath another—will cause things to melt. And so will a collision that compresses and thickens terrane. After a continent-to-continent collision, the crust might double; a batholith will come up within thirty million years. In deep burial, the heat from such radioactive and universal elements as uranium, potassium, and thorium is trapped. The heat increases until the rocks melt themselves and their surroundings. Granite should be forming under Tibet at present, where India has hit the Eurasian Plate in a collision that is not yet over. Under California, both thickened crust and plate-under-plate subduction contributed to the making of the batholith, at first after Sonomia came in and sutured on and deformed itself, and again after Sonomia was hit from the west and further deformed.

The Sierra batholith is melted crust of oceanic origin as well as continental. Most of the world's great batholiths are not quite true granite but edge on down the darkening spectrum and, strictly speaking, are granodiorite. Too strictly for me. But that is the rock of the High Sierra, which almost everyone refers to as granite.

After the batholith came nothing during the many millions of years of the Great Sierra Nevada Unconformity. At any rate, nothing from those years was left for us to see. The rock record jumps from the batholith to the andesite flows of recent time, patches of which Moores pointed out from the lookoff at Emigrant Gap. A few million years ago, when lands to the east of us began to stretch apart and break into blocks, producing the province of the Basin and Range, the Sierra Nevada was the westernmost block to rise, lifting within itself the folds and faults of the Mesozoic dockings, the roots of mountains that had long since disappeared. The chronology at Emigrant Gap ends with the signatures of glaciation on the new mountains—the bestrewn boulders and dumped tills, the horns, the arêtes, the deep wide U of the Bear Valley.

I remarked that geologists are like dermatologists: they study, for the most part, the outermost two per cent of the earth. They crawl around like fleas on the world's tough hide, exploring every wrinkle and crease, and try to figure out what makes the animal move.

Moores said he begged to differ. He said the whole earth is involved in plate tectonics. The earthquake slips of subducting plates could be read as deep as four hundred miles, and seismic data were now indicating that the plates' cold ocean-crustal slabs may descend all the way to the core-mantle boundary. Bumps on the core may be related to the activity of hot spots like Hawaii, Yellowstone, and Iceland. He said he wouldn't call that dermatology.

Since Moores had learned geology in the late nineteen-fifties and early nineteen-sixties, when the theory of plate tectonics was still in a formative unheralded stage, I asked him what he had been taught. How had his teachers at the California Institute of Technology explained—in what is now known as the Old Geology—the building of mountains, the rise of volcanoes, the construction of North America west of Salt Lake? And how large had the transition been from what he learned then to what he knew now?

He said that the science's understanding of mountain-building mechanisms took its first great step forward in the second half of the nineteenth century, when James Hall, the state geologist of New York, conceived of the geosynclinal cycle, and so put in place the geology that prevailed until 1968, when plate tectonics was nailed to the church door. Since mountain belts tended to rise at the margins of continents and to contain, among other things, folded marine sediments and intruding batholiths, Hall imagined the long wide seafloor trough, the deep dimple, in which vast amounts of sediment would pile up and where magmas would intrude until the material was ready to rise as mountains. That is what Moores was taught.

The geosyncline, like any admirable and serviceable fiction, contained a lot of truth. From stratigraphy to structure, geology was understood in terms of geosynclines for about a hundred years. You found gold with your knowledge of geosynclines. You found silver, antimony, and oil. You started conceptually with a geosyncline and projected events forward in time until you saw the geosyncline shuffled up in the mountains before you. Or you started with the mountains, disassembled them in your mind, and made palinspastic reconstructions, backward in time, as far as the geosyncline. The entire procedure—from the making of rock to the making of mountains to the destruction of mountains to the making of fresh formations of rock—was the geosynclinal cycle.

Inevitably, the concept was improved, refined, unsimplified. The archetypical geosyncline was deep in the middle and shallow at the sides, and grew different kinds of rocks in various places. The German tectonicist Hans Stille proposed the names miogeosyncline and eugeosyncline for the shallows and the deeps. The vocabulary was universally accepted. Miogeosynclines were the source of shallow-water sediments (limestone, for example) and no volcanics. In the eugeosynclines, volcanism occurred, and deepwater sediments, like chert, collected. In the twentieth century, as the science matured and thickened, mio- and eu- became inadequate to prefix all the differing synclinal scenes that new generations of geonovelists were describing. The germinant term was soon popping like corn. The professional conversation came to include parageosynclines, orthogeosynclines, taphrogeosynclines, leptogeosynclines, zeugogeosynclines, paraliageosynclines, and epieugeosynclines.

Moores had entered Caltech in 1955. "In the Old Geology, one learned of the eugeosyncline and miogeosyncline of western North America, which started in the late Precambrian and went through the Cretaceous," he said. "Rock deformed by orogeny—folding and thrusting—from the center of the eugeosyncline out toward the continental shelf. The mechanism was 'orogenic forces.' Here in the Sierra, for example, you had a eugeosyncline and a miogeosyncline, and the eugeosyncline was thrust on the miogeosyncline. And that was the Golconda Thrust. No one knew how this 'orogeny' happened."

If California rock was disassembled on paper and palinspastically reassembled as the original geosyncline, there were shallow-water sediments followed by deepwater material, but there was no other side. "That was never explained," Moores went on. "Also, the geosynclinal cycle was said to be about two hundred million years. In the Overthrust Belt in Montana, forty thousand feet of Precambrian sediment had been thrust over Cretaceous sediment. As students, we wondered why all that Precambrian was still there. What had the source geosyncline been doing sitting there for a billion years when the cycle was two hundred million? There was no answer."

Hall's idea was not preposterous. It was incomplete. There was, after all, marine rock in mountains. Between the geosyncline and the mountains, though, something was missing, and what was missing was plate tectonics.

We continued west from Emigrant Gap through cuts in unsorted glacial till, buff and bouldery, and past the many blue doors of the pink garage of the Transportation Department's mountain center for snowplows and road maintenance, situated, with its cavalry of trucks, within a slowly moving earthflow, a creeping descent of unstable moraine, a sedate landslide. "The engineers strike again," Moores said, but in scarcely three miles his contempt went into a subduction zone, melted, and came back up as appreciation for a long high competent roadcut that exposed bright beds of rhyolite tuff. Twenty-nine million years ago, this air-fall ash came out of a volcano in what is now Nevada, he said, as he pulled over to the side of the road, got out, and put his nose on the engineered outcrop. While he examined the tuff through his hand lens, an eighteen-wheeler that had also come from Nevada was smoking down the mountain grade. Its brakes were furiously burning, and emitted a dark cloud. Long after the truck had gone, the cloud hung stinking in the air. The ash had been launched in several eruptive episodes. Blown west, it had landed hot, and had welded solid in successive bands. Here, more than sixty miles from the source volcano, a single ash fall was more than a metre thick. The ash had settled, of course, horizontally. Having risen with the Sierra, it was now tilting west. We descended past the four-thousand-foot contour, moving on among volcanic rocks five times older than the tuff and of more proximate origin: rock of the Sonomia Terrane altered in the heat and pressure of the assembly of California and weathered along the interstate into an abstract medley of red and orange and buff and white.

Now thirty miles west of Donner Summit, we were well into the country rock of California gold—the rock that was there when, in various ways, the gold itself arrived. The most obvious place to look for it was in fluviatile placers—the rubble of running streams. In such a setting it had been discovered. *Placer*, which is pronounced like Nasser and Vassar, was a Spanish nautical term meaning "sandbank." More commonly, it meant "pleasure." Both meanings seem relevant in the term "placer mining," for to separate free gold from loose sand is a good deal easier than to crack it out of hard rock. Some of the gold in the running streams of the western Sierra was traceable to the host formations from which it had eroded—traceable, for example, to nearby quartz veins that had grouted ancient fissures. Within two years of the discovery of gold in river gravel, gunpowder was blasting the hard-rock fissures. Into the quiet country of the low Sierra—between the elevations of one thousand and four thousand feet—gold seekers spread more rapidly than an explosion of moles. Their technology was as rampant as they were, and in its swift development anticipated the century to come. In 1848, the primary instrument for mining gold was a sheath knife. You pried yellow metal out of crevices. Within a year or two, successively, came the pan, the rocker, the long tom, and the sluice—variously invented, reinvented, and introduced.

There was also a third source of gold. It was found in dry gravel far above existing streams—on high slopes, sometimes even on ridges. The gravel lay in discontinuous pods. Geologists, with their dotted lines, would eventually connect them. In cross section, they were hull-shaped or V-shaped, and in some places the deposits were more than a mile wide. They had the colors of American bunting: they were red to the point of rutilance, and white as well, and, in their lowest places, navy blue. They were the beds of fossil rivers, and the rivers were very much larger than the largest of the living streams of the Sierra. They were Yukons, Eocene in age. Fifty million years before the present, they had come down from the east off a very high plateau to cross low country that is now California and leave their sorted bedloads on a tropical coastal plain. Forty million years later, when the Sierra Nevada rose as a block tilting westward, it lifted what was left of that coastal plain. It included the beds of the Eocene rivers, which were fated to become so celebrated that they would be known in world geology less often as "the Eocene riverbeds of California" than as, simply, "the auriferous gravels." Fore-set, bottom-set, point bar to cutbank, under the suction eddies—gold in varying assay was everywhere you looked within the auriferous gravels: ten cents a ton in the high stuff, dollars a ton somewhat lower, concentrated riches in the deep "blue lead."

To separate gold from gravel, you wash it. But you don't wash a bone-dry enlofted Yukon with the flow of little streams bearing names like Shirt Tail Creek. Mining the auriferous gravels was the technological challenge of the eighteen-fifties. The miners impounded water in the high country, then brought it to the gravels in ditches and flumes. In five years, they built five thousand miles of ditches and flumes. From a ditch about four hundred feet above the bed of a fossil river, water would come down through a hose to a nozzle, from which it emerged as a jet at a hundred and twenty miles an hour. The jet had the diameter of a dinner plate and felt as hard. If you touched the water near the nozzle, your fingers were burned. This was hydraulic artillery. Turned against gravel slopes, it brought them down. In a contemporary account, it was described as "washing down the auriferous hills of the gravel range" and mining "the dead rivers of the Sierra Nevada." A hundred and six million ounces of gold—a third of all the gold that has ever been mined in the United States—came from the Sierra Nevada. A quarter of that was flushed out by hydraulic mining.

The dry bed of an Eocene river carries Interstate 80 past Gold Run. The roadside records the abrupt change. As if you were swinging off a riverbank and dropping into the water, you go out of the metavolcanic rock and into the auriferous gravels. We stopped, stood on the shoulder, and looked about a hundred feet up an escarpment that resembled an excavated roadcut but had not been excavated by highway engineers. It was capped by a mat of forest floor, raggedly overhanging. The forest, if you could call it that, was a narrow stand of ponderosas, above an understory of manzanita with round fleshy leaves and dark-red bark. The auriferous gravels were russet, and were full of cobbles the size of tomatoes—large stones of long transport by a most impressive river.

To the south, across the highway, the scene dropped off into a deep mountain valley. The near end of the valley was three hundred feet below the trees above us. The far end of the valley was nearly twice as deep. A mile wide, this was a valley that had not been a valley when wagons first crossed the Sierra. All of it had been water-dug by high-pressure hoses. It was man-made landscape on a Biblical scale. The stand of ponderosas at the northern rim was on the level of original ground.

The interstate was on a bench more than halfway up the gravel. Above us, behind the trees, were the tracks of the Southern Pacific. In the eighteen-sixties, when the railroad (then known as the Central Pacific) was about to work its way eastward across the mountains, it secured the rights to this ground before the nozzles reached it. Moores and I made our way up to the tracks, where the view to the north was over a hosed-out valley nearly as large as the one to the south, and bordered by white hydraulic cliffs. The railroad, with the interstate clinging to its hip, ran across a septum of the old terrain, an isthmus in the excavation, an unmined causeway hundreds of feet high made of gravel and gold.

This was the country of Iowa Hill, Lowell Hill, Poverty Hill, Poker Flat, Dutch Flat, Red Dog, You Bet, Yankee Jim's, Gouge Eye, Michigan Bluff, and Humbug City. It was the country of five hundred camps that sprang up for many dozens of miles to the north and south of the present route of I-80. For a year or two, it had been a center of world news, and for some decades had clanged with industry. Now, in the dry air, nothing was stirring, not even a transcontinental freight. But looking down the two sides into the artificial valleys you could almost hear the water jets and the caving slide of gravel. Poverty Hill yielded four million dollars' worth of gold. You Bet yielded three. Humbug City got its name from a lack of confidence in the claims there, but when five million dollars came out of forty million cubic yards of flushed-away ground the name was changed to North Bloomfield. The water-dug valleys below the ground where we stood had yielded six million dollars in gold.

Yankee Jim was an Australian. A red-dog bank was a savings-and-loan ahead of its time. It issued notes in excess of its ability to redeem them. Across most of the Sierra, Interstate 80 runs close by the line of two counties—Placer and Nevada—which together produced five hundred and sixty million dollars in gold. Translated into modern values, that would be five billion. Yankee Jim was hanged in his eponymous town.

The ancient riverbed beneath us evidently passed through Gold Run, picked up a fossil tributary coming in through Dutch Flat, and went off to the northwest via Red Dog and You Bet. Before human beings appeared on earth, glacial ice and modern streams and other geologic agents had obliterated large parts of the Eocene river system. People had come near eliminating the rest. "Man is a geologic agent," Moores said, with a glance that swept the centennial valleys. Erosion occurs, for the most part, in what geologists call catastrophic events—hurricanes, rockslides, raging floods—and in that category full credentials belonged to hydraulic mining, for scouring out and taking away thirteen thousand million cubic yards of the Sierra.

I remember Moores rapping his geologic hammer on an outcrop of olivine in northern Greece. He was drawn to the rock for academic reasons, but he remarked that it might be gone before long, because of its use in a brick that is resistant to very high temperatures. I asked him how he felt about being in a profession that calls attention to the olivine that people tear up mountainsides to take away. He said, "Schizophrenic. I grew up in a mining family, a mining town, and when I got out of there I had had it with mining.

Now I am a member of the Sierra Club. But you have to face the fact that if you are going to have an industrial society you must have places that will look terrible. Other places you set aside—to say, 'This is the way it was.' "

I remember him referring to the same disease in response to my asking him, one day in Davis, what effect his professionally developed sense of geologic time had had upon him. He said, "It makes you schizophrenic. The two time scales—the one human and emotional, the other geologic—are so disparate. But a sense of geologic time is the most important thing to get across to the non-geologist: the slow rate of geologic processes—centimetres per year—with huge effects if continued for enough years. A million years is a small number on the geologic time scale, while human experience is truly fleeting—all human experience, from its beginning, not just one lifetime. Only occasionally do the two time scales coincide."

When they do, the effects can be as lasting as they are pronounced. The human and the geologic time scales intersect each time an earthquake is felt by people. They intersect when mining, of any kind, begins. After 1848, when the two time scales intersected in the gold zone of the western Sierra, California was populated so rapidly that it became a state without ever being a territory. As the attraction diminished, newcomers ricocheted eastward, in sunburst pattern—to Idaho, to Arizona, to Nevada, New Mexico, Montana, Wyoming, Utah, Colorado—finding zinc, lead, copper, silver, and gold, and transmogrifying the West in a manner more pervasive than the storied transition from bison to cattle. The event of 1848 in California led directly to the discovery of gold in Australia (after an Australian miner who rushed to the Sierra saw auriferous facsimiles of New South Wales). By 1865, at the end of the American Civil War, seven hundred and eighty-five million dollars had come out of the ground in California, making a difference—possibly *the* difference—in the Civil War. The early Californian John Bidwell, an emigrant of 1841, expressed this in his memoirs:

It is a question whether the United States could have stood the shock of the great rebellion of 1861 had the California gold discovery not been made. Bankers and business men of New York in 1864 did not hesitate to admit that but for the gold of California, which monthly poured its five or

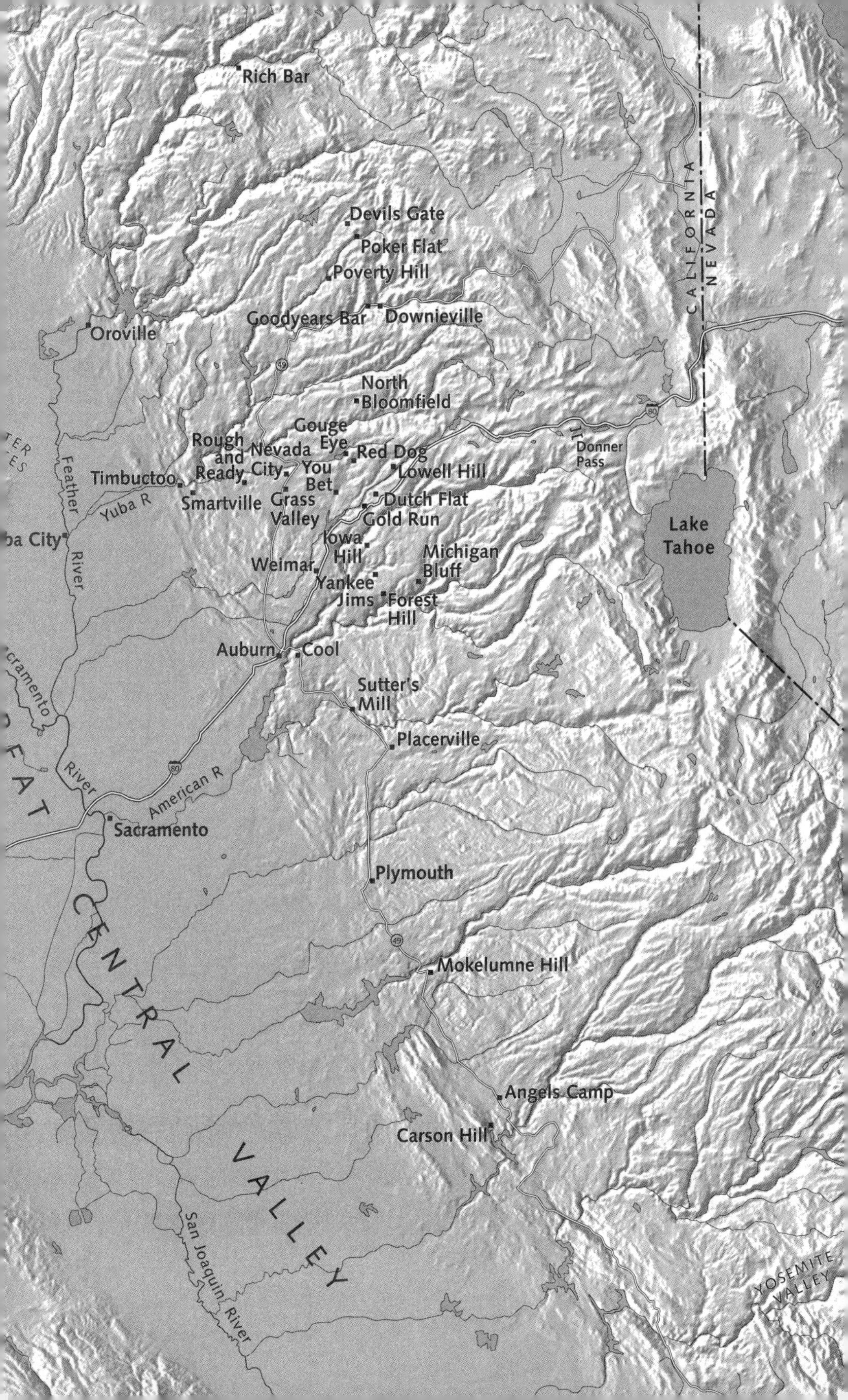

Rich Bar
Devils Gate
Poker Flat
Poverty Hill
Goodyears Bar
Downieville
Oroville
49
North Bloomfield
80
CALIFORNIA
NEVADA
Gouge Eye
Rough and Ready
Nevada City
Red Dog
You Bet
Lowell Hill
Donner Pass
Timbuctoo
Smartville
Grass Valley
Dutch Flat
Gold Run
Feather
Yuba R
River
Iowa Hill
Michigan Bluff
Lake Tahoe
Weimar
Yankee Jims
Forest Hill
Auburn
Cool
Sutter's Mill
Placerville
River
80
American R
Sacramento
Plymouth
49
Mokelumne Hill
CENTRAL VALLEY
Angels Camp
Carson Hill
San Joaquin River
YOSEMITE VALLEY

six millions into that financial center, the bottom would have dropped out of everything. These timely arrivals so strengthened the nerves of trade and stimulated business as to enable the government to sell its bonds at a time when its credit was its life-blood and the main reliance by which to feed, clothe, and maintain its armies. Once our bonds went down to thirty-eight cents on the dollar. California gold averted a total collapse and enabled a preserved Union to come forth from the great conflict.

Moores and I returned to the shoulder of the interstate and walked along the auriferous escarpment. The stream-rounded gravels, asparkle with quartz, are so compactly assembled there that they suggest the pebbly surface of a wide-wide screen. One does not need a director, a film, or rear projection to look into the bright stones and see the miners in motion: the four thousand who are in the region by the end of '48, the hundred and fifty thousand who follow in the years to 1884. With the obvious exception of the natives, no one is as sharply stricken by the convergences of time as Johann Augustus Sutter. He has come into a scene in which gold is unsuspected—this blue-eyed, blond and ruddy, bankrupt Swiss dry-goods merchant, with his broad-brimmed hat and his broader belly and his exceptionally creative dream. He is thirty-six years old. He envisions a wilderness fiefdom—less than a kingdom but more than a colony—with himself as a kind of duke. On a ship called Clementine, he arrives in Monterey in 1839, accompanied by ten Hawaiians and an Indian boy once owned by Kit Carson and sold to Sutter for a hundred dollars. The Mexican government, which seeks some sort of buffer between coastal California and the encroaching United States, grants him, on an incremental schedule, a hundred thousand acres of land. Sailing around San Francisco Bay, he spends a week hunting for the mouth of the Sacramento River. Soon after he finds it, there is a collection of Hawaiian grass huts on what is now Twenty-seventh Street, in a section of downtown Sacramento still called New Helvetia. He has cannons. He builds a fort, with walls three feet thick. He does not overlook a dungeon. A roof slopes in above peripheral chambers to frame a parade ground of two acres. Sutter's goal is to develop an independent agricultural economy, and he prospers. He has a gristmill. He brings in cattle and builds a tannery. He hires weavers and makes textiles. He widens his fields of grain, and draws plans for a second gristmill. He attracts many people. He issues passports.

One of the attracted people is James Wilson Marshall, of Lambertville, New Jersey, a mechanic-carpenter-wheelwright-coachmaker who is experienced as a sawyer. Sutter sees possibilities in cutting lumber and floating it to San Francisco. Meanwhile, he needs boards for the new gristmill. He sends Marshall up the American River to a small valley framed in canyons and backed by a mountainside of sugar pines. Like many handsome moments in western scenery, this one is prized by the natives, who think it is theirs. A bend in the river touches the mountains. Marshall lays a millrace across the bend.

He sees "blossom" in the stream gravel and remarks that he suspects the presence of metal.

A sawyer asks him, "What do you mean by 'blossom'?"

Marshall says, "Quartz."

As the sawmill nears completion, its wheel is too low. Water is ponding around it. The best correction is to deepen the tailrace down through the gravel to bedrock. Yalesumni tribesmen help dig out the tailrace, where, early in the morning of January 24, 1848, Marshall picks up small light chips that may not be stone.

Having some general knowledge of minerals, I could not call to mind more than two which in any way resembled this—*sulphuret of iron*, very bright and brittle; and *gold*, bright, yet malleable; I then tried it between two rocks, and found that it could be beaten into a different shape, but not broken.

He sets it on glowing coals, and he boils it in lye. The substance shows no change.

Carrying a folded cloth containing flakes the size of lentils, Marshall journeys to New Helvetia, and insists that he and Sutter talk behind a locked door. Sutter pours aquafortis on the flakes. They are unaffected. Sutter gets out his Encyclopedia Americana and looks under G. Using an apothecary's scales, he and Marshall are soon balancing the flakes with an equal weight of silver. Now they lower the scales into water. If the flakes are gold, their specific gravity will exceed the specific gravity of the silver. Underwater, the scales tip, and Marshall's flakes go down.

Sutter at once can see the future and is dismayed by the look of it. Who will work in his sawmill if gold lies in the stream beside it? Who will complete the gristmill? What will become of his New Helvetia, his field-and-forest canton, his discrete world, his agrarian dream? He and Marshall agree to urge others to keep the discovery a local if not total secret until the mills are finished.

Coincidentally in Mexico (that is, only five days after Marshall's visit to Sutter), Nicholas P. Trist, American special agent, who has defied orders recalling him to Washington and pressed on with negotiations, successfully concludes the Treaty of Guadalupe Hidalgo. For fifteen million dollars, Mexico, defeated in battle, turns over to the United States three hundred and thirty-four million acres of land, including California.

Sutter writes a twenty-year lease with the Yalesumni for the land around the sawmill. He agrees to grind their grain for them and to pay them, in clothing and tools, a hundred and fifty dollars a year. Seeking a validation of the lease, Sutter sends an envoy to Monterey—to Colonel Richard Mason, USA, military governor of California. The envoy sets on a table a number of yellow samples. Mason calls in Lieutenant William Sherman, West Point '40, his acting assistant adjutant general.

Mason: "What is that?"

Sherman: "Is it gold?"

Mason: "Have you ever seen native gold?"

Sherman: "In Georgia."

Sherman bites a sample. Then he asks a soldier to bring him an axe and a hatchet. With these he beats on another sample until —malleable, unbreakable—it is airy and thin. Sherman learned these tests in 1844, when he was twenty-four, on an investigative assignment in Georgia having to do with a military crime.

Mason sends a message to Sutter to the effect that the Indians, having no rights to the land, therefore have no right to lease it.

In an April memorandum, the editor of the *California Star* says, in large letters, "HUMBUG" to the idea that gold in any quantity lies in the Sierra. Six weeks later, the *Star* ceases publication, because there is no one left in the shop to print it. Thousands come through Sutter's Mill and spread into the country. On the American River at the discovery site, Marshall tries to charge tithes, but the forty-eighters ignore him and overrun his claims. They stand hip to hip like trout fishermen, crowding the stream. Like fishermen, too, they move on, restlessly, from cavern to canyon to flat to ravine, always imagining something big lying in the next pool. Indians using willow baskets wash sixteen thousand dollars. People are finding nuggets the size of eggs. "There is a chance now for every white man now in the country to make a fortune," says a letter written to the *New York Herald* on May 27, 1848. One white man, in some likelihood Scottish, is driven insane by the gold he finds, and wanders around shouting all day, "I am rich! I am rich!" Two miners in seven days take seventeen thousand dollars from a small gully.

In June, Colonel Mason travels from Monterey to San Francisco and on to New Helvetia to see for himself what is happening in the foothills. He takes Sherman with him. They find San Francisco "almost deserted," its harbors full of abandoned ships. Ministers have abandoned their churches, teachers their students, lawyers their victims. Shops are closed. Jobs of all kinds have been left unfinished. As Mason and Sherman cross the Coast Ranges and the Great Central Valley, they see gristmills and sawmills standing idle, loose livestock grazing in fields of ripe untended grain, "houses vacant, and farms going to waste." It is as if a devastating army had traversed a wide swath on its way to the foothills from the sea.

Sutter, in the shadow of his broad-brimmed hat—his silver-headed cane tucked under his arm—warmly greets the military officers. It is scarcely their fault that two thousand hides are rotting in the vats of his abandoned tannery, that forty thousand bushels of standing wheat are disintegrating on the stem, that work has ceased on the half-finished gristmill, that the weavers have abandoned their looms, that strangers without passports—here today, gone tomorrow—have turned his fort into a boarding house and taken his horses and killed his cattle. To short-term profit but long-term disaster his canton is doomed. His dream is drifting away like so much yellow smoke. We could follow him to his destitute farm on the Feather River and on to the East, where he dies insolvent in 1880, but better to leave him on July 4, 1848, sitting at the head of his table in the storehouse of his fort, host of a party he is giving to celebrate—for the first time in California—the independence of the United States. He has fifty guests. With toasts, entertainment, oratory, beef, fowl, champagne, Sauternes, sherry, Madeira, and brandy, he presents a dinner that costs him the equivalent of sixty thousand dollars (in the foothills' suddenly inflated prices converted into modern figures). In no way has he shown resentment that rejection has met his appeal to secure his claims in a discovery of sufficient magnitude to pay for a civil war. Seated on his right is Richard Barnes Mason. Seated on his left is William Tecumseh Sherman.

By the end of 1848, a few thousand people, spread out over a hundred and fifty miles, have removed from modern stream placers ten million dollars in gold. The forty-eighters have the best of it in 1849, because the forty-niners are travelling most of that season, at the end of which fifty thousand miners are in the country. There are a hundred and twenty thousand by the end of 1855. The lone miner all but disappears. To stay abreast of the sophisticating technologies, individuals necessarily form groups. Groups are crowded out by corporations. More and more miners make less and less money, until many independents are living hand-to-mouth and their way of life is called subsistence mining. Watching companions die of disease in Central American jungles or drown in Cape storms, they have travelled thousands of miles in pursuit of a golden goal that has now turned into "mining for beans." Always, though, there are fresh stories going east—stories that would cause almost anyone to start thinking about trying the overland route, the isthmus, the Horn.

In a deep remote canyon on the east branch of the north fork of the Feather River, two Germans roll a boulder aside and under it find lump gold. Another couple of arriving miners wash four hundred ounces there in eight hours. A single pan yields fifteen hundred dollars. The ground is so rich that claims are limited to forty-eight inches square. In one week, the population grows from two to five hundred. The place is named Rich Bar.

At Goodyears Bar, on the Yuba, one wheelbarrow-load of placer is worth two thousand dollars.

From hard rock above Carson Creek comes a single piece of gold weighing a hundred and twelve pounds. After black powder is packed in a nearby crack, the blast throws out a hundred and ten thousand dollars in gold.

A miner is buried in Rough and Ready. As shovels move, gold appears in his grave. Services continue while mourners stake claims. So goes the story, dust to dust.

From the auriferous gravels of Iowa Hill two men remove thirty thousand dollars in a single day.

A nugget weighing only a little less than Leland Stanford comes out of hard rock in Carson Hill. Size of a shoebox and nearly pure gold, it weighs just under two hundred pounds (troy)—the largest piece ever found in California. Carson Hill, in Calaveras County, is in the belt of the Mother Lode—an elongate swarm of gold-bearing quartz veins, running north-south for a hundred and fifty miles at about a thousand feet of altitude. There are Mother Lode quartz veins as much as fifty metres wide.

American miners come from every state, and virtually every county. Others have arrived from Mexico, India, France, Australia, Portugal, England, Scotland, Wales, Ireland, Germany, Switzerland, Russia, New Zealand, Canada, Hawaii, Peru. One bloc of several thousand is from Chile. The largest foreign group is from China. Over most other miners, the Chinese have an advantage even greater than their numbers: they don't drink. They smoke opium, certainly, but not nearly as much as the others like to think. The Chinese miners wear outsized boots and blue cotton. Their packs are light. They live on rice and dried fish. Their brothels thrive. They are the greatest gamblers in the Sierra. They make Caucasian gambling look like penny ante.

Some of the early gold camps are so deep in ravines, gulches, caverns, and canyons that the light of the winter sun never reaches the miners' tents. If you have no tent, you live in a hole in the ground. Your backpack includes a blanket roll, a pick, a shovel, a gold pan, maybe a small rocker in which to sift gravel, a coffeepot, a tobacco tin, saleratus bread, dried apples, and salt pork. You sleep beside your fire. When you get up, you "shake yourself and you are dressed." You wear a flannel shirt, probably red. You wear wool trousers, heavy leather boots, and a soft hat with a wide and flexible brim. You carry a pistol. Not everyone resembles you. There are miners in top hats, miners in panama hats, miners in sombreros, and French miners in berets, who have raised the tricolor over their claims. There are miners working in formal topcoats. There are miners in fringed buckskin, miners in brocaded vests, miners working claims in dress pumps (because their boots have worn out). There are numerous Indians, who are essentially naked. There are many black miners, all of them free. As individual prospecting gives way to gang labor, this could be a place for slaves, but in the nascent State of California slavery is forbidden. On Sundays, while you drink your tanglefoot whiskey, you can watch a dog kill a dog, a chicken kill a chicken, a man kill a man, a bull kill a bear. You can watch Shakespeare. You can visit a "public woman." The *Hydraulic Press* for October 30, 1858, says, "Nowhere do young men look so old as in California." They build white wooden churches with steeples.

In four months in Mokelumne Hill, there is a murder every week. In the absence of law, lynching is common. The camp that will be named Placerville is earlier called Hangtown. When a mob forgets to tie the hands of a condemned man and he clutches the rope above him, someone beats his hands with a pistol until he lets go. A Chinese miner wounds a white youth and is jailed. With a proffered gift of tobacco, lynchers lure the "Chinee" to his cell window, grab his head, slip a rope around his neck, and pull until he is dead. A young miner in Bear River kills an older man. A tribunal offers him death or banishment. He selects death, explaining that he is from Kentucky. In Kentucky, that would be the honorable thing to do.

Some miners' wives take in washing and make more money than their husbands do. In every gold rush from this one to the Klondike, the suppliers and service industries will gather up the dust while ninety-nine per cent of the miners go home with empty pokes. In 1853, Leland Stanford, twenty-nine years old, opens a general store in Michigan Bluff, about ten miles from Gold Run. John Studebaker makes wheelbarrows in Hangtown.

Stanford moves to Sacramento, where he sells "provisions, groceries, wines, liquors, cigars, oils & camphene, flour, grain & produce, mining implements, miners' supplies." Credit is not in Stanford's vocabulary. Miners must "come down with the dust." They come down with the dust to Mark Hopkins, a greengrocer who, sensing greater profit in picks, shovels, and pans, goes out of produce and into partnership at Collis P. Huntington, Hardware. They come down with the dust to Charles Crocker, Mining Supplies. When the engineer Theodore Judah comes down from a reconnaissance of the

Sierra with the opinion that a railroad can be built across the mountains, these merchants of Sacramento have the imagination to believe him, and they form a corporation to construct the Central Pacific. The geologic time scale, rising out of the ground in the form of Cretaceous gold, has virtually conjured a transcontinental railroad.

It leaves Sacramento in 1863, and not a minute too soon, for in a sense—which is only a little fictive—it is racing the technology of mining. As the railroad advances toward Donner Pass at the rate of about twenty miles a year, the miners are doing what they can to remove the intervening landscape. Their ability to do so has been much accelerated in scarcely a dozen years. This is the evolution of technique:

Prospectors find the fossil rivers within two years of James Wilson Marshall's discovery, and soon afterward vast acreages are full of holes that seem to have been made by very large coyotes. In the early form of mining that becomes known as coyoting, you dig a deep hole through the overburden and lower yourself into it with a windlass. You hope that your mine will not become your grave. You dig through the gravel to bedrock, then drift to the side. Some coyote shafts go down a hundred feet. One goes down six hundred. When water first arrives by ditch and flume, it not only washes excavated pay dirt but is allowed to spill downslope, gullying the gravel mountainsides and washing out resident gold. This is known as ground-sluicing, gouging, booming, or "picking down the bank." Even now the terrain is beginning to reflect the fact that these visitors are not the sort who carry out what they carry in. Jack London will write in "All Gold Canyon":

> Before him was the smooth slope, spangled with flowers and made sweet with their breath. Behind him was devastation. It looked like some terrible eruption breaking out on the smooth skin of the hill. His slow progress was like that of a slug, befouling beauty with a monstrous trail.

So far, the technology is not new. From high reservoirs and dug canals, the Romans ground-sluiced for gold, as did Colombian Indians before 1500, and people in the eighteenth century in the region known as the Brazils. In the words and woodcuts of *De Re Metallica* (1556) the Saxon physician Georg Bauer, whose pen name was Georgius Agricola, comprehensively presented gold metallurgy, from panning and sluicing to the use of sheepskin:

> Some people wash this kind of sand in a large bowl which can easily be shaken, the bowl being suspended by two ropes from a beam in a building. The sand is thrown into it, water is poured in, then the bowl is shaken, and the muddy water is poured out and clear water is again poured in, this being done again and again. In this way, the gold particles settle in the back part of the bowl because they are heavy, and the sand in the front part because it is light. . . . Miners frequently wash ore in a small bowl to test it.

> A box which has a bottom made of a plate full of holes is placed over the upper end of a sluice, which is fairly long but of moderate width. The gold material to be washed is thrown into this box, and a great quantity of water is let in. . . . In this way the Moravians, especially, wash gold ore.
>
> The Lusitanians fix to the sides of a sluice, which is about six feet long and a foot and a half broad, many cross-strips or riffles, which project backward and are a digit apart. The washer or his wife lets the water into the head of the sluice, where he throws the sand which contains the particles of gold.

The Colchians placed the skins of animals in the pools of springs; and since many particles of gold had clung to them when they were removed, the poets invented the "golden fleece" of the Colchians.

(Translation by Herbert Clark Hoover and Lou Henry Hoover, 1950.)

California's momentous innovation in placer mining comes in 1853, after Edward E. Matteson, a ground-sluicer, is nearly killed when saturated ground slides down upon him and knocks his pick from his hand. Matteson thinks of a way to dismantle a slope from a safe distance. With his colleagues Eli Miller and A. Chabot, he attaches a sheet-brass nozzle to a rawhide hose and bombards a hill near Red Dog with a shaped hydraulic charge. That first nozzle is only three feet long and its jet at origin three-quarters of an inch in diameter. Soon the nozzles are sixteen feet long, and are called dictators, monitors, or giants. They require ever more ditches and flumes. In the words of *Hutchings' California Magazine*, "The time may come when the whole of the water from our mountain streams will be needed for mining and manufacturing purposes, and will be sold at a price within the reach of all." Where two men working a rocker can wash a cubic yard a day, two men working a mountainside with a dictator can bring down and drive through a sluice box fifteen hundred tons in twelve hours, and this is the technology that the railroad is racing to the ground at Gold Run.

Although the nozzle has the appearance of a naval cannon, it is mounted on a ball socket and is so delicately counterbalanced with a "jockey box" full of small boulders that, for all its power, it can be controlled with one hand. Every vestige of what has lain before it —forest, soil, gravel—is driven asunder, washed over, piled high, and flushed away. At a hundred and twenty-five pounds of pressure per square inch, the column of shooting water seems to subdivide into braided pulses hypnotic to the eye, and where it crashes at the end of its parabola it sounds like a storm sea hammering a beach. In one year, the North Bloomfield Gravel Company uses fifteen thousand million gallons of water. Through the big nine-inch nozzles go thirty thousand gallons a minute. Benjamin Silliman, Jr., a founding professor of the Sheffield Scientific School, at Yale, writes in 1865, "Man has, in the hydraulic process, taken command of nature's agencies, employing them for his own benefit, compelling her to surrender the treasure locked up in the auriferous gravel by the use of the same forces which she employed in distributing it!"

To get at the deepest richest gravels, which lie in the hollows of bedrock channels, the miners dig tunnels under the beds of the fossil rivers. When they reach a point directly below the blue lead, they go straight up into the auriferous gravel, where they set up their nozzles and flush out the mountain from the inside. At Port Wine Ridge, Chinese miners make a tunnel in the gravel fifteen miles long. Surface excavations meanwhile deepen. Twelve million cubic yards of gravel are washed out of Scotts Flat, forty-seven million cubic yards of gravel out of You Bet and Red Dog, a hundred and five million cubic yards out of Dutch Flat, a hundred and twenty-eight million out of Gold Run. After a visit to Gold Run and Dutch Flat in 1868, W. A. Skidmore, of San Francisco, writes, "We will soon have deserted towns and a waste of country torn up by hydraulic washings, far more cheerless in appearance than the primitive wilderness of 1848." In the middle eighteen-sixties, hydraulic miners find it profitable to get thirty-four cents' worth of gold from a cubic yard of gravel. In a five-year period in the eighteen-seventies, the North Bloomfield Gravel Company washes down three and a quarter million yards to get $94,250. Soon the company is moving twelve million parts of gravel to get one part of gold.

As the mine tailings travel in floods, they thicken streambeds and fill valleys with hundreds of feet of gravel. In their blanched whiteness, spread wide, these gravels will appear to be lithic glaciers for a length of time on the human scale that might as well be forever. In a year and a half, hydraulic mining washes enough material into the Yuba River to fill the Erie Canal. By 1878 along the Yuba alone, eighteen thousand acres of farmland are covered. Mud, sand, cobbles—Yuba tailings and Feather River tailings spew ten miles into the Great Central Valley. Tailings of the American River reach farther than that. Broad moonscapes of unvegetated stream-rounded rubble conceal the original land. Before hydraulic mining, the normal elevation of the Sacramento River in the Great Valley was sea level. As more and more hydraulic detritus comes out of the mountains, the normal elevation of the river rises seven feet. In 1880, hydraulic mining puts forty-six million cubic yards into the Sacramento and the San Joaquin. The muds keep going toward San Francisco, where, ultimately, eleven hundred and forty-six million cubic yards are added to the bays. Navigation is impaired above Carquinez Strait. The ocean is brown at the Golden Gate.

In the early eighteen-eighties, a citizens' group called the Anti-Debris Association is formed to combat the hydraulic miners. On June 18, 1883, a dam built by the miners fails high in the mountains—apparently because it was insufficiently engineered to withstand the pressure of high explosives. Six hundred and fifty million cubic feet of water suddenly go down the Yuba, killing six people and creating a wasteland much like the miners'. On January 9, 1884, a United States circuit court bans the flushing of debris into streams and rivers. Although the future holds some hydraulic mining—with debris dams, catch basins, and the like—it is essentially over, and miners in California from this point forward will be delving into hard rock.

Edward E. Matteson—of whom the *Nevada City Transcript* said in 1860, "His labors, like the magic of Aladdin's lamp, have broken into the innermost caves of the gnomes, snatched their imprisoned treasures, and poured them, in golden showers, into the lap of civilized humanity"—spends the last days of his life at Gold Flat, near Nevada City, working as a nighttime mine watchman and a daytime bookseller. Even in the high years of his invention, he never applied for a patent. From 1848 onward, James Wilson Marshall has been literally haunted by the fact of his being the discoverer of California gold. William Tecumseh Sherman will remember him as "a half-crazy man at best," an impression that Marshall confirms across the years as he claims to consult with spirits, asking them where he might again find gold. Newcomers to California in mid-century believe that Marshall really does have some sort of divine intuition, and—to his bitter annoyance—follow him wherever he goes. With respect to further gold strikes, nowhere is where he goes. Drinking himself to heaven, he drips tobacco juice through his beard. It stains his shirt and dungarees. Looking so, he makes a visit home. From his family's house, on Bridge Street in Lambertville, he goes up into the country toward Marshall's Corner and the farmhouse where he was born, prospecting outcrops of New Jersey diabase, hoping to discover gold. He picks up rock samples. He carries them to a sister's house and roasts them in the oven.

At the end of the twentieth century, the small farmhouse where he was born is still standing. Part fieldstone, part frame, it has long since been divided into three apartments, enveloped in a parklike shopping center called Pennytown. A boldly lettered sign on a screen door indirectly recalls Marshall's compact with Sutter. It says, "Don't Let the Cat Out."

Beside I-80, Moores inserted a knife in the auriferous gravels and pried loose a few rounded stones. He carved them to demonstrate their softness, and said, "They are practically clay. They have weathered so much they could be in Georgia."

In the nineteenth century, some of the nuggets found in the auriferous gravels were electrum—a natural pale-yellow alloy of gold and silver. Other nuggets were full of mothy cavities, where something had been eaten away—quite possibly silver. This was some of the first evidence that California enjoyed a coincidence of golds, for electrum was not characteristic of the hard-rock gold of the Sierra. The gravels had brought those nuggets from somewhere else. Rock soft enough to carve with a knife would disintegrate if it were tumbling in the bed of a stream; therefore, it had softened after it arrived. Because gold changes shape so easily, the mothy pitting of nuggets necessarily occurred after transport, too.

That the auriferous deposits were Eocene was affirmed by the fossil plants among them. The gravels themselves, with channel deposits six hundred feet deep and stones the size of basketballs, described the power of the river that brought them, the Himalayan loft of its headwaters. Fossils of subalpine Eocene vegetation have since been found in central Nevada.

"There is gold in the Carson Range, east of the Sierra, that is like the nuggets that were found in these gravels," Moores said. "The source of some California gold is probably under Nevada now."

If something as crazy-sounding as that had been said to miners in the eighteen-fifties, the miners in all likelihood would not have been surprised, for they were familiar with geologists, and geologists were not their heroes. In 1852, at Indian Bar, a miner remarked to a doctor's wife, "I maintain that science is the blindest guide that one could have on a gold-finding expedition. Those men, who judge by the appearance of the soil, and depend upon geological calculations, are invariably disappointed, while the ignorant adventurer, who digs just for the sake of digging, is almost sure to be successful." The doctor's wife, Louisa Amelia Knapp Smith Clappe, is probably the most interesting writer who was on the scene in the early days of the gold rush. Indian Bar was close by Rich Bar, where the two Germans in the deep canyon rolled a boulder and found lump gold. The doctor and his wife became resident there in 1851. She wrote letters to a sister in Amherst, Massachusetts, which have been preserved in publication under her pseudonym, Dame Shirley. At times, she may be even more purple than the interior of the Rich Bar saloon, but when she speaks of "the make-shift ways which some people fancy essential to California life," she is hitting for distance. She speaks of "red-shirted miners . . . reclining gracefully . . . in that transcendental state of intoxication, when a man is compelled to hold on to the earth for fear of falling off." She speaks of "the Irishman's famous down couch, which consisted of a single feather laid upon a rock." And she has thoughts to add about geologists:

Wherever Geology has said that gold *must* be, there, perversely enough, it lies not; and wherever [geology] has declared that it could *not* be, there has it oftenest garnered up in miraculous profusion the yellow splendor of its virgin beauty. It is certainly very painful to a well-regulated mind to see the irreverent contempt, shown by this beautiful mineral, to the dictates of science; but what better can one expect from the "root of all evil"?

There were prospectors in the Sierra who wore over their hearts a device they called a gold magnet, explaining that in the presence of gold the magnet tingled and shocked. There were prospectors who carried forked hazelwood rods that were said to point to gold as if it were water. The miners had as much respect for them as they had for the geologists. Over their shoulders as they took off up the canyons the miners liked to say, "Gold is where you find it." As early as 1849, the *Sacramento Placer Times* remarked:

> The mines of California have baffled all science, and rendered the application of philosophy entirely nugatory. Bone and sinew philosophy, with a sprinkling of good luck, can alone render success certain. We have met with many geologists and practical scientific men in the mines, and have invariably seen them beaten by unskilled men, soldiers and sailors, and the like.

All that notwithstanding, the legislature of the new State of California created in 1860 a state geological survey, and recruited the Yale-trained and already distinguished Josiah D. Whitney to be the state geologist. Nearly everybody imagined that Whitney would investigate and catalogue places in California where the earth could be turned for a profit. Instead, he gave them paleontology, historical geology, igneous petrology, stratigraphy, structure, tectonics. He gave them the minutest points of mineralogy, and he gave them the global setting. He gave them academic geology in the form that can least be turned into capital—the disciplines that lead to understanding of the history and composition of the planet. California fired him. They fired him in the modern sense that after a few years he was defunded. His name rests on the highest mountain in the Sierra Nevada.

By the erosive scenes at Iowa Hill, Poverty Hill, Forest Hill, North Bloomfield, Michigan Bluff, Gold Run, You Bet, Dutch Flat, Poker Flat, Downieville, and Smartville—the major Eocene-river deposits—Josiah Whitney was not appalled. He liked the hydraulic diggings. They flushed away the soft stuff and exposed solid rock, the better for geologists to see.

Moores cast a final glance over the man-made valley by Gold Run, and said, "It's not all that bad. Some places like this do not look bad. They are spaced out. They are not the English industrial Midlands. I like to drive cars. I like to move rapidly from place to place. There is a price we pay. If people wish to eschew all that, let them walk. When they get rid of their cars and their hi-fi sets, their credibility will rise."

He lingered long enough for a change of mood. His voice resumed at a lower and softer register. "In a couple of hundred years we are doing a good job of extracting minerals deposited over billions of years. High-grade gold deposits are just gone. Ditto copper. The U.S. has had it. There just won't be any more until we go through a few million more years of erosion, allowing the geologic processes of secondary enrichment to take place. Meanwhile, technology must extract lower- and lower-grade resources. We don't realize what we're doing."

I said I thought that we knew what we were doing and didn't give a damn.

He said, "Americans look upon water as an inexhaustible resource. It's not, if you're mining it. Arizona is mining groundwater."

Soon we were dropping toward two thousand feet, among deeply weathered walls of phyllite, in color cherry and claret—the preserved soils of the subtropics when the unrisen mountains were a coastal plain. Geologists call it lateritic soil, in homage to the Latin word for brick. All around the Sierra, between two and three thousand feet of altitude, is a band of red soil, its color deepened by rainfall that leaches out competing colors and intensifies the iron oxide. Not only phyllites but also mica schists, shales, tuffs, and sandstones in the roadcuts were red. When the road dipped far below the rooflike plane of the western Sierra Nevada, the dissected inclines around us had the appearance of red mountains covered with manzanita.

At Weimar, a little off the highway and close to the two-thousand-foot contour, was a narrow band of serpentine, the California state rock. Moores said, "Worldwide, there is an association between serpentine and gold-bearing quartz, as there is here, in the belt of the Mother Lode. Gold-quartz deposits and serpentines just go together. Where there's a hard-rock mine, serpentine will not be far away. The relationship between serpentine and quartz-vein gold is not well understood, but the miners talked about it. It was a fact of their life." On the geologic map, the serpentines showed up as strings and pods in a rich wisteria blue, like some sort of paisley print, trending north-south, signing the Mother Lode.

Also accompanying the Mother Lode was a family of major faults, confined to a zone that was scarcely fifteen miles wide but extended, both to the north and to the south of Interstate 80, more than a hundred miles. Three of the faults crossed the highway in and close to Auburn, about twenty miles below Gold Run and thirty-five above Sacramento. Auburn, once known as Rich Dry Diggings, is now the seat of Placer County. Gold was found there in a living stream less than four months after Marshall's discovery at Sutter's Mill, and mined hard-rock ore was still being stamped to powder at Auburn well into the twentieth century. The placer discovery was in Auburn Ravine, which the interstate touches as it passes through the town and under the Southern Pacific. In 1849, Auburn was as far uphill as you could haul things conveniently from Sacramento in a wagon. For people and pack mules, it was the trailhead to the burgeoning mines. Within a few years, Auburn had become known as the Crossroads of the Mother Lode. In masonry walls of block schist, in windows arched with sawed soapstone, something is left in Auburn of the Roaring Fifties.

In Auburn Ravine, a couple of hundred yards below the railroad overpass and exactly twelve hundred feet above sea level, the interstate had been cut through charcoal-gray rock that had very evidently been damaged by a great deal more than human engineering. We pulled over as soon as the shoulder was wide enough, and walked back to have a look. We walked past talc schists and sheared serpentines and integral blocks of volcanic rock separated by shear zones. The cut that had caught Moores' attention was ten feet high and nondescript, below gray pines and trees of heaven. It was tight to the interstate, and tandem trailers were screaming past us. A billboard across the road said, "Placer Savings, It's the Extras That Count." Picking and prying at the Sierra Nevada a roadcut at a time, Moores had crossed the mountains showing all levels of absorption and excitement. In the presence of unusual rock, he variously fizzes and clicks. Now, as he leaned into this outcrop with his lens, he began to do both.

It was fine-grained diabase, in magnification asparkle with crystals—free-form, asymmetrical, improvisational plagioclase crystals bestrewn against a field of dark pyroxene. It was a much finer diabase than you would find in, say, the Palisades Sill, across the Hudson River from Manhattan. It had cooled and frozen more rapidly, but it derived from a chemically identical magma—that is to say, *essentially* identical, there being no exact copy in geology except a Xerox of your last mistake. Had this magma been extruded into air or water it would have become basalt, but—like granite, diorite, gabbro—it had chilled and formed its crystals in the absence of both. There was a signal difference, however—far beyond cooling rates or chemical composition—between this diabase and the rock of the Palisades Sill or any magma that intrudes and then hardens as a single body. To see the difference, you did not need to make a thin section—a tiny slice of rock for a microscope slide. You did not even need the hand lens. This rock had been assembled in vertical laminations like successive layers of wallboard. It had frozen not all in one piece but in continual fashion, layer after layer—a history that could be read from one lamination to the next, like bar codes indefinitely extended. Moores, ebullient, said, "We're in Fat City." Lens to eye and leaning into the outcrop, this professed and practicing agnostic said, "God, it's fantastic! God Almighty! This is a jackpot, a tremendous bit of serendipity. We've struck gold."

Given the fact that we were at twelve hundred feet in the western foothills of the Sierra Nevada and in close proximity to serpentine and quartz, I could be forgiven if at first I took him literally. Yet all that glistered in this outcrop was pyroxene. Gold is where you find it, though, and for Eldridge Moores this indeed was gold. Unlike all the other rock we had seen as we traversed the mountains, or were likely to see in most of the aerial world, this rock in its origin was not of any continent. It was not from slope, not from shelf, not from lake, stream, or land. It had no genetic relationship to continental rock. Like a blue-water fish on a farmhouse platter, it had been moved a great distance. Only a meteorite could have been more out of place.

Nineteenth-century geologists would have called this rock augite porphyrite; the miners would have called it blue diorite or slate. It was rock of the ocean crust. Formed at spreading centers, ocean crust gradually turns cold as it travels away from the hot rift of its beginnings toward the deep trenches where nearly a hundred per cent of it is consumed. Down the vertical column from salt water to mantle rock, ocean crust has varying components, of which these laminations are the clearest record of lateral movement. A layer at a time, the fluid rock is driven upward in the spreading center, solidifies, and takes its place in the long march. Most of this happens in the mid-oceans, in the world system of separating boundaries of plates. It also happens in the short, isolated, and slice-like spreading centers that develop near island arcs. In all geology where rock forms in successive layers, the layers are initially horizontal—with this one exception. The laminations of the ocean crust form vertically, and remain vertical as they move to become the floors of abyssal plains and until they disappear into trenches. In Moores' words, "This is the only situation where age progression goes sideways."

Although the rock in this outcrop had obviously been shattered by a very great tectonic force—and although it had to some extent been recrystallized as well in the attendant heat and pressure—neither its disfigurement nor its metamorphosis had masked its structure. The laminations—known in geology as sheeted dikes—were as narrow as ten centimetres and as wide as eighty. By looking closely at their edges, you could all but see the spreading center that the accumulating rock had slowly moved away from. Layer after layer was glassy along its right-hand edge. The magma had cooled quickly there after touching solidified rock. The spreading center, therefore, had been to the left. After a new lamination of magma touched hard rock and turned marginally to glass, the rest of the lamination froze more slowly, forming the fine crystals. Some layers had glassy margins on both sides. They had split the weak center of previous and still-cooling layers. In a minor and local way, they corrupted the chronology.

When seismology first revealed the dimensions of the ocean crust, it proved to be surprisingly thin—about fifteen thousand feet thin—with remarkable uniformity all over the world. The sediments upon it, generally speaking, are not much more than a veneer. Rock of the ocean crust—departing from spreading centers with bilateral symmetry, ultimately disappearing in the subduction zones—is everywhere younger than most rock of the continents. The oldest known continental rock was discovered east of Great Bear Lake, in the Canadian Northwest Territories, in 1989, and has a uranium-lead age of 3.96 billion years. The earth itself, according to radiometrics, is six hundred million years older than that. The oldest ocean-crustal rock that has yet been found in any seafloor in the world is early-middle Jurassic—a hundred and eighty-five million years old. That is less than one-twentieth the age of the oldest continental rock and one-twenty-fifth the age of the earth itself. From spreading to subduction, from creation to extinction, the ocean crust completely cleans house in fewer than two hundred million years. A lithospheric plate will typically include both continental rock and ocean crust, but trenches get rid of the ocean crust while the continents stay afloat. Since rock of the sort that Moores and I were looking at does not form on continents and will not be found under a Hudson Bay, a Sea of Okhotsk, or any epicontinental sea, what was it doing in Auburn, California, more than five hundred miles from the nearest abyssal floor?

Moores did not have to be asked, for if he had a tectonic and petrologic specialty this was it. He had travelled the earth to see this kind of rock. Where you found it up on dry land, it proclaimed an event in the making of new country, in the mobile history of plates. It was not a signature after a fact but a precursory signing in. In its transportation from the deep and its emplacement on a continent, it was not merely a clue but an absolute statement that scenery had been shifted in an operatic manner.

Toward the end of the middle Jurassic—in the high noon of dinosaurs, about a hundred and sixty-five million years ago—an island arc like the Aleutians or Japan had moved in from the western ocean and docked here. This was the third terrane at this latitude: the one that followed Sonomia and smashed into it with crumpling, mountain-building effects that propagated eastward turning soils into phyllites, sandstones into quartzites, siltstones into slates—the metamorphics we had seen up the road. In aggregate, the three terranes extended the continent by at least four hundred miles. The third one, suturing here, had doubled the width of what is now California.

The sheeted diabase that we found in Auburn—shattered so grossly in the collision—was a part of the ocean crust at the leading edge of the third terrane. As the island arc drifted eastward and the continent westward, nearly all the intervening ocean crust was consumed, but some broke off and came to rest on the continental margin, announcing the collision.

North-south, the third terrane probably came near to being a thousand miles long. What remains of it is closer to a hundred. Its width, including the part that is under the Great Valley, is about a hundred miles, too. This ten-thousand-square-mile piece of ground, named for a gold camp some twenty-five miles north of Auburn, is known in geology as the Smartville Block.

If you look at a map of the Mother Lode and lode-gold belts related to it—a narrow band, north-south, lying under Grass Valley, Forest Hill, Placerville, Plymouth, Mokelumne Hill, Angels Camp, Carson Hill—you are, for practical purposes, looking at a map of the Smartville suture. As a geologically immediate result of the collision, the nearby rock developed the numerous high-angle faults that now appear on the geologic map along the Mother Lode. The voluminous magmas of the batholith came into the country. Water moving down through the faults would have circulated close to—or actually in—the magma, dissolving high-temperature gold compounds, and carrying them upward to precipitate the gold in fissures. In this manner, the Smartville Block, docking in the Jurassic, not only doubled the size of central California but created its Mother Lode.

If you could pull up an acre of abyssal plain anywhere in the world—lift into view a complete column of the ocean floor, from the accumulated sediments at the top to mantle rock at the base—you would find the sheeted dikes about halfway down. In contrast to the rock columns you find all over the continents—giddy with time gaps among lithologies of miscellaneous origin and age—this totem assemblage from the oceans tells a generally consistent story. At its low end is peridotite, the rock of the mantle, tectonically altered in several ways on departure from the spreading center. Above the mantle rock lie the cooled remains of the great magma chamber that released flowing red rock into the spreading center. The chamber, in cooling, tends to form strata, as developing crystals settle within it like snow—olivine, plagioclase, pyroxene snow—but above these cumulate bands it becomes essentially a massive gabbro shading upward into plagiogranite as the magmatic juices chemically differentiate themselves in ways that relate to temperature. Just above the granites are the sheeted dikes of diabase, which kept filling the rift between the diverging plates. Above the sheeted dikes, where the fluid rock actually entered the sea, the suddenly chilled extrusions are piled high, like logs outside a sawmill. Because these extrusions have convex ends that bulge smoothly and resemble pillows, they are known in geology as pillow lavas. Above the pillows are the various sediments that have drifted downward through the deep sea: umbers, ochres, cherts, chalk. Unlike the rest of the crust-and-mantle package, the sediments may hint at the surrounding world. Water that gets down through all this and into the mantle rock—at the spreading center or anywhere else—will change the nature and appearance of that rock. Through an alteration of minerals, the rock takes on a silky lustre and a very smooth texture, becomes fibrous, and develops color—occasional streaks and spots of white, but mainly chrome green, myrtle green, Nile green, in patterned shapes within the mantle black. Because the patterns strongly suggest the skin of a snake, this rock has been known—for nearly six hundred years in the English language—as serpentine. Geologists—in their strange, synecdochical way—have named the entire oceanic assemblage for this one component rock. But not directly. In their acute sense of time, they were not content to settle for a term of Latin derivation. Instead, they extracted from a deeper stratum ὄφις—*ophis*—the Greek word for snake. From the mantle upward, the complete column of ocean-floor rock is collectively known in geology as an ophiolite. The generally consistent differences within it are the ophiolitic sequence.

On the American River under the bluffs of Auburn, in 1852, a single pan of gravel might be worth a hundred dollars. In 1857, after the lone miners had worked the place over, the American River Ditch Company built a dam there, to impound water for hydraulic mining. The dam eventually crumbled. The dam site did not. As environmentalists have discovered to their eternal chagrin, a dam site is a dam site forever, no matter what the state or the nation may decide to do about it in any given era. On present road maps of

California, that part of the American River is marked "Auburn Dam and Reservoir (Under Construction)."

The dam site is scarcely a mile from the shattered ophiolite of Interstate 80, so Moores and I went to see how the dam might relate to the Smartville collision, and we have returned there since. The river's deep canyon is walled with sheared foliated rock—broken, disrupted, deformed lithologies, *Bruchgeitrochen*, tortured rocks—as one would expect of a place where an oceanic island arc had sutured onto the continent. There were sheeted dikes, serpentines, plagiogranites, gabbros, and other items from the ocean suite. The type of dam chosen in 1967 by the Department of the Interior's Bureau of Reclamation was a thin arch of concrete rising six hundred and eighty-five feet from channel to crest. Its purpose was to store winter runoff for use in summer, supplementing the storage behind Folsom Dam, fifteen miles downstream. The new reservoir, Lake Auburn, would reach twenty miles into the Sierra, filling two forks of the river—up the North Fork past Codfish Creek and Shirt Tail Creek beyond Yankee Jim's almost to Iowa Hill, and up the Middle Fork over New York Bar and Murderers Bar and the Ruck-a-Chucky Rapids to Volcanoville. The lake would cover ten thousand acres and be twice as deep as the Yellow Sea.

When Moores and I first visited the site, in 1978, it resembled one of the huge excavations flushed out by the hoses of hydraulic mining. Benched roadways descended switchbacks a thousand feet down the canyon walls. A cofferdam had tucked the river to one side. Reaching eleven hundred and fifty feet across the canyon floor lay the white concrete of the dam's base. From the outset, the construction project had had to deal with the inconvenience of the faulting that had followed the arrival of the Smartville Block. The dam site was squarely in the suture zone. Under the dam's foundation ran a fracture known to engineers as the F-1 Fault. A tectonic event on the scale of an arc-to-continent docking will not result in every fissure's being filled with quartz and gold. Countless empty cracks remain. In order to secure the dam's basement, the Reclamation engineers had performed what they described as "dental work," a "root canal." They had sealed in the Smartville fault zone with three hundred and thirty thousand cubic yards of grout.

Moores remarked, "If you want to find a fault in California, look for a dam."

Scarcely had the dental work been completed when, in 1975, an earthquake struck near Oroville, forty-five miles up the Smartville Block. Its Richter magnitude was 5.7. Near Oroville on the Feather River was the eighth-largest dam in the world. It had been completed in 1968, only seven years before the earthquake, and gradually its reservoir had impounded forty-six hundred million tons of water, or enough to put a lot of pressure on the rocks below. It was a gravity dam, broad and squat, an earth-fill dam, and what it had been filled with was seventy-eight million cubic yards of hydraulic-gold-mine tailings. Absorbing the earthquake, the dam just sat there, holding its lake. However, the United States Geological Survey (a sibling agency of Reclamation within the Department of the Interior) quietly noted that twenty-five per cent of reservoirs of comparable depth had, by their sheer weight, triggered earthquakes. The earthquake at Oroville had been five times larger than the maximum earthquake that the dam at Auburn was designed to withstand.

This collection of facts soon assembled itself in the editorial offices of the *Sacramento Bee*. On its front page the *Bee* envisioned an earthquake that would knock out Auburn Dam, releasing water that would in less than two hours stand in Sacramento twenty feet deep. In the words of a former Assistant Secretary of the Interior, this would be "the worst peacetime disaster in American history." Estimates were that a quarter of a million people would drown.

The federal government considers faults inactive if they haven't jumped for a hundred thousand years. The latest known movement along the F-1 Fault in Auburn was in the Jurassic—a hundred and forty million years before the present—but that did not soothe Sacramento. Work on Auburn Dam was suspended. In 1978, Moores and I found the site silent, dry, reliquary. It looked that way many years later, when we went there again. Geologic time and human time seemed to have met and parted.

Mountain lions go through the dam site. Bears. Feral goats. The project is dormant and appears dead, or vice versa, depending on how the beholder eyes it. The bureau keeps a skeleton crew there, each of whom speaks of the dam in the future positive.

You ask at what altitude the lake level was to be.

Response: "The lake level will be eleven hundred and thirty feet."

You ask if the boat ramp would have been paved.

Response: "Yes, that will be paved."

The gravel boat ramp, several hundred yards long, descends a steep slope and ends high and nowhere, a dangling cul-de-sac. The skeletons call it "the largest and highest unused boat ramp in California." Houses that cling to the canyon sides look into the empty pit. They were built around the future lakeshore under the promise of rising water. You can almost see their boat docks projecting into the air. Thirty-three hundred quarter-acre lots were platted in a subdivision called Auburn Lake Trails.

Moores wanted to know if a geology student from Davis might be permitted to study the rocks that had been exposed during construction.

"Fine, but we would not want the student's conclusions to inconvenience the dam."

The dam has cost several hundred million dollars so far. The bureau spends a million dollars a year maintaining the site while nothing happens.

We looked for a cup of coffee in Cool, California, after crossing the American River on a seven-hundred-foot-high bridge. Not particularly long, the bridge was built so high in order to clear the lake that wasn't there. From houses in Cool, picture windows framed the lake air. There were numerous for-sale signs. Mother Lode Realty. Cool was a placer-mining camp of the eighteen-fifties. In Cool Quarry, marine limestone is mined now—a lenticular pod, a third of a cubic mile, shoved into California by the arriving Smartville Block. If you had lived on the moon then, as a full earth came into the sky you would have seen two large continents (Laurasia and Gondwanaland), one above the other, surrounded and divided by ocean. West to east, the dividing seas were the incipient Caribbean (Central America was not there), the incipient Atlantic, and—from Gibraltar through China—the long water known in geology as Tethys. Worldwide, fossils from that time are described as Tethyan. Tethys, mother of rivers, was the consort of Oceanus. Cool was named for Aaron Cool. In the limestone pod in Cool, California, caught up in the docking of the Smartville Block, are Tethyan fusilinids and Tethyan corals.

As an overriding plate scrapes the plate below, it acts like the blade of a bulldozer and piles up sand, seashells, cherts, phyllites—

whatever happens to be there. Impressive amounts of material can be accreted in this manner. As the Philippine Plate has scraped westward, overlapping the Eurasian Plate, an accretionary wedge has risen as Taiwan. In an arc-to-continent collision that is reducing the distance between Taipei and Beijing, Taiwan is the first piece of the West Luzon Island Arc to reach the Eurasian slope. In the mélange of rocks in the Taiwan accretionary wedge are not only sands, seashells, cherts, and phyllites but enough scraped-up ocean-floor debris—sheeted dikes, pillow lavas, gabbros, serpentines—to be known as the East Taiwan Ophiolite. A large and intact package of ocean crust-and-mantle can be expected to follow, as one did when Smartville began to close with North America.

The Smartville Block pushed before it not only the limestone of Cool and the schists and serpentines of Auburn Dam but also the red-weathered phyllites and the argillites and cherts we had seen along the interstate as we descended toward Auburn. These and a great miscellany of additional rocks were Smartville's mélange, its accretionary prism—highly foliated, sheared, broken, disrupted, deformed—caught up in the Smartville suture, the docking of arc and continent.

There was, of course, a subduction zone—a trench—between the arc and the continent as they drew together, and in the collision it disappeared. It was actually stuffed shut, according to present theory. First, ocean crust-and-mantle of the North American Plate went down the trench. Eventually, the continental rock itself reached the trench and jammed it, like a bagel in a toaster. Continents are too light and thick to be subducted, and where they arrive at trenches the trenches cease to function. Australia has jammed the trench to the north of it with such force that it has produced New Guinea. As Moores envisions the Jurassic event in California, a large overlap of Smartville ocean crust (the upper plate) was left lying on the North American continental slope after subduction stopped. The region cooled in the postcollisional stillness. The lower, west-moving plate was no longer descending, dragging everything downward. Isostasy, the force that lifts light objects when other forces cease to hold them down, began to work on the combined terranes. Lifting them, it broke off a large piece of the Smartville ocean crust-and-mantle and carried it into the air in what would eventually become the foothills of the Sierra Nevada.

As a geological and geophysical specialty, the study of ophiolites is only a few years old, and therefore provokes argument on almost any question raised—from the environment of the origin of the rock itself to the putative method by which it made its way from extra-continental depths to dry land. The emplacement story for the Smartville Block was worked out by Eldridge Moores.

Descending westward, just below Auburn, you cross the thousand-foot contour, and the Great Central Valley comes into view, running flat out of sight to the horizon. Sacramento is down there, and, fifteen miles farther, Davis. It is an abrupt, absolute change of physiographic worlds, where the mountains hinge. The Smartville Block extends under the valley and ends beneath the Coast Ranges.

After the trench at Auburn disappeared, another one had to develop, as the trailing edge of Smartville became the new front of the North American Plate, moving west. To balance the earth's books by consuming an amount of ocean crust comparable to what is made at spreading centers, subduction zones develop when and where they are needed. Across geologic time, subduction zones have come and gone quite often. On one side of the Smartville Block the new subduction must have been developing even while the old subduction died on the other. "For sure," Moores said. "You've got to produce that convergence somewhere. You don't cut things off. The subduction here at Auburn was Permian to early Jurassic. The new subduction zone to the west was operating by the late Jurassic, and it operated all through Cretaceous and into Tertiary time. The volcanoes came up in the Sierra, and the main batholith formed. I think the geology is really neat. The new plate margins produced their own accretionary prism, piling up out there to the west of us to become the rest of California."

Moores had become aware of the unusual rocks in the Sierra foothills only four years before I met him. On a spring day in 1974, he and his family left Davis for a camping trip in the mountains, and instead of following their usual route, up the Feather River, they took a right at Yuba City and went into the canyon of the Yuba. The theory of plate tectonics was six years old. Eight more years would pass before the term "exotic terrane" would be coined. Geologists, caught up in the fresh intensity of a scientific revolution, were still at the beginnings of seeing the world anew. Few people then envisioned the western United States as a collection of lithospheric driftwood. Moores, however, had worked for some years in Macedonia, struggling to understand the large unrooted masses of mantle rock that lay there on the surface as mountains, and to relate it to the stratiform gabbros, plagiogranites, and sheeted dikes nearby. With Fred Vine, now of the University of East Anglia, he had also worked in the Troodos Mountains of Cyprus, where sheeted diabase ran on for seventy miles, and where the gabbros, the granites, and capping marine sediments were present as well. Moores and Vine decided that the whole assemblage had somehow been removed from a blue-ocean setting and emplaced upon the African slope. In 1971, they published in the *Philosophical Transactions of the Royal Society (London)* an establishing paper, the significance of which was expressed in its title: "The Troodos Massif, Cyprus, and Other Ophiolites as Oceanic Crust."

Between the Great Valley and the High Sierra, the Yuba drainage lay in the geographic center of what was not yet known as the Smartville Block. Moores wasn't there to do geology. All he was trying to do was to go uphill out of the oak woodland and into the ponderosas, a task to which his micropowered microbus was almost unequal. It chugged, and—among the brown grasses of a dry country—the rock at the roadside passed slowly. At work or play, a geologist always drives like Egyptian painting—eyes to the side. Near the tributary Dry Creek, about ten miles north of the Yuba, the microbus ascended a particularly formidable incline between high roadcut walls of a dark and igneous rock. Moores did not stop, but the hair on his arm may have moved. He remembers feeling that he could have been in Macedonia. More emphatically, in Cyprus.

On the geologic map of California, the lithology of that area was vaguely described as "Jura-Triassic metavolcanic rocks." The eye sees what it is trained to see. Or, according to a maxim that Moores often quotes, "the eye seldom sees what the mind does not anticipate." When Moores returned to that roadcut with a lens in his hand, he did not need it. He found—one standing beside the next beside the next beside the next—classic sheets of sparkling diabase, each with a glassy margin, all in prime condition, undeformed.

The sheeted dikes that he would later find on Interstate 80 at Auburn were deformed in the Smartville suture almost beyond recognition. These in the Yuba Valley were well back of Smartville's impacted leading edge. When he and I stopped there once, he said, "You could look at hand specimens of rocks taken from this cut and not be able to say if they were from Cyprus, Pakistan, Oman, New Guinea, Newfoundland, or California. Only by dating and by detailed chemical analysis of minor elements can you tell them apart." From Auburn Dam north to Oroville, there is a continual exposure of forty miles of these sequential laminations of seafloor spreading.

I asked him if he could say from what distance and what western compass point the Smartville Block had come.

"No," he said. "We have tried doing paleomagnetic work, but so far it's inconclusive. The Smartville arc probably developed not more than a thousand kilometres from North America. It probably came from the northwest."

In altitude, the Smartville Block rises roughly from zero to five thousand feet. To move about on its small roads is to move among zones of time that are not all written in rock. U-2s drop out of the stratosphere and hang-glide into Beale Air Force Base while you walk up a dry streambed at Oregon House, where placers were first panned in 1850, in a valley of sheeted diabase that was injected into the floor of the North Pacific Ocean about a hundred and sixty million years ago and emplaced on California five million years after that. The ophiolite—the whole vast assemblage of transported deep-ocean rock—now rests on California like a ship stuck in sand, listing thirty degrees to the west. The ophiolite tilts more steeply than the slope of the Sierra. Therefore, as you climb the modern foothills, geologically you go downsection, ever deeper into the former seafloor, from the spreading rift to the granites and gabbros of the magma chamber that fed it, and on to the cumulate layers of heavy crystals that settled on the mantle at the boundary that is known in the science as the Moho. On up the mountains (and farther downsection) are scattered serpentines derived from the mantle itself. Where the rock has not been folded, the descent of the ophiolite is as clear as it is complete. It is the perverseness not of the science but of the earth itself that you often go downsection when you are going uphill.

In the other direction, the ocean-crustal rock that lay above the sheeted dikes is in the foothills also. Below Timbuctoo Bend on the Yuba, for example, a green ledge of what appear to be massed satin pillows reaches into the fast clear river. Notwithstanding the boulder fields of hydraulic-mining tailings which cover the floodplain and extend out of sight downstream, it is a place of such appeal that you reflexively reach for a fly rod, or look around for a place to pitch a tent. Whereas the pillows at Auburn are so extensively crushed that only a specialist can reassemble them in the eye, the pillows of Timbuctoo are rotated but undamaged. Each about two feet in diameter, they are simple in form, elegant, ovate—a spread of huge caviar. Nowhere on land, Moores said, will you see pillow lavas more perfect than these.

Timbuctoo was a placer camp in 1849. When Moores and I first stopped there, in the nineteen-seventies, we read these words on a wall of a roofless masonry building:

GOLD DUST BOUGHT
WELLS FARGO & BROTHERS

And, barely legible in fading paint, "Stewart Brothers have for sale dry-goods, boots and shoes, ready-made clothing, groceries and provisions." That structure was all there was of Timbuctoo, where twelve hundred people lived in the eighteen-fifties. Now there is even less. The Wells Fargo building has collapsed. One masonry corner juts chimneylike above the rubble. The words are gone, and only a hill beside the river retains an indelible scar, torn out by hydraulic mining as if by rapid landslide.

In or close to the magma chambers under oceanic spreading centers, seawater, which has descended through fissures, dissolves metals (copper, silver, iron, magnesium, gold); it carries the metals upward and precipitates them on top of the new rock. If the new rock, in its migration, happens to end up on a continent, the metal comes with it. In the suturing process, faults form. Deeply circulating groundwater redissolves the metals and redeposits them in quartz veins in the faults. In this way, Moores said, it made sense to imagine that the Mother Lode gold of California came in from the deep ocean riding the Smartville Block.

Smartville, California, is a living town, 95977. It is just uphill from Sucker Flat and a mile from Timbuctoo. Sucker Flat was named for Illinois. The miners looked upon Illinois as "the Sucker State." Jim Smart, of Smartville, was not from Illinois, and he was smart enough not to be a miner. He ran a hotel. Never mind that the Sucker Flat Channel, between Timbuctoo and Smartville, yielded two and a half million dollars in gold. There is a white wooden church in Smartville, its paint peeling. There are two gas pumps under a sign that says "Bait." There are slopes of brown grass under blue oaks—and the houses of a hundred and fifty people. Roadcuts in Smartville are full of bulging pillows.

In the roadcuts of Rough and Ready, nine miles from Smartville, we saw massive gabbros. Rough and Ready was founded by forty-niners, and its citizens soon voted to secede from the Union. If geologists of the nineteenth century and the first three-quarters of the twentieth century failed to see the intrinsic bond between the gabbros of Rough and Ready and the pillow basalts of Smartville and Timbuctoo (and the relationship of those rocks to the diabase and plagiogranite and serpentine nearby), Moores was sympathetic. He said, "If you found a headlight, a hubcap, a brake drum, and a radiator, you would say, 'Ah, pieces of an automobile.' But if you had never seen those things assembled you would not relate them to each other. The ophiolitic sequence is one of the most classic things of importance to the plate-tectonics story. Its emplacement is evidence of the spreading process and of the subduction process, not to mention the consumption of vast amounts of ocean crust. There are ten thousand square miles of ground here with no one arguing about the fact that it's island-arc material. The original arc may or may not have been the size of Japan, or the Philippines, or the Marianas, or the Antilles, or the Aleutians. But it was surely an island arc, and its arrival is signalled by these ophiolitic rocks. If ophiolites are found at the suture of the Urals—as they are—it means that there was a sea between Siberia and Europe. The sea was consumed. Ophiolites were emplaced. And the Urals are welded in a Permo-Triassic suture. Before the Permo-Triassic, in the sea and the islands, the Gulag Archipelago was real."

Moores was doing postdoctoral work at Princeton University, in the middle nineteen-sixties, when he first heard about what was described to him as "a wonderful complex on Cyprus," but when he tried to look it up he could find no data. He requested, from Nicosia, memoirs of the Cyprus Geological Survey. This was at the time when the theory of plate tectonics was in its coalescing phase. The term itself did not yet exist. Oceanic spreading centers were known, and the consumption of ocean crust in subduction zones was beginning to be understood. The idea of a suite of rock called an ophiolite had been around in the science for many years but had not yet been widely accepted or related to the new theory. As the story of plate tectonics further unfolded, the story of the ophiolitic sequence would accompany it like an echo.

As it happened, Fred Vine was also working at Princeton at the time—Vine, who had co-authored with his Cambridge University colleague Drummond Matthews the paper that contributed the manifesting insight into the movement of the ocean crust and placed Vine and Matthews among the handful of people in various parts of the world who collectively brought about the plate-tectonics revolution. When Moores' packet arrived from Cyprus, Moores already had behind him three years' experience in Macedonia among the lower components of ophiolitic rock, but he had not sensed their origins. He had not imagined that they had formed in one milieu and been transported to another. Like most geologists, he thought of them as rocks formed from magmas that had welled up under Greece. Opening a geologic map of Cyprus, he saw diabase dikes all running in the same direction; he saw mantle-derived serpentine at the bottom of the section and basaltic pillows at the top. This time, his imagination made the jump. Unfolding the map in front of Fred Vine, he said, "How does this look for oceanic crust formed at a spreading center?"

In the science, the ancestral glimmerings of that intuition were nearly a century old. By the eighteen-eighties, geologists had begun to reflect on a common association of serpentine, gabbro, diabase, and basalt; and in *The Face of the Earth*, Eduard Suess, of Vienna, observed that all these "green rocks," as he called them, could characteristically be found in folded-and-faulted mountains. He said that they had formed in the geosynclines from which the mountains derived.

In 1892, when the German geologist Gustav Steinmann visited San Francisco, Berkeley's Andrew Lawson took him to the north side of the bridgeless Golden Gate to see the rocks of Marin. (Moores came upon this remarkable account while he was reading Steinmann in German in 1967.) Steinmann had wandered the Apennines and Alps noticing serpentines, pillow lavas, and radiolarian chert—always in that order, upsection. Now, in Marin, he said to Lawson, "These rocks are the same." There were solid cliffs of red chert, and pillow lavas as well. On the San Francisco side of the strait was a headland of serpentine. Steinmann commented that since the chert was stratigraphically at the top of the sequence the whole assembly must have come from deep sea. In 1905, Steinmann published a definitive study of the same three rock types threading through the Alps. They became known in the science as the Steinmann Trinity.

After the German meteorologist Alfred Wegener presented his hypothesis of continental drift (1912), no one connected it to the Steinmann Trinity. For about four decades, the two ideas hung on to the trailing edge of the science, with no one suspecting that Wegener's idea would effloresce as a scientific paradigm, or that the ophiolite story (the advanced edition of the Steinmann Trinity) would, thereafter, provide the chapter headings to the plate-tectonic history of the world.

There were hints. In 1936, Harry Hess, of Princeton, whose "History of Ocean Basins" would—in 1960—introduce the spreading seafloor and begin the new tectonic story, gave a paper in Moscow in which he related Alpine peridotites to island arcs and called this "a contribution to the ophiolite problem." The ophiolite problem was manifold. Most geologists did not accept the association of such disparate rocks. The few who did asked elemental and unanswerable questions: Are the rocks in their original place? If not, where have they come from and how did they move? Peridotite, which in alteration becomes serpentine, is now seen to be of the deepest origin of any rock found at the earth's surface. It is now thought to be rock of the earth's mantle, and few disagree. Hess had studied two belts of Alpine-type peridotites in the Appalachians. He asserted that they were not just garden-variety igneous rocks that had come up as magma under existing Appalachians but rock that had intruded much earlier, on the edges of a geosyncline. Hess went on to say that peridotites seemed to be introduced in the initial phases of mountain building—and not thereafter. He said they were cold, intact, and solid as they came up in the rising mountains—that they had been, in other words, tectonically emplaced. Hess was describing a collision between a continent and an oceanic subduction zone, but in 1936 he didn't know it. In Moores' words, "This is the clearest example I know of of a guy saying the right thing for the wrong reason. Hess said this was the most important event in the forming of the Appalachians, this 'intrusion' in the Ordovician. It was. But plate tectonics did it."

By 1955, the burst of data that had come with the ocean-exploration programs of the Cold War had yielded—in the work of Russell Raitt, of the Scripps Institution of Oceanography, and Maurice Ewing, of Columbia University—the first descriptions of ocean crust to be derived from seismic refraction. Everywhere, the ocean crust seemed to be some sort of package—a once molten but nonetheless zoned assembly, in three general bands.

The German geologist W. P. de Roever was the first to imagine rock of the earth's mantle in the thin air of the Alps. In a 1957 paper, he said that the Alpine peridotites appear to be solid intrusions; they have deformed fabrics; they appear to be coming up from the mantle. For "solid intrusions" you could read "emplacements from elsewhere." In their deformed fabrics you could see that they had been moved. How they had been moved remained unclear.

In 1959, Jan Brunn, who had mapped the Vourinos ophiolitic complex in Macedonia, published an abstract in French in which he compared ophiolites to the Mid-Atlantic Ridge. He was the first person ever to suggest that the ophiolites of the aerial world were similar to crust found at mid-ocean ridges. He compared ophiolitic rocks to dredged rocks. All this was before seafloor spreading was recognized, and no one noticed the work of Brunn. But it was Brunn who took the Steinmann Trinity and the ophiolitic sequence out of the Oz of geosynclines and placed them in the center of the widening oceans.

Over the next nine years (1960–68) appeared the twenty-odd

scientific papers reporting the plate-tectonics story: that plates are essentially rigid, and deform at their boundaries; that all plates include ocean crust, and generally a very large amount of it (the continents are passengers on the plates); that new seafloor moves away from a spreading center until it goes down into a trench to be consumed; that plates sliding past each other (as at San Francisco) do so in strike-slip sporadic jumps; that ocean crust colliding with continental crust can pry up something like the Andes; that continental crust hitting continental crust will build Himalayas, Urals, Appalachians, and Alps.

While these novel facts were still for the most part unknown, Moores, in Macedonia in the early nineteen-sixties, was becoming thoroughly at home with the petrology and structure of the Vourinos Complex, about thirty miles west of Mt. Olympus. Brunn's ideas notwithstanding, Moores thought of the Vourinos rocks as homegrown—in his words, "a partially molten diapiric blob." Diapirs are the bodies of rock that, balloonlike, rise, crashing their way into the country rock above them. Harry Hess, one of Moores' supervisors, went to Greece to inspect his work. Hess was already in the process of abandoning the geosyncline and shedding the Old Geology like old skin. He decided that the Vourinos rock had formed in an oceanic environment, and probably at a spreading center. Moores, clinging to what he had been taught (by, among others, Hess), decided that Hess was crazy.

Moores' conservatism can be understood in the light of the disregard in which ophiolites were generally held in the United States at that time. When a graduate student at Stanford, doing field work not far from the university, suggested that the local sediments had been deposited on an ophiolitic complex, the Geology Department specifically forbade him to use the word "ophiolite" in a Ph.D. thesis. The professors explained that ophiolites were a wild European idea, clearly wrong—and, in any case, not applicable to California.

The concept of the spreading seafloor had gained acceptance by 1966, when Moores first saw the geologic map of Cyprus. Moores and Vine prepared to go there, but had to wait, because of the political tension in the six-year-old country. They worked there in 1968 and 1969. The evidence was convincing that Cyprus was essentially a piece of ocean crust, thrust up in some way and now aerially exposed. Moores and Vine's paper, which has influenced all subsequent understanding of ocean crust, was the first to establish ophiolites as ocean-floor remnants, for the most part formed by spreading processes. This odd collection of rocks ranging from water-cooled lavas to mantle blocks, so difficult to explain in continental settings, could now be seen not as ordinary igneous formations but as tectonic features that had moved from one place to another in the course of epic alterations of landscape.

In Pacific Grove, California, at the end of 1969, a Penrose conference on "The Meaning of the New Global Tectonics" drew structural geologists from all over the world. William Dickinson, of Stanford, dismantled the geosyncline and assigned its parts to various aspects of plate tectonics—collisions, island arcs, abyssal plains, mélanges, trenches, transform faults. Moores describes the conference as "a watershed of geology—that was when people really began to realize how important plate tectonics was." Listening to Dickinson, he thought of all the ophiolitic and volcanic-island rock that he was seeing in the Sierra Nevada and the Coast Ranges, and it occurred to him that these mountain systems could be understood in terms of island arcs accreting. The arrival times of ophiolites could date successive mountain-building events. In his words, "Recognition of the fact that ophiolite emplacement had to be by the collision process meant that you could explain the western part of the United States by that kind of sequence. That idea came to me on the last morning of the Penrose conference in 1969, and I wrote it down: the idea that you could explain the progressive eugeosyncline-miogeosyncline development and the progressive orogenies that you see in western North America as a series of island-arc complexes—the things we call terranes—that have collided with the continent. I was so excited I could hardly sit still for several days—in fact, for a couple of weeks after that. I came back to Davis all bubbling over." Before long, he had sent a paper to *Nature*. It appeared in 1970 and was the first to suggest the collisional assembling of California and of vast related portions of the North American Plate.

The idea that California is in large part a collection and compaction of oceanic islands was a reverberation of an ancient myth as well as a development in a science. For at least two thousand years, people described certain undiscovered islands with a force of imagination that became belief. In the Middle Ages and the Renaissance, such islands appeared on global maps, and when navigation revealed that they were not where they were said to be cartographers moved the islands to new locations in unexplored seas—the Fortunate Isles, for example, and the Seven Cities of Cíbola, and the Lost Atlantis. One such island was California, a utopia of the western Atlantic Ocean. It is described as follows in *Las Sergas de Esplandián*, a Spanish romance published in 1508:

> Know, then, that, on the right hand of the Indies, there is an island called California, very close to the side of the Terrestrial Paradise, and it was peopled by black women, without any man among them, for they lived in the fashion of Amazons. They were of strong and hardy bodies, of ardent courage and great force. Their island was the strongest in all the world, with its steep cliffs and rocky shores. Their arms were all of gold, and so was the harness of the wild beasts which they tamed and rode. For, in the whole island, there was no metal but gold.

The wild beasts were griffins—half lion, half eagle—which the women rode through the air into battle, and which they trapped as fledglings. To the griffins they fed the voyaging men who came their way, and their own male infants. Ruling California was the "mighty Queen, Calafia . . . the most beautiful of all of them, of blooming years." (Translation by Edward Everett Hale, 1864.)

With an eye on the new tectonics, the paleontologist James W. Valentine, of the University of California, Davis, worked out a curve of the distribution of marine-invertebrate families across Phanerozoic time (the past five hundred and forty-four million years). He saw the diversity of creatures expand, decline, rise again. The thought occurred to him that if numerous small continents were spread around the world in fair proximity to the equator one would expect the diversity of life to be very high, whereas if continental masses should happen to be clustered (and especially if they were clustered around a pole) diversity should be very low. A typical Valentine graph showed creatures from continental shelves starting out at a very low level of diversity in late Precambrian and early Cambrian time, coming up to a high level in mid-Paleozoic time, then crashing at the end of the Permian, and then rising again. Valentine showed his graphs to Moores, and said, "I wonder if you can explain these patterns of diversity in terms of continental drift and redistribution of land."

Among the results of this dialogue were papers by Valentine and Moores in *Nature* (1970) and the *Journal of Geology* (1972). The title of the second one was "Global Tectonics and the Fossil Record." The concept of continental drift had always implied the preexistence of a supercontinent, which Wegener had called Pangaea. After all, if Australia and Africa and the Americas and Eurasia spread apart from their obvious fit, they had to have been together in the first place. The first place—according to the newly determined vectors of the lithospheric plates—was two hundred million years ago, when Pangaea began to split into Laurasia and Gondwanaland. In the second place, they split farther, to sketch the present globe. Valentine's diversity patterns were in harmony with this story: the greater the breakup of landmasses, the more diverse the fossil families. Moores looked at mountain belts that had come into existence hundreds of millions of years before the dispersal of Pangaea. If the new theory worked, it would work not only forward in time but backward. Gradually, he reassembled Paleozoic and Precambrian continents—not only the continents that came together to create Pangaea but also the continents that came together to create an earlier supercontinent, which no one had thought of before. Moores and Valentine called it Protopangaea, or Pangaea Minus One. More widely, it has become known in the science as Rodinia (from a Russian word meaning "motherland"). Since agglomerations and dispersals of terrane seemed to be cyclical in nature, there may have been who knows how many supercontinents before Rodinia. The new theory was like a stable and inventive structure built in the mind of a composer in advance of a composition, but there were those who didn't like what they heard. "Global Tectonics and the Fossil Record" was attempting to demonstrate how continental drift affects evolution, and—heaven knows—it succeeded in enraging no small number of geologists and paleontologists, who felt that Moores, especially, and other "plate-tectonics boys" were ignoring phenomena like ocean currents in their headlong lust to write all aspects of the geologic pageant into the plate-tectonic model. But in 1978 Patrick

Morel and Ted Irving, of Canada's Department of Energy, Mines, and Resources, presented paleomagnetic evidence for Pangaea Minus One.

Moores and Valentine also sensed a relationship between plate-tectonic history and the history of the level of the sea. Moores explained, "If you look at the stratigraphic record on platforms such as the midcontinent of the United States, you see times of high stands of the seas, when the continent was nearly submerged, and times of really low stands of the seas. Could that be related to continental drift? Take an average ocean basin and add a hot and voluminous spreading ridge. You will diminish the volume of the ocean basin and force water up over the continents. Conversely, if ridges die for some reason—lose their heat and collapse, or otherwise disappear—that will increase the volume of ocean basins and allow seawater to drain off the continents. The ocean's transgressions and regressions seem to represent seafloor spreading going on or not going on. Others worked on this before we did, but we extended it back to the Cambrian-Precambrian boundary. In the geologic record, you see a great regression of seas in late Precambrian time, then transgression in Cambrian and Ordovician time, then regression in Permo-Triassic time, and then the transgression of Cretaceous time. The late-Precambrian regression coincides with Pangaea Minus One. As it split apart, with spreading centers forming all over the place, the Cambro-Ordovician transgression occurred, and when the smaller continents came back together there was the Permo-Triassic regression, and when they split apart again came the famous Cretaceous transgression, when Colorado was underwater."

Hugh Davies, of the Papuan Geologic Survey, published cross sections of New Guinea that showed a huge ophiolite dipping northward. The Indo-Australian Plate, like a shovel, had lifted this piece of Pacific mantle and crust. The ophiolitologists Robert Stevens, John Malpas, and Harold Williams, of Memorial University of Newfoundland, described a series of ophiolites in Newfoundland as remnants of an arc that collided with North America in Ordovician time, signing in the Taconic Orogeny. Eli Silver, of the University of California, Santa Cruz, working in Indonesia, traced a large ophiolite there from its emplaced position on a submerged microcontinent northward into an ocean basin.

As comparable research went on around the world, a question

that attended all of it was "What ancient geography can be deduced from ophiolites?" and spectacular deductions continued to be forthcoming. The ophiolite suite seemed not only to spell out in detail the process of formation of oceanic lithosphere but to record plate collisions of which all other evidence was long since gone. Vanished oceans were recalled, and vanished plates inferred, as continents were deconstructed and continents were reconstructed. An ocean (or oceans) that once existed where the Atlantic is now had closed from north to south in Cambro-Devonian time, and the Permian disappearance of the ocean where the Urals are now had completed Pangaea. The Pacific Plate, now the largest in the world, did not exist then, but a spreading ridge, propagating westward, split Pangaea into Laurasia and Gondwanaland in late Triassic and early Jurassic time, and opened the Tethys Ocean. The early central Atlantic was a part of Tethys. As oceans came and went and continents evolved, island arc after island arc had been swept into larger masses—a story that could suggest that the first dry lands of Genesis were arcs accreting in a globe-girdling sea.

As with so many things that were obscure or mystifying before the arrival of plate tectonics, the discovery of the origin of ophiolites was something like the discovery that the rock you have been using for twenty-five years as a doorstop is actually the Stone of Scone. Suddenly, for example, the concept unmentionable in a Stanford doctoral thesis was helping to tell the story of California as it had not been understood before. Moores, reflecting on this, said to me once, "If the story of California sounds fantastic, with all its accreting arcs and mélanges coming from the western sea, just look at a map of the southwest Pacific—look at the relationship between Australia and Indonesia right now."

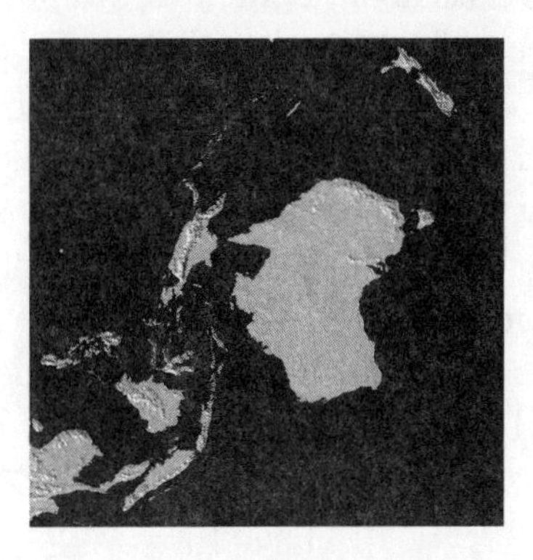

As fresh attention accrued to the ophiolitic suite where—in completeness—it was found on land, the actual dimensions of the ocean lithosphere could be measured part by part from top to bottom, clarifying the results of seismic refraction. After the sequence has formed at spreading centers—where it is in large part liquid and is much swollen with heat—it cools and thins as it moves away, and after travelling twenty million years and seven hundred miles is a deep cold slab:

A few tens of metres of ocean sediments drift down upon the deep cold slab, settling on top of

a kilometre or so of pillow lavas, under which is

a kilometre or so of sheeted dikes, under which is

a kilometre or so of plutonic rock (plagiogranite, gabbro), under which is

a kilometre or so of plutonic rock in which cumulate crystals settled in layers upon a distinct chamber bottom—

the Moho—

under which is a kilometre or so of mantle rock, some of which was melted in the spreading center and some of which is peridotite in its solid original form and if water has reached it is serpentine.

When Moores talks about ophiolites with fifth-graders in Davis, he sketches the ocean floor much as he has for me, simplifying the vertical sequence and presenting an idealized rock column that closely resembles the one above. It is more than just a useful model. In its general way, it is accurate. But as a description it is only somewhat more encompassing than to say that Herman Melville wrote a novel about a one-legged madman in vengeful pursuit of a whale. Remarkable as ocean lithosphere may be for its worldwide youth and repeated character, it is not nearly as simple as that summary outline. In Moores' words, "Where ophiolites are made, in spreading centers, hot fluids are mixing, cooling, and so forth. We are not talking about clear sedimentary layers; ophiolite contacts are gradational. Moreover, there are various types of ophiolites: some are from the basins behind or in front of island arcs, some from the intersections of spreading centers and transform faults. There is undersea 'weathering.' Parts of the sequence erode, more sediment comes down, and as a result there are hiatuses in seafloor rock, just as there are in rock that accumulates on land. In Italian ophiolites, the diabase is missing, and so is the gabbro. The serpentine is full of calcite—it's called ophicalcite, very beautiful white and green and red stone, expensive building stone. There is some gabbro in Elba. But there are no sheeted-dike complexes in Italy. Obviously, Italian ophiolites formed in a different ocean environment, and what that may have been is not well understood. Oceanic crust is not a simple three-layered thing, as geophysics is telling us now. Geophysicists are unable to produce a consistent model. Nature is messy."

One day, on a field trip we made to Cyprus, Moores did a long and detailed inspection of an outcrop he had not studied before, and figured out a chain of magmatic events in which layered gabbro had come first and a plagiogranite sill had intruded below the gabbro—an inversion of the usual sequence. "This reminds us not to take the chart of the ophiolites simply," he said. "Layered gabbro may be lower than plagiogranite in the master chart, but here we see a plagiogranite sill under the layered gabbro. It came in later. Things don't always happen in the earth as they do on charts."

Not to confuse me but just to give me a reality shot, he sketched out what he described as "an expanded ophiolitic assemblage"—an elaboration of the picture I had written on my palm. He repeated the generalized column with enough added detail to suggest the actual complexity of the rock that lies below the oceans. I could forget it as soon as I read it, but at least I ought to sense the tangle of nature, and thus the nature of science.

Where you find ophiolites on land, you might find at the top of the sequence the shallow-water limestones of the sea from which they emerged, or even laterites from soils formed in air as the ophiolite was lifted by the continental margin.

The deepwater sediments that drifted down upon the moving lithospheric plate may be chalk (as in Cyprus) or the product of volcanoes (as in the Smartville Block) or chert (as in Italy or Greece). They tell you something about the oceanic environment through which the lithosphere travelled.

Beneath the massive pillow lavas are

more pillow lavas, shot through with diabase dikes that came up in molten state with enough pressure to continue past the

zone of massive sheeted dikes. If the ophiolite were an animal, these originally vertical laminations would be its brain. Taken together, they are the tape measure and chronometer of the seafloor. As each new dike forced its way into the complex, the seafloor spread that much. New dikes would intrude every fifty to a hundred years, often splitting the previous dike up the middle. The average width of the split dikes would be about seventy centimetres. They recorded absolutely the episodic widening of the ocean lithosphere, in contrast to the arithmetical notion of geophysicists, who suggest that seafloors spread continuously at a quotient number of centimetres per year.

Among the plagiogranites are large diabase dikes (feeders of the sheeted complex)

and large diabase dikes are among the massive varitextured gabbros as well.

The stratiform gabbros are generally cyclic: plagioclase and pyroxene cumulate above olivine and pyroxene cumulate with intercumulate plagioclase (and sometimes olivine and plagioclase) above olivine cumulate with some chromite in it, a combination known as dunite. You will not at once recognize all these things where an ophiolite has made an appearance in a roadcut or a mountain cliff, but at least try to remember that somewhere within this zone is the geophysical Moho—

and at the bottom of this zone is the petrologic Moho.

I have to interrupt him. Two Mohos? How could there be two Mohos? *The* Moho, as fifth-graders can tell you, is where the crust ends and the mantle begins—about five kilometres down from the ocean bottom and thirty-five kilometres below most places on continents and as much as sixty kilometres below deeply floating mountains. "Moho" is a geophysical term, coined in honor of the Croatian seismologist Andrija Mohorivičić, who in 1909 discovered the crust-mantle boundary. When geophysicists examine their autodriven strip-chart recordings, which look like close-up photographs of matted gray hair, they see what Mohorivičić saw. They see seismic waves speeding up when they hit the olivine-rich cumulates, and they call that the change from crust to mantle. Geologists looking at ophiolites in geographical settings see mantle material below the olivine-rich cumulates and discern that the mantle supplied the olivine that went into the rock above it and now devilishly accelerates the geophysicists' seismic waves. Below the olivine-rich cumulates geologists see what they regard as the true transition from crust to mantle, and they call it the petrologic Moho. Geophysicists insist their machines can't be wrong, the nature of the rock notwithstanding. Moores says to remember, however, that Moho means "Mohorivičić discontinuity" and the seismic discontinuity is decidedly where geophysicists say it is. The discontinuity is seismic and is recorded on paper, but the crust-mantle boundary, which is lower, is recorded in the rock. Nature, in this case, is not messy or confused. The science is—for the time being. Like the early cartographers piecing together the face of the globe, geologists and geophysicists are now trying to map places that no human being will ever see but which are features of the earth no less than Scotts Bluff or the Shetland Islands, and were features of the earth when Scotts Bluff and the Shetland Islands were other rock in other places, and will be features of the earth when Scotts Bluff has totally disintegrated and the Shetland Islands are under the sea. The two Mohos are like a camera's divided range finder trying to close. The two Mohos are an imperfectly mapped frontier

under which, in the ophiolitic sequence, the kilometre or so of mantle rock is peridotite, a general term for rock composed mainly of olivine and a little pyroxene. Depending on the kind of pyroxene and the amount of pyroxene that is in it, peridotite is also called harzburgite and lherzolite and dunite, and, if water has reached it, serpentine. Peridotite that has moved in its solid original form is called tectonite, and presumably extends to the bottom of the lithospheric plate. One of the thickest measurable sections of mantle rock emplaced on any continent today is in Macedonia, and is about seven kilometres from top to bottom, a considerable weight to lift from the mantle into the air.

In this young field within earth science, the two Mohos are scarcely the sole battleground. Because ophiolites develop not only at mid-ocean ridges but also in the small spreading centers associated with island arcs and sometimes along oceanic transform faults, arguments about their origins can be intense. There is considerable agreement that the Smartville ophiolites relate to the arrival of an exotic island arc, and that the Bay of Islands ophiolite complex, in Newfoundland, formed at a mid-ocean spreading center. But Moores thinks Cyprus formed in mid-ocean, and most ophiolitologists do not. Italy's ophiolites, with their missing parts, seem to some, but not others, to be fragments of oceanic transform faults. The Papuan ophiolites of New Guinea are so complicated that they seem—to some workers—to have come not only from a mid-ocean spreading center but also from behind an island arc.

Moores says that ophiolites are more important as models for the mechanism of spreading than they are as relics of the environments where they were made. The elapsed time between formation and emplacement is measurable by various methods, and some people argue that emplacement seems to have happened too quickly—an average of thirty million years—for most of the world's ophiolites to have derived from mid-ocean centers. Moores says he has no argument with people who hold that most ophiolites form near continents—in fore-arcs or back-arcs—but he insists that there is enough time to bring them in from mid-ocean ridges as well. Geologists argue about the chambers of magma under spreading centers, and whether the chambers were continuous over time or were punctuated units that crystallized and were followed by new chambers that in turn crystallized, and so forth. They wonder, above all, why only a few ophiolites are more than a thousand million years old, while the earth itself is four and a half times older. For the last thousand million years they can work out the tectonic history—the shifting shapes of continents, the rise of long-gone mountains—from the ophiolites that were left behind. Before that, in the early Proterozoic and the Archean Eon, what was going on? Was something different going on? Something other than plate tectonics?

The long argument over how ophiolites are emplaced has provided workers with a more immediate distraction. In 1971, R. G. Coleman, of the United States Geological Survey, proposed that where ocean crust slides into a trench and goes under a continent, a part of the crust—i.e., an ophiolite—is shaved off the top and ends up on the lip of the continent. He called this "obduction." In 1976, David Elliott, of Johns Hopkins, decided that rocks could not stand the roughshod tectonics that Coleman had proposed—that the ophiolites would be shattered into countless parts and would just not make it up there. Elliott proposed gravity sliding—ophiolites as toboggans coming to rest. But something would have to lift the seafloor and break some off before it could slide. And what, for example, could have lifted the Macedonian mantle more than forty-five thousand vertical feet? The idea that enjoys the widest acceptance was around before either of the others, but no one much noticed, because it came from a graduate student and a postdoctoral fellow. In 1969, Peter Temple and Jay Zimmerman had proposed that the emplacement of an ophiolite might occur when a continental margin goes down under ocean crust, jams a trench, and then isostatically lifts the ocean crust.

Their proposal, derived from seismic data, recognized that the subduction of lithospheric plates was far more varied than people had supposed. Not only did ocean floor dive under continents but also—and much more commonly—it dived under other ocean floor, like two carpets overlapping. The lower slab, after melting, rose through the upper one as a volcanic-island arc. Now the island arc begins to move with the plate on which it rests. Plate motions shift. New trenches form. In back-arc basins, new ocean crust is made. Some island arcs go in one direction for a while and then reverse themselves. They choke a trench, say, and then go the other way, eating up their own crust as they go. Some of the crust might get emplaced as an ophiolite. The Marianas back-arc basin is spreading now, and so is the Lau-Havre Basin behind the Tonga-Kermadec arc, and so is the basin behind the South Sandwich Islands. Between Indonesia and the Philippines are two trenches that are eating their way toward each other and if nothing stops them will destroy each other. Some people—Moores among them—think that in similar fashion off Jurassic California two trenches were active simultaneously, the easterly one dipping to the east and the westerly one dipping west, and both were destroyed during the Smartville emplacement. In *Geology* for December, 1983, the volcanologist Alex McBirney, after dealing with the increasingly complicated attempts to relate igneous rocks to plate tectonics, closed with a vision of the decade to come. He said, "I predict that our present confusion about igneous rocks will rise to undreamed-of levels of sophistication."

Within half a decade, it was decided that the Smartville arc had formed where a spreading center developed in a transform fault or a fracture zone, after which the vector of the plate changed. When plate motions change, transform faults may turn into subduction zones or spreading centers. Forty-three million years ago, for example, the Pacific Plate changed its heading from north to northwest. In the hot-spot track of the Emperor and Hawaiian seamounts, the change is recorded in a pronounced bend—from north-south to northwest-southeast—dating exactly to that time. As transform faults turned into trenches all over the ancestral Pacific, the Tonga-Kermadec arc was created, and the Aleutian Islands and the Marianas. Something like this seems to have happened a hundred and sixty million years ago, creating the Smartville island arc.

That, Moores said, is where things are today—in the understanding of an ophiolite and its history in the world. If I was confused, so were geologists, not to mention geophysicists. "It has taken people a lot of time to stop thinking of these things as locally derived igneous rocks, and to begin thinking of them as transported tectonic features." If the Smartville Block was bewildering in the third and fourth dimension, I should at least reflect on the complexity of the surface, where Yuba City is the county seat of Sutter County, Marysville is the county seat of Yuba County, Auburn is the county seat of Placer County, Placerville is the county seat of El Dorado County, and El Dorado is the county seat of nowhere.

Heaping Pelion upon Ossa, Moores reminded me that there was a possible ophiolite higher in the Sierra and that in many ways it perplexed the relative simplicity of the story he had been telling. Known in geology as the Feather River peridotite, it was a large body of serpentine and related rocks that we had seen near the interstate just below Dutch Flat, where we went into a canyon one day whose walls, soft to the touch, were snakeskin black and green. The Feather River peridotite defied explanation, because it was much older than the terranes on either side of it. If this region of the western United States consisted of accreted terranes, progressively younger as you moved west, what was the Feather River peridotite doing there between Sonomia and Smartville? "The dates on it range from the Devonian through the Permian," Moores said. "We don't know how it got here. It was metamorphosed at six hundred degrees centigrade and twenty kilometres down. Must have been an extraordinary thrust that brought it up from twenty kilometres. Some of the rocks it deforms on its eastern contact are Triassic in age. It is older than Sonomia but was emplaced later. It is older than Smartville and lies east of Smartville, but—who knows how or why?—it was emplaced later. We have the story of several successive terranes neatly tying onto North America, and then we discover that the Feather River peridotite is between No. 2 and No. 3, that it is much older than any rock in No. 2, and that its emplacement is younger than the arrival of No. 3. The Feather River peridotite was about two hundred million years old when emplaced here. What does that mean? That it's anomalously old? That it was picked up first by an island arc and ultimately emplaced twice? Is it, in the first place, an ophiolite? What else can it be? It includes serpentine, unserpentinized peridotite, metagabbros, other amphibolites of diverse parentage, and, possibly, deformed sheeted dikes. What does that suggest?"

To Steve Edelman, a young structural geologist and tectonicist trained at Davis and the University of South Carolina, that suggested a genuine ophiolite, however baffling its history might be, and he believed he had found the sheeted dikes that would seal the discussion. He was the one geologist who had ever come out of the Sierra mentioning this possibility. One day in 1989, Edelman got off a train in Davis (being between jobs and between grants, he could not afford to fly) and started for the mountains to enrich his field research. He asked Moores to go with him, and I went along.

Edelman, whose beard is red, looked like a tennis player prepared to serve with a rock hammer. He was wearing a pink eyeshade in foam plastic, an aqua T-shirt, shorts, Adidas shoes, and white socks with red and blue bands. We went down a very steep canyonside more than five hundred vertical feet into the narrow defile of Slate Creek, close to the mining camps of Grass Flat, French Camp, and Yankee Hill, and closer to Devils Gate. Slate Creek was a place of spare light, where miners had removed a hundred thousand ounces of gold. In California canyons as remote as this, gold has more recently been grown in the form of plants with spiky leaves of the sort that were painted by Henri Rousseau. In a statewide effort to stop the industry, the government in Sacramento had organized something called CAMP—Campaign Against Marijuana Production. CAMP narcs were said to be posing as geologists when they roamed the wild country. A geology graduate student from the University of Texas, doing field work in the northern Coast Ranges, had been murdered, presumably by marijuana growers—shot in the back of the head. The killers have never been caught. A man in the California Division of Mines and Geology whose work frequently takes him into the Sierra foothills goes into every bar in every old mining camp and forest hamlet within a radius of five or ten miles and tells everyone present what he is doing and where. Out in the crops and outcrops, when he encounters people they know about him.

To the nineteenth-century miners a lot of rock was slate. Slate Creek was flowing over a beautiful gray diabase. I remarked that it was one of the clearest streams I had seen anywhere south of the Brooks Range. Moores marvelled, too, saying that he could see "shear-sense indication in porphyroclasts at the bottom of the creek—it's that clear."

This seemed to please Edelman, but not greatly. He had asked Moores to come to Slate Creek to see if he could agree that the diabase, as old and recrystallized and murkily deformed as it was, showed any pentimento of the laminations and chilled margins of sheeted ophiolitic dikes.

Seeing nothing in his lens that impressed him as he moved from ledge to ledge upstream, Moores struggled to be cooperative. He said, "It's not a good stretch. We have to get around the next bend. . . . I think I see some folding, some layering. Maybe it's gabbroic."

Edelman seemed to speed up a bit.

"Maybe that is a pillow there," Moores said, hopefully. But doubt was the pillowcase.

Now Edelman was on the run. He promised better things ahead. We had not yet reached the exposures he wanted us to see. Around the next bend, he closely scrutinized a big outcrop that jutted into the stream. He said that in his judgment it was a set of sheeted dikes. What did Moores think?

Moores leaned in to the rock, lingered, drew back, and said, "It's O.K. if you're a believer. It's not a skeptic's outcrop."

"And this one?"

"Only slightly more convincing than the last."

Edelman was right, though. The farther we moved upstream, the more the gray rock seemed to exhibit the features he had asked Moores to witness. But the weathering and deformation were such that the laminations were not easy to discern. Moores, with his leather cases and leather pouches lined up on his belt, his wide-brimmed fedora shading his beard, appeared to have been there since 1849.

"I don't think a Howard Day type skeptic is going to believe this," he said, mentioning a metamorphic petrologist at Davis.

"No," Edelman said. "I guess not."

Moores said, "Fortunately, there's only one of those in the world."

Edelman said, "I know it gets good up here ahead of us. Suspend your judgment until you see the good stuff—until you see the well-defined sheeted dikes."

A little farther on, Moores paused at an outcrop for a long close look, with and without a lens. Minutes went by. Slate Creek went by, making the only sound. At last, Moores said, "That's a good one. That would convince Howard Day."

The next ledge was even better. Moores examined it for some time. Edelman's expression was full of sniffed victory. Moores said, "I have no doubt that that is a chilled margin. And that. And that." Looking into Edelman's eyes, he added, "But I'm a believer."

Another few yards, another rock face, and Moores said, "If you were an ophiolite and someone took you down to six hundred degrees and twenty kilometres, maybe this is what you would look like.

If this is not a dike complex, what else could it have been? Nothing, is my answer."

Edelman said, "So. Do you believe in the Devils Gate Ophiolite?"

Moores said, "I believe in the Devils Gate Sheeted Dike Complex."

Edelman: "You happy?"

Moores: "Yes."

Edelman: "Another ophiolite in the Sierra."

A lot of geology is learned now from seismic waves and satellites, and pieced together on printout paper in artificial light. Neither a seismometer nor a satellite was ever going to see what Edelman had seen.

Moores said, "If you took Japan and its old ocean crust and collided it with the state of Washington, and the old crust was shoved over Washington, you might have something like the Feather River peridotite. If this is an ophiolite, it is bigger than the Troodos of Cyprus. It does not appear to be a part of Sonomia. It could be a subsidiary of Smartville. You don't know what it's doing here. You just know it's big and it's important."

If the Feather River peridotite inconveniences the exotic-terrane story, it is well that it should, he said. The old picture of the western margin of North America is gone, but not long gone. The present description of assembling terranes is so new—and calls for so much working out—that it hardly requires acid-free paper. Did a large pre-assembled terrane—the Stikine Superterrane, of which Smartville would be a part—dock in the Jurassic? Or did the United Plates of America, as they have been called, arrive separately? "Most of the ophiolites from the Brooks Range on down through central British Columbia into the western United States and in Baja California and Costa Rica are Mesozoic in age," Moores said. "They appear to represent some sort of island-arc complex that collided with western North America in mid-Jurassic time. Singly or collectively? That's hard to say. You can make it the one or the other. We can't work that out now. The timing seems to be different. It seems to be lower Cretaceous in the Brooks Range, mid-Jurassic in British Columbia, mid-Jurassic in the Sierra Nevada, and somewhat younger as one goes farther south. That could be a single, ragged-edged collision. Or it could be several terranes coming in. We're not prepared to say."

There remain in the world, of course, geologists who are not prepared to say that exotic terranes exist in such prodigious quantity. Homicidal in their sarcasm, they still like to assert that their less conservative colleagues are prone to name new terranes for any change of lithology, at any formation boundary—in fact, at any place, however small, where it is easier to claim that something is exotic than to figure out the relationships of present and missing parts. In the geology of such people, it is said, a microterrane is a field area. A nanoterrane is an outcrop. A picoterrane is a hand specimen. A femtoterrane is a thin section.

Although the outer shell of two-thirds of the earth is rock of the ocean crust, it is so inaccessible to the field geologist that to study even the fragments that have broken off on continental margins requires prodigious travel. Moores has pursued his specialty from Oman to Yap to Tierra del Fuego to Pakistan, and routinely has returned to the eastern Mediterranean—most of all to Cyprus. One autumn in the nineteen-eighties, when I was working in Switzerland, I flew to Cyprus to watch him do his geology there.

He met me at Larnaca, we drove north, and within the hour were tectonically deconstructing a huge broiled fish and drinking dark Cypriot wine. Politically, Cyprus was in Asia, Moores said, but geologically it belonged to no continent. It rested on the lip of Africa but was not African. It was not Eurasian. In the lowercase and literal sense, it was mediterranean. Long after the last supercontinent began rifting and its new internal shorelines bordered the Tethys Ocean, what was to become the foundation rock of Cyprus welled up as magma in a Tethyan spreading center. At that time, about ninety million years before the present, the Eurasian and the African sides of Tethys were twice as far apart as they are now. They continued to separate for another ten million years, and then plate motions changed. As the North Atlantic began to open, Africa began to move northeast, closing with Eurasia, as it continues to do. Ongoing results include the Alps, the Carpathians, the Caucasus, the Zagros—in Moores' words, "a big slug of deformation that's throwing up mountains everywhere." By late Miocene time, geologically near the present, not much remained of Tethys in that part of the world except the Mediterranean Sea, the Black Sea, and the southern Caspian Sea.

It was in the late Miocene that Africa levered the Tethyan floor, and broke off Cyprus. The ophiolite was thickly covered with chalk, which had settled upon the pillow lavas as a clean lime ooze, unadulterated with sediment from any landmass—as clear an indication as you could ever hope to find that Cyprus was a piece of remote deep ocean.

For seeing and touching ocean crust—for leaning against it with a hand lens, for removing small cores to study their remanent magnetism, for mapping the varying rocks from zone to zone—there was no example in the world as well preserved as Cyprus. Shaped like a razor clam, the island had a long foot that reached up in the general direction of Turkey, which was fifty miles away. This northeastern extremity was a long low range of mountains, whose geologic history was not well understood: it seemed to be in some sense accretionary, perhaps a collection of island fragments from a confused Eurasian sea, and thus to belong—in a tectonic sense—to Turkey, which seized it by force in 1974, and established it as the Turkish Federated State of Cyprus. The Turkish Federated State of Cyprus had so far been recognized only by geologists. The thick body of the island, which rose higher than the White Mountains of New Hampshire, was the ophiolite—the exposed and integral lithospheric crust that was without continental affiliation and was the heart and substance of the independent Republic of Cyprus.

Each day, we drove out of Nicosia down the great treeless plain of the Mesaoria, on our way to the high Troodos. Trending east-west, the plain divided the island's most prominent lithologies. Off to our right about ten miles was the low silhouette of the north-coastal range, and ahead to the left were the Troodos. Through the Mesaoria ran a boundary drawn by the United Nations which was known among Greek Cypriots as the Green Line. Turks called it the Attila Line. Along it, over the plain, ran United Nations sentry boxes, like populated fence posts, farther than the eye could see. All north-south roads, even unpaved ruts, were barricaded and marked with warnings and instructive signs: "HALT!"

The feel of day in the Mesaoria was like crouching too close to a campfire. Up in the Troodos were groves of shade. Under Aleppo pines, the air was as cool as deep water, and about as still. There were whole ridgelines of sheeted diabase, weathered out in silver blades—like thousands of playing cards in one standing deck—recording in subcenturies the spreading of ten million years. There were cliffs of chalk, and bulging extrusions of pillow basalt. There were layered cumulates in massive black gabbro. There were serpentine peaks. The highest peak was Mt. Olympus. In the Hellenic world are enough Mt. Olympuses to suggest tract housing for redundant gods. It is a godly talent that geologists have: not only to see ocean lithosphere in mountain crests but to feel comfortable in the knowledge that some of the lowest rock in the ophiolitic sequence is the highest rock of a place like Cyprus—with nothing overturned. A sawyer would also understand this—and almost anyone who could look at woodwork and see the original trees. The Cyprus ophiolite—great slab of the ocean—was bent upon the slope of Africa. It was draped, hung, arched, folded—not quite like Dali's watches, but the image would do. Water entering peridotite to make serpentine had swelled the whole affair, and then erosion had taken over, finding the serpentine within the crown of the arch and variously stripping the other stuff off the top and down the sides until the serpentine stood highest and the ophiolitic sequence (reading upward) went down the mountains in successive steps, ending in peripheral cliffs of chalk. In some of the higher country, Moores chipped away at interbanded cumulates (so-called magmatic sediments) from the deepest pools of gabbro, and among them found a trace current, which he described as "a stream channel in the magma chamber." Rapping it with his hammer, he said, "Shows you which way is up."

Among stone cabins whose metal roofs were weighted with stone, we hammered stone. We moved among ripening apples, Lombardy poplars, and red-roofed white villages spread on the dry mountains. We went into deep canyons. A good deal less of the massif was under forest than under grapes. From the high ridges with comprehensive views, the mountains looked like coalesced football stadiums, vineyards terraced into the sky. Above Palekhori, we picked and ate fourteen grapes, doubling the annual number destined for fresh consumption. At a table under a tree in Palekhori, we drank coffee that was good enough to eat. It was brewed in a briki scarcely larger than a gill. It came with a glass of cool water. "Palekhori" means old village. It is in the sheeted diabase, whose stately laminations are so distinct that you can all but see their lateral motion. Moores and Vine had found their best evidence of seafloor spreading at these altitudes. More recently, Moores had been studying fault blocks in the Troodos. In various tiltings of the sheeted diabase, he had seen that the rift valleys of ocean spreading centers tend to break into blocks as they widen—a version, on a small scale, of the faulting characteristic of the Connecticut Valley, for example, or the Newark Basin, or the Culpeper Basin, or the great western province of the Basin and Range.

The mid-ocean ridges run around the world very much like the stitching on a baseball, not in simple lines but in oscillating offset segments. Such a pattern evidently accommodates a sphere. In any case, it is what the earth looks like where it is pulling itself apart. The ocean ridges of the world jump from rift valley to transform fault to rift valley to transform fault, everywhere they go. The rift valleys are typically about forty miles long, and are offset, also about

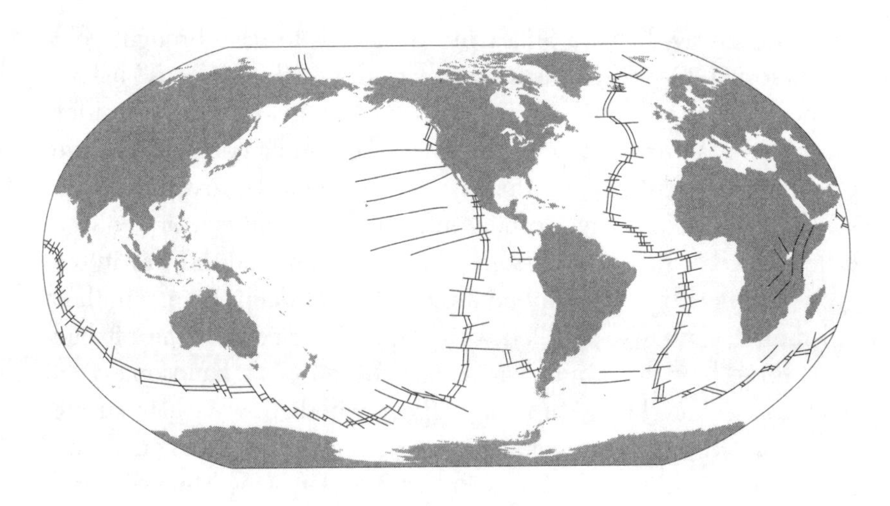

forty miles, by the transform faults. Moores had found this pattern in the high Troodos. In rock beside a road, he pointed out sandstone sediments that fell into a bathymetric depression where a transform fault intersected a spreading center. A goatherd walked by—blue shirt, soft olive hat, a stick bag on his back. "*Yasas!*" he said, in greeting, above a din of goats. Nothing in his face suggested that he found it at all strange to come upon two men with the beards of Greek Orthodox priests squinting into coarse rock with the lenses of jewellers.

In a remote mountain valley, we walked into a box canyon and saw feeder dikes five metres wide that had driven upward through pillow lavas to break into the ocean and form more pillow lavas. Moores and Vine were working there in 1968 when a Cypriot dressed in a business suit materialized before them. He said that he was just out for a drive and had seen their vehicle, and he asked what they were doing. He also said that he was Minister of the Interior. They told him that they were drilling into the rock to remove small cylindrical cores for paleomagnetic data, and in the course of conversation they also told him that their work was in part supported by the National Science Foundation, in Washington, D.C. The minister said, amiably, "Do you mean to tell me that your government pays you to come over here and drill holes in my island?" He stayed long enough to learn that his island was the keystone of the sea. A year and a half later, he was assassinated.

In the way that the Smartville arc seems to have brought gold to California, Cyprus brought copper to the world. "Cyprus" means copper. Whether the island is named for the metal or the metal for the island is an etymology lost in time. The mining geologist George Constantinou, director of the Cyprus Geological Survey, took us south from Nicosia one morning into the hill country near a village called Sha. On soil eroded from pillow basalt, he led the way into a grove of pines that surrounded a pit forty feet deep. Like countless old mines, it was partly filled with water. He said that the pit had been there four thousand years. Constantinou was a handsome man of alternately bright and brooding aspect, with light wavy hair, strong features, and such commanding stage presence that I imagined him as an actor. I imagined him as Prince Hamlet, King Henry V, and Archie Rice, because his physical resemblance to Laurence Olivier was so close it was unnerving. Of this small excavation framed by Aleppo pines he spoke with resonance as well as reverence. Cypriots thirty-five centuries before Christ had walked into this pine grove, and others like it, and had found native metallic copper lying on the surface, he said. Pine resins in the groundwater had mixed with copper sulphate and reduced the copper to metal.

When Cyprus was spreading in the Tethyan floor, seawater descended through fissures and—close to or within the magma—picked up quantities of dissolved copper, and lesser amounts of mercury, manganese, tin, silver, and gold. Like the black smokers active now in the Red Sea and the Gulf of California, hot brine plumes rose through the Cypriot rock and precipitated metals and metallic compounds on the pillow lavas. From everywhere in the ancient world, people turned to Cyprus for weapons-grade copper. The swords, spears, and shields of innumerable armies were made from Cypriot copper. Before long, though, the resin-reduced metals were gone. More than a millennium passed before the Cypriots learned that dark earths where the metals had been were not a whole lot less cuprous than the metal itself. Rainwater—rare in Cyprus on the human scale but continual in geologic time—had removed lighter materials and had concentrated the copper minerals malachite and azurite in an upper zone of extreme high assay. Geologists of the twentieth century would describe such a concentration as a supergene enrichment. The ancients somehow discovered that if they mixed the cuprous earth with umber, and then heated the mixture, molten copper would flow. There was plenty of umber close at hand. Umber is an oxide of manganese and iron. In spreading environments on ocean floors today, umber is piling up on the pillow lava in large dark-chocolate cones beside the black smokers, as it did on the pillows of the nascent Cyprus.

In 2760 B.C., smelting began in Cyprus, Constantinou told us. And in the following centuries Cyprus became an island of seven kingdoms. Slag heaps developed in forty places. The *Iliad* is populated with warriors armed in bronze. Bronze is copper hardened by adding some tin, and the copper would have come from Cyprus. (Copper was mined on Cyprus for nearly two thousand years before the lifetime of Homer.) In 490 B.C., Darius the Persian attacked Greece with forty thousand soldiers who carried bronze shields and bronze javelins. The Phoenicians also mined copper on Cyprus, the Romans as well. The ancients stripped the supergene, and other rich ores, down to the water table, where they had to stop. The Republic of Cyprus once used ancient slag for roads, but the old slag heaps are now protected as monuments.

After the mine at Sha, we drove on ancient slag through roadcuts of pillow basalt, west-northwest. There were orchards of carobs, figs, and pistachios, and an understory of prickly pears. This was not Lawrence Durrell's north-coastal range of "silk, almonds, apricots, oranges, pomegranates, quince." It was an interior country of buff eucalyptus, of thousand-year-old olive trees fluted at the base. Nearly every farmhouse was white, and most had sky-blue shutters—the colors not of Cyprus but of Greece. On virtually all rooftops were boxy solar water heaters, like raised sarcophagi. With few exceptions, they carried advertising.

On the Mesaoria, we passed new, isolated towns—one-story clusters of temporary housing, marginally superior to internment camps, built to shelter Greek Cypriot refugees from communities north of the Attila Line. We left the highway and went into the new and extremely narrow streets of Peristerona, which seemed less a town than a military barracks. Many of its people were natives of Katokopia, scarcely three miles away, but Katokopia was lost to them, beyond the Turkish line. Among these was Anastasia Constantinou, the mother of the director of the Geological Survey. She was elderly, tall, dressed in black, obviously pleased to see her son no matter what he might bring with him. The hour was near noon, the air somewhat humid and above a hundred degrees. There was a small greenhouse full of gardenias, camellias, and azaleas. The geologist went among them with a mist propagator. He opened folding chairs on a concrete terrace with a view across the treeless plain to the marble mountains of Kyrenia—the dark wall of the forbidden north. His mother set on a table bowls of boiled rice flour that had lightly thickened as it cooled. Sprinkled with sugar, it was very cold, standing in rose water. In that volcanic heat, it had four times the effect of a cold fruit soup, twice the effect of gazpacho. If you closed your eyes, you saw pools among gardens descending into pools.

At Skouriotissa, southwest of Peristerona, the concurrence of geologic time and human time had been long enough to approach a record. A very large working strip mine there had been in operation for four thousand three hundred years. The slag, piled in pyramids, represented all that time. Constantinou said that there were at least two million tons of ancient slag in and around Skouriotissa. "Skouria" means slag. The massive copper-bearing sulphide ores of Cyprus have a very characteristic sugary structure, he said, resulting in an incompetent rock that was always easy to mine. "The ancients were excellent geologists. They knew the geology of the ore bodies of Cyprus. I am an exploration geologist with a Ph.D. I don't think we will find any ore body the ancients did not know about."

The earliest known smelting of copper was in China. Did the Cypriots figure it out for themselves or learn how to do it from others? In ancient manuscripts, Constantinou said, there is no insight that helps to answer that question. Where you find copper, you will find iron. The umber of Cyprus is more than half iron. "It seems logical to expect that Cypriot umber was the world's first iron source, that iron was invented on Cyprus. The ancients used gabbro to grind the ore. They lined their furnaces with serpentine."

They fired their furnaces with Aleppo pines, and other conifers—the ancient forests of Cyprus. To smelt one pound of copper from sulphide required three hundred pounds of charcoal. From the earliest beginnings of the mining until the last years of the Roman Empire, about two hundred thousand tons of copper were smelted on Cyprus. That used up fifty-eight thousand square miles of pinewood forest, on an island whose total area is thirty-six hundred square miles. The forest had to be rejuvenated sixteen times for copper alone, not to mention the fleets of ships that were made on Cyprus, or the firing of the island's world-renowned kilns.

"For lack of wood, Oman, Iran, Saudi Arabia, Egypt, and Israel all had short-lived mines," Constantinou said. "The Troodos gave water here to support trees. But sixty million tons of charcoal made from 1.2 billion cubic metres of wood is no joke. Sometimes I close my eyes and see that ancient scene. I get crazy. I see all those people, tens of thousands of people, carrying ore, carrying wood."

Before Moores went back to California and I to Switzerland, I accompanied him on a brief reconnaissance in Macedonia. While Cyprus was surely, as he had described it, one of the best-developed ophiolite complexes in the world, the sequence there did not include a large percentage of mantle rock. In a typical slice of ocean lithosphere, the mantle rock is nearly twenty per cent, bottoming at the asthenosphere, the lubricious zone in the mantle which allows plates to glide. In Cyprus, the serpentinized mantle rock of the Troodos was a relatively small part of the peridotite that had once been included in the package. The rest was buried or lost. If you wanted to sense what was missing, you could do so in Macedonia, where seven vertical kilometres of exposed mantle constitute one of the thickest measurable sections of mantle rock emplaced on any continent. Moores said, "Presumably, it goes to the bottom of the plate."

In the disjunct cacophony of the airport in Athens—through a sea of Arabs wearing kaffiyehs and tobes and running shoes, of Ethiopians with wallets the size of magazines, of Panasonic briefcases turned up full—we found Hertz. We were soon at the foot of the Acropolis, establishing our bearings. Moores told me to notice the red shales and red cherts around the Theatre of Dionysus, on the low ground, and—over his shoulder as he rapidly climbed—to watch for the contact where the lithology changes. It would have been hard to miss. The freestanding, high-standing Acropolis is an almost pure massive limestone, sitting on the cherts and shales. Our way was blocked by a ticket booth. We paid a hundred drachmas. American college students were all over the summit. In the frugal shade of the Parthenon, Moores had the look and certainly the sound of a free-lance English-speaking guide. He mentioned Ictinus, Callicrates. American students leaned in to listen. He lauded the durable Phidias. He moved to the south side of the building, and students followed. He mentioned that limestone is soluble in water. Therefore, it includes caves. In caves within this hill, gods were thought to reside. Grottoed limestone will impound water. If you were seeking refuge, or a place to endure a siege, you would choose a hill like this one. We were looking south over the Stoa of Eumenes to the shore of the Saronic Gulf. From runways there, 747s were rising. They seemed in no hurry to go away. They seemed to hang like barrage balloons. Moores said, "After the battle of Salamis, ships beached themselves by the airport. As caves in limestone enlarge, their roofs eventually collapse."

He mentioned the Parthenon's historical stratigraphy—temple, church, mosque—and the erosional forces that had brought the building to its present condition: rain, acid rain, smog, gunpowder. In 1687, the Parthenon was in use as a Turkish powder magazine. Venetians bombarded it, and the powder exploded. The event was geomorphologically catastrophic. For two thousand years the Parthenon had stood there uneroded, until that night in 1687. "There is no mortar in the Parthenon," Moores added pensively. "It is all marble, and held together by gravity. And it's gone through earthquakes, too. The geology is not well worked out here. The general story is that the Acropolis is a klippe, resting on the red cherts and shales. It is not a deep block."

A klippe is a remnant of a nappe. A nappe is a large body of rock that has been moved—by gravity, by thrust-faulting, or by any other mechanism—some distance from its place of origin. If you liked, you could call Cyprus a special kind of nappe. Moores gestured to the east, across the white city and its numerous hills, to the serrated profile of the Hymettus Range, less than ten miles away. "It is thought that the Acropolis came from there," he said. "There are problems with the idea, but it is distinctly possible."

In the heat and pressure of a collision or some other tectonic event, limestone softens, recrystallizes, and hardens as marble. The Hymettus Range is for the most part marble. Its limestone first collected on the floor of Tethys, and was later folded in collisional mountains compressed by Africa moving northeast. Marble quarries in the range had been there for something like three thousand years. The Parthenon came out of a quarry at the foot of Mt. Hymettus.

If the Acropolis is a klippe, the Acropolis itself came away from the Hymettus Range, in Eocene time, and travelled overland to Athens. About fifty million years later, in late Holocene time, the Parthenon followed, in carts.

"An alternative possibility," Moores said, "is that the Acropolis is a large block in a mélange with a matrix of red cherts and shales—an accretionary wedge tectonically scraped from Tethys when the seafloor was subducting in the Mesozoic."

The American students were looking at one another, and Moores was becoming self-conscious. A guide he may have been, but the language he was speaking was, to the students, local. "I have always thought it sacrilegious to come here and do geology," he said. The tone was apologetic but not sincere. His next words were "I think the shales would correlate with the Olonos-Pindos deepwater sediments, which extend from the Peloponnesus through western Greece."

As one might expect in a marine country sitting on a microplate caught in the crunch between Africa and Europe, ophiolitic fragments of varying age are strewn about Greece like amphora handles. As Moores drove up the broad avenue of Vasillisis Sofias and into Sintagma Square, he remarked that the silver that financed Athens came out of a metamorphosed ophiolite at Laurium, near Cape Sounion, the southern tip of Attica. He parked beside the Bank of Greece. Standing at a teller's cage on a green-and-black floor of polished serpentine, he changed dollars to drachmas.

North through Attica we moved swiftly, as if we were in a light plane flying at an altitude of three feet. A few miles west of the North Euboean Gulf, Moores pulled over and stopped at what appeared to be a long high roadcut. It was actually a limestone cliff, of the Parnassus Range, which rose steeply behind it. "There are bits and pieces of ophiolite on top of the Parnassus," Moores said.

"This is Thermopylae." The broad coastal plain to the east was full of olives and cotton and was large enough to accommodate a very large army. In 480 B.C., however, the coastal plain was not there, and water lapped close to the rock. There was insufficient room for the attacking Persian Army. Leonidas, king of the Spartans, defended the narrow margin of land with Parnassus ridges at his back. He was defeated after the Persians learned of a route around the ridges. "Hot springs were at the foot of the mountains then," Moores said. "They are long gone. The coastal plain came up recently—at some point in the past twenty-five hundred years—as a result of an earthquake."

In the nineteen-sixties, in his long and lonely field seasons in Macedonia, Moores read his way in several senses into the country. He learned to speak the language. His interests were well spread across a couple of hundred million years. He asked me if I knew John Cuthbert Lawson's *Modern Greek Folklore and Ancient Greek Religion*, and described it as "polytheistic mysticism with a superficial patina of Christianity." Not being a geologist, I took his word for it. Farther north, he said:

——*This is Lamia. A few kilometres outside Lamia, during the Second World War, Greek partisans shot three German soldiers. Germans stopped the first hundred and thirty-eight people who came down the road, and killed them. The German rule was fifty Greeks for one German.*

——*Over there to the east, fifteen miles, is Volos, where Jason and the Argonauts sailed from.*

——*Pharsala is about twenty-five miles west of us. Near Pharsala, Caesar defeated Pompey on alluvium at the western margin of the Tsangli ophiolite. Pharsala itself is on serpentine. If you see a black church, you are probably looking at serpentine. That's particularly so in Italy. In Florence, the dark rock in the walls of the Duomo is serpentine—and the Giotto campanile and the baptistery with the Ghiberti doors. In Istanbul, the dark columns in the Hagia Sophia are serpentine.*

We came into a vast horizon of land so flat it seemed unnatural. It seemed flatter than the Great Central Valley of California, if that is possible. Solitary oaks were widely spaced above cotton, wheat, barley—a tree on each twenty acres, more or less. In a manner that called up ritual, tall telephone poles stalked across the two-dimensional landscape. Woodhenge. A huge Pleistocene lake had been here. In the stillness of its depths, smooth silts collected:

——*This is the Plain of Thessaly. In the eighth century* B.C., *Greek tribes settled here, chasing off the inhabitants. The fugitives went into the hills. They were fine horsemen. They raided the Greek settlements on horseback. The legend of the centaurs may have come from this. At night, you couldn't tell man from horse.*

——*Do you see those switchbacks climbing out of the plain? The Greeks used to survey a road by putting a hundred kilos on the back of a burro and sending him uphill. They followed the burro with a road.*

Three coastal mountains now formed the eastern skyline: Mt. Pelion, Mt. Ossa, and Mt. Olympus. In an island universe of Mt. Olympuses, there is one Mt. Olympus. This one. Sealed in its own integument, at the moment it was eighty per cent cloud. We could see only the base. Of the country more immediately around us, Moores said:

——*This is the Pelagonian Massif, a Mesozoic microcontinent that was thrust over a dome of younger sediment. The dome is marine rock—shallow-water limestone and flysch. From the highest part of the dome, the Pelagonian rock has worn away, and in that window stands Mt. Olympus, ten thousand six hundred feet. I think you need technical equipment to get to the summit.*

——*This is the Vale of Tempe, where the Muses came down off Olympus and played in the waters.*

We were soon in highland Macedonia, with wide views over red-and-white villages to far-distant mountains: west into the Pindus

Range, north into Albania. At five thousand feet, the modest summits of Macedonia—Vourinos, Flambouron—were about as high as the Adirondacks of northern New York. Vourinos and Flambouron and the country roundabout had risen a great deal more than that. Somehow they had been lifted forty-five thousand vertical feet, and were almost pure mantle.

The Vourinos Complex—this mantle peridotite included—was Tethyan seafloor that formed in a spreading center in Jurassic time. It was emplaced on the Pelagonian microcontinent in lower Cretaceous time, and later broken into four major fault blocks. The narrative was straightforward and fairly simple, but of course it had not been helpful to Moores as he began work here in 1963, because plate tectonics and emplaced ophiolites were uncoined terms and the narrative did not exist. He was further inconvenienced by the ravages and deceptions of erosion, which had caused the lowest rock in the sequence to stand highest in the country. Moreover, the sheeted dikes of Vourinos diabase were misleadingly parallel to the sedimentary beds that had formed above them. After years of living with this terrane and taking it apart in his mind, Moores had come to realize that soon after the dikes formed at a spreading center they had been rotated ninety degrees from their original plane. In recent time, the four fault blocks had tilted over as well, and were like broken segments of a fallen ancient column. On paper, Moores had brought them into spatial coherence. He had worked here intensively for three years, and continually after that. He had recognized the contrast between the mantle rock and the magmatic rocks above it, and had become convinced that these parts taken together were in turn parts of a larger sequence.

Now, on the road to Skoumtsa, in the valley of a small clear stream, he was standing on the contact between the ophiolite and the Pelagonian rock on which it had been emplaced. Pelagonian limestone was overlain by serpentine that had been sheared up badly and turned into a messy schist as it skidded to a tectonic stop. The mantle peridotite had been serpentinized there where it scraped upon the continent, but as we moved away from that boundary the peridotite became purer to the point of zero serpentinization. The rock had never been magma. We were seeing (he presumed) the earth's mantle in an essentially unaltered state.

Since the ophiolitic column was lying on its side, we could advance overland on the face of the earth and lithologically descend deeper and deeper into the mantle. We did this on a very rocky and narrow road, some of which had been cut by engineering. After crossing the stream, we paused to look at a dynamited outcrop—a rough texture, dark green, knobby with pyroxene in a smooth matrix of olivine. "This is solid mantle," he said. "It's just about as fresh as you'll ever find, with the exception of mantle material in a diamond pipe. We are roughly five kilometres below the petrologic Moho, and that would have been about ten kilometres below the Tethyan ocean bottom, and fifteen kilometres below Tethyan sea level. Some people like to think that this rock slid by gravity into its position on the Pelagonian continental platform. From fifteen kilometres below sea level? How do you do *that* with a gravity slide?"

Now and again in that country we saw, in the peridotite, bands and lenses of dunite, which generally contains chromite, the source of chrome. The chromite looked like small, spattered blobs of tar, not like the grille of a Jaguar. The village of Khromion was ten kilometres away. During the Second World War, Germans used Greeks as forced labor there. Chrome was essential for the high-strength steel in the armor of tanks; and chrome, as it happened, sharpened the effect of armor-piercing shells. Peridotite with dunite in it is softened, more susceptible to erosion. "You can almost map the rock by looking at the relative roughness of the terrain," Moores said. "The smooth and grassy swales have dunite under them. The pure peridotite is in the rough ridges."

For a couple of years, Moores worked alone in Macedonia, attracting the attention of nothing much but mastiffs, which appeared out of nowhere. The mastiffs were protective of sheep, and hostile to geologists and wolves. In 1965, Moores turned up in the country with a wife (Judy had just graduated from Mount Holyoke), and later on with the Volkswagen bus and Geneva, Brian, and Kathryn. This attracted the attention of old Macedonian women, who lived on the Vourinos Ophiolite, and would appear out of nowhere. They admired the Moores children so much that they spat on them. This was a custom that warded off the evil eye. Judy learned to snatch baby Kathryn away from crones who adored her. In the Volkswagen one day, Judy turned around and saw Brian spitting on his teddy bear.

In Davis in 1989, when Kathryn had just turned sixteen she fell sick with acute mononucleosis, and her throat was so choked by swollen tonsils that she was completely unable to speak. The antidotal spit was recalled at the kitchen table. Kathryn wrote on a slip of paper: "Maybe it only lasts sixteen years."

It was in Macedonia that I asked Moores how he felt about being in a profession that had identified the olivine that people would be ripping the mountainsides to take away, and he said, "Schizophrenic. I grew up in a mining family . . . Now I am a member of the Sierra Club."

If you ask someone in Arizona where Crown King is, the usual answer is a shrug. Someone going home to Crown King would turn off the Flagstaff–Phoenix highway, raise a plume of dust on Bloody Basin Road, and go west-southwest toward mountains. The elevation of the basin is thirty-five hundred feet. The ridgeline ahead is seven thousand feet, and while you lurch and rattle toward it—as Moores and I did recently in a rented pickup equipped absurdly with cruise control—a good deal of time goes by but the mountains seem no closer. Moores said he remembered his father whistling a Schubert serenade on Bloody Basin Road. Moores was hearing it now, as he always had when crossing this particular stretch of country, although he had not been there in twenty years. In early slanting light, fields of prickly pears flashed like silver dollars.

We went through a one-house town. On the seat between us was an Exxon map of Arizona. "That was Cordes," I said. "Why is that place called Cordes?"

Moores said, "Because that was Bill Cordes' house."

The dust behind us thickened. We were now on the old Black Canyon Highway. When Moores was teen-aged, this unpaved and bridgeless thoroughfare lined with mesquite, cat's-claw, and paloverde was the main route from Prescott to Phoenix. We passed five buzzards eating a rabbit. While distance compiled, the mountains continued their retreat. After leaving Black Canyon Highway, we threw even more dust on saguaro cactus, agave, cholla, and ocotillo. We went through Cleator, a town that somehow managed to seem smaller than Cordes. There was a gas pump dating from the twenties which seemed to have died in the thirties. Beside it—in the open air—was a radio that had last heard the Blue Network. If you squinted hard enough into the cactus, you could see Dorothea Lange changing film. After more long miles, we began to climb, and now—in direct proportion to the gradient—the route evolved from an ordinary unpaved right-of-way into one of the oddest and certainly one of the costliest dirt roads in America. Its humble surface passed through roadcuts of exceptional engineering. It went through narrow defiles past vertical walls of competent granite blasted by construction crews in the century before. A railroad had once climbed three thousand feet there, its purpose being to help dismantle the mountains themselves, to ease down from Crown King in gondolas inexhaustible ores of hard-rock gold. "This railroad was an incredible feat of engineering, resulting in futility," Moores said. "The ore just wasn't there. Mine promoters are a breed apart. Their mentality is 'Of course it's there.' When I was ten, I heard a promoter say, 'We have a thousand tons of ore blocked out. When we get going, we're going to process a hundred tons a day.' He didn't stop to calculate that a hundred per cent of his ore would be gone in ten days. Mine promoters will believe anything, and so will their backers. The money came from New York, principally. The promoters were always looking for people with more money than smarts."

A voice said, "They looked in the right place."

Some of the deep cuts in the granite were cul-de-sacs, and the dirt road turned sharply before them. Trains had entered them, and then backed up across a switch and on toward the higher ground. Some cuts were low-sided, like the sunken lanes of England. In the course of the climb, the chaparral of the Sonoran Desert gave way to forest of ponderosa pine. The chaparral ran highest up south-facing slopes. We bypassed a tunnel, its ends now mostly caved in. When Moores and his three sisters were children, their grandfather convinced them that the tunnel was the home of a monster known as the Geehan. Long before they were born—after the mining scheme failed and the rails were removed—the Geehan moved in.

Crown King was at six thousand feet, in a subsummit swale—

a few dozen buildings, spread through a mile of forest. When the air was still, the people could hear vehicles far down the switchbacks, climbing. They could tell from the rattles who was approaching. In the Crown King that Moores returned to now, not a great deal had changed. The same old welcome sign stood on the outskirts:

THE FIRE DANGER TODAY IS
EXTREME

The main street was a rocky swath of white granitic dust. Two pickups were parked by a retaining wall below the veranda of the general store. An apple tree thirty feet high grew out of the veranda. In a window, a red neon ring circled the word "Lite" under the painted words "U.S. Post Office." There was an old anonymous gas pump, still dispensing gas. It served the mountain. A white pole beside it flew the American flag.

Roads threading the declivities above Crown King led to various mines—gold mines, mainly, and silver and zinc—one of which served as the town well. The air and the forest were so dry, the community so high, that a ready source of water was beyond imagining. Yet water emerged from fissures at the Philadelphia Mine. In these exceptionally arid mountains, monsoonal rains arrive in August. Moores remembered an August day when four and a half inches fell in a single hour.

Crown King now had four telephone lines, four parties on each line. In the nineteen-forties, when Moores was growing up, there was one line. When you turned the crank of one of the four phones in town, three other people picked up phones to hear your outgoing call. I asked him what else especially came back to him about those years. He said, "The smell of warm pine needles and the sound of the wind."

Under the ponderosas, on the dry needles, were granitic boulders. Playing hide-and-seek, he had hidden behind the boulders. When he and his friends played baseball, the bases were rocks. One year, there were six students in the Crown King school—grades 1 through 8. Usually, there were ten or fifteen. Moores climbed half a mile from home to school, sometimes in fairly deep snow. We went up there now, and found that a room was being added to the building, and thus it would become a two-room school. Academic privies had served when Moores was a child. They had been replaced by indoor plumbing. The wood stove was gone. In the ceiling, the chimney hole was covered with a board. The old classroom was otherwise the same, with its tongue-and-groove walls of horizontal boards, its long span of lead pipe supporting a curtain so that one end of the room could serve as a stage. There was a personal computer on the teacher's desk. The Lombard piano was an upright that Moores remembered. Playing a few bars on it, he found it "more or less in tune."

When Moores was a child, there was a piano in Crown King that belonged to his grandmother Annie Moores. She was from San Francisco, where, as a girl, she had routinely gone to the opera and returned home to play the scores from memory. After her husband became a miner and they began a life of moving from one remote mine to the next, her piano went with her.

It helped that he owned trucks. From the railroad at Flagstaff to the Colorado River he had trucked the steel of the Navajo Bridge. His name was Eldridge Moores, and he was primarily a small-scale pick-and-shovel all-around hard-rock miner whose body temperature became progressively higher in the presence of lead, zinc, copper, silver, and gold. In the middle nineteen-thirties, he was mining copper in the Verde River drainage when he decided to pick up the piano and haul it to Crown King. His son, Eldridge Moores, father of Eldridge Moores, worked for his father, Eldridge Moores. His son's wife, Geneva Moores, had a piano as well, and the two families in concert, in two Ford trucks, doremifasoled up the mountain.

Eldridge III—the future tectonicist, ophiolitologist, structural geologist, editor of *Geology*—was born in 1938. One of his earliest memories is of his father and grandfather saying that no one seriously engaged in mining would seek or follow the advice of a geologist. They said it often. Typically, Eldridge's father would say, "Huh—geologists. They think they can see through solid rock."

Five miles from Crown King and a thousand feet higher, Eldridge's grandfather tunnelled into solid rock. The family's Gladiator Mine was just below the ridgeline of the mountains—a ten-man operation that grossed about a million dollars in ten years. Gladiator was a gold mine with enough lead and zinc to be declared a strategic industry in the Second World War, so it was not shut down, as most gold mines were. The shafts and adits (tunnels) went into the mountain several hundred feet to the stopes—chambers with five-foot ceilings, angled with the gold vein at sixty degrees. They loosened the ore with pneumatic drills, and took out fifty thousand tons, getting half an ounce of gold per ton.

The main shaft was now covered with chicken wire and surrounded by rusting debris—a contusion in the mountain a century old. "You rely on the competence of the rock to keep the chambers open," Moores said. "You play it by ear. You develop a sense of what the good ore looks like. If it had a lot of flashing sulphides, it was ore." The sulphides were galena (lead) and sphalerite (zinc). "The lead and the zinc betrayed the gold. They were only two or three per cent of the rock, but they were the clue."

Soon after the war, Eldridge's grandfather moved on to something else, and his father, having found a place below the summit where the gold vein outcropped, started a new adit there. The mine was called War Eagle. Modestly, it would support his family, with three hundred ounces a year. He went at it first with a pick and shovel, then with a hand-driven bull prick, and finally with a pneumatic drill, as the rock, ever farther from weather, became fresher and harder. In ten minutes, he could shovel a ton of rock, enough to fill an ore cart. "It was backbreaking work," Moores said. "But my father had a tough back." When we looked into the small, cave-like mouth of War Eagle, Moores said, "There's the vein. That little fault zone—that's what it is. I was here when he started the mine. The rusty streak in the rock marks the vein he was following. These old miners had a very good sense of where things would be in rocks. They would look at an outcrop, see a streak of iron oxide, and say, 'Ah, yes, this must be the vein.' They ranked geologists with garbagemen and dogcatchers. Most of the geologists they met were starving third-rate consulting geologists who came into small-mining areas looking for money. The tectonic and chemical models weren't in place yet, so the understanding of ore deposits wasn't very good. The miners had an intuitive feel for where things would be that was probably better than what a geologist coming in cold could give them. My dad sure thought that geologists were a worthless bunch, basically—people who came into the country to write reports for mining companies telling them what they wanted to hear."

Helping at War Eagle, Moores as a boy shoved the hand-propelled ore carts from the working face to the ore bin, outside the mine. He emptied the carts on the grizzly, an inclined steel grid. The smaller stuff fell through, into the bin. (Ore-vein rock, generally weaker than the rock around it, tended to break into small pieces.) The larger chunks rolled onto a steel platform. By hand, he sorted them, choosing what he thought was good ore, and heaving rejects aside. From the ore bin, the rock went down chutes into trucks. After the war, his father had acquired an International ten-wheeler with six-wheel drive, capable of hauling twelve tons. When Moores was learning to drive, his version of the family sedan with the automatic shift in the supermarket parking lot on Sunday morning was an International ten-wheeler between Crown King and the summit. The mine's driveway, five miles long, had been engineered and was maintained by his father. Not the least of its features was a railless plunge on the outboard side. As we inched along it in the cruise-control pickup, Moores said, "This was my first driving experience, this road. In that International, I was one scared little teen-ager." As he practiced, he dragged a road grader behind the truck. His father was back there, working the grader.

Sometimes he rode with his father to Phoenix in the big truck, on a dirt road flanked by desert, first encountering pavement in country that is now city, seven miles north of the state capitol. The hot truck stank of transmission oil. When the truck quit, they revived it. Sometimes they would be stranded for hours by a flash flood.

The family lived for some years in the most imposing house in Crown King, which they rented for twenty-five dollars a month. It is squarish, board-and-batten, on a rocky platform behind a retaining wall. A refrigerator reposed on the front porch. Moores called that "an emblem of rural Arizona." When he was nine, his family bought another board-and-batten house, with a yard of dry needles and granite boulders. It was light blue now, with white trim and a tin roof, and suggested, in its setting, the quarters of a forest ranger. There were two bedrooms and three daughters, so his father brought a shack from the mine and set it on a poured slab, six feet by nine, and that was his son's freestanding bedroom. The main house was heated by unvented butane. The well went dry for several months each year. Eldridge's father put a tank of water in a dump truck, and raised the body. The family had its own water tower.

At the age of ten or eleven, Eldridge noticed with some interest that two of the large rocks close to the house were different in color, yet each was called granite. He was somewhat puzzled, but his hair did not stand on end. This was the one touch of geological curiosity that he felt throughout his youth in the mining camp. In his portable bedroom, there was nothing that even vaguely resembled a mineral collection. Where a budding herpetologist might have a closet full of snakes, a chemist a set of volatile powders, a cosmologist a wheel of stars, Eldridge had musical instruments. When his father opened War Eagle, Eldridge's interest was keen: "I wanted them to find the good stuff, because it put shoes on the feet. But I wasn't curious about the vein."

Any rock that was hard and dark was blue diorite to the miners; anything platy was schist; everything else was granite. One did not have to go to Caltech to learn this geology. Endlessly, his father and other miners talked about the provender of rock. They sat on their porches in front of the refrigerators and reminisced about mining camps, mining failures, and yields in ounces per ton of ore. Eldridge's mind was elsewhere. Even before he entered his teens, he dreamed of places far from the ridge. He would forever remember Carl Vanlaningham, a friend of his father, remarking one day, with a glance around town, "Optimism is highest at the beginning. A mining camp has nowhere to go but down." Eldridge as a child had sensed this in a general and pervasive way. One day when he was accompanying his parents from one switchback to the next on the interminable road to Crown King, he suddenly burst out, "I've had it! If I never do another thing, I'm going to go out of here and stay out of here." His parents looked sad. He was ten years old.

He finished eighth grade and enrolled at North Phoenix High School when he was twelve. His father had built a house on the outskirts of Phoenix to accommodate his children's education. Eldridge's mother stayed with them. After she became the teacher at the Crown King school, his sister Carolyn (two years older than he) was in charge of the household in Phoenix. Moores' developing opinion of developing Phoenix was a good deal lower than his opinion of Crown King. As he would explain in later years, "there's something wrong with a place that looks to Miami for its cultural leadership."

In high school, his principal interests were music and history. Teachers urged the cello on him, because his hands were large. He has nearly perfect pitch. "If you ask me to sing a note, I can get within a few cycles per second. If I'm listening to a piece on the radio, I can tell you what key it's being played in." In Crown King, while his father struck notes on the family piano Eldridge with his back turned could identify the notes. Eldridge has said of his father, "Music meant a lot to him, too. But, as he saw things, men didn't go into music. They had to go into something that was practical." To fulfill the high school's requirements, Eldridge also studied science and math. He was not inwardly driven toward a scientific or technological career, but he was already being nudged in that direction by what he has described as "a regional force." He explains, "In that part of the United States, it was assumed that any bright high-school student would go into science or engineering." His grade-point average was almost unimprovable. North Phoenix High School took particular pride in steering its best students toward the California Institute of Technology. Caltech offered him a larger scholarship than any other school to which he applied. Tractably, compliantly—sixteen years old—he went to Caltech.

Academically, he was at home there. For him, Caltech proved to be no more formidable than the one-room school in Crown King. Eventually, however, the day arrived when he had to choose a major—to decide, in effect, what he wanted to do with his life—and he experienced a sort of intellectual ambush. To his considerable surprise, he came to realize that there was only one discipline at Caltech that appealed to him strongly enough for such a commitment, and the discipline was geology. Under all his early indifferent attitudes, not to mention his avowals to escape from Crown King, there had obviously lain ambivalence. Evidently, what he thought he hated he did not altogether hate. He wondered still about those colors in granite. He may not have cared how the gold got out of the mountains, but he did want to know how the mountains came there to receive the gold. He remembers looking out through classroom windows, seeing the San Gabriel Mountains, and wishing he were up there. In the years ahead of him, he decided, he would like to combine history with science, and to travel the world out-of-doors. Those prerequisites could be combined, and restated as a single word. So he majored in geology. Today, if you look at him closely through your hand lens and ask him why he did that, he will give a little shrug and say, "I grew up in the mountains."

He will also say, "I had a hard time coming to grips with going into geology. While I was in graduate school, I still wondered if I could make a career in music."

At the mouth of the Gladiator Mine, we set up a bench with a discarded plank, resting it on discarded ores. Not many feet away were irises planted by his grandmother Annie Moores, whose house had once stood beside them. Just below the ridgeline, looking east, we ate sandwiches and spread on the ground before us the geologic map of Arizona. Six million acres of the original were also spread before us. We could see the Superstition Mountains, east of Phoenix. We could see the Four Peaks of Mazatzal Land. We could see a hundred miles. I remembered him once shaking his head with amazement at the joy experienced by a paleontologist who, in ten hours bent over in a blistering Wyoming gully, might find a couple of sharks' teeth. "I'm a ridge man, not a ravine man," Moores remarked at the time. "I like to get up and look out." Now in the foreground three thousand feet below us was the valley we had crossed in coming to Crown King. The axis of the valley, he said, running his finger along a black line on the map, was the Shylock Fault—"a major zone of tectonism that is reminiscent of major mélange zones that characterize consuming plates." Like the rock of the mine, the rock of the valley was Precambrian. He had found there—in sequence—serpentines, gabbros, and basalts. Pillow basalts. All this had suggested to him "plate activity in the Precambrian"—a collision, a docking, an addition to the continent arriving. He had published an abstract on the subject. Precambrian tectonics are, in their great antiquity, extremely difficult to read, but the rock of the valley suggested to him that he grew up in exotic terrane.

The granite of Crown King, he said, was technically quartz monzonite—a granite sibling, almost a twin. Where it contained a little iron, it would be pink. It had come into the earth molten, as a pluton, about seventeen hundred million years ago, and the country rock it had intruded was the rock here at the mine. (We had crossed the contact driving up from Crown King.) The rock at the mine was a rusty-looking, darkish, metavolcanic metasediment, two thousand million years old. Kicking at it, Moores said, "I would rank these rocks as not particularly easy to work with."

I thought he was referring to the picks, the shovels, the pneumatic drills—the backbreaking labors of his father and his grandfather. But he meant the geology.

"It's taken me a long time to get this stuff out of my system," he continued. "Metal deposits are telling you something interesting about tectonic systems." Thus, by the front door, his interests returned to the place where he developed. The livelihood of his family had depended on the yield of this rock. The relationship of an Arizona pluton to the rock in which it had ballooned as magma was now heating his imagination, as, before, it had heated his grandfather's and his father's.

They died in 1949 and 1979. "I don't have gold fever," Eldridge said. "When I get it, I stamp it out. I avoid the study of ore deposits, except as they are a scientific subject. Small mining is dirty, dangerous, boring, and dismal. After a while, watching people do it gets to you. The prospect of doing that made me want to get out, get away. I developed a hankering for places that were dust-free. I promised myself I was going to live in a place that was green and cool. Basically, this kind of mining is a futile operation. No one ever gets rich. It's got to be something in your blood."

I said, "Wouldn't you say it's in *your* blood?"

He said, "Sort of like an antibody."

From the Auburn suture of the Smartville Block, where you glimpse for the first time (westbound) the Great Central Valley of California, the immense flatland runs so far off the curve of the earth that its western horizon makes a simple line to the extremes of peripheral vision. In California's exceptional topography—with its crowd-gathering glacial excavations, its High Sierran hanging wall, its itinerant Salinian coast—nothing seems more singular to me than the Great Central Valley. It is far more planar than the plainest of the plains. With respect to its surroundings, it arrived first. At its edges are mountains that were set up around it like portable screens.

While looking out on the Great Central Valley one time, Moores described a winter day when he took off from Newfoundland in a snowstorm and flew to Toronto, landed there in strafing sleet, and flew on to Chicago (horizontal sleet), and then on across the high blizzarded plains and the Rockies (whited out) and the snowed-under Basin and Range to descend over the snowpacked Sierra, and bank toward a landing in a world that was apple green. The Great Central Valley was cool and moist and not cold and frozen, and for him, now, it was home.

It's not always cool, or—heaven knows—moist. Trunks of fruit and nut trees are painted white to keep them from being sunburned. Summer days commonly rise above a hundred, but the air will fall toward the fifties at night.

The ground surface is so nearly level that you have no sense of contour. A former lakebed can be much the same, where sediments laid in still water have become a valley floor. Such valleys tend to be intimate, however, while this one is fifty miles wide and four hundred miles long. It is not a former lake, although in large part it is a former swamp. Geology characteristically repeats itself around the world and down through time, but—with the possible exceptions of the Chilean Longitudinal Valley and the Dalbandin Trough in Pakistan—the Great Central Valley of California has no counterpart on this planet.

Engineers designing roads in the valley are frustrated by the lack of topographical relief. They have nothing to cut when they're in need of fill. If a new highway must cross over something, like a railroad track, the road builders go back half a mile or so and sink the highway into the earth in order to dig out enough dirt to build ramps to a bridge jumping the railroad. From the middle of this earthen sea, the flanking mountains are so low and distant that the slightest haze will give you the feeling that you are out of sight of land.

Over open ocean, the number of miles you can see before your line of sight goes off the curve of the earth is roughly equal to the square root of your eye level in feet. If your eyes are forty-nine feet up, you can see seven miles to sky. The formula is of very little use on land but is practical in this valley. When Darwin, off the Beagle, was travelling in Argentina, he sensed the subtle contours of the pampas:

> For many leagues north and south of San Nicolas and Rozario, the country is really level. Scarcely anything which travellers have written about its extreme flatness, can be considered as exaggeration. Yet I could never find a spot where, by slowly turning round, objects were not seen at greater distances in some directions than in others; and this manifestly proves inequality in the plain.

He could have done the same in the Great Central Valley. Much of it is near sea level, but it does rise. North from Interstate 80, the valley rises steadily, with a grade so imperceptible it is measured by laser. Moores compares this rise to the inclined sides of a mid-ocean ridge. With respect to the abyssal plains, the ridges rise about six thousand feet, but over so great a distance that, in his words, "if you were put down on a mid-ocean ridge flank and told to walk toward the ridge crest you would not know which way to go." The grade of a fast-spreading ocean ridge, like the East Pacific Rise, is about the same as the grade of the Great Central Valley between Sacramento and Redding.

You watch magpies over the valley, larcenous even in flight. You watch crop dusters—buzzing up and down, up and down, like trapped houseflies. Rice is sown from the air. Where DDT was once laid down in an aerosol, fish—as living insecticides—are now dropped from airplanes. They live in the checks, as paddies are called, and eat larvae. In this essential flatness, where there is no visible relief, instruments can find a minuscule astonishing topography; as if on a smooth board very slightly weathered, the unevenness of the land is discovered. Your eye can't discern it, but if you were a rice rancher you would be dealing with it every day. Rice seedlings need to stand in enough water to cover them but not enough to kill them, and this delicate margin has caused rice fields —from the air—to reveal the structure of the valley.

When the rice was first planted, vehicles bearing surveyors' rods drove in circles around fixed transits to discover the valley's contour lines, and along them berms were nudged up a few inches to contain

the shallow water. Sinuously paralleling the contours, the berms made the rice fields look, from the air, like supermagnified topographic maps. They showed tectonism expressible in centimetres. They showed the noses of anticlines, the troughs of synclines, microfolds, depressions—all too minimal to be detected by the human eye. Even very shallow water would run off these surfaces in every direction. So the rice plantations were terraced—each check, in altitude, scantly different from the next. Just as volcanologists use lasers to sense expansion in eruptive ground, rice ranchers have in recent years surveyed their rice fields with rotating beams. Contours are adjusted by the laser-controlled blades of earthmoving triplane levellers. The newer berms tend to be straight and less indicative of the geologic structures beneath them. There are five hundred thousand acres of rice in California. The climate is much the same as the climate in Egypt, which has the highest-yielding acreage of any rice country in the world. In the office of Jim Hill, who teaches rice at Davis, a sign says "Have a Rice Day."

The Great Central Valley is drained by two principal rivers, one flowing south and the other north. They meet in the valley and discharge themselves together into San Francisco Bay. The north-flowing river is the San Joaquin. The south-flowing river is the Sacramento, with its tributary the Feather River, which is dammed to reserve the snowmelt of the Sierra Nevada, not only to flood the rice fields and irrigate the other crops of the valley but also to travel six hundred miles in a life-support tube that is taped to the nose of Los Angeles. Rivers with common deltas are rare in the world. It would be difficult to name more pairs of them than the Kennebec and the Androscoggin, the Ganges and the Brahmaputra, the Tigris and the Euphrates, the Sacramento and the San Joaquin.

The floodplains of the Sacramento and the San Joaquin are dozens of miles wide. Before they were drained, and checkered like kitchen floors in shades of patterned green, they were known as the tulares or the tules or the tule swamps, in honor of the tough bulrushes that (pretty much alone) were able to survive the scouring inundations of spring. Sacramento stands only a little higher than the tules. To the conventional wisdom that one ought never to build on a floodplain, California has responded with its capital city. Old houses of Sacramento are in a sense upside down. Long exterior stairways lead to porches and entrance halls on upper floors. Whole neighborhoods are on stilts.

Emerging from the Sierra foothills among the blue oaks of the grass woodland, Interstate 80 rests on gravelly loams until it reaches the silty clays and humic gleys of Sacramento. Now I-80 has become a long elevated causeway that reaches across rice and sugar beets, marsh grass and milo, as if it were in search of Key West instead of San Francisco. In years of exceptional floods, fifty million acre-feet of water have come under the causeway. Some row crops and tree crops are on the deeper lighter soils near the river—the natural levees, where floodwaters give up the larger part of the material they carry. Eventually, imperceptibly, the ground beyond the floodplain goes up a few inches onto the outer lenses of an alluvial fan, a fine loam that has spilled eastward from the Coast Ranges like thin paint. The difference it makes is widely expressed in field crops, truck crops, orchards. In soil taxonomy, there are ten groups in the world. Nine are in this valley. Each is suited to a differing roster of crops. Plums, kiwis, apricots, oranges, olives, nectarines—it is the North American fruit forest. In some parts of the valley, roots are inhibited by a zone of hardpan. The soil's B-horizon, firmly cemented with silica, is like a concrete floor. Farmers used to bore holes in it and grow trees in the holes. Now they use diesel-powered earth rippers. Especially, this is the valley of beets and peaches, grapes and walnuts, almonds and cantaloupes, prunes and tomatoes. That is to say, the Great Central Valley of California grows more of each of those things than are grown in any other state of the union.

In 1905, the College of Agriculture of the University of California, in Berkeley, set up an experimental farm in Davis, Yolo County, in the valley's center. In 1925, the farm itself became an agricultural college. In 1959, it became a general campus in the state's university system. The livestock-judging pavilion is now a Shakespearean theatre. Under skyscraping water towers, the ground-hugging university is of such breadth and grandeur that it has its own beltway. It may have more bicycles than Shanghai. But Davis is still the main agricultural research center in California, and just outside the glassy postmodern geology building are sties containing massive monolithic pigs.

From the geology-building roof, Moores has looked across the immense flat ground and picked out features of Yosemite, a hundred miles to the southeast. Fifty miles southwest, he sees the detached oddness of Mt. Diablo, protruding beside the Coast Ranges. Looking the other way, he has seen Lassen—a white cone on the northern horizon, a hundred and forty miles distant—and, on nearer ground, the Sutter Buttes, a recent volcanic extrusion that has left a ring of jagged hills standing in the valley like a coronet set on a table. These landmarks embrace more than six million acres, or—in one look around—a landscape larger than Massachusetts. It is less than half of the valley.

When Moores arrived, in the nineteen-sixties, ten thousand people lived in Davis. Although the number has increased fivefold, Davis is still a quiet town, still a field station. From its shaded streets, crops and orchards reach out in all directions. To an open-air market in Davis on Saturdays farmers bring their lime thyme, their elephant-heart plums, their lemon cucumbers, bitter melons, and peaches the size of grapefruit. They bring tomatoes that are larger than the peaches, and tomatoes of every possible size down to tomatoes the size of pearls. Yolo County grows about as many tomatoes as Florida does. Yolo County grows ketchup in the form of "processing" tomatoes that could sit on a tee and be driven two hundred yards.

The Mooreses' turn-of-the-century farmhouse is "in the county," down a long thoroughfare of black walnuts at the edge of town. Their street is named Patwin, for the tribe that preceded the farms and the walnuts. The house faces north, across tomato fields. From the east windows you can sometimes see the low line of the Sierra, and from the west windows the Coast Ranges, but there is no sense of valley. The word seems misapplied. As the edges of a flat so vast, those montane curbs fail to suggest the V that a valley brings to mind.

When Moores looks out upon landscapes, he sees beneath them other landscapes. Like most geologists, he carries in his head a portfolio of ancient scenes, worlds overprinting previous worlds. He sees tundra in Ohio, dense forestation on New Mexican mesas, the Persian Gulf in the Painted Desert. Once, after a day in the Sierra, while he was sitting outside at home beneath what appeared to be a chandelier of apricots, I asked him to describe this one particle of the planet, his own backyard, at differing times. He responded in the present tense, as geologists often do, while his narrative went backward, scene by scene—episodically, stratigraphically—disassembling and dissolving California.

In the late-middle Pleistocene, when pulses of alpine ice are appearing on the Sierra, this place would be much the same, on the natural levee of a creek, surrounded not by fruit trees but by swamps. Valley oaks are on the dry ground, inches above the swamps. The mountains—east and west—are even lower. Mt. Diablo is not there. It has "scarcely begun to grow."

Three million years before the present—in the Piacenzian age of late Pliocene time—neither the Coast Ranges nor the Sierra is above the horizon. From the hinge under the Great Central Valley, the Sierra fault block begins to rise. The tectonic behavior of the Coast Ranges is different. Sluggishly, they come up from the deep. They have no integral structure. They are a fragmentary mass, a marine clutter. They will be known in geology as the Franciscan mélange. Appearing first as islands, the Franciscan pushes against the level sediments of the coastal plain and bends them upward until they are nearly vertical. Up through the mélange come volcanoes that spew lava and tuffaceous ash in and around the Napa Valley. There are active volcanoes on the crest of the Sierra. And between the nascent ranges the Sutter Buttes erupt.

I have a question. Why all this Fourth of July geology as recently as three million years ago, when all we have in these latitudes now are run-of-the-mill earthquakes?

Because in the Pliocene a triple junction of lithospheric plates is just off San Francisco, Moores replies. A subduction zone is dying out as its trench turns into the San Andreas Fault. The volcanism relates to that. For tens of millions of years, a lithospheric plate of considerable size lay between North America and the Pacific Plate. It is known in geology as the Farallon Plate. By the late Pliocene, this great segment of crust and mantle, possibly at one time a tenth of the shell of the earth, had in large part been consumed. Fragments of it remain: in the north, the Juan de Fuca Plate, whose subduction under North America has produced Lassen Peak, Mt. Shasta, Mt. Rainier, Mt. St. Helens, Glacier Peak, and the rest of the volcanoes of the Cascades; in the south, the Cocos Plate and the Nazca Plate, whose subduction has created Central America and elevated the Andes. For tens of millions of years, the Farallon Plate went under the western margin of North America, while North America gradually scraped off the Franciscan mélange of coast-range California. To the west, under the ocean, was the spreading center that divided the Farallon Plate from the Pacific Plate. As the Farallon Plate, moving eastward, was consumed, the spreading center came ever closer to California. At Los Angeles and Santa Barbara, the Pacific Plate first touched North America, twenty-nine million years before the present. Where it touched, the trench ceased to function, the spreading center ceased to function, and the plate boundary became a transform fault. It was only a few miles long at first, but steadily the great fault propagated from Los Angeles and Santa Barbara to the north and to the south, shutting the trench like a closing zipper. The triple junction of the Farallon Plate, the Pacific Plate, and the North American Plate migrated northward with the northern end of the fault. And so, in the Pliocene, three million years ago, the triple junction was off San Francisco. The volcanoes in the Sierra were the dying embers of Farallon subduction. The volcanoes in the Napa Valley and adjacent coastal ranges were a result of the new fault pulling the earth apart at kinks and bends. The eruption of the Sutter Buttes almost surely relates to the dying subduction or the new plate motions but is, as they say, not well understood. Now, in the Holocene, the triple junction is still moving north. For the moment, it is at Cape Mendocino, where the San Andreas ends and what is left of the Farallon Trench continues. That is how things appear, anyway, in present theory.

Six million years before the present, in the late Miocene, Moores and his apricot tree would be in or beside a saltwater bay that covers most of the Great Central Valley. It is full of tuna and other large fish, because an upwelling of cold water (like the upwelling in the Humboldt Current off modern Peru) has filled the bay with nutrients. There is no Golden Gate. The bay's outlet is at Monterey. A terrane is moving along the west side of the San Andreas Fault. Carrying with it the sites of San Diego, Los Angeles, Santa Barbara, San Luis Obispo, Big Sur, Monterey, and Salinas, it will someday be known as Salinia.

In the Eocene, fifty million years ago, Moores' backyard in Davis is mud at the bottom of the Farallon Ocean, some thirty miles offshore, on the continental shelf. As Eocene rivers pour into these waters—having advanced their gravels from Tibet-like altitudes and across the low country that will one day rise as the Sierra—they cut submarine canyons through the future Great Valley. The rock that preserves this story is a marine shale, loaded with shelf creatures of Eocene age. Below Davis, in the Great Valley Sequence of sediments, it lies about twenty-five hundred feet down.

In the Cretaceous, some eighty or ninety million years ago, Moores' address is a precariously inclined deep-sea fan—a spilling of sediment down the continental slope toward the trench where the Farallon Plate is disappearing. About fifty miles wide, the trench lies in the space that will one day separate San Francisco and Fairfield. As the slab of the Farallon Plate melts beneath North America, it contributes to the magmas of the great batholith and the superjacent volcanoes of the ancestral Sierra Nevada.

At the end of the Jurassic, about twenty million years after the docking of the Smartville Block, another island arc comes in and docks against Smartville, more or less directly under Moores and his tree. Geology will call it the Coast Range Ophiolite, and it will lie under forty thousand feet of Great Valley sediments and be warped into the coastal mountains. One of its large fragments will end up in the Oakland hills.

When Smartville docks, in the Jurassic, its individual islands possibly resemble Hokkaido, Kyushu, and Honshu. The trench closes east of Sacramento and a new one opens west of Davis and begins to consume the Farallon Plate. The downgoing slab of the Farallon Plate depresses the region and creates the structural basin that will fill up with sediments and become the Great Central Valley. Since there will be no Sierra Nevada and no Coast Ranges for nearly a hundred and fifty million years to follow, the result will be a valley that is not a conventional river valley but a structural basin filled to the brim with sediments that (almost wholly) do not derive from the mountains around it.

Before Smartville, blue ocean—extracontinental, abyssal ocean. In the earliest Triassic, the site of Davis is far out to sea. The continental shelf is back in Idaho and Nevada. North America in these latitudes has been growing. Two terranes have already come in. But here at the dawn of the Mesozoic the continent has not yet received so much as a hint of California.

The Napa Valley is thirty-five miles due west of Davis—an easy run for a field trip, a third of it flat and straight. The occasions have been several, not to mention spontaneous, when Moores and I have made westering traverses, collecting roadside samples of rock and wine.

After the level miles of field crops and fruit trees and almond groves, the ground suddenly and steeply rises in oak-woodland hills, so brown and dry for much of the year that geologists working among them can accidentally start fires with sparks from their hammers. Putah Creek, the stream that has spread its fine silts to Davis, is here a kind of door to the Coast Ranges, spilling forth their contents, coarsely bedded. Among the stream's cutbank gravels are layers of air-fall tuff that descended from the coast-range volcanoes of the Pliocene, and conglomerates that contain serpentine pebbles, peridotite pebbles, chert pebbles, graywacke pebbles, volcanic pebbles—the amassed detritus of several geologies, suggesting the commotion in the rock to come. Also present are fine-grained remnants of extremely fluid basalts that burst out in the northwest in middle Miocene time, covered areas the size of Iceland in a single day, and are thought to have been the beginnings of the geophysical hot spot that has since migrated to Yellowstone. The Columbia River flood basalts reached their southern extremity here.

As we go up the stream valley and arrive at the shore of Lake Berryessa, we pass through huge roadcuts of sedimentary rock whose bedding planes, originally horizontal, have been bent almost ninety degrees and are nearly vertical. Reaching for the sky in distinct unrumpled stripes, the rock ends in hogbacks, jagged ridges. Cretaceous in age, these are the bottom layers of the Great Valley

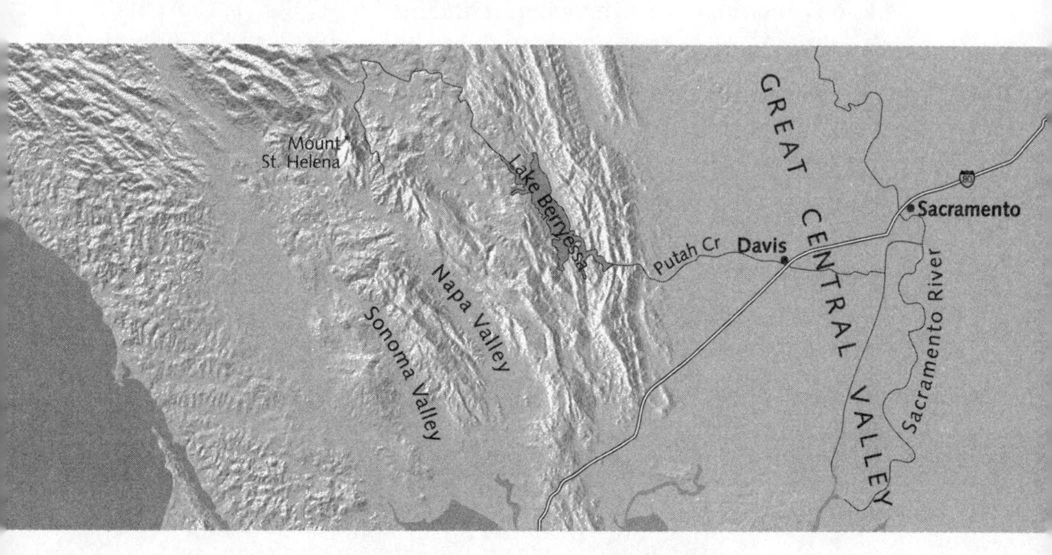

Sequence, bent high enough to resemble the bleaching ribs of a shipwreck. They are some of the strata that were folded against the Franciscan mélange when it rose (or was pushed) to the surface as the latest addition to the western end of the continent. In the heat and pressure of the Farallon Trench, the Franciscan sediments had been metamorphosed to varying extents, with the result that when they ultimately appeared on the surface they were miscellaneous and heterogeneous well beyond the brink of chaos. This lithic compote is the essence of the Coast Ranges. You leave the precise bedding planes and jagged ridgelines of the Great Valley Sequence and enter a country of precipitous nobs and rootless outcrops resting in scaly clay. In its lumpiness it resembles a glacial topography magnified many times. If the Great Valley Sequence can be compared to regimental stripes, the Franciscan is paisley.

"Look at this munged-up Franciscan glop!" Moores exclaims.

Narrow thoroughfares twist among the giddy hills. Ink Grade Road. Dollarhide Road.

"Look at that mélange! Holy moly, look at the lumps!"

Between the grinding lithospheric plates, the rock of this terrain was so pervasively sheared that a roadcut in metabasalt looks like green hamburger. We clearly see its contact with the scaly clay.

"That clay is the matrix of the Franciscan, in which blobs of various material are everywhere contained, and that is the guts of the Coast Range story. The metabasalt is a tectonic block in the matrix. You can see why people who tried to map stratigraphy went crazy. Imagine—before plate tectonics—the aching problems that this fruitcake, this raisins-in-a-pudding kind of stuff, produced. It doesn't fit the stratigraphic rules we all grew up on. It was assumed that you had a stratigraphic sequence here, and for years people tried unsuccessfully to explain these places in terms of eroded and deformed stratigraphies. In 1965, Ken Hsü proposed the mélange idea. But he suggested that the mélange had come here by gravity —that it had slid off the Sierra. No one had the idea of underthrusting—what we now see as the subduction of one plate beneath another, with all this miscellaneous material being scraped together and otherwise accumulating at the edge of the overriding plate. In 1969, Warren Hamilton, of the U.S.G.S., published a paper on the underflow of the Pacific crust beneath North America in Cretaceous and Cenozoic time. He presented the paper at the Penrose conference on the new global tectonics. Suddenly, people had a new view of the Franciscan. They said, 'Oh, that must be a berm resulting from subduction.' And the whole story broke open."

The Franciscan mélange contains rock of such widespread provenance that it is quite literally a collection from the entire Pacific basin, or even half of the surface of the planet. As fossils and paleomagnetism indicate, there are sediments from continents (sandstones and so forth) and rocks from scattered marine sources (cherts, graywackes, serpentines, gabbros, pillow lavas, and other volcanics) assembled at random in the matrix clay. Caught between the plates in the subduction, many of these things were taken down sixty-five thousand to a hundred thousand feet and spit back up as blue schist. This dense, heavy blue-gray rock, characteristic of subduction zones wherever found, is raspberried with garnets.

In a 1973 paper by Kenneth Jingwha Hsü appears a sentence describing the Franciscan mélange—this five-hundred-mile formation, the structural nature of which he was the first to recognize—in terms that could be applied to almost any extended family sitting down to a Thanksgiving dinner:

> These Mesozoic rocks are characterized by a general destruction of original junctions, whether igneous structures or sedimentary bedding, and by the shearing down of the more ductile material until it functions as a matrix in which fragments of the more brittle rocks float as isolated lenticles or boudins.

Hsü was born in China and began to use his umlaut as a tenured professor at the Swiss Federal Institute of Technology.

The mélange above Auburn, which collected against North America before the arrival of the Smartville Block, tells the same sort of story as the Franciscan, with the difference that the rock in the Sierra mélange has been almost wholly recrystallized, as a result of the collisions that completed California. Kodiak Island and the Shumagin Islands are accretionary wedges, too—shoved against Alaska by the north-bound Pacific Plate. The Oregon coast is an accretionary wedge (the Juan de Fuca Plate versus the North American Plate), complicated by a chain of seamounts that have come drifting in, making, among other things, Oregon's spectacular sea stacks. The outer islands of Indonesia are accretionary berms like the California Coast Ranges (the Indo-Australian Plate versus the Eurasian Plate), not to mention the Apennines of Italy, the north coast of the Gulf of Oman, and the Arakan ranges of Burma.

Now and again in the Coast Ranges you see ophiolite pillows on top of the mélange—a typical relationship, since the mélange forms at the edge of the overriding plate and the ophiolite is already on the overriding plate, having been previously emplaced there. Ocean-crustal detritus is widespread and prominent among the rocks of the Franciscan, but the Coast Range Ophiolite, in more concentrated form, is in the eastern part of the mountains, where it has been bent upward with the overlying Great Valley sediments, and pretty much shattered. Between Davis and Rutherford is a block of serpentine—disjunct, floating in the Franciscan—that underlies the bowl of a small mountain valley. The serpentine has weathered into soil, now planted to vines. These are some of the few grapes in California that are grown in the soil of the state rock. Moores is predisposed toward the wine. To him, its bouquet is ophiolitic, its aftertaste slow to part with serpentine's lingering mystery. To me, it tastes less of the deep ocean than of low tide. The stuff is fermented peridotite—a Mohorivičić red with the lustre of chromium.

The winery is in the deep shade of redwoods on a tertiary road. It makes only ten thousand gallons and has been in one family for a hundred years. The cave is in Franciscan sandstone. The kegs, tanks, and barrels are wood. Outside the cave, we stand on a wooden deck looking into a steep valley through the trunks of the big trees. Passing a glass under his nose, Moores remarks that the aroma is profound and reminds him of the wines of Cyprus. There is an intact ophiolite on the side of Mt. St. Helena, at the northwest end of the Napa Valley, he tells me—an almost complete sequence, capped with sediments but lacking pillows. There's a complete sequence on the east side of Mt. Diablo. "If you mapped the Coast Range Ophiolite, it would go from Oregon all the way down, in discontinuous blobs, plus the shards you see around San Francisco and elsewhere—rocks of the ophiolitic suite that just lie around as broken pieces, like the block that is under these grapes, and cannot be read in sequence." When Moores was first in California, he happened upon a report about mercury deposits at the north end of the Napa Valley. It mentioned "gabbro . . . along the contact between serpentine and volcanics." Moores got into his van, went to the Napa Valley, and looked. He then interested Steven Bezore, a graduate student, in working there. Bezore's master's thesis was the first demonstration of an ophiolitic complex in California, and led to the recognition of the Coast Range Ophiolite. After the winery, we stop at a crossroads store, Moores explaining that he requires coffee "to back-titrate the wine."

The descent is deep to the floor of the Napa Valley, which is flat. For a Coast Range valley, it is also spacious—as much as three miles wide. Vines cover it. Up the axis runs the two-lane St. Helena Highway, which seems to be lined with movie sets. This road is the vague but startling equivalent of the Route des Vins from Gevrey-Chambertin to Meursault through Beaune. The apparent stage sets are agricultural Disneylands: Beringer's Gothic half-timber Rhine House, Christian Brothers' Laotian Buddhist monastic château, Robert Mondavi's Spanish mission. Most offer tours, and wines to sip. As a day progresses, tongues thicken on the St. Helena Highway, where the traffic begins to weave in the late morning and is a war zone by midafternoon. The safest sippers are in stretch limos, which seem to outnumber Chevrolets.

Most valleys in the Coast Ranges are smaller and higher than this one, their typical altitude at least a thousand feet. The southern end of the Napa Valley, being close to the San Francisco bays, is essentially at sea level. The valley floor rises with distance from the water, but not much. St. Helena, in the north-central part of the Napa Valley, has an elevation of two hundred and fifty-five feet. It is surrounded by mountains that are comparable in height to the Green Mountains of Vermont or the White Mountains of New Hampshire. Why this deep hole in such a setting?

The San Andreas family of faults is spread through the Coast Ranges, and outlying members are beneath the Great Valley. Where a transform fault develops a releasing bend—which is not uncommon—the bend will pull apart as the two sides move, opening a sort of parallelogram, which, among soft mountains, will soon be vastly deeper than an ordinary water-sculpted valley. In the Coast Ranges, most depressions are high and erosional. Some are deep tectonic valleys that are known in geology as pull-apart basins. In the Napa region, Sonoma Valley, Ukiah Valley, Willits Valley, and Round Valley are also pull-apart basins. Lake Berryessa lies in a pull-apart basin, and so does Clear Lake.

Where pull-apart basins develop—stretching and thinning the local crust, drawing the mantle closer to the surface—volcanic eruptions cannot be far behind. In the Pliocene, after the Farallon Trench at this latitude ceased to operate and the San Andreas family appeared, the Napa basin had scarcely pulled itself apart before the fresh red rhyolite lavas came up and air-fall tuffs poured in. The Coast Ranges were aglow with sulphurous volcanism, its products hardening upon the Franciscan. The nutritive soils derived from these rocks prepared the geography of wine.

The rocks are known in geology as the Sonoma Volcanics. Napa and Sonoma are Patwin Indian names: "Napa" means house; "Sonoma" means nose or the Land of Chief Nose. The rocks are the Land of Chief Nose Volcanics. Chief Nose was a Tastevin before his time. The heat of the volcanics lingers in the mud baths and hot springs of Calistoga. The heat lingers under cleared woods near Mt. St. Helena, where small power stations dot the high ground like isolated geothermal farms.

As the new fault system wrenched the country, fissures opened, and hot groundwater burst out in the form of geysers and springs. They precipitated cryptocrystalline quartz and—in this matrix—various metals. Some gold. More silver. Near the surface, easiest to mine, were brilliant red crystals of cinnabar (mercuric sulphide). Mercury will effectively pluck up gold from crushed ores. In the nineteenth century, the Coast Ranges were tunnelled for mercury. It was carried across the Great Central Valley and used in the Sierra. The gold of the Coast Ranges was in those days insignificant but is more than significant now. In the nineteen-eighties, the Homestake Mining Company dug two open pits in ridges north of the Napa Valley. In surface area, they aggregate roughly a square mile. The gold is too fine to be seen through a microscope but is nonetheless there in sufficient concentration to be dissolved economically with cyanide. Discoveries of submicroscopic gold in California, Nevada, and elsewhere put the United States in a position to surpass South Africa in gold production by the turn of the twenty-first century—news that geologists regard as only slightly less astounding than the landings on the moon. Homestake's underground mine in the Black Hills of South Dakota is about a century old, and at latest count was eight thousand and fifty feet deep—the deepest mine in the Western Hemisphere. Homestake has produced more gold than any corporation in North America. With these new claims in the Coast Ranges, the company announced that it had more than doubled its reserves.

In 1880, Robert Louis Stevenson—aged thirty, newly married, consumptive—fled the "poisonous fog" of San Francisco and went into the mountains above Calistoga, where he and his American bride and her twelve-year-old son spent the summer squatting in an empty cabin at a closed-down mine called Silverado. From their high bench among rusting machinery and rubbled tailings, they looked down into the green rectangles of the Napa Valley.

The floor of the valley is extremely level to the roots of the hills; only here and there a hillock, crowned with pines, rises like the barrow of some chieftain famed in war.

Stevenson had more than a passing sense of the geology.

Here, indeed, all is new, nature as well as towns. The very hills of California have an unfinished look; the rains and streams have not yet carved them to their perfect shape.

Hot Springs and White Sulphur Springs are the names of two stations on the Napa Valley railroad; and Calistoga itself seems to repose on a mere film above a boiling, subterranean lake.

He began making notes for what became "The Silverado Squatters" and various settings for later work. He described the summit of Mt. St. Helena as "a cairn of quartz and cinnabar." He noted that Calistoga was a coined name. A Mormon promoter had been thinking of America's premier spa. Fortunately, his idea failed to travel, or there would be a Nevastoga, a Utastoga, a Wyostoga. Rattlesnakes resounded in the air like crickets. For a couple of months, Stevenson didn't know what he was hearing.

The rattle has a legendary credit; it is said to be awe-inspiring, and, once heard, to stamp itself forever in the memory. But the sound is not at all alarming; the hum of many insects, and the buzz of the wasp convince the ear of danger quite as readily. As a matter of fact, we lived for weeks in Silverado, coming and going, with rattles sprung on every side, and it never occurred to us to be afraid. I used to take sun-baths and do calisthenics in a certain pleasant nook among azalea and calcanthus, the rattles whizzing on every side like spinning-wheels, and the combined hiss or buzz rising louder and angrier at any sudden movement; but I was never in the least impressed, nor ever attacked. It was only towards the end of our stay that a man down at Calistoga, who was expatiating on the terrifying nature of the sound, gave me at last a very good imitation; and it burst on me at once that we dwelt in the very metropolis of deadly snakes, and that the rattle was simply the commonest noise in Silverado.

Without so much as a warning rattle, the owner of the Silverado Mine turned up one day, discovering and embarrassing the illegal squatter.

I somewhat quailed. I hastened to do him fealty, said I gathered he was the Squattee. . . .

Stevenson's summer was four years after the battle of the Little Bighorn. The West was that Old. Yet he counted fifty vineyards in the Napa Valley. Farmers had been in the valley for nearly half a century. In the eighteen-thirties, George Yount, of North Carolina, had been converted to Catholicism and had had himself baptized Jorge Concepción Yount in order to obtain a Mexican land grant of almost twelve thousand acres. An English surgeon to whom the Mexicans also gave a Napa Valley land grant named his place Rancho Carne Humana. In 1876, the Beringer winery was founded by Germans from Mainz. In approximate replication of their ancestral home, they built Rhine House in 1883. The stretch limos park there now, beside wide lawns under tall elms. Off the jump seats come people who go inside and lay down forty dollars for the magnum opus *Beringer: A Napa Valley Legend*. Leafing through the book, Moores picks up the information that the foundation and first story of Rhine House are limestone. He goes outside and squints at the house through his ten-power Hastings Triplet. "Jesus Christ!" he says. For Moores, this is new ground. He has never before seen limestone that came out of a volcano. "It's poorly welded volcanic ash with lots of big vesicles, pumice lapilli," he goes on. "It's friable volcanic ash! A welded tuff! An ignimbrite!"

Louis Martini's cement-block roadhouse, south of St. Helena on the way to Rutherford, is a low, clean-lined, postroad-modern building that lacks windows and has a long portico and a few wrought-iron lamps. Its architectural statement is upper-middle prime rib. Among the building's tiled rooms are showcases of Martini wines and a long dark bar. No one hurries anyone away, and in the cool quiet we sample half a dozen bottles, talking geology with our noses in the outcrop. Louis Martini's wines are straightforward, stalwart, allusive, volcanic. They are prepared to travel—like the terrane they derive from, and like the first Martini (who emigrated from Italy in 1894), and, according to Moores, like Italy itself, which departed from Europe in the Jurassic but later went home. Italy became a prong of Africa, he says, cupping his hand and orbiting a cabernet sauvignon. Italy left Europe, joined Africa, and later smashed back into Europe in the collision that made the Alps. The quarried Tuscan serpentines in the walls of the Duomo and the Giotto campanile are particles of the ophiolites that underscore this story.

Martini's pinot noir has the brawny overtones of an upland Rioja, the resilient spring of an athletic Médoc. Moores wonders if I have noticed that "the claret coast of France" and the Cantabrian coast of northern Spain seem to suggest an open bivalve, with Bordeaux at the hinge. In the early Cretaceous, when the Atlantic was young and narrow, there was no water between western France and northern Spain; the hinge was closed. The whole of Iberia got caught up in the spreading, and was perhaps yanked by Africa as Africa moved northeast. A rift opened, and widened, and became the Bay of Biscay. In a comparatively short time, the Iberian Peninsula swung ninety degrees and assumed its present position.

During the zinfandel, Moores summarizes the United Kingdom as "the remnants of a collision that occurred at the end of the Silurian." Mélanges resembling the Franciscan were caught in it, he says—for example, Caernarvonshire and Anglesey, in Wales. Collisional ranges appeared, later to be dismembered by the opening of the ocean. In France, the Massif Central is actually a continuation of the northern Appalachians. The southern Appalachians go up to New Jersey and then jump to North Africa as the Atlas Mountains and then to the Iberian Plateau and to the Pyrenees, which were later enhanced by compressions that developed as Spain swung around.

During the Napa Valley Reserve Petite Sirah, I mention the Brooks Range, where I have recently been.

The Brooks Range, Moores says, is a sliver of exotic continental material that came in from above Alaska, hit a subduction zone, and put ophiolite sequences along what is now the south slope. In the collision that followed, the exotic sliver was folded into mountains.

"When was that?"

"I forget. In the Jurassic, probably, or the early Cretaceous."

The Seward Peninsula—where Nome is, in west-central Alaska—is a piece of Jurassic blue schist surrounded by ophiolitic rock, but no one knows where the Seward Peninsula came from. For that matter, he adds, there is no certainty about where any of Alaska came from. It seems to consist entirely of exotic pieces that drifted to North America in Mesozoic time. South of the Denali Fault, which runs east-west and is close to Mt. McKinley, is the huge terrane that geologists call Wrangellia. It was an island arc, developed over an ocean plateau. Moores describes Mt. McKinley as "a bit of granite" that came up into Wrangellia after it arrived. Not long ago, Japan was attached to Asia. It drifted away. Japan is coming toward North America one centimetre a year. It may be a part of Alaska in eight hundred million years.

There is a shift change at Louis Martini's. One hostess replaces another. The new one says to her departing colleague, "Be careful out there. It's intense. They're driving all over the road."

That conversation in Louis Martini's winery occurred in 1978, when the theory of plate tectonics was ten years old and people who talked the way Moores was talking were widely considered daft. I may have thought it was the wine, but I was not in a position to know. Over the years, Moores and I have returned so often to the subject of world ophiolites and global tectonics—as they have recorded and described the changing face of the planet—that what follows is a sampling of all such dialogues, which I have compiled in the hope of reflecting, through his remarks, some of the geologic thinking of the nineteen-seventies, eighties, and nineties.

In order to move from place to place and let time float free, it would be well to bear in mind that the plate-tectonics narrative of the past fifteen hundred million years principally describes the assembling and disassembling of two supercontinents—Rodinia and Pangaea. Of the mountain ranges of Rodinia we have nothing today but evidential roots, attended by some ophiolites that speak of the collisions that built those Precambrian mountains. After Rodinia breaks up, about six hundred million years ago, its fragments result in a map of the world so different from the present one that it could be a map of a different planet; Kazakhstan, for example, is contiguous with Norway and New England. By two hundred and fifty million years before the present, the scattered continents and microcontinents have reassembled as Pangaea, whose sutures are today expressed in dwindled but palpable topographic relief (the Urals, the Appalachians). While Pangaea in turn disassembles, in the Mesozoic, not only is the Atlantic Ocean born but all over the world recognizable pieces of dispersing land move in the direction of the present map.

Wherever tectonically emplaced ophiolites happen to be, they lead to local geographic histories within the general story of the successive supercontinents. The presence of an ophiolite is a notation that while something is added to a continental margin an ocean basin of unknown size disappears. It could be Pacific-size. Moores is planning a book relating ophiolites to their origins. Chapter 1 might develop his analogy between the great complexity of islands north of modern Australia and the loose landmasses that once cluttered the Farallon Ocean off western North America and are now consolidated as California and other additions to the continent. North of modern Australia is a confused piece of the globe, made so by the encroaching motions of the Indo-Australian, Eurasian, and Pacific plates. They have broken the crust between them into microplates that look like the results of severe impact on a hard-boiled egg. The small pieces continue to be rigid, and remain in place, but the shell is shattered. The Philippine Plate, largest of the microplates, is surrounded by ocean trenches six and seven miles deep. On the east, Pacific crust is going under the Philippine Plate. On the west, Asian crust is going into the Manila Trench, where melting has produced the West Luzon arc and where Taiwan is docking with mainland China. That much is straightforward compared with the many smaller microcontinents and minor ocean basins that are also in the region, where a subduction zone is apparently in the process of flipping over, another is bending back upon itself, and another has curled around almost far enough to meet itself in a circle. This is a carnival of plate tectonics—of numerous island-arc-to-island-arc collisions and continent-to-island-arc collisions. As in the story of western North America, some arcs seem to be joining one another before attaching to a continent.

Ocean crust of the Indo-Australian Plate descending into the Java Trench has resulted in the arc from the Andaman Islands to the Banda Sea: Sumatra, Java, Bali, others. Where the Australian continental shelf has already jammed the trench and has picked up the Papuan ophiolites, it has buckled its own Australian sediments sixteen thousand feet into the air, making most of New Guinea. Moores thinks that the subduction zone will flip over now, and the Pacific Plate will begin to slide under Australia. In that event, he says, "Australia will keep going and will pick up the Philippines and every intervening island and then go after Japan on Japan's way east."

"You're saying that a north-dipping subduction zone will swing like a pendulum and become a south-dipping subduction zone? That is possible?"

"That seems to be what's happening. That's what you see in the seismicity. There's nothing magical or indelible about the present plate margins. Consuming margins, especially, can change their nature very readily."

In the Sierra Nevada near the Mother Lode, where the geology suggests to plate theorists that a pair of ocean trenches came together in the Jurassic, evidently there was no spreading center, and the trenches just ate up the crust between them, leaving undigested the accretionary phyllites, cherts, argillites, and limestones that lie uphill from Auburn. To many people, the idea of unspreading seafloor being consumed from two sides by converging trenches has seemed especially farfetched. In the late nineteen-seventies, however, a pair of active trenches doing exactly that was discovered in the Celebes Sea. They are moving toward each other. The intervening crust is disappearing. With depth finders and seismographs, geologists can see this happening, but they can't explain it.

Tracing a finger northward on a geologic map of the world, Moores follows ophiolites from Beijing to Siberia. There are several parallel strings of them, connecting two Precambrian continental blocks. The place-names on his map are written in Cyrillic characters, because it is a Russian map, but the rocks are readable, in the international colors and symbols of the science. "These sutures tell you that China used to be separated from Siberia by two or three oceans," Moores says. "They disappeared in the Paleozoic."

North of China, the Verkhoyanski Mountains make a sinuous track through Siberia to the Arctic Laptev Sea. Landmasses extend two thousand miles east of the Verkhoyanskis and four thousand west, yet the mountains contain rocks derived from a spreading center in a vanished ocean. The Verkhoyanski ophiolites are early Cretaceous in age—at least a hundred million years younger than the sutures in China, which were involved in Pangaea's assembling. The Verkhoyanski collision occurred after the supercontinent started to break up. As the Atlantic Ocean widened and the North American Plate moved west and the Eurasian Plate moved east, the two landmasses eventually touched each other, nearly halfway around the world, and made the Verkhoyanski Mountains. This was the plate boundary where Asia and North America actually came together. The Chukchi Sea and the Bering Sea, which separate Alaska and Siberia, are merely water lying on the North American continent.

Moores moves west to the Urals, which are flanked with ophiolites emplaced in Silurian time, in the middle Paleozoic, at the edge of the ocean that separated Asia from Europe. The ensuing collision did not begin until the Mississippian period, a hundred million years later. The fact that so much time passed between the emplacement of the ophiolites and the continent-to-continent collision means that a lot of seafloor was consumed, at least enough for an ocean a thousand miles wide. Russian geologists call this the Paleoasian Ocean. When the collision finally came, it completed Pangaea, two hundred and fifty million years ago.

Putting my hand on Spitsbergen and the rest of the Svalbard archipelago—ten degrees from the North Pole—I ask him, "What are they?"

With a sweeping move down the Atlantic he connects their story to Alabama. Among the various ocean basins that disappeared while Pangaea coalesced, the most intensively studied is the one that geologists have named for Iapetus, the father of Atlas: the ocean basin—or group of ocean basins—that lay between continental landmasses that are now substantial parts of Europe, Africa, and North America. Iapetus appears to have been larger than the modern

Atlantic. Five hundred million, four hundred million, and three hundred million years ago, as Iapetus gradually closed, the lands on either side in no way resembled the modern configurations of Europe and North America, but they were composed of rock we see in those places now. In the Iapetus Ocean, or oceans, were arcs and trenches, spreading centers, microplates, subduction zones, strike-slip faults, a mess of islands. Much of this seems to have resembled the Farallon Ocean off California in Mesozoic time, and the southwest Pacific today. The collisions that eradicated Iapetus and made a kind of headcheese of the intervening islands began more or less at Spitsbergen, and—roughly, sporadically—crunched their way south. In the terms of the Old Geology, this was the series of mountain-making episodes that were known as the Caledonian, Taconic, Acadian, and Alleghenian orogenies. The trail of these events was blazed with ophiolites. Ophiolitic emplacements in Newfoundland, Quebec, and Vermont, for example, signal the docking of an island arc—the event long known as the Taconic Orogeny. The consequential mountain building in New England and much of eastern Canada was thought—in the early days of plate tectonics—to be the result of a continent-to-continent collision. "The Taconic Orogeny is a collision of ophiolitic terrane with the North American continent, full stop," Moores says. "It is an oceanic terrane—and not yet Europe—colliding with North America." In the assembling of New England, at least two more arcs followed. These Paleozoic additions to the eastern seaboard are remarkably analogous to the assembling of California an era later.

When continents collided, Africa docked with, among other places, the Old South. About a hundred and fifty million years later, when Africa departed, it apparently left a large piece of Florida, which is now covered with what Moores calls "a lot of modern limestones that developed on top of the Appalachian suture, which can be traced seismologically under northern Florida and off into the continental shelf."

Hesitating, I say to him, "Florida is covered with marine sand on top of limestone on top of Paleozoic rocks. The Paleozoic rocks derive from Africa. That is what you are saying?"

"That's right. Southern Florida is a piece of Africa which was left behind when the Atlantic opened up."

"People in Florida say that southern Florida is Northern and northern Florida is Southern."

"Civilization reflects geology."

Southeastern Staten Island is a piece of Europe glued to an ophiolite from the northwest Iapetus floor. Nova Scotia is European, and so is southeastern Newfoundland. Boston is African. The north of Ireland is American. The northwest Highlands of Scotland are American. So is much of Norway.

While the Atlantic continues to open, the western Tethys, or Mediterranean Sea, is closing like a pair of tongs, hinged in the Atlantic crust roughly a thousand miles west of Casablanca. During the initial spreading of Tethys and later in the narrowing and even the lengthening of Tethys as effects of the forces that opened the Atlantic, the Mediterranean seafloor has been such a battleground that ophiolitic pieces of it are scattered around the basin like shell cases. Introducing chapters of the Mediterranean story, they are all through the Alps, Corsica, the Apennines, the Carpathians, the Dinarides, the Balkans, the Hellenides, Crete, the Cyclades, and the western part of Turkey.

The Mediterranean is full of tectonic rubble, no other single example being as large or as destructive as the Italian microcontinent, also known as the Adriatic Plate. Its western boundary is a subduction zone off the Campanian coast whose melt has become Vesuvius, and whose compressional distortions have become the Apennines. The boundaries of the Italian microcontinent run north into Switzerland, northeast down the Rhine to Liechtenstein, east to include the Austrian Alps and Vienna, then south through Zagreb and Sarajevo and past the Vourinos Ophiolite in Macedonia to the central Peloponnesus, and back to the boot of Italy. In the Jurassic, the Italian microcontinent made its attempt to become a permanent part of Africa. As Africa moved northeast with the opening of the Atlantic, it returned Italy to sender, picking up the Tethyan ocean crust that became the ophiolites of the Alps. The Alpine collision began in the Eocene, about fifty million years ago, and has not completely stopped.

Muscat, in Oman, sits at the base of peridotite cliffs and is the only capital city in the world hewn into rock of the earth's mantle. Almost all of northern Oman is ophiolite, lifted by the shelf of Arabia in the closing of Tethys.

Four large parts of Africa, dating from the Archean Eon, are more than three thousand million years old: the West African Craton, the Congo Craton, the Zimbabwe Craton, and the Kalahari Craton. Long defined in geology as continental basements, continental shields, or continental cores, cratons are the ancient fundament to which younger and more legible rocks adhere. Plate tectonics now suggests (not to everybody) that the older parts of continents were themselves assembled much as the younger parts have been. For example, the African cratons are separated by belts of deformation that occurred after the Archean Eon but still in deep Precambrian time. Perhaps the deformed rocks are suture belts where preexisting oceans disappeared. If plate tectonics was functioning then as it functions now, the crusts of those vanished African oceans consisted of rocks of the ophiolitic sequence. There are late Precambrian ophiolites in the Kalahari Desert of southwest Africa, in Sudan, in Egypt, in Arabia, and at Bou Azzer, in the western Sahara. Moores says that Sudan, Egypt, and other parts of Africa are full of exotic terranes—island arcs that were, as he puts it, "crunched in there in the late Precambrian." He continues, "Geologists have long seen Africa as having developed in place, but the story must be wrong. Little is known about the mobile tectonics there in Precambrian time. I think it's kind of an unknown frontier of the ophiolite story."

The mobile tectonics of more recent years are a good deal easier to see. If you look on a world map at Antarctica, South America, Africa, and Australia, you virtually see them exploding away from one another. You can reassemble Gondwana in your mind and then watch it come apart. In the Cretaceous, Africa and South America start to separate from Antarctica, and from each other. India is still part of southern Africa, but soon it breaks away. Australia remains attached to Antarctica until the Eocene, when it breaks away, forms its own plate, and heads north. India and Australia move separately for a time but then weld themselves together to become a single plate.

Madagascar begins to separate from Africa soon after India does, but India leaves Madagascar far behind. The Seychelles move away from Africa in the same manner—rifting obliquely, opening a small ocean basin with an active spreading center whose stairlike geometry (as in the Gulf of California) consists of short ridges connected by long transform faults. When the spreading stops, Madagascar and the Seychelles rejoin the African Plate.

I ask Moores why transform faults, like the San Andreas Fault, are so few and far between on land, whereas ocean floors are full of them.

He says, "They are rare on land because when they do appear in such a setting they rapidly take the land away and turn into marine transform faults. A transform fault carried India away from Africa. Look at the east coast of Madagascar. It's a long straight line, where India departed. Look at the corresponding part of India, the Malabar Coast below the Western Ghats. It's a long straight line. The Salinian Block, in California—with San Diego, Los Angeles, and so forth—will go on out to sea to the northwest, away from North America. The fault-divided halves of New Zealand's South Island, which were once apart, will come apart again. Equatorial Africa and northeast Brazil slid away from each other and developed an intervening sea."

I ask him why spreading centers and subduction zones take up most of the length of the world's plate boundaries and transform faults take up so little.

"Because the earth is a globe," he answers. "The curves of the spreading centers and the curves of the subduction zones will meet, or nearly meet. Where they don't meet, as in California, you find a transform fault."

When India separated from Africa, India's geographical center was more than a thousand miles farther south than the present position of the Cape of Good Hope. Three thousand miles of Tethys Ocean separated India from Tibet. India moved northeast as rapidly as any drifting continent in the calculable history of plate motions. At least one island arc lay in its path, and maybe several. There were microcontinents, too.

In and around the Himalaya are well-preserved ophiolitic sequences that describe the disappearance of that part of Tethys. These include ophiolites of Pakistan. Until the 1971 war, they were known in geology as the Hindubagh Complex. They are now called the Muslimbagh Complex. They run from the Indus Gorge east along the Indus River and the Tsangpo and Brahmaputra rivers for two thousand kilometres, acquiring other local names along the way. This continuous belt of ophiolites consists of ocean crust that formed at a spreading center late in the Cretaceous and was emplaced on the northern margin of India in Paleocene time, when India had completed about half of its journey north. India evidently reached a trench with an island arc behind it, choked the trench, and picked up the ocean crust from the peripheries of the arc. That, in any case, is how Moores tells the story. He likens the ophiolite to a cow on a cowcatcher in front of an old western train. When those ophiolites were emplaced and India swept up the island arc (about sixty million years before the present), Australia was still on its own plate. After the two plates joined, in the Eocene, the whole enterprise that geologists now refer to as the Indo-Australian Plate continued to move northward, gathering islands, for a few tens of millions of years. Then, with India as its hammerhead, it struck the Asian mainland. Moores thinks that the collision has scarcely begun.

This most emphatic of all contemporary continent-to-continent collisions is often described as a head-on crash, as if it had occurred within the ticks of human time. As India moved north, its highest rate of speed was a hundred and forty-two miles per million years. The present rate of compression is about a quarter of that, or two inches a year. If this could be recorded in stop-action photography, like the boiling swirls of cumulus clouds or the unfolding of a rose, it could indeed express itself kinetically. But two inches a year is an encounter so slow that a word like "collision" distorts its scale.

While India was closing with Tibet, it buckled the intervening shelf, raising from the sea a slab of rock more than a mile thick, a part of which is now the top of Mt. Everest. From the depths of lithification to the rock's present loft, it has been driven upward at least fifty thousand feet. In the tectonic history of the globe, we have no idea how many times something of this proportion has happened. The probability is that it has happened often.

The boundary between the Eurasian Plate and the plate that carries India and Australia seems pretty obvious, but actually it cannot be narrowly defined. It is not as simple and precise as the Indus suture, where ophiolites are embedded along the northern slopes of the higher mountains, and it cannot be limited to the Great Himalaya Range itself, though that appears to be a clear partition between the hitter and the hit. Across two thousand miles, from the Ganges River north to Lake Baikal, the boundary between the Indo-Australian Plate and the Eurasian Plate is indistinct. It was once described as a separate plate—the unfortunately named China Plate. The whole zone is seismically active. It contains the highest large plateau in the world. The Indian collision has produced additional mountain ranges north of the Himalaya that are comparable in altitude to the Andes. Included also is the Sinkiang Depression, where collisional downbending has put the ground below sea level. Essentially, all of China is a part of the plate boundary, and in all of China there are very few rocks that are undeformed. Chinese geologists travelling in America incessantly snap pictures of simple flat-lying sediments—a geological basic that they have seldom seen. The crust under the Tibetan Plateau is twice as thick as most continental crust. The Indo-Australian Plate, pressing northward, seems to have caused this. To accommodate two inches of relentless annual advance, various things have to bulge or give way. The mountains have risen. The plateau has thickened. But these two changes have not been enough to account for the total compression. A growing number of geologists, following work that is being done by a group of French tectonicists, are beginning to agree that a large part of Southeast Asia has also been forced to one side. Where Burma meets India, the high ranges bend almost at a right angle and go off to the southeast. This is the Burma Syntaxis (the term refers to a bend in a mountain chain), and near it are the beginnings of a whole series of great rivers—the Brahmaputra, the Mekong, the Irrawaddy, the Salween—initially in parallel valleys, veining Southeast Asia from the Bay of Bengal to the South China Sea. Controlling these valleys are long strike-slip faults, in motion like the San Andreas. The French tectonicists are proposing that Vietnam, Laos, Thailand, Cambodia, Burma—the whole of Indochina—slid southeastward among those strike-slip faults, like a great terrestrial hernia. On a relief map you can see India ramming Asia and squeezing all that country out to the southeast. As the mechanism has gained acceptance among tectonic theorists, it has become known as continental escape.

I ask Moores if he thinks there's a chance that plate tectonics may someday seem to have been a rational fiction, as the geosynclinal cycle does now.

"For parts of the world, maybe so," he says. "Whatever is going on in central Asia is no one's idea of plate tectonics. But as an explanation for eighty per cent of the surface processes of the earth, plate tectonics is in, firm." Repeating the words of the volcanologist Alex McBirney, he says, "Remember, 'In the next ten years, our confusion will reach new heights of sophistication.' "

(Or, in words dubiously attributed to Mark Twain: "Researchers have already cast much darkness on the subject, and if they continue their investigations we shall soon know nothing at all about it.")

The thought occurs to me, not for the first time, that I am following a science as it lurches forward from error to discovery and back to error. In my effort to describe some of the early discoveries of plate tectonics, I must also be preserving some of the early misconstructions.

"Inevitably," Moores agrees. "That is the nature of science, and geology is surely no exception." His mentor Harry Hess, a combat veteran with the rank of rear admiral in the Naval Reserve, once told him, "Geologists make better intelligence officers than physicists or chemists, because they are used to making decisions on faulty data."

The data that sketch departed geographies are actually numerous. Where pictures are clearest, the data cross-check with confirming frequency. For example, the ophiolitic narrative will conform with ancient latitudes preserved in the remanent magnetism of rock. The fossil record must not disagree. Where strike-slip faults have sliced a landscape and carried two sides apart, matchups can be traced in time and space. Sedimentary sequences, blue-schist belts, batholithic belts, thrust belts, and mélanges will orchestrally tell what happened. If they are not synchronous, it didn't happen.

That Asian plate boundary two thousand miles wide untidies the theory of plate tectonics more than any other place in the world with the probable exception of the American West. Moores is among the growing number of geologists who believe that Salt Lake City is on the eastern side of a muddled and dishevelled boundary between the North American and Pacific plates. Not long ago, when the boundary was utterly different—when an ocean trench off North America was consuming the Farallon Plate—the Rocky Mountains appeared, from Alaska to Mexico. In a major way, they defy expla-

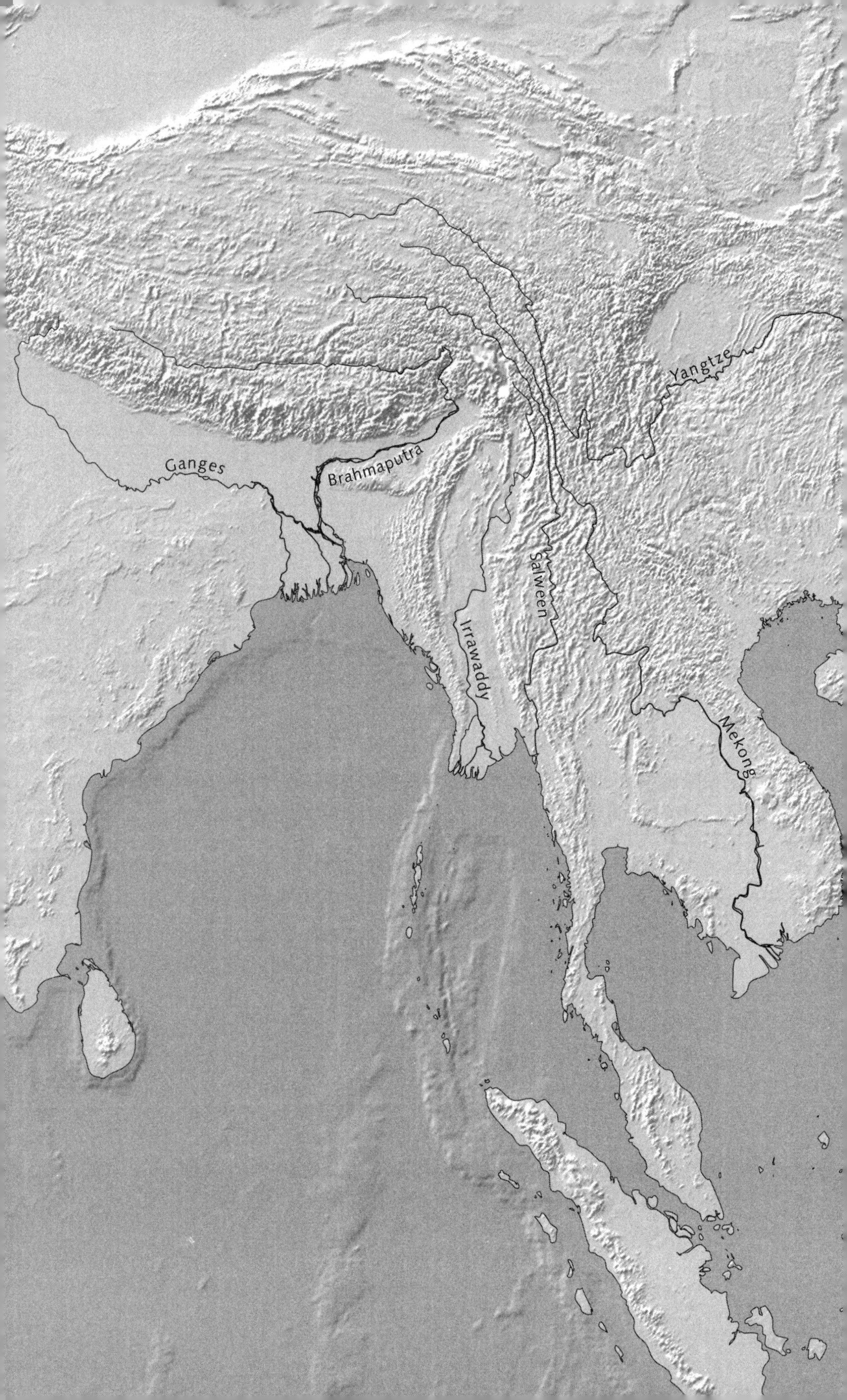
Ganges
Brahmaputra
Yangtze
Salween
Irrawaddy
Mekong

nation. As you look at the world map and see India hitting Asia, with the Himalaya and all the additional deformation to the north to show for it, you might begin to wonder why there is no India against the west coast of the United States. Obviously, the Smartville Block and the other accreted arcs added something to the mountains' compression, but the impact of those terranes could not have been sufficient to deform a third of a continent. Force from the west evidently made the mountains. But what force? If it was a colliding landmass, where has it gone?

If you run your eye up the coast a couple of thousand miles, you see protruding from North America a body of land at least as conspicuous as India. Staring at the map one day, I ask Moores if he is ready for some academic arm waving on a windmill scale.

"Shoot," he says.

"Why isn't Alaska the missing India of North America?"

"It is."

"Why didn't Alaska—not all at once but in parts and in successive collisions—strike North America at the California latitudes and then take off on transform faults and slide north to where it is now?"

"That's exactly what Alaska has done. When the Brooks Range swung into place, none of the rest of Alaska was there. Below the Brooks Range, all of Alaska consists of exotic terranes. They seem to have come from the Southern Hemisphere, and even from the western Pacific. Where each and every part originated we have no idea, but a lot of it collided down here in California and then went north on transform faults. Sonomia, Smartville, and so forth are probably fragments of terranes that in large part broke away and went north."

From radiometric dating, paleomagnetism, matching fossils, and connectable orogenies, George Gehrels, of the University of Arizona, and Jason Saleeby, of the California Institute of Technology, have proposed that the Alexander Terrane of southeastern Alaska, which includes Juneau and Sitka, drifted ten thousand miles from eastern Australia to Peru and then north to its present position. Vancouver Island seems to have followed; its paleomagnetism indicates that it came from the latitude of Bolivia and arrived in the Eocene.

Such travels seem modest in comparison with a vision that

Moores has of the world half a billion years earlier. In 1991, he published a paper claiming that Antarctica and western North America were once conjunct. This was during the existence of Rodinia. Tectonicists have for some time agreed that something must have rifted away from western North America in latest Precambrian time—that the craton cracked and separated, much as the Nubian-Arabian Craton is rifting now and making the Red Sea. In proposing Antarctica as the other side of the rift, Moores traces Precambrian lithologies of eastern Canada down through Alabama, Texas, and Arizona into Queen Maud Land, East Antarctica. Ian Dalziel, of the University of Texas, who had made a field trip with Moores to Antarctica, took Moores' proposal and extended its Precambrian juxtapositions, reconstructing the whole of Rodinia. According to Dalziel and Moores, if you had journeyed due north from Morocco you would have crossed western Africa and gone into Venezuela and through Brazil and Chile to West Virginia and on through Arizona into Antarctica, beyond which lay Australia. When they published their conclusions, *Time*, *Science News*, the *New York Times*, the *Los Angeles Times*, the *San Francisco Chronicle*, the *Washington Post*, and countless other publications adorned the story with maps beside which the cartographic efforts of the fifteenth century seem to be precision documents.

Running a finger down through Mexico and into Guatemala almost to Honduras, Moores says that an east-west fault zone there is laced with ophiolitic rock and seems to have been the southern extreme of Paleozoic North America. After Rodinia dispersed, what lay beyond Guatemala was open ocean. When Pangaea later coalesced, north and south touched there.

One of the places where the breaking up of Pangaea may have begun, in the Mesozoic, was not far away. If you reassemble the Triassic terranes around the Atlantic, the dikes radiate from a point in the Bahamas—suggesting that a hot spot that was centered there broke open a large part of Pangaea. As the supercontinent rifted, a large oceanic gap reopened between North and South America—in effect, a western extension of Tethys. It contained no Antilles, Lesser or Greater—none of the present island arcs or subduction zones of the Caribbean. It was blue, abyssal ocean. Its bottom was ocean crust-and-mantle—the ophiolitic sequence. Evidently, the Caribbean Plate, bringing arc sequences with it, came drifting in from far to the west to take up the position it holds today. It appears to have collided with the Bahama platform, and perhaps with the shelf of North America, in latest Cretaceous time. Ocean crust chipped off its edges as it fitted into place. All around the basin, island arcs came up above sea level, and fragments of ocean crust came up with them, as the ophiolites of northern Venezuela, of Cuba, of Hispaniola, of western Puerto Rico. The island of Margarita, near the Venezuelan coast, appears to be one large ophiolite. Off the Cayman Trench, in mid-Caribbean, the ocean crust is thicker than ocean crust commonly is. Thick basalts seem to have poured out over the original ophiolitic sequence in some sort of mid-plate volcanic event. "The only other place we know about where that sort of thing occurred in that time is out in the western Pacific, in the Nauru Basin," Moores continues. "Some people have suggested that these two things may be related. If you do a reconstruction of the East Pacific Rise and the plates in the Pacific, you can bring the Caribbean back into contact with the Nauru Basin in early Cretaceous time." If that is where the Caribbean Plate came from, the distance that it travelled eastward is eight thousand miles.

The gap between the Americas was subsequently filled in two stages. Not long after the arrival of the Caribbean Plate, Honduras and Nicaragua drifted in from who knows where. Ultimately, Panama came in from the Pacific, about seven million years ago. The collision calls to mind a cork going into a bottle, but was not so neat a fit. Panama was a part of an island arc, and the ophiolite that preceded it runs from Costa Rica to the Colombian Cordillera Occidental and on down to the Gulf of Guayaquil, indicating that the Choco Terrane, as the arc is called, was nearly a thousand miles long.

In Colombia's Cordillera Central, one range east of the Choco Terrane, are older ophiolites that report like a tree ring the continent's growth in successive laminations. Moores has been in Brazil among Precambrian ophiolites that seem to describe the basement of the continent collecting. In southern Patagonia, he has traced a line of ophiolites, called the Rocas Verdes Complex, from the fiftieth parallel to Tierra del Fuego. The Atlantic island of South Georgia, two thousand miles east of Tierra del Fuego, is part ophiolite and is believed to be a travelled piece of the Rocas Verdes Complex.

"Travelled?"

"Probably on a transform fault," Moores says.

In other words, South Georgia was once the easternmost part of Tierra del Fuego, and it took off.

Another part of Tierra del Fuego appears to have departed but returned. That, at any rate, is what Moores and others concluded in 1989 during a geologic voyage there. The Straits of Magellan lie across an ophiolite complex that they interpret as the back-arc basin behind a piece of South America that moved west before changing its mind. Perhaps South Georgia will come back, too.

Down the high Andes from Ecuador to fifty degrees south are no known ophiolites. This is a surprising interruption in a story that touches every other plate-boundary mountain range or former-plate-boundary mountain range in the world. After being tucked into the tectonics all the way from Alaska to Ecuador, ophiolites disappear. The gap in which they are missing is equal to the distance from the equator to Seattle. In ophiolite tectonics, perhaps the largest question at present is: What has happened in the central Andes?

In his office in Davis, with the Americas spread out on paper before him, Moores greets the question laconically. He says, "It makes you wonder."

I remark, "It's the one place on the western margin of the Americas where you don't find ophiolites providing some sort of story about exotic terranes—about the Smartville Block joining California, about Caribbean crust coming several thousand miles to its present position. Your intuition must be telling you something."

Moores says, "There are two possibilities: one, it is different from everywhere else that we know of; two, the evidence is hidden or is not preserved. My guess—this will make Andean geologists cringe—is that much of the Mesozoic volcanics one sees in the western Andes and the coastal ranges of Chile and Peru represents something that was at one time exotic to South America. My guess is that there's a suture. We just haven't found it. It may well be on the Chilean-Argentine frontier."

"So you're talking about an ophiolite announcing a terrane that came drifting in from wherever to fill the four thousand miles from the equator to fifty south, and you don't even know where the ophiolite is?"

"That's right."

"So you're . . ."

"Doing violence to the geology as it's known."

"The understanding of plate-tectonic history as it appears to be developing for the Pacific margin of the two Americas from Alaska to Tierra del Fuego includes lots of evidence for such terranes attaching themselves all the way, with the single exception of . . ."

"Zero to fifty south."

"And you think that . . ."

"Chile is exotic terrane."

"Chile and western Peru are from somewhere off in the South Pacific?"

"It's not a credible position to defend. But that's what I think."

"Where are the ophiolites on the east side of the Andes?"

"They haven't been found."

We go out of Davis one morning past a sign that says "OPEN TRENCH," and head for San Francisco on Interstate 80. In the median are dense effusions of oleander, in blossom pink and white, beguiling the westbound traffic with the pastel promise of California. A pickup in front of us is carrying an all-terrain vehicle studded with flashing sequins. As we move out to pass, the pickup abruptly moves out and blocks us. The pickup has California plates and a bumper sticker. "Don't Like My Driving? Dial 1-800-EAT-SHIT."

By the edge of the western hills, Mt. Diablo stands up like a hat on a table. All the way from Sacramento to the Coast Ranges you can see it from the highway. Moores calls it a piercement structure. It is a balloonlike mass of Franciscan mélange that has been squeezed up through valley sediments as if from a pastry sleeve. It is nearly four thousand feet high. Its topographical base is at sea level, beside the common delta of the great conjoining rivers, where a canal with no locks takes oceanic merchant ships across the Great Central Valley to Sacramento.

To climb Mt. Diablo, you go up through Pittsburg out of

Honker Bay. In the tranquillity of its oak woodland (the mountain is a park) you see the rhythmically bedded red cherts characteristic of the Franciscan. You see its featureless, unbedded sandstones. You see elements of the Coast Range Ophiolite. Moores calls the mountain "low-cost fruitcake, stirred up even more than most Franciscan." Interlayered volcanics and cherts are folded and refolded there like the leaves of a croissant. From the summit, you can see more than a hundred miles to the High Sierra or look down into a quarry cut in sheeted diabase from some ocean's spreading center who knows where.

The geology along the interstate west-southwest of Davis recapitulates the traverse to the Napa Valley, but in less obvious fashion, for the hills are lower; the delta and the bays are in a structural depression. In the swells of the oak woodland, the bent-upward sediments of the Great Valley are under the straw-brown grass. Where the grass is thin, you can see the bedding. Near Cordelia, Green Valley Creek comes in from the north, following the Green Valley Fault, which is actively slipping. Here in the hill country between Fairfield and Vallejo, the lovely hummocky topography is the result of creeping landslides, earthflow, solifluction. The solifluction scars are like stretch marks, waves advancing downhill. This is not the sort of place to site a house, but it is the sort of place where houses are sited. Here geology in motion is just another factor in daily life. When a classified ad in a local paper says "Owner Suddenly Called East Must Sell," the possibility is not inconceivable that the house is a sled.

On the top of Sulphur Springs Mountain, before Vallejo, the highway is in a benched throughcut a hundred feet high. On I-80, this is the beginning of the Franciscan. The Coast Range Ophiolite is here as well. Some of the rock is serpentine. We pull over, and walk the cut from one item to another in the mélange. After the serpentine comes a black wall that reflects light as if it were made of obsidian. Moores opens a pocketknife and easily carves the rock. He studies it in his hand lens. He says it is "the decrepitated matrix of the Franciscan"—the scaly clay in which are embedded all the continental shards and abyssal sediments, the bits of seamounts and ocean crust, the litter of half the world.

The interstate descends toward Vallejo, toward Benicia, toward

Suisun and San Pablo bays. When Robert Louis Stevenson first saw Vallejo, he described the community as "a blunder." Vallejo was twice, briefly, the capital of California. Benicia was Mrs. Vallejo. Between the two bays, we cross Carquinez Strait and immediately pass through the soft marine sediments of the largest roadcut on Interstate 80 between the Atlantic and Pacific oceans. Its vertical dimension is three hundred and six feet, but the sediment is so weakly cemented that the two sides lie open like butterfly wings and are thus immense. After the Farallon Trench quit, these marine deposits settled upon the Franciscan mélange. Throughout the Coast Ranges, the mélange has an icing of sediment, acquired while it was still underwater.

Running through Richmond over landfilled marshes beside San Pablo Bay, we cross the Wildcat Fault and, moments later, the close and parallel Hayward Fault. Off to our left, southeast, runs the long escarpment of the Berkeley Hills. Their steepness breaks at the Hayward Fault, below which is gently sloping ground. The obeliscal Berkeley Campanile, of the University of California, stands like a marker close by the fault, whose itinerary on the campus goes right through Memorial Stadium. The Hayward Fault is like one of the yard lines.

The road veers west, and we are suddenly high above water, on the upper deck of the San Francisco–Oakland Bay Bridge. Left and right of us are fifty miles of safe anchorage, waters that are in places twelve miles across. It strains credulity that in two centuries of nautical exploration this most prodigious harbor in North America did not reveal itself to a single ship. So far as is known, no ship passed through the Golden Gate until 1775—a modern date, even in California time.

As a geologic feature, the bay is the youngest thing in sight—younger than the rivers that feed it. During the glacial pulses of the Pleistocene, when so much water rested on the continents as ice and sea level was lower than it is now by a couple of hundred feet and shorelines worldwide were far outboard of present beaches, the Sacramento–San Joaquin flowed through the Golden Gate and forty or fifty miles west before it emptied into the sea. When the ice melted, the sea came up, and drowned innumerable river valleys—drowned the Susquehanna and made the Chesapeake, drowned the

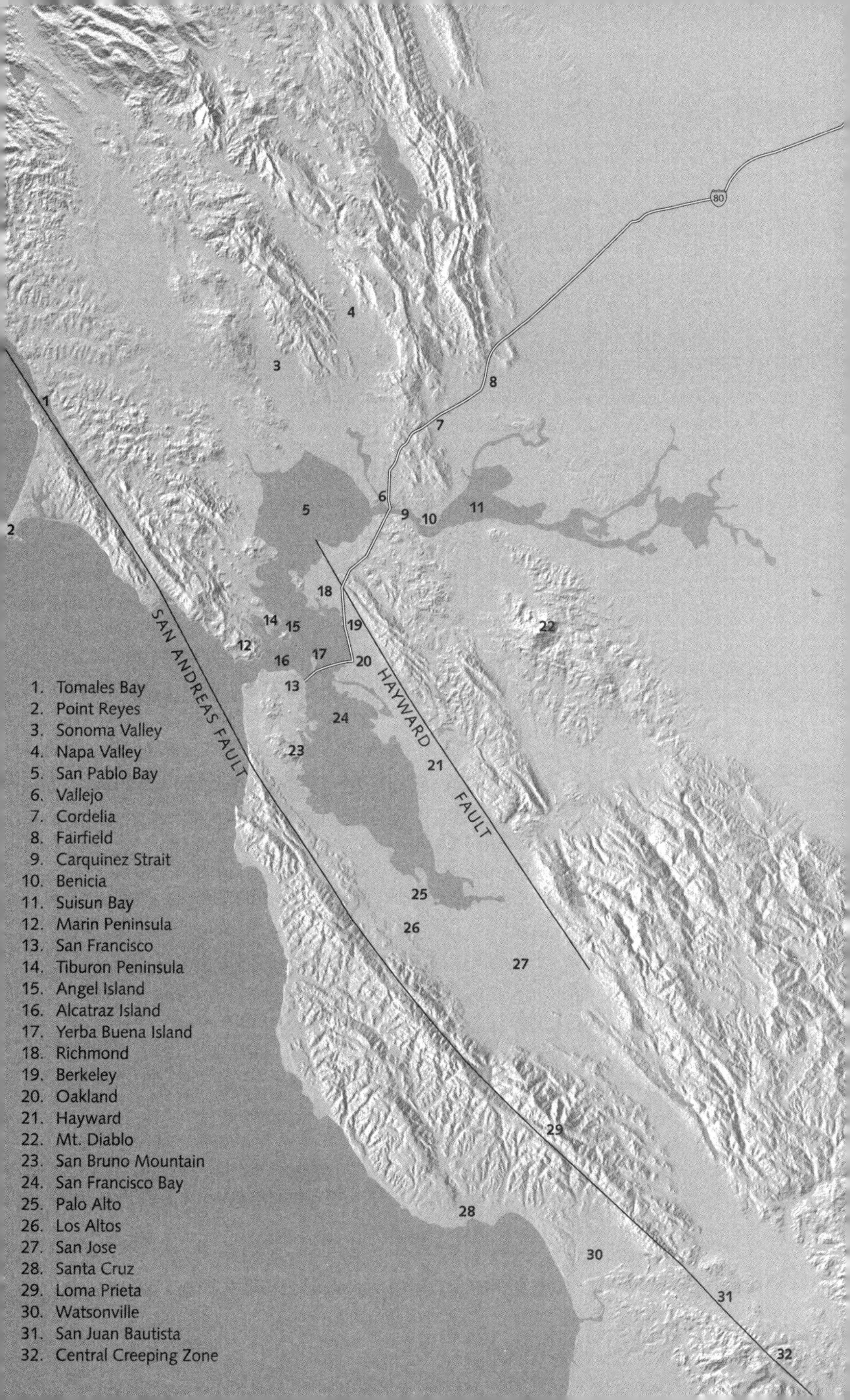
80
SAN ANDREAS FAULT
HAYWARD FAULT
1. Tomales Bay
2. Point Reyes
3. Sonoma Valley
4. Napa Valley
5. San Pablo Bay
6. Vallejo
7. Cordelia
8. Fairfield
9. Carquinez Strait
10. Benicia
11. Suisun Bay
12. Marin Peninsula
13. San Francisco
14. Tiburon Peninsula
15. Angel Island
16. Alcatraz Island
17. Yerba Buena Island
18. Richmond
19. Berkeley
20. Oakland
21. Hayward
22. Mt. Diablo
23. San Bruno Mountain
24. San Francisco Bay
25. Palo Alto
26. Los Altos
27. San Jose
28. Santa Cruz
29. Loma Prieta
30. Watsonville
31. San Juan Bautista
32. Central Creeping Zone

Delaware to Trenton, drowned the Hudson to Albany, drowned the Sacramento–San Joaquin from the Golden Gate through the Coast Ranges and into the Great Central Valley, filling the Bay Area's bays.

If Alcatraz, Angel Island, Yerba Buena, and so forth were elsewhere in the Coast Ranges, they would be the summits of mountains, and not islands in a bay. Something has depressed the bay region, incidentally allowing the rivers to get out of the Great Valley. In breadth and depth, the depression is unlike any other among the mountains of the California coast. Asked to explain, Moores can only speculate. Possibly the depression is just a system of erosional valleys, at the business end of the state's great watershed. Maybe it's a synclinal fold—a compressional trough, made as a side effect of San Andreas motion. Maybe it's a huge pull-apart basin. Maybe it's all three. Of his guesses, he likes best the pull-apart basin.

"What's under the bay?"

"Volcanics, conglomerates, glaucophane schist, sandstone, serpentine, chert—Franciscan. This is the place it was named for, and this is the tectonic product. The Franciscan is under the city of San Francisco and is the basement of the bay. Franciscan rocks in the Bay Area are from more parts of the subduction complex than you are likely to find in such concentration anywhere else on the coast."

The bridge arrives at Yerba Buena Island, and a tunnel runs through it: through sandstone of the Franciscan, outcropping above the entrance and derived from some continent somewhere, perhaps from our own—indurated American sediment that slid down the slope to the trench and there became incorporated into the Franciscan. In the light again and on the western span of the bridge, the run is short to the white city.

Not just any city is the topic of a serious tourist guidebook called *A Streetcar to Subduction*, a copy of which Moores and I have with us. Written by the geomorphologist Clyde Wahrhaftig, of Berkeley, it is a shining little book of geological field trips on public transportation.

Not just any city can claim to have formed in a trench where the slab of a great ocean dived toward the center of the earth, where large pieces of varicolored country came together, and where competent rock was crushed to scaly clay. After the churning stopped and the whole mixture was lifted into the weather, the more solid chunks very soon stood high and the softer stuff washed down. In Ina Coolbrith Park, we climb to the top of Russian Hill. Lining the path are large sharp pieces of red Franciscan chert, but Russian Hill is sandstone and the chert is decorative. It has come in trucks. Moores trains his binoculars on Alcatraz. He has read the geology of Alcatraz. He says it is a sequence of quartzofeldspathic sandstone and shale. In the compote of the Franciscan, Alcatraz is lying on its side. From Russian Hill he can see that—in the bedding planes. The rock of Alcatraz is continental in origin. There is no telling what continent, or when it escaped. The hill we stand on is also quartzo-feldspathic sandstone and is believed to be of the same provenance as Alcatraz. Behind us is the summit of Nob Hill. Russian Hill and Nob Hill are actually the mammary climax of the same small mountain. Some tectonicists look upon this sandstone as being so discretely integral that it deserves nomenclatural distinction. They see it as a picocontinent that drifted into the Franciscan. In their terminology, Nob Hill is part and parcel of the Alcatraz Terrane.

Because the mélange contains rocks from all over the Farallon rim, it lends itself to reclassification by those who prefer to describe it not as a tectonic unit but as a boneyard of exotica. To them the Bay Area is not so much the type locality of the Franciscan as a tessellation of six miniterranes. Over the water past Alcatraz we see Angel Island and the southern tip of the Tiburon Peninsula, which, with scattered bits along the Hayward Fault in Richmond, Berkeley, and Oakland, have been called the Yolla Bolly Terrane. The Marin Headlands Terrane includes not only southern Marin County but much of the city of San Francisco. There is a San Bruno Mountain Terrane, a Permanente Terrane, and, where doubt retains a foothold, an Unnamed Terrane.

We cross the Golden Gate, take the first exit, and curl downhill in overshadowing roadcuts of radiolarian chert. Radiolaria are creatures that live near the tops of warm oceans and look like microscopic sea urchins. After they die, their external skeletons go to the bottom to be radiolarian ooze, which lithifies as a very hard, very beautiful rock—a wine-red cryptocrystalline quartz of arrowhead quality. In fifty per cent of the world ocean, radiolarian chert lies like an enamel on the ophiolitic sequence. Where the sequence goes, so goes the chert.

Where you find chert from the open ocean, if you keep going downsection, basalt will soon follow, Moores remarks. Walking downsection, he finds the contact of red upon black. He says that the basalt was a seamount, one of the countless submarine volcanoes that ride the ocean floor.

In the early nineteen-thirties, the north pier of the Golden Gate Bridge was sunk two hundred feet through the red chert and into the basalt below it. At the north pier, the competence of the rock was never a problem. Support for the south pier, in San Francisco, was a good deal less promising. The south pier had to stand, in water, on sinuous, slippery serpentine, the state rock. Also under the ocean, two miles away, was the San Andreas Fault. The serpentine was thought to be potentially unstable, so it was hollowed out, like a rotten molar. The hollow was a little more than an acre, and ten stories deep. It was to be filled with concrete to anchor the bridge. While it still lay open and dry, within coffering walls thirty feet thick, the structural geologist Andrew Lawson, of Berkeley, was lowered in a bucket to inspect the surface of the bedrock. With his pure-white hair, his large frame, his tetragrammatonic mustache, Lawson personified Higher Authority. The stability of the serpentine had been called into question and made a public issue not only by a mining engineer but by the world-renowned structural geologist Bailey Willis, of Stanford, who predicted disaster. Lawson regarded Willis' assessment as "pure buncomb." Getting out of his bucket a hundred and seven feet below the strait, Lawson found that "the rock of the entire area is compact, strong serpentine remarkably free from seams of any kind." He wrote in his report, "When struck with a hammer, it rings like steel."

About the proximity of the great fault, Lawson had realistically observed (during the design phase) that an earthquake strong enough to knock down the bridge would also raze the city. He went on to say, "Though it faces possible destruction, San Francisco does not stop growing and that growth necessarily involves the erection of large and expensive structures."

In June, 1935, when the south tower stood nearly complete—seven hundred and forty-seven feet high, with no cables attached—it began to sway in a middle-energy earthquake. A construction worker named Frenchy Gales—as quoted in John van der Zee's *The Gate*—continues the story:

It was so limber the tower swayed sixteen feet each way. . . . There were twelve or thirteen guys on top, with no way to get down. The elevator wouldn't run. The whole thing would sway toward the ocean, guys would say, "Here we go!" Then it would sway back, toward the Bay. Guys were laying on the deck, throwing up and everything. I figured if we go in, the iron would hit the water.

The iron did not hit the water. Charles Ellis, the chief designer, had likened his developing bridge to a hammock strung between redwoods. Addressing the National Academy of Sciences in 1929, he said, "If I knew that there was to be an earthquake in San Francisco and . . . this bridge were built at that time, I would hie me to the center of it, and while watching the sun sink into China across the Pacific I would feel content with the thought that in case of an earthquake I had chosen the safest spot in which to be."

Moores and I recross the bridge. South of the south pier, we go down a steep trail to the Pacific beach that comes up from Point Lobos. There are serpentine outcrops above the beach—large blocky hunks in the scaly Franciscan matrix. The closer we come to the pier of the bridge, the more serpentine we see. "It could be part of the basement of the Marin Headlands Terrane," Moores says. "The seamount, with this serpentine as a basement, is a really far-travelled piece. It's old and equatorial. It began life way out in the central ocean. What a pile of hash-mash to put a bridge on!" The serpentine is massive, soft and soapy, threaded with asbestos, and below the great bridge it stands up in cliffs. In high positions are concrete gun emplacements built for five-inch pillar-mounted rifles, for six-inch disappearing rifles. There are nudes on the beach. Other men on the beach are sitting upright in little pillboxes that shelter them from the wind. The traffic on the bridge hammers the expansion joints and sounds like the firing of distant guns.

In section more than a mile wide, the serpentine traverses San Francisco from the Golden Gate Bridge to the old naval shipyard on the bay. It underlies all or parts of the Presidio, the University of San Francisco, the Civic Center, Mission Dolores, Haight-Ashbury, Pacific Heights, Hayes Valley, Bayshore. Potrero Hill is serpentine. On Candlestick Hill, near the naval shipyard, we find pillow lavas and red chert. Climbing Billy Goat Hill, above Thirtieth and Castro, we see more pillows under more red chert. Behind a grammar school in Visitacion Valley, at the south end of the city, we scramble up a dark hill of gabbro. Around a pond in McLaren Park, nearby, are diabase boulders.

Chert, pillow lavas, diabase, gabbro, serpentine—item for item, the city seems to be in part composed of the ophiolitic sequence. But here in the Franciscan it is not a sequence. "These rocks are not necessarily from one ophiolite," Moores comments. "They are probably from all over the globe—bits of ocean crust, of varying age and provenance, collected in the mélange."

It was Andrew Lawson who, in 1895, named the rocks Franciscan. He assumed that they were a conventional formation with traceable stratigraphy—with an eroded structure that could nonetheless be deciphered and spatially reconstructed. One might as well empty a cement mixer and try to number the pebbles in the order in which they entered the machine. On the other hand, Lawson came uncannily close to describing the ophiolitic sequence and to seeing it as ocean crust, and in this respect he was more than half a century ahead of his time. Of the San Francisco sea cliffs that are now described as pillow basalts, Lawson wrote that they resemble "an irregular pile of filled sacks, each having its rotundity deformed by contact with its neighbor." In 1914, enumerating the igneous rocks of the Franciscan, he came closest to describing the ocean-crustal sequence: "The igneous rocks are genetically allied peridotites, pyroxenites, and gabbros, the first named preponderating and being generally very thoroughly serpentinized; and the second spheroidal and variolitic basalts and diabases." He mentioned sandstone that had been deposited "upon the sinking bottom of a transgressing sea," and added that "cherts followed." Lawson's sinking bottom was the cooling ocean slab, on its way to the trench.

There are hillsides in downtown San Francisco that are too steep for cars, cable cars, or human locomotion short of rope-and-piton climbing. If you look at a street map, you will see a hiatus in Green Street near the Embarcadero. A few feet west of Sansome, Green ends at the edge of a vacant lot. No one can build there, because the vacant lot is vertical, a cliff of solid rock. "Grungy-looking sandstone," Moores remarks, glancing upward. "Can't see any bedding. It's marine sandstone that went into the trench." Houses line the top of the cliff. An apartment building with cantilevered balconies seems to hang over the edge. Somewhere up there, Green Street continues west.

Filbert Street, a couple of blocks away, is similarly interrupted. The escarpment is a little less sheer. A roadway is out of the question, but the street turns into a staircase. Houses are on both sides of the steps, some of them dating from the mid-nineteenth century. A few are hutlike, with stovepipe chimneys and sagging windows. It is reported that in April, 1906, when the temblor destroyed the municipal water system and the helpless city was being razed by fire, these houses were saved by sheets, blankets, tablecloths, and bedspreads soaking wet with wine. The pale painters and coughing poets who once lived here are gone now, replaced by readers of *Barron's Weekly*. Their cliff dwellings are charming past the threshold of envy, and they look ten miles over water.

We climb three hundred and eighty-seven risers, not to mention intercalated terraces and ramps. We are climbing, actually, Telegraph Hill. Moores refers to it as "a single large thick turbidite bed." In the rock behind the stairway, marine sandstones are interbedded with shales. The sands and muds probably derived from North America but slid out far enough to get into the trench.

At the summit of the hill is Coit Tower, and we go on up that as well—for a dollar, in an elevator. Coit Tower is two hundred and ten feet tall, and its observation terrace is five hundred feet above the sea. The view incorporates the city, the bays, the shoreline suburbs. We look down upon absurdly straight thoroughfares roller-coasting the precipitous hills. Moores says, "The hills have risen rapidly and have therefore eroded steeply. They're still rising rapidly. San Francisco streets were drawn on paper, without regard to geology or topography. There is one reaction. You laugh."

Skyscrapers ascend the apron of Nob Hill. South, from the tower, the view passes over them as if they were stalagmites. The eye is stopped by the San Bruno Mountain ridgeline, altitude thirteen hundred feet, fencing off the peninsular city. San Bruno Mountain, like Telegraph and Nob and Russian hills, is a large loose piece of marine sandstone within the Franciscan mélange. In the opposite direction, the view crosses the Golden Gate, passes over the Marin Headlands, and is again stopped, this time by Mt. Tamalpais, another block of float sandstone, twice as high as the San Bruno ridge. Francis Drake, the English pirate, two years shy of being knighted by the queen, spent the winter of 1579 camped beside a Pacific beach close to the base of Mt. Tamalpais. Neither he nor anyone in his crew went up the mountain to look around, to discover the three hundred and fifty square miles of protected water close to their encampment. The probable explanation is fog: the cold and almost quotidian sea fog that will overlap the coastal land when the air of California is otherwise cloudless; the fog that fosters the growth and survival of redwoods; the fog that conceals the Golden Gate Bridge and brings out the sounds of tubas.

In the summer and fall of 1769, sixty-four Spanish soldiers walked north four hundred miles from San Diego in search of Monterey Bay. Having no more to go on than a navigator's description a hundred and sixty-six years old, they failed to recognize Monterey Bay; and they kept on walking, another hundred miles, until they came up against a large coastal mountain. On a clear day, they climbed it. They were fourteen miles from the Golden Gate, but they could not see water there. All they could see was the ocean to their left and, ahead of them, an endless reach of mountains. Forty miles up the coast they could see Point Reyes. Several soldiers were sent ahead to blaze a trail to Point Reyes and prove that it was not Monterey. Their exact route is not clear. It is more than probable that they went far enough to be stopped by the deathly currents of a narrow strait set against unfriendly cliffs, and that—from one of the numerous hills—they became the first Europeans to look upon San Francisco Bay. Soldiers back at the main encampment, nearly starving, climbed San Bruno Mountain hunting deer, and saw the bay.

We, on the tower, have the city of San Francisco spread around us, whereas the soldiers, from similar positions, looked out upon a confused topography of deep swales, creased gullies, and high dunce-cap hills. The hills were all but treeless. They were matted with bush monkeyflowers, woolly painted cups, coffeeberries, Christmasberries, bush lupine, and poison oak. In the swales and gullies were wax myrtle, arroyo willows, coast live oaks, creek dogwood. Large parts of the future city were covered with marching dunes, restrained by dune tansy, coyote shrubs, and sand grass. The scouts probably shrugged. In any case, their mission had failed.

Seven years after the discovery of the bay, thirty-three families with the intent of settling beside it trekked eight hundred miles north from the part of Mexico that now is Arizona. Their leader, Juan Bautista de Anza, went ahead of them and examined the terrain. On top of the serpentine cliffs quite close to what is now the southern approach of the Golden Gate Bridge, he erected a cross. This was to be the site of the citadel (*presidio*). He erected another cross a few miles away, beside a small lake. This was to be the site of the mission. When the settlers arrived, they pitched their tents by the lake. Their seventh day of residence was the Fourth of July, 1776.

San Francisco is on the North American side of the San Andreas Fault, barely. The fault comes in from the ocean at Mussel Rock and goes straight down the San Francisco Peninsula, southeast. It runs close beside Skyline Boulevard above South San Francisco, San Bruno, Millbrae, Burlingame. The "skyline" is one of several Coast Range ridges that are separated from others by the depression of the bay. Five miles south of the city proper, Moores and I left Skyline Boulevard one December day and climbed through a subdivision to an elevation that was close to a thousand feet. We walked a hundred yards or so to a lookoff above a finger lake. It was three miles long and a tenth as wide, very straight, trending north forty degrees west. It lay in what resembled a small rift valley. Called San Andreas Lake, it lay in the trace of the fault. Here, actually, was where the fault was given its name—by Andrew Lawson, in 1895, who thought he was describing a local feature when in fact it extended more than seven hundred miles. Rock of the fault zone, frequently mashed, erodes easily, and the erosion leaves a groove in the terrain. The groove ran on as far as we could see, and included a second and much longer lake, called Crystal Springs Reservoir. Both lakes were man-made—a term, in this milieu, that women might be pleased to accept. In California, the San Andreas Fault is used as a place to store drinking water. In ditches and pipes the water travels a hundred and fifty miles from Hetch Hetchy Reservoir, in the Sierra Nevada, which lies in a valley adjacent to and equal to Yosemite. In 1913, environmentalists led by John Muir lost America's first great conservation battle when the valley of the Hetch Hetchy was dammed. Above San Andreas Lake, we could see over a flanking ridge and down through a deep pool of sky to San Francisco International Airport, at the edge of the bay. One after another—up through the pool slowly—747s were rising.

The San Andreas and Crystal Springs reservoirs were built before the turn of the century. On April 18, 1906, when the fault-zone surface in northern California was ruptured for nearly three hundred miles, the parallel sides of the reservoirs slid in different directions. The motion here was eight feet. The Pacific side moved north. The rupture went straight up both lakes, but the dams did not break. Nor have they since. They held in March, 1957, when a 5.5 temblor epicentered near Mussel Rock tore open a smaller reservoir. And they held in the Loma Prieta earthquake, of October, 1989.

"There's a seismic gap in here that did not get filled by that last event," Moores remarked. He meant that in 1989 in this stretch of the San Andreas Fault—well north of the epicenter—the two sides did not move.

The idea of the seismic gap first occurred to the seismologist Akitsune Imamura, in Tokyo, more or less at the time of the great San Francisco earthquake of April, 1906. As he studied Japanese earthquake records, which went back hundreds of years, Imamura arranged them graphically in zones of time and place. Where he found quiescent stretches—unfilled areas of his charts—he could see that they had been temporary, as pressure built to fill them. He could see that Tokyo—for what was then the time being—was in a large quiescent zone. In 1912, he began warning the public that the Tokyo gap was soon to be filled. He said that its size suggested to him a severe shock. Essentially, no one was interested. Imamura repeated his warnings for eleven years. The response remained as empty as the gap. In 1923, a hundred and forty thousand people died as Imamura's gap, in a couple of minutes, closed.

In 1906, when fewer than a million people lived near San Francisco Bay, an estimated three thousand died as a result of the earthquake. The population now exceeds six million, and the much publicized fact that the region is traceried with active faults—that the San Andreas system is not just one trace but a whole family of faults in a stepped and splintering band a great many miles wide—has done nothing to discourage the expanding populace from creating new urban shorelines and new urban skylines and so crowding the faults themselves that the faults' characteristic landforms are obscured beneath tens of thousands of buildings and homes. In addition to troughs and sag ponds, the motion of transform faults produces scarps, scarplets, saddles, notches, kerncols, kernbuts, and squeeze-up blocks. Streambeds are offset. Alluvium is ponded. Undrained depressions form, and parallel ridges and shutter ridges. Springs appear, and oases. Scarce has the geology made these features afresh when earthmovers move in to move them out, preparing the ground for structures even less permanent. One has to travel to remoter parts of the San Andreas Fault to see its full range of geomorphic features. In San Mateo County, the first county south of the city, nearly all such features have been obscured or destroyed by housing since 1945. Greater San Francisco, the most beautiful urban landscape in the United States, just will not be inconvenienced by a system of sibling faults. Less than a year after the major earthquake of 1989, a modest (fourteen-hundred-square-foot) two-bedroom house in the Marina, the most devastated residential district in San Francisco, could be had for five hundred and sixteen thousand dollars, a fall of barely ten per cent from pre-earthquake prices.

Visiting California after the 1906 earthquake, Harry Fielding Reid, of Johns Hopkins University, conceived the theory of elastic rebound, which is also known as the Reid mechanism. It describes the mechanics of fault motion. It preceded by sixty years the larger story of where such motions can lead. In a couple of hundred miles of the San Andreas fault trace, nature's hints to Harry Reid were not faint. He saw offset crop rows, tree lines, and fences. He found tunnels, highways, and bridges misaligned. Reid decided that elastic strain must have accumulated for years in the rock below until a moment came when the strain surpassed the strength of the rock, causing an abrupt slip, which released the stored energy.

Because the slip followed the direction—the strike—of the rupture, the San Andreas was a strike-slip fault. It was also known as a wrench fault. Not until the discovery of plate tectonics would it also be called a transform fault. The magnitude of the slip—the jump—diminishes with distance from the epicenter. In Marin County in 1906, a dirt road was severed where it crossed the Olema Valley at the head of Tomales Bay. It sprang apart twenty feet. That, or something near it, was the earthquake's maximum jump. The epicenter was underwater, not far away. The shaking lasted a full minute in 1906—four times as long as the shaking in the event of October, 1989, which released about one thirty-fifth as much energy.

Tomales Bay, long and narrow, resembles San Andreas Lake, and is also in a trough directly on the fault. Standing on its shore, you are impressed not only by its fjordlike dimensions but even more by the complete dissimilarity of the two sides. A tan cotton sock on one foot and a green wool sock on the other could not represent a greater mismatch. On the east side of Tomales Bay are bald unpopulated hills, straw brown in most seasons, and a scatter of lone oaks. Over the west shore of the bay—above the small riparian towns—is a dark-green vegetated ridge, a comparative jungle, which expresses the geology beneath it. The rock of the east shore is Franciscan mélange, and presents at its surface a typical Coast Range demeanor. The west side is, for the most part, granite. In age, the two formations are millions of years apart, but, more to the point, they are different in provenance as well. The granite on the west side of Tomales Bay, like the granite under the sea off Mussel Rock, broke away from the southern Sierra Nevada and has travelled north along the fault at least three hundred miles, an earthquake at a time.

Similar offsets line the great fault. A Cretaceous quartz monzonite on the east side of the fault in San Bernardino County mirrors a Cretaceous quartz monzonite on the west side of the fault near San Luis Obispo. An Eocene sandstone on the east side of the fault near San Luis Obispo appears to be the same as an Eocene sandstone on the west side of the fault in the Santa Cruz Mountains. The volcanics of Pinnacles National Monument, on the west side of the fault at the latitude of Monterey, evidently broke away from a chemically identical formation that remains on the North American side of the fault some two hundred and fifty miles south. (Wedged into a tight canyon at Pinnacles with his entire family one day,

Moores found himself intoning, "Fault, don't move now.") A Cretaceous gabbro on the east side of the fault in Santa Barbara County closely matches a Cretaceous gabbro on the west side of the fault in Mendocino County, three hundred and sixty miles away. Included in the gabbros in both places are bits of rare purple amygdaloidal andesite. On the peninsula south of San Francisco is a piece of a structural basin that seems to have broken away from the San Joaquin Basin, two hundred miles down the fault. In the nineteen-sixties, when all this motion was beginning to be understood as the steady movement of the Pacific and North American plates sliding past each other, a foundation hole was dug for a nuclear power plant exactly in the trough of the San Andreas Fault at Bodega Head, fifty miles up the trace from San Francisco. Half the hot fuel rods would have ended up in the Tropics and the other half in Alaska, but environmentalists halted the project. The Salinian Terrane, sliding past North America with San Diego aboard—and Big Sur and Salinas and Santa Cruz—will, in time, carry Los Angeles to San Francisco. Meanwhile, a part of northern Salinia is the Point Reyes Peninsula—the granite west of Tomales Bay. Looked at from the air, the Point Reyes Peninsula seems about as disjunct from the rest of California as Saudi Arabia is from Africa, and for the same reason: a boundary of lithospheric plates.

Along the San Andreas Fault, the average annual rate of slip is enough to transport something one nautical mile in sixty thousand years. Locally, the jump in 1906 represented roughly two hundred years. In some parts of the fault system, the motion assumes an almost steady creep, but, over all, most of the slip is staccato in time and occurs in elastic rebounds. The strain is essentially constant as the Pacific Plate tugs northwest. In response, earthquakes occur annually in the tens of thousands, most of them below the threshold of human sensitivity. Where the two sides of the fault are most tightly locked, the strain builds highest before it goes. The event of 1906 was what is now known as a large plate-rupturing earthquake. Vertically, the earth broke all the way down to the lower crust. Laterally, it opened the surface like a zipper—from the epicenter northward, where the fault trace for the most part lies just offshore and parallel to the coast, and southward, throwing up what appeared to be a plowed furrow through the rifted hills of Marin, tearing the seafloor where the fault passes west of the Golden Gate, opening the cliff at Mussel Rock, splitting the San Francisco Peninsula, and stopping near San Juan Bautista, east of Monterey Bay. Only once in the historical record has a jump on the San Andreas exceeded the jump of 1906. In 1857, near Tejon Pass outside Los Angeles, the two sides shifted thirty feet.

Kerry Sieh, a San Andreas specialist at Caltech, has dug trenches in numerous places across the fault zone near Los Angeles in order to examine the evidence in the exposed sediments. He has established that twelve great events have occurred on the south-central San Andreas Fault in the past two millennia, with intervals averaging a hundred and forty-five years. The Tejon event of January 9, 1857, is the most recent. One does not have to go to Caltech to add a hundred and forty-five to that.

In 1992, the United States Geological Survey completed a series of studies of the fault segment near San Francisco, and concluded that earthquakes on the order of magnitude of the 1906 event—it has been estimated at 8.3—probably recur every two hundred and fifty years. To the human eye, such a number appears in dim light. In a country where people get up in the dead of night to see what has happened on the Tokyo market, who is worried about two hundred and fifty years? When your complete range of concern begins with your grandparents and stops with your grandchildren, one of your safest bets is elastic rebound. In the prodigious roster of earthquakes on the San Andreas Fault, nearly all of them affect no one. The plates drift, the people with them. Fourteen times a year, an earthquake on the order of the 1989 event near Loma Prieta in the Santa Cruz Mountains occurs somewhere in the world. That might seem to thicken the risk. But not much. In California, only thirteen events have occurred at that level since 1769. So why not move in, spread out, build up, lay back, occupy this incomparable terrain?

It is said that if a cow lies down in California a seismologist will know it. In Iceland, which is seismically one of the most active countries in the world, there are fewer than thirty monitored seismographs. In California, there are seven hundred. Among countries of the world, only Japan has more seismographs than California. To track plate motions on both a large and a local scale, geologists also use very-long-baseline interferometry, in which the arrival times of noise from quasars at widely separated stations on earth are used to measure distances (even very long distances) with an error margin of less than a centimetre. (They can measure the actual distance that Africa and South America move apart in one calendar year.) The seismographs, the V.L.B.I.s, and other devices enabled Lynn Sykes and Stuart Nishenko, of Columbia University's Lamont-Doherty Earth Observatory, to predict in 1983 that "the segment of the San Andreas fault from opposite San Jose to San Juan Bautista, which ruptured less than 1.5 m in 1906 and which probably also broke in 1838, is calculated to have a moderate to high probability of an earthquake of magnitude 6¾ to 7¼ during the next 20 years." In 1989, after that particular stretch of the San Andreas Fault produced a magnitude 7.1 temblor, television interviewers rolled their eyes toward soundproof ceilings when geologists told them that predictions could not be more exact than a time frame of twenty years. This was too much for the anchor flukes, who think in airtime. But a ratio of one moment to twenty years actually represented an amazing juxtaposition of human time and geologic time. To predict a major earthquake within twenty years was like shooting out a candle flame at five thousand yards.

Sykes and Nishenko also noted that the segment of the San Andreas Fault between Tejon Pass and the Salton Sea was another place quite likely to experience a major event. In the lower part of that region, east of Los Angeles and San Diego, there has not been a great earthquake—8.0 and larger—in historic time. A loud announcement that such an event may be forthcoming was made by the Joshua Tree, Landers, and Big Bear earthquakes of 1992, which occurred fairly near that part of the San Andreas and effectively increased the stress upon it. The combined power of Landers and Big Bear, which came on the same day, caused oil to flow heavily from mountain seeps a hundred and fifty miles west, further endangering the endangered threespine stickleback (a fish). Astonishingly, the Landers earthquake and one that occurred only a few weeks earlier at Cape Mendocino rank among the twelve largest earthquakes that have happened in California in the twentieth century. With three earlier and lesser earthquakes, Joshua Tree and Landers form a straight line pointing north. They have broken open a new fault. Like a river seeking a straight path, the San Andreas seems to want to shift direction and go north through the Mojave Desert and up the east side of the Sierra Nevada (a probability mentioned to me in 1978 by Ken Deffeyes). The Landers earthquake, as if to emphasize the significance of the new fault vector, reached 7.3 and would have been extremely destructive had it happened in a setting other than a desert. Even before 1992, the accumulated strain on the nearby San Andreas was thought to be enough to open a plate-shattering rupture two hundred miles long. In 1981, the Federal Emergency Management Agency published a warning that a temblor of 8.3 or better was likely to occur on that part of the San Andreas before the end of the century. The agency said that property losses would amount to roughly twenty billion dollars, large numbers of people would be hospitalized, and as many as fourteen thousand would be dead. In 1982, the California Division of Mines and Geology chimed in with a special publication describing the same putative earthquake as "an event expected to take place during the lifetime of many of the current residents of southern California," and going on to say that "two of the three major aqueduct systems which cross the San Andreas fault will be ruptured and supplies will not be restored for a three- to six-month period."

In the same year, the California Division of Mines and Geology issued a companion publication called *Earthquake Planning Scenario for a Magnitude 8.3 Earthquake on the San Andreas Fault in the San Francisco Bay Area.* The scenario predicted that the Bay Bridge would withstand the shaking. So would the Golden Gate. Of a viaduct running through the Marina the scenario said it would "collapse." Of the Nimitz Freeway in Oakland the scenario said, "The hydraulic fills used to construct miles of freeway along the east shore of the Bay in Alameda County may liquefy during heavy shaking, with long sections becoming totally impassable. . . . The elevated section through downtown Oakland is expected to be extensively damaged." The earthquake of the prediction was thirty-five times as intense as the earthquake that actually came.

By the time of these scenarios, the rock offsets along the San Andreas had been explained, and the role of earthquakes at the plate boundary was understood. In 1906, the great earthquake was an unforeseeable Act of God. Now the question was no longer *whether* a great earthquake would happen but *when*. No longer could anyone imagine that when the strain is released it is gone forever. Yet people began referring to a chimeric temblor they called "the big one," as if some disaster of unique magnitude were waiting to happen. California has not assembled on creep. Great earthquakes are all over the geology. A big one will always be in the offing. The big one is plate tectonics.

At one time and another, for the most part with Moores, I have travelled the San Andreas Fault from the base of the Transverse Ranges outside Los Angeles to the rocky coast well north of San Francisco. In clear weather, a pilot with no radio and no instrumentation could easily fly those four hundred miles navigating only by the fault. The trace disappears here and again under wooded highlands, yet the San Andreas by and large is not only evident but also something to see—like the beaten track of a great migration, like a surgical scar on a belly. In the south, where State Route 14 climbs out of Palmdale on its way to Los Angeles, it cuts across the fault zone through two high roadcuts in which Pliocene sediments look like rolled-up magazines, representing not one tectonic event but a whole working series of them, exposed at the height of the action. On the geologic time scale, the zone's continual agitation has been frequent enough to be regarded as continuous, but in the here and now of human time the rift extends quietly northward through serene, appealing country: grasses rich in the fault trough, ridges intimate on the two sides—a world of tight corrals and trim post offices in towns that are named for sag ponds.

Farther north, it loses, for a while, its domestic charm. Almost all water disappears in a desert scene that, for California, is unusually placed. The Carrizo Plain, only forty miles into the Coast Ranges from the ocean at Santa Barbara, closely resembles a south Nevada basin. Between the Caliente Range and the Temblor Range, the San Andreas Fault runs up this flat, unvegetated, linear valley in full exposure of its benches and scarps, its elongate grabens and beheaded channels, its desiccated sag ponds and dry deflected streams. From the air, the fault trace is keloid, virtually organic in its insistence and its creep—north forty degrees west. On the ground, standing on desert pavement in a hot dry wind, you are literally entrenched in the plate boundary. You can see nearly four thousand years of motion in the bed of a single intermittent stream. The bouldery brook, bone dry, is fairly straight as it comes down the slopes of the Temblor Range, but the San Andreas has thrown up a shutter ridge—a sort of sliding wall—that blocks its path. The stream turns ninety degrees right and explores the plate boundary for four hundred and fifty feet before it discovers its offset bed, into which it turns west among cobbles and boulders of Salinian granite.

You pass dead soda ponds, other offset streams. The (gravel) road up the valley is for many miles directly on the fault. Now and again, there's a cattle grid, a herd of antelope, a house trailer, a hardscrabble ranch, a fence stuffed with tumbleweed, a pump in the yard. A daisy wheel turns on a tower. Down in the broken porous fault zone there will always be water, even here.

With more miles north come small adobes, far apart, each with a dish antenna. And with more miles a handsome spread, a green fringe, a prospering ranch with a solid house. The fault runs through the solid house. And why should it not? It runs through Greater San Francisco.

Of the two most direct routes from southern to northern California, always choose the San Andreas Fault. If you have adequate time, it beats the hell out of Interstate 5. Nearly always, some sort of road stays right in the fault zone. Like a water-level route through rough country, the fault is a place to find gentle grades and smooth ground. When the fault makes minor turns, they are nothing compared to the bends of a river. With more distance north, the desert plain yields to hay meadows and then to ever lusher country, until vines are standing in the fault-trace grabens and walnuts climb the creaselike hills. Ground squirrels appear, and then ever larger flocks of magpies, and then cottonwoods, and then oaks in thickening numbers, and velure pastures around horses with nothing to do. In age and rock type, the two sides of the fault are as different as two primary colors. Strewn up the west side are long-transport gabbroic hills and deracinated ranges of exotic granite. Just across the trough is Franciscan mélange—stranger, messier, more interesting to Moores.

Near Parkfield, you cross a bridge over the San Andreas where Cholame Creek runs on the fault. The bridge has been skewed—the east end toward Chihuahua, the west end toward Mt. McKinley. Between Cholame and Parkfield, plate-shattering ruptures have occurred six times since 1857, an average of one every twenty-two years, and the probability that another would occur before 2003 had been reckoned at ninety-eight per cent. Thirty-seven people live in Parkfield. If the population is ever to increase, seismologists will be the first to know it, for the valley here is wired like nowhere else. Parkfield has attracted earthquake-prediction experts because the brief interval time on this segment of the fault suggests that if they monitor this place they may learn something before they die. Also, the Parkfield segment has—in Moores' words—"relatively simple fault geometry." And the last three earthquakes have had a common epicenter and have been of equal magnitude.

An average of one plate-shattering earthquake every twenty-two years works out to forty-five thousand per million years. The last big Parkfield event was in 1966. It broke the surface for eighteen miles. Words on the town's water tower say "Parkfield, Earthquake Capital of the World, Be Here When It Happens." The actual year doesn't matter much. The instrumentation of Parkfield assumes that a shock is imminent. Its purpose is not to confirm the calculated averages but to develop a technology of sensing—within months, days, hours, or minutes—when a shock is coming. Even a minute's warning, or five minutes', or an hour's, let alone a day's, could (in highly populated places) save many lives and much money. Accordingly, the Cholame Valley around Parkfield—between Middle Mountain, to the north, and Gold Mountain, to the south—has been equipped with several million dollars' worth of strain gauges, creepmeters, earth thumpers, laser Geodimeters, tiltmeters, and a couple of dozen seismographs. It is said that the federal spending has converted the community from Parkfield to Porkfield. Some of the seismographs are in holes half a mile deep. Experience suggests that rocks creep a little before they leap. The creepmeters are sensitive to tens of millionths of an inch of creep.

If ever there was a conjectural science, it is earthquake prediction, and as research ramifies, the Tantalean goal recedes. The maximum stress on the San Andreas Fault—the direction of maximum push—turns out to be nearly perpendicular to the directions in which the fault sides move, like a banana peel's horizontal slip when pressure comes upon it from above. A fault that moves in such a manner must be weak enough to slide—must be, in a sense, lubricated. Among other things, the pressure of water in pores of rock in the walls of the fault has been mentioned as a lubricant, and so has the sudden release of gases that may result from shaking. Such mechanisms would tend to randomize earthquakes, diminishing the significance of mounting strains and temporal gaps. Those who practice earthquake prediction will watch almost anything that might contribute to the purpose. A geyser in the Napa Valley inventively named Old Faithful seems to erupt erratically both before and after large earthquakes that occur within a hundred and fifty miles—an observation that is based, however, on records kept for not much more than twenty years. In 1980, the United States Geological Survey began monitoring hydrogen in soils. Two years later, near Coalinga, about twenty miles northeast of Parkfield, the hydrogen in the soil was suddenly fifty times normal. It appeared in bursts, and such bursts became increasingly numerous in April, 1983. In May, 1983, a 6.5 earthquake occurred on a thrust fault under Coalinga. Releases of radon are watched. So are patterns and numbers of microquakes, especially those that are known as the Mogi doughnut. In the mid-nineteen-sixties, a Japanese seismologist noticed on his seismograms that microquakes occurring in the weeks before a major shock sometimes formed a ring around the place that became the epicenter. Mogi's doughnut is a wonderful clue, but—like hydrogen bursts and radon releases—before most major shocks it fails to appear.

People who live in earthquake country will speak of earthquake weather, which they characterize as very balmy, no winds. With prescient animals and fluctuating water wells, the study of earthquake weather is in a category of precursor that has not attracted funds from the National Science Foundation. Some people say that well water goes down in anticipation of a temblor. Some say it goes up. An ability to sense imminent temblors has been ascribed to snakes, turtles, rats, eels, catfish, weasels, birds, bears, and centipedes.

Possible clues in animal behavior are taken more seriously in China and Japan than they are in the United States, although a scientific paper was published in *California Geology* in 1988 evaluating a theory that "when an extraordinarily large number of dogs and cats are reported in the 'Lost and Found' section of the *San Jose Mercury News*, the probability of an earthquake striking the area increases significantly."

Earthquake prediction has taken long steps forward on the insights of plate tectonics but has also, on occasion, overstepped. Until instrumentation is reliably able to chart a developing temblor, predictors obviously have a moral responsibility to present their calculations shy of the specific. The mathematical equivalent of a forked stick will produce such absurdities as the large earthquake that did not occur as predicted in New Madrid, Missouri, on December 2, 1990. A U.S.G.S. geologist and a physicist in the United States Bureau of Mines whose research included (among other things) the study of rocks cracking in a lab predicted three great earthquakes for specific dates in the summer of 1981, to take place in the ocean floor near Lima. The largest—9.9—was to be twenty times as powerful as any earthquake ever recorded in the world. A few hundred thousand Peruvians were informed that they would die. Nothing happened.

As Louis Agassiz discovered, if you set stakes in a straight line across a valley glacier and come back a year later, you will see the curving manner in which the stakes have moved. If you drive fence posts in a straight line across the San Andreas Fault and come back a year later, almost certainly you will see a straight line of fence posts—unless your fence is in the hundred miles north of the Cholame Valley. There the line will be offset slightly, no more than an inch or two. Another year, and it will have moved a little more; and a year after that a little more; and so forth. In its seven hundred and forty miles of interplate abrasion, the San Andreas Fault is locally idiosyncratic, but nowhere more so than here in the Central Creeping Zone. Trees move, streams are bent, sag ponds sag. In road asphalt, echelon fractures develop. Slivers drop as minigrabens. Scarplets rise. The fault is very straight through the Central Creeping Zone. It consists, however, of short (two to six miles), stepped, parallel traces, like the marks made on ice by a skater. Landslides occur frequently in the Central Creeping Zone, obscuring the fresh signatures of the creep.

"The creep is relatively continuous for a hundred and seventy kilometres here and seems to account for nearly all of the movement," Moores remarked. "Creep is rare. Most fault movement is punctuated. The creep produces numerous small earthquakes. There are actual 'creep events,' wherein as much as five hundred metres of the fault zone will experience propagating creep in one hour." There were many oaks and few people living in the creep zone. The outcrops on the Pacific side of the fault sparkled with feldspar and mica—the granitic basement of the Gabilan Range. More than three thousand feet in elevation and close against the fault trace, the Gabilan Range creeps, too.

Jumping and creeping, the San Andreas Fault's average annual motion for a number of millions of years has been thirty-five millimetres. The figure lags significantly behind the motion of the Pacific Plate, whose travels, relative to North America, go a third again as fast. In the early days of plate tectonics, this incongruous difference was discovered after the annual motion of the Pacific Plate was elsewhere determined. The volcanic flows that crossed the San Andreas and were severed by the fault had not been carried apart at anything approaching the rate of Pacific motion. This became known as the San Andreas Discrepancy. If the Pacific Plate was moving so much faster than the great transform fault at its eastern edge, the rest of the motion had to be taken up somewhere. Movements along the many additional faults in the San Andreas family were not enough to account for it. Other motions in the boundary region were obviously making up the difference.

With the development of hot-spot theory (wherein places like Hawaii are seen as stationary and deeply derived volcanic penetrations of the moving plates) and of other refinements of data on vectors in the lithosphere, the history of the Pacific Plate became clearer. About three and a half million years ago, in the Pliocene epoch, the direction in which it was moving changed about eleven degrees to the east. Why this happened is the subject of much debate and many papers, but if you look at the Hawaiian Hot Spot stitching the story into the plate, you see, at least, that it did happen: there is an eleven-degree bend at Pliocene Oahu.

The Pacific Plate, among present plates the world's largest, underlies about two-thirds of the Pacific Ocean. North-south, it is about nine thousand miles long, and, east-west, it is about eight thousand miles wide. What could cause it to turn? Various events that occurred roughly three and a half million years ago along the Pacific Plate margins have been nominated as the cause. For example, the Ontong-Java Plateau, an immense basaltic mass in the southwest Pacific Ocean, collided with the Solomon Islands, reversing a subduction zone (it is claimed) and jamming a huge slab of the Pacific Plate under the North Fiji Plateau. The slab broke off. Suddenly released from the terrific drag on its southwest corner, the rest of the great northbound plate turned eleven degrees to the northeast. A number of coincidental collisions along the plate's western margin may have contributed to the change in vector. Additional impetus may have been provided by the subduction of a defunct spreading center at the north end of the plate. The extra weight of the spreading center, descending, may have tugged at the plate and given it clockwise torque. Whatever the cause, it's not easy to imagine a vehicle that weighs three hundred and forty-five quadrillion tons suddenly swerving to the right, but evidently that is what it did.

The tectonic effect on North America was something like the deformation that results when two automobiles sideswipe. Between the Pacific and North American plates, the basic motion along the San Andreas Fault remained strike-slip and parallel. But as the Pacific Plate sort of jammed its shoulder against most of California a component of compression was added. This resulted in thrust faults and accompanying folds—anticlines and synclines. (Petroleum migrated into the anticlines, rose into their domes, and was trapped.)

Earlier—about five million years before the present—the ocean spreading center known as the East Pacific Rise had propagated into North America at the Tropic of Cancer, splitting Baja off the rest of the continent, and initiating the opening of the Gulf of California. The splitting off of Baja was accompanied by very strong northward compression, which raised, among other things, the Transverse Ranges above Los Angeles, at the great bend of the San Andreas. That the Transverse Ranges were rising compressionally had been obvious to geologists long before plate tectonics identified the source of the compression. But not until the late nineteen-eighties did they come to see that compression as well as strike-slip motion accompanies the great fault throughout its length, as a result of the slight shift in the direction of the Pacific Plate three and a half million years ago. All these compressional aspects taken together—anticlines, synclines, and thrust faults in a wide swath from one end of California almost to the other—account for some of the missing motion in the San Andreas Discrepancy. The Los Angeles Basin alone has been squeezed about a centimetre a year for two million

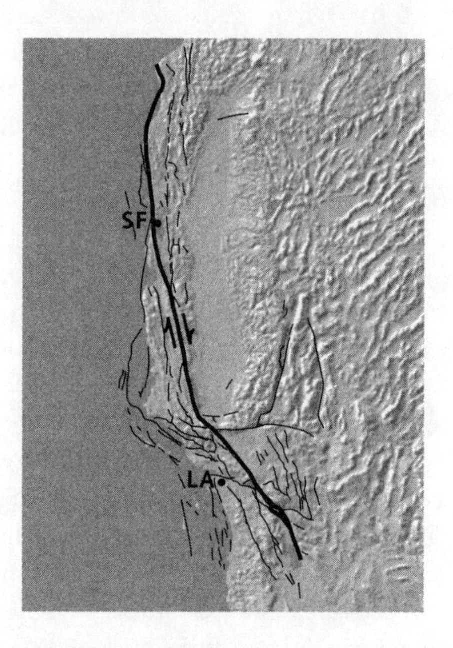

two hundred thousand years. The sites of Laguna Beach and Pasadena are fourteen miles closer together than they were 2.2 million years ago. This has happened an earthquake at a time. For example, both the Whittier Narrows earthquake of 1987 and the Northridge earthquake of 1994 lessened the breadth of the Santa Monica Mountains and raised the ridgeline.

The Whittier Narrows hypocenter was in a deeply buried fault in a young anticline. Such faults tend to develop about ten miles down and gradually move toward the surface. Northward for five hundred miles, young anticlines on the east side of the San Andreas

Fault are similar in nature—the products of deep successive earthquakes. Most are recently discovered, and many more, presumably, remain unknown. They make very acute angles with the fault, like the wake of a narrow boat. When a temblor goes off like a hidden grenade, geologists often have not suspected the existence of the fault that has moved. The 6.5 earthquake at Coalinga in 1983 was that kind of surprise. It increased the elevation of the ridge above it by more than two feet.

In 1892, a pair of enigmatic earthquakes shook Winters, which is near Davis, in the Great Central Valley. Evidently, the earthquakes occurred on the same sort of blind thrust that is under Coalinga, but the Winters thrust is of particular interest, because it is east of the Coast Ranges and fifty miles from the San Andreas. Yet it is apparently a product of the newly discovered folding and faulting that everywhere shadow the great fault. The Central Valley of California is about the last place in the world where virtually any geologist would look for an Appalachian-style fold-and-thrust belt. Without shame, Moores sketches one on a map of California; it goes up the west side of the valley almost all the way from the Tehachapi Range to Red Bluff and reaches eastward as far as Davis. He and his Davis colleague Jeff Unruh have been out looking for tectonic folds in the surreally flat country surrounding the university. This is a game of buff even beyond the heightened senses of the blind. They have found an anticline—an arch with limbs spread wide for many miles and a summit twenty-five feet high. They call it the Davis Anticline. It is a part of what Moores likes to describe as "the Davis campus fold-and-thrust belt." He is having fun, but the folds are not fictions. The anticline at Davis has developed in the past hundred thousand years. It is rising ten times as fast as the Alps.

On perhaps the weirdest geologic field trip I have ever been invited to observe, he and Unruh went out one day looking for nascent mountains in the calm-water flatness of the valley. There were extremely subtle differentiations. Moores said, "We are looking here on the surface for something that is happening five kilometres down—blind thrusts. Compressional stress extends to the center of the valley."

"Topography doesn't happen for nothing," Unruh said. "Soil scientists have long recognized that these valley rises are tectonic uplifts. Soils are darker in basinal areas. There's a fault-propagation fold in this part of the valley."

Moores later wrote to me:

We continue to gather evidence. We have seen two seismic profiles that show a horizontal reflection, presumably a fault, that extends all the way from the Coast Ranges to the Sacramento River. Jeff has been working at stream gradients. The rationale is that where there is a sharp change in gradient on a flood plain there is a reason, and the reason here is uplift. The analysis fits the two areas of acknowledged uplift west of Davis pretty well, and seems to indicate a new north-trending zone of uplift that goes right through Davis itself. Maybe there was a reason why the Patwin Indians selected this particular spot on the banks of Putah Creek for this village, after all. It was a high spot in a swamp, and it was high because it is coming up!

The compressive tectonism associated with the plate boundary contributes to the total relative plate motion, but not much: the overall average is less than a centimetre a year. And that does not nearly close the numerical gap. Surprisingly, the rest of the missing motion seems to come from the Basin and Range, the country between Reno and Salt Lake City, wherein the earth's crust has been stretching out and breaking into blocks, which float on the mantle as mountains—where the stretching has increased the width of the region by sixty miles in a few million years. Very-long-baseline interferometry has shown that the Basin and Range is spreading about ten millimetres a year in a direction west-northwest. This supplies enough of the total plate-boundary motion between the Gulf of California and Cape Mendocino to make up the difference in the San Andreas Discrepancy. If some Pacific Plate motion is coming from Utah, Utah is a part of the plate boundary.

The westernmost range of the Basin and Range Province is the Sierra Nevada, which has risen on a normal fault that runs along the eastern base of the mountains. The fault has experienced enough earthquakes to give the mountains their exceptional altitude. The most recent great earthquake there was in 1872. In a few seconds, the mountain range went up three feet. In the same few seconds, the Sierra Nevada also moved north-northwest twenty feet. That would help to fill in anybody's discrepancy.

Perhaps a sixth of the total motion between the plates is contributed by the other faults in the San Andreas family. Each is strike-slip, active, right-lateral—that is, viewed from one side of the fault, the other side appears to have gone to the right.

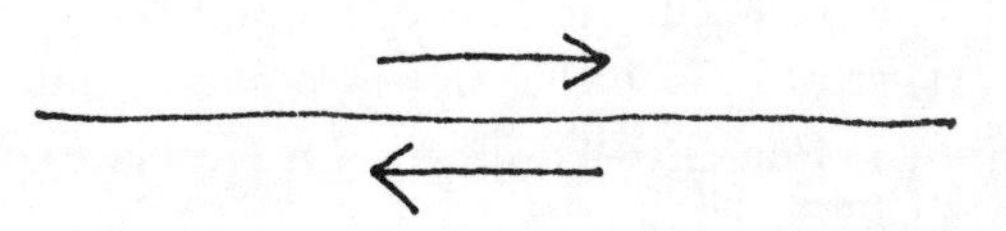

In a general way, you can demonstrate their relationship to one another with a deck of cards. Hold the deck, side up, between the palms of your hands, and slide the hands, pulling the right side toward you, pushing the left side away, and keeping pressure on the deck. The cards will respond by slipping, sticking, locking, sliding. Some may slide more than others. There may even develop a primary break. In any case, the fifty-one slips between the cards are, as in California, a family of right-lateral strike-slip faults. If one has moved more than the others, in effect you may have cut the cards, and you could call that cut the San Andreas Fault. But all the cards, to varying extents, have contributed slip to the total motion.

Moores believes that the plate-vector change three and a half million years ago is what probably created so large a grouping of boundary faults. Parallel and subparallel to the San Andreas, they have been likened to the tributaries of a river or the branches of a tree. But they are not dendritic. Often they do not conjoin. They are more like the checks that appear in dry timber. They all trend northwest. Many of them have varying local names, because the field geologists who did the naming, as much as a hundred years ago, did not suspect their continuity. The actively slipping Green Valley Fault, which intersects I-80 at Cordelia, in the eastern Coast Ranges, continues to the south as the Concord Fault. I-680, branching off, follows the fault and stays right on it. Down through the Napa Valley and under San Pablo Bay and through Berkeley and Oakland and Hayward and farther south than San Jose runs a continuous fault that is in segments named Healdsburg, Rodgers Creek, and Hayward. Portentous microquakes on the Rodgers Creek Fault have suggested to the United States Geological Survey the possibility of an earthquake equal in intensity to the 1989 wrench on the San Andreas near Loma Prieta.

The Calaveras Fault runs close to the Hayward Fault and extends somewhat farther south, like the Sargent Fault, the Wildcat Fault, the Busch Ranch Fault. The Antioch Fault is in the Great Central Valley. In seven hundred miles of splintery faults, the ones I have mentioned are all in the San Francisco Bay Area east of the San Andreas. West of it—on the San Francisco Peninsula or under the ocean—are the Pilarcitos Fault, the La Honda Fault, the Hosgri Fault, the San Gregorio Fault. The San Gregorio Fault extends from San Francisco to Big Sur, south of Monterey. Its longest historical jump is thirteen feet. It has produced great or major earthquakes on an average of once every three hundred years—nothing to be concerned about, unless it is your year. In the San Gregorio fault zone, San Mateo County has forty-acre zoning.

A comparable cross-sectional anatomy of the San Andreas system could be described for any latitude from Cape Mendocino to the Salton Sea, with a long list of names of contributive strike-slip faults. Within the system, the twentieth-century earthquake second in severity occurred in 1952 on the White Wolf Fault, near Bakersfield. In southern California, the belt is as much as a hundred and fifty miles across—three times as wide as it is at San Francisco, and a good deal more complex.

The Hayward Fault alone has contributed more than a hundred miles of offset. Running southeast from San Pablo Bay, north of San Francisco, to the latitude of Santa Cruz, it disappears near Gilroy, not far from the San Andreas Fault at San Juan Bautista. In many places, such as Berkeley, the Hayward Fault has Jurassic rock of the Franciscan mélange on one side of it and Cretaceous rock of the Great Valley Sequence sort of dredged up on the other side. A return specialist in the football stadium going eighty yards through a broken field will gain or lose about fifty million years. In a split, unpredictable second, he can be tossed out of bounds by a shift of the sod beneath him. The Hayward Fault runs not only through Memorial Stadium but also through or very near the Alameda County hospital, the San Leandro hospital, and California State University, Hayward. The Hayward Fault also ran through the California School for the Deaf and Blind, but the State became nervous, moved the school to another site, and then filled up its old dorms with Berkeley undergraduates.

The Hayward Fault separates the Cretaceous Berkeley Hills from the Jurassic university campus and the Holocene alluvium of the flat ground near San Francisco Bay. To no small extent, the Hayward Fault has created the Berkeley Hills, which are an obvious fault scarp, the result of a vertical component in an otherwise strike-slip motion. The change is abrupt from gentle slope to steep escarpment because the fault is so active and the hills are so young. Large earthquakes occurred on the Hayward Fault in 1836 and 1868—not the sort of information that is likely to plumb the tilt in a laid-back sophomore at Berkeley. However, a U.S.G.S. Miscellaneous Field Studies map of predicted maximum ground-shaking from large earthquakes on the San Andreas Fault and the Hayward Fault shows three areas of A-level intensity—characterized as "very violent"—for a Hayward earthquake: the environs of lower University Avenue in Berkeley; the blocks just east of Lake Merritt in downtown Oakland; and the Warren Freeway in Piedmont. The Warren Freeway uses the Hayward Fault in the way that water uses a riverbed.

The Geological Survey sees a sixty-seven-per-cent chance of another major San Francisco Bay Area earthquake on either the San Andreas or the Hayward Fault before the year 2020, with probabilities leaning toward the Hayward, because a jump there is so long overdue. If it should equal the intensity of several nineteenth-century shocks, the new one—according to estimates by the Federal Emergency Management Agency and the California Division of Mines and Geology—would result in as many as four thousand five hundred deaths, a hundred and thirty-five thousand injuries, and forty billion dollars' worth of damage. The Geological Survey adds a comment to these possibilities: the center of San Francisco "is as close to the Hayward Fault as it is to the San Andreas Fault."

San Lorenzo Creek, coming out of the hills into the outskirts of Hayward, bends sharply right when it hits the fault, flows northwest on the fault for more than a mile, and then takes a left and heads southwest to the bay. Hayward is about fifteen miles down the trace from Berkeley, and if ever there was a type locality to define the geological meaning of "type locality" Hayward is the place. Rarely is a fault zone so sharply drawn. Through part of the town runs a steep emphatic bench, resembling a sloped medieval wall, that is the product of the fault beside it. On D Street between Mission and Main, the curb, running east, makes a right-lateral bend to the south, and then continues east. When Moores and I were on D Street not long ago, a pressure ridge had buckled the sidewalk. The fault went through the service department of Boulevard Buick. There were fresh patches in the sidewalk outside. Between C and D, a building that had long housed the municipal government—and had been designed and constructed as the Hayward City Hall—stood precisely on the fault, which was tearing the building apart as if it were a tuft of cotton. Tiles and plaster had showered the bureaucrats until they fled. The bureaucracy included a department established to deal with emergencies confronting the city. At 934 C Street was a store that declared itself to be the "Hayward Sewing Center, Alterations." The alterations included a canted curb, a pressure ridge in the sidewalk, and a wall in the act of bulging toward Mexico City. There was a bend in the long wall of the Action Signs Building, at 22534 Mission. The fault had offset the Spoiled Brat Parking Lot. In a municipal parking lot nearby, the lines of meters took right-lateral bends and then resumed their original direction. Robert's School of Karate—in motion relative to the antique shop next door—had recently moved south an inch. A sign in the window said "KARATE KENPO GUNG-FU BOXING—The Only School in Hayward Teaching Street Fighting." Sidewalks were patched on every east-west street. The occupied house at 923 Hotel Avenue had been so torqued by the fault that one of its walls was concave. Looming over it—fifty feet high—was the Hayward escarpment.

In the country south of Hayward, we saw numerous streams that came down through the hills, made a right turn at the fault line, and then, after a bit, rediscovered their offset beds. We saw a deep ravine bent like a crochet hook. Like the San Andreas, the Hayward Fault seems to be stuck hard in some places (Berkeley) and in others (mainly in the south) to be creeping. A culvert curves like macaroni. A water tunnel begins to leak. Railroad tracks move. In Fremont, Moores and I climbed over some walls and fences in order to get up onto the ballast and squint down the rails of the Union Pacific, which, where it crossed the Hayward Fault, was bent into an echelon of kinks.

In 1986, a small earthquake on the Quien Sabe Fault, a deeply hidden blind thrust near the south end of the Hayward Fault, sent out elastic waves that shook open a twenty-thousand-gallon vat of cabernet sauvignon in an Almadén winery twenty-five miles from the epicenter. The vat was thirty feet high. A thunderous winefall flooded an office, broke out of the building, and poured down the road. This seems to have done it for Almadén. The Quien Sabe Fault? Like almost everybody else, Almadén had never heard of it, but the Quien Sabe Fault added insult to chronic injury. Ever since the winery was built, it had slowly been coming apart. Whereas the little Quien Sabe Fault was twenty-five miles away, the San Andreas Fault happened to run right through the building. The road outside was called Cienega, meaning "swamp," as in reed-filled sag pond, an example of which was beside the winery. *Almadén* is a Spanish geological term meaning "mine," and this location—about twelve miles south of Hollister, in the Central Creeping Zone—was no place to mine wine. Nevertheless, like so many parts of the San Andreas trace, it was an intimate and lovely valley, full of walnut orchards and olive groves and horse pastures and signs of anxious warning: "CAUTION: CHILDREN WALKING TO SCHOOL."

This was the place where slow tectonic creep, also called aseismic slip, was first observed. The winery stands quiet now under Atlas cedars. Almadén has shut it down. Isabel Valenzuela, a caretaker who had left Mexico two weeks before, gave Moores and me a tour in Spanish. In the gloom of the great space, casks were ranked, and not a few were broken—cracked, like immense standing eggs. Through the floor a wide crack ran from one end to the other of the long rectangular building. Outside was a bronze plaque mounted on a freestanding wall of unreinforced masonry. It bore the words "San Andreas Fault has been designated a Registered Natural Landmark."

I am from the Northeast and have never felt a destructive earthquake. The closest one to my home came in 1980, when a temblor centered in Cheesequake, New Jersey, caused no damage. My wife and I were in San Francisco in mid-October, 1989, and departed by train from Oakland, heading north. A niece in Berkeley drove us to the station and had difficulty finding it, in dark streets among warehouses on low flat bayfill ground. Up one street and down the next she searched—back and forth, around and under the Nimitz Freeway. The earthquake came more than a hundred hours after we left—an inexpressibly short span on the geologic scale, an irrelevant long one on ours. I wish I could say that I felt in my neuroplasm that it was time to go, but I was not a missing cat.

A few weeks later, when I was back in California, Moores and I were approaching the Santa Cruz Mountains from the south, and we stopped off at the Spanish mission in San Juan Bautista. Where scouts discovered two wells of water only twenty paces apart, the mission was established in 1797. Down the middle of a broad plain ran a sharply defined escarpment about fifty feet high, like the one in Hayward. It was a single long step, steeply inclined, like a grandstand. A modern geologist seeing such a break with springs or shallow wells beside it would be wondering not "Why is it there?" but "How frequent is the slippage upon it?" Moores remarked, "A fault is a good place for a well. It's brecciated. There's lots of porosity. Aquifers on either side are truncated, and spill into the fault." The Franciscans built their mission on the top edge, the brink—not near, but on, the San Andreas. By October, 1800, they had begun or completed eight buildings, of adobe roofed with sedge, when earthquake after earthquake—as many as six shocks a day—brought much of the mission down.

The cool cloisters look immovable, the chapel looks repaired. Its plans (unique among the missions) called for three aisles divided by columns, but the colonnades were finished as walls. The view from the scarp is much the same: the fault is about all that has moved. In April, 1906, when San Juan Bautista was the southern end of the plate-shattering rupture, there was some destruction. Now, in 1989, there was no sign of the shaking of a few weeks before. Damage had been essentially nil in San Juan Bautista and everywhere in the fault zone to the south.

The fault scarp beside the chapel was actually used as a grandstand for dog races on a dirt track below, but the fans lost interest, and vines have grown over the seats. Beneath the bells of the mission, we sat in the bleachers and talked among the crawling vines, right on the fault, looking east toward distant mountains across the unaltered plain. Like almost all topography, it seemed to be immutable.

"People look upon the natural world as if all motions of the past had set the stage for us and were now frozen," Moores remarked. "They look out on a scene like this and think, It was all made for us—even if the San Andreas Fault is at their feet. To imagine that turmoil is in the past and somehow we are now in a more stable time seems to be a psychological need. Leonardo Seeber, of Lamont-Doherty, referred to it as the principle of least astonishment. As we have seen this fall, the time we're in is just as active as the past. The time between events is long only with respect to a human lifetime."

(In a 1983 paper called "Large Scale Thin-Skin Tectonics," Seeber addressed the possibility that crustal deformations have occurred on an areal and cataclysmic scale never imagined or described. "Our direct view of geologic phenomena has been severely limited by the relatively short span of history and by the relatively small vertical extent of outcrops," he wrote. "In many respects we only have a two-dimensional snapshot view of the geologic process. Moreover, the interpretation of geologic data was probably influenced by the psychologic need to view the earth as a stable environment. Manifestations of current tectonism were often perceived as the last gasps of a geologically active past. Thus, subjected to the principle of least astonishment, geologic science has always tended to adopt the most static interpretation allowed by the data.")

Earthquakes in the six-to-seven range occurred at least once a decade in the San Francisco Bay Area between 1850 and 1906. Afterward, that segment of the San Andreas Fault was essentially quiet for more than eighty years, a not very significant exception being the 5.5 event near Mussel Rock in 1957. While strain accumulated along the San Francisco Peninsula, whole lifetimes passed, so the principle of least astonishment, which works to a fare-thee-well in a place like New York, seemed to be working even here. In recent days, of course, the newspapers had been full of comment suggesting that least astonishment was no longer a principle in the Bay Area. Withal, there was an undercurrent of implication that it had not died and would come back.

Jerry Carroll, in the *San Francisco Chronicle*:

> There is no greater betrayal than when the earth defaults on the understanding that it stay still under foot while we go about the business of life, which is full enough of perils as it is.

Stephanie Salter, *San Francisco Examiner*:

A traumatic experience . . . started in the depths of the earth and wreaked damage all the way to the depths of the psyche. . . . Or maybe the truth is, earthquake time is the most real time of all, a time when all the bull ceases and the preciousness of life is understood most acutely.

Herb Caen, the venerable columnist of the *San Francisco Chronicle*, who had seen his share of accumulating strain:

[This is] a headstrong, careless city dancing forever on the edge of disaster. . . . We realize afresh the joys and dangers of living here, and we reaffirm our belief that it is worth the gamble, however great. . . . We have been validated as San Franciscans.

A few miles up the trace, we looked across the Pajaro Valley at the high notch where the fault slices into the southern extremities of the Santa Cruz Mountains. The east side was Oligocene shale, Moores said, and on the west side was a quartz diorite of the mid-Cretaceous. The two formations differed by at least sixty million years and by who knows how many miles of sliding offset. It was as if an apple and a pear had both been vertically sliced in half, and two of the differing halves had been placed together to make the mountains. The lookoff where we stood was at the northern end of the Gabilan Range. The Pajaro River, narrow and slow, ran westward toward the ocean through a topographic gap that punctured the mountains. Indians of the eighteenth century informed the arriving Spanish that the small stream in that huge gap had not always been so modest, that it had once been the outlet of the interior rivers, also draining the bays—the role now played by the Golden Gate. The Indians were right, Moores said. Never mind that they may not have suspected that the whole of the coastal country was moving northwest, occluding what lay to the east. Geologists have described the Pajaro Valley as "one of the most seismically active regions in the coterminous United States." On April 18, 1906, a freight train crossing the valley was thrown off the tracks.

The San Andreas Fault was exceptionally smelly where it crossed the Pajaro River and went into the Santa Cruz Mountains.

Highway 129 follows the right bank there. From the fault zone—a landslide of sedimentary hash about a hundred and twenty yards wide—a dozen sulphur springs were pouring. Galvanized pipes had been driven into the springs, causing the water to spout onto the roadside and color it yellow-cream.

Just to the west was Watsonville, which looked like a French battle town in 1944. The eighty-six-year-old St. Patrick's Church, until recently a steepled brick structure with four spires, stood in its own red scree. That it stood at all was remarkable. Its crosses were aslant, its buttresses denuded, its brick gone in swatches from the walls. Like cartoon lightning, jagged cracks descended through the brick that remained. Two spires were stretched out in the church parking lot. Where Highway 1 crossed Struve Slough on concrete columns, the columns had punctured the pavement and now protruded upward like standing stones. Wide acreages in the town center had been bulldozed bare and brown. Ford's Department Store (1851) was totally destroyed. Countless buildings were shuttered with plywood or wrapped in chain-link fencing. There were fissures in cement-block walls. Two hundred and fifty houses were off their foundations, many of them crushed like foam cups. There were tents. There were cellar holes where houses had been. In a single long moment at the edge of town, a million apples had fallen to the ground.

California building codes that involve seismic requirements were first written in 1933. They covered school buildings and nothing else. San Francisco did not extend such codes to other structures until the late nineteen-forties, when they appeared in the laws of virtually all communities around the bays and of many around the state. While most buildings are still "pre-code," what is most remarkable is how effective the codes have been. In the 1989 Loma Prieta earthquake, sixty-two people died. In an earthquake of similar magnitude in Armenia in 1988, fifty-five thousand died. In Mexico City in 1985, ten thousand died. In the Iranian earthquake of 1990, fifty thousand died. The difference may lie partly in luck, in site, in relative intensity, but largely it lies in building codes, and the required or suggested strengthening of existing structures. Certain vulnerabilities notwithstanding, California seems to know what it is up against, and what to try to do about it. Never mind that in October, 1989, twenty-one thousand homes and commercial buildings were cracked, crumpled, or destroyed, and nature's invoice for a few moments of shaking was six billion dollars.

During the previous summer, there had been a 5.2 earthquake in the Santa Cruz Mountains, and a 5.1 quake the year before. These could be looked upon as precursors, but precursors never become such until a large jump follows, and are therefore useless as warnings. A score of 5.2 on the Richter scale is made by an earthquake with seven hundred times less energy than the one that shattered Watsonville. Richter was a professor at Caltech. His scale, devised in the nineteen-thirties, is understood by professors at Caltech and a percentage of the rest of the population too small to be expressible as a number. Another professor at Caltech in Richter's time—and someone who manifestly understood the principles involved—was Beno Gutenberg, who provided the data from which the scale was made. The data applied only to southern California; subsequently, Gutenberg and Richter jointly developed the worldwide scale, which has since been variously refined. Gutenberg did not see or hear well and was understandably reluctant to deal with reporters. He generally asked his young colleague Charles F. Richter to explain the scale to them. Since I have no idea how the scale works, let me say only that it is a mathematically derived combination of three scales parallel to one another: a magnitude scale flanked by scales of amplitude and distance. (Amplitude is the height of the mark an earthquake produces on a seismogram.) Where a line drawn between amplitude and distance crosses the central scale, it registers magnitude. With each rising integer on the magnitude scale, an earthquake's waves have ten times as much amplitude and thirty times as much energy. Richter always insisted that it was the Gutenberg-Richter scale.

There is a swerve in the San Andreas Fault where it moves through the Santa Cruz Mountains. It bends a little and then straightens again, like the track of a tire that was turned to avoid an animal.

Because deviations in transform faults retard the sliding and help strain to build, the most pronounced ones are known as tectonic knots, or great asperities, or prominent restraining bends. The two greatest known earthquakes on the fault occurred at or close to prominent restraining bends. The little jog in the Santa Cruz Mountains is a modest asperity, but enough to tighten the lock. As the strain rises through the years, the scales of geologic time and human time draw ever closer, until they coincide. An earthquake is not felt everywhere at once. It travels in every direction—up, down, and sideways—from its place and moment of beginning. In this example, the precise moment is in the sixteenth second of the fifth minute after five in the afternoon, as the scales touch and the tectonic knot lets go.

The epicenter is in the Forest of Nisene Marks, a few hundred yards from Trout Creek Gulch, five miles north of Monterey Bay. The most conspicuous nearby landmark is the mountain called Loma Prieta. In a curving small road in the gulch are closed gates and speed bumps. PRIVATE PROPERTY. KEEP OUT. This is steep terrain—roughed up, but to a greater extent serene. Under the redwoods are glades of maidenhair. There are fields of pampas grass, stands of tan madrone. A house worth two million dollars is under construction, and construction will continue when this is over. BEWARE OF DOG.

Motion occurs fifty-nine thousand eight hundred feet down—the deepest hypocenter ever recorded on the San Andreas Fault. No drill hole made anywhere on earth for any purpose has reached so far. On the San Andreas, no earthquake is ever likely to reach deeper. Below sixty thousand feet, the rock is no longer brittle.

The epicenter, the point at the surface directly above the hypocenter, is four miles from the fault trace. Some geologists will wonder if the motion occurred in a blind thrust, but in the Santa Cruz Mountains the two sides of the San Andreas Fault are not vertical. The Pacific wall leans against the North American wall at about the angle of a ladder.

For seven to ten seconds, the deep rockfaces slide. The maximum jump is more than seven feet. Northwest and southeast, the slip propagates an aggregate twenty-five miles. This is not an especially large event. It is nothing like a plate-rupturing earthquake. Its upward motion stops twenty thousand feet below the surface. Even so, the slippage plane—where the two great slanting faces have moved against each other—is an irregular oval of nearly two hundred square miles. The released strain turns into waves, and they develop half a megaton of energy. Which is serious enough. In California argot, this is not a tickler—it's a slammer.

The pressure waves spread upward and outward about three and a half miles a second, expanding, compressing, expanding, compressing the crystal structures in the rock. The shear waves that follow are somewhat slower. Slower still (about two miles a second) are the surface waves: Rayleigh waves, in particle motion like a rolling sea, and Love waves, advancing like snakes. Wherever things shake, the shaking will consist of all these waves. Half a minute will pass before the light towers move at Candlestick Park. Meanwhile, dogs are barking in Trout Creek Gulch. Car alarms and house alarms are screaming. If, somehow, you could hear all such alarms coming on throughout the region, you could hear the spread of the earthquake. The redwoods are swaying. Some snap like asparagus. The restraining bend has forced the rock to rise. Here, west of the fault trace, the terrain has suddenly been elevated a foot and a half—a punch delivered from below. For some reason, it is felt most on the highest ground.

On Summit Road, near the Loma Prieta School, a man goes up in the air like a diver off a board. He lands on his head. Another man is thrown sideways through a picture window. A built-in oven leaves its niche and shoots across a kitchen. A refrigerator walks, bounces off a wall, and returns to its accustomed place. As Pearl Lake's seven-room house goes off its foundation, she stumbles in her kitchen and falls to the wooden floor. In 1906, the same house went off the same foundation. Her parents had moved in the day before. Lake lives alone and raises prunes. Ryan Moore, in bed under the covers, is still under the covers after his house travels a hundred feet and ends up in ruins around him.

People will come to think of this earthquake as an event that happened in San Francisco. But only from Watsonville to Santa Cruz—here in the region of the restraining bend, at least sixty miles south of the city—will the general intensity prove comparable to 1906. In this region are almost no freeway overpasses, major bridges, or exceptionally tall buildings. Along the narrow highland roads, innumerable houses are suddenly stoop-shouldered, atwist, bestrewn with splinters of wood and glass, even new ones "built to code." Because the movement on the fault occurs only at great depth, the surface is an enigma of weird random cracks. Few and incongruous, they will not contribute to the geologic record. If earthquakes like Loma Prieta are illegible, how many of them took place through the ages before the arrival of seismographs, and what does that do to geologists' frequency calculations?

Driveways are breaking like crushed shells. Through woods and fields, a ripping fissure as big as an arroyo crosses Morrill Road. Along Summit Road, a crack three feet wide, seven feet deep, and seventeen hundred feet long runs among houses and misses them all. Roads burst open as if they were being strafed. Humps rise. Double yellow lines are making left-lateral jumps.

Cracks, fissures, fence posts are jumping left as well. What is going on? The San Andreas is the classic right-lateral fault. Is country going south that should be going north? Is plate tectonics going backward? Geologists will figure out an explanation. With their four-dimensional minds, and in their interdisciplinary ultraverbal way, geologists can wiggle out of almost anything. They will say that while the fault motion far below is absolutely right lateral, blocks of rock overhead are rotating like ball bearings. If you look down on a field of circles that are all turning clockwise, you will see what the geologists mean.

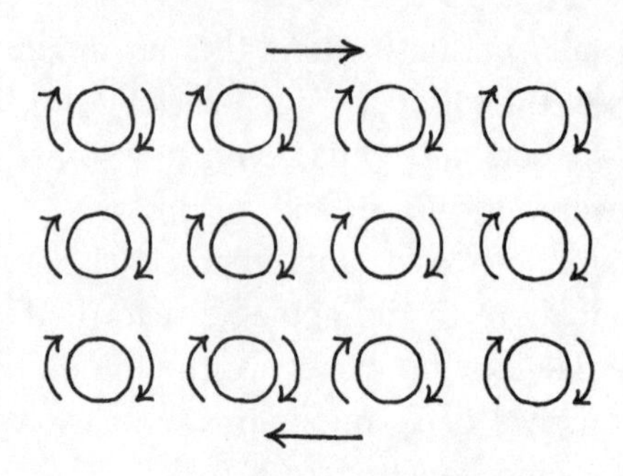

Between one circle and the next, the movement everywhere is left lateral. But the movement of the field as a whole is right lateral. The explanation has legerdemain. Harry Houdini had legerdemain when he got out of his ropes, chains, and handcuffs at the bottom of the Detroit River.

All compression resulting from the bend is highest near the bend, and the compression is called the Santa Cruz Mountains. Loma Prieta, near four thousand feet, is the highest peak. The words mean Hill Dark. This translation will gain in the shaking, and appear in the media as Dark Rolling Mountain.

At the University of California, Santa Cruz, three first-year students from the East Coast sit under redwoods on the forest campus. As the shock waves reach them and the trees whip overhead, the three students leap up and spontaneously dance and shout in a ring. Near the edge of town, a corral disintegrates, horses run onto a highway, a light truck crashes into them and the driver is killed. Bicyclists are falling to the streets and automobiles are bouncing. Santa Cruz has been recovering from severe economic depression, in large part through the success of the Pacific Garden Mall, six blocks of old unreinforced brick buildings lately turned into boutiques. The buildings are contiguous and are of different heights. As the shock waves reach them, the buildings react with differing periods of vibration and knock each other down. Twenty-one buildings collapse. Higher ones fall into lower ones like nesting boxes. Ten people die. The Hotel Metropol, seventy years old, crashes through the ceiling of the department store below it. The Pacific Garden Mall is on very-young-floodplain river silts that amplify the shaking —as the same deposits did in 1906.

Landslides are moving away from the epicenter in synchrony

with the car alarms. As if from explosions, brown clouds rise into the air. A hundred and eighty-five acres go in one block slide, dozens of houses included. Hollister's clock tower falls. Coastal bluffs fall. Mountain cliffs and roadcuts fall.

The shock waves move up the peninsula. Reaching Los Gatos, they give a wrenching spin to houses that cost seven hundred and fifty thousand dollars and have no earthquake insurance. A man is at work in a bicycle shop. In words that *Time* will print in twenty-four-point type, he will refer to the earthquake as "my best near-death experience." (For a number of unpublished fragments here, I am indebted to editors at Time Warner, who shared with me a box-ful of their correspondents' files.)

Thirteen seconds north of the epicenter is Los Altos, where Harriet and David Schnur live. They grew up in New York City and have the familiar sense that an IRT train is passing under their home. It is a "million-dollar Cape Cod," and glass is breaking in every room. This is scarcely their first earthquake.

David: "Why is it taking so long?"

Harriet: "This could be the last one. Thank God we went to *shul* during the holidays."

The piano moves. Jars filled with beans shatter. Wine pours from breaking bottles. A grandfather clock, falling—its hands stopping at 5:04—lands on a metronome, which begins to tick.

The shock reaches Stanford University, and sixty buildings receive a hundred and sixty million dollars' worth of damage. The university does not have earthquake insurance.

The waves move on to San Mateo, where a woman in a sixteenth-floor apartment has poured a cup of coffee and sat down to watch the third game of the World Series. When the shock arrives, the apartment is suddenly like an airplane in a wind shear. The jolt whips her head to one side. A lamp crashes. Books fall. Doors open. Dishes fall. Separately, the coffee and the cup fly across the room.

People are dead in Santa Cruz, Watsonville has rubble on the ground, and San Francisco has yet to feel anything. The waves approach the city directly from the hypocenter and indirectly via the Moho. Waves that begin this deep touch the Moho at so slight an angle that they carom upward, a phenomenon known as critical reflection. As the shaking begins in San Francisco, it is twice as strong as would generally be expected for an earthquake of this magnitude at that distance from the epicenter.

Two men are on a motor scooter on Sixteenth Street. The driver, glancing over his shoulder, says, "Michael, stop bouncing." A woman walking on Bush Street sees a Cadillac undulating like a water bed. She thinks, What are those people *doing* in there? Then the windows fall out of a nearby café. The sidewalks are moving. Chimneys fall in Haight-Ashbury, landing on cars. In Asbury Heights, a man is watering his patch of grass. He suddenly feels faint, his knees weaken, and his front lawn flutters like water under wind. Inside, his wife is seated at her seven-foot grand. The piano levitates, comes right up off the floor as she plays. She is thinking, I'm good but not this good. A blimp is in the air above. The pilot feels vibration. He feels four distinct bumps.

In Golden Gate Park, high-school girls are practicing field hockey. Their coach sees the playing field move, sees "huge trees . . . bending like windshield wipers." She thinks, This is the end, I'm about to fall into the earth, this is the way to go. Her players freeze in place. They are silent. They just look at one another.

In the zoo, the spider monkeys begin to scream. The birdhouse is full of midair collisions. The snow leopards, lazy in the sun with the ground shaking, are evidently unimpressed. In any case, their muscles don't move. Pachy, the approximately calico cat who lives inside the elephant house, is outside the elephant house. She refused to enter the building all day yesterday and all day today. When someone carried her inside to try to feed her, she ran outside, hungry.

At Chez Panisse, in Berkeley, cupboard doors open and a chef's personal collection of pickles and preserves crashes. The restaurant, renowned and booked solid, will be half full this evening. Those who come will order exceptionally expensive wine. Meanwhile, early patrons at a restaurant in Oakland suddenly feel as if they were in the dining car of a train that has lurched left. When it is over, they will all get up and shake hands.

In the San Francisco Tennis Club, balls are flying without being hit. Players are falling down. The ceilings and the walls seem to be flowing. Nearby, at Sixth and Bluxome, the walls of a warehouse are falling. Bricks crush a car and decapitate the driver. Four others are killed in this avalanche as well.

In the hundred miles of the San Andreas Fault closest to San Francisco, no energy has been released. The accumulated strain is unrelieved. The U.S. Geological Survey will continue to expect within thirty years an earthquake in San Francisco as much as fifty times as powerful. In the Survey's offices in Menlo Park, a seismologist will say, "This was not a big earthquake, but we hope it's the biggest we deal with in our careers." The Pacific Stock Exchange, too vital to suffer as much as a single day off, will trade by candlelight all day tomorrow.

Passengers on a rapid-transit train in a tube under the bay feel as if they had left the rails and were running over rocks. The Interstate 80 tunnel through Yerba Buena Island moves like a slightly writhing hose. Linda Lamb, in a sailboat below the Bay Bridge, feels as if something had grabbed her keel. Cars on the bridge are sliding. The entire superstructure is moving, first to the west about a foot, and then back east, bending the steel, sending large concentric ripples out from the towers, and shearing through bolts thicker than cucumbers. This is the moment in which a five-hundred-ton road section at one tower comes loose and hinges downward, causing one fatality, and breaking open the lower deck, so that space gapes to the bay. Heading toward Oakland on the lower deck, an Alameda County Transit driver thinks that all his tires have blown, fights the careening bus for control, and stops eight feet from a plunge to the water. Smashed cars vibrate on the edge but do not fall. Simultaneously, the Golden Gate Bridge is undulating, fluctuating, oscillating, pendulating. Daniel Mohn—in his car heading north, commuting home—is halfway across. From the first tremor, he knows what is happening, and his response to his situation is the exact opposite of panic. He feels very lucky. He thinks, as he has often thought before, If I had the choice, this is where I would be. Reporters will seek him later, and he will tell them, "We never close down." He is the current chief engineer of the Golden Gate Bridge.

Peggy Iacovini, having crossed the bridge, is a minute ahead of the chief engineer and a few seconds into the Marin Headlands. In her fluent Anglo-Calif she will tell the reporters, "My car jumped over like half a lane. It felt like my tire blew out. Everybody opened their car doors or stuck their heads out their windows to see if it was their tires. There were also a couple of girls holding their chests going oh my God. All the things on the freeway were just blowing up and stuff. It was like when you light dynamite—you know, on the stick—it just goes down and then it blows up. The communication wires were just sparking. I mean my heart was beating. I was like oh my God. But I had no idea of the extent that it had done."

At Candlestick Park, the poles at the ends of the foul lines throb like fishing rods. The overhead lights are swaying. The upper deck is in sickening motion. The crowd stands as one. Some people are screaming. Steel bolts fall. Chunks of concrete fall. A chunk weighing fifty pounds lands in a seat that a fan just left to get a hot dog. Of sixty thousand people amassed for the World Series, not one will die. Candlestick is anchored in radiolarian chert.

The tall buildings downtown rise out of landfill but are deeply founded in bedrock, and, with their shear walls and moment frames and steel-and-rubber isolation bearings, they sway, shiver, sway again, but do not fall. A woman forty-six floors up feels as if she were swinging through space. A woman twenty-nine floors up, in deafening sound, gets under her desk in fetal position and thinks of the running feet of elephants. Cabinets, vases, computers, and law books are flying. Pictures drop. Pipes bend. Nearly five minutes after five. Elevators full of people are banging in their shafts.

On the high floors of the Hyatt, guests sliding on their bellies think of it as surfing.

A quick-thinking clerk in Saks herds a customer into the safety of a doorjamb and has her sign the sales slip there.

Room service has just brought shrimp, oysters, and a bucket of champagne to Cybill Shepherd, on the seventh floor of the Campton Place Hotel. Foot of Nob Hill. Solid Franciscan sandstone. Earthquakes are not unknown to Shepherd. At her home in Los Angeles, pictures are framed under Plexiglas, windowpanes are safety glass, and the water heater is bolted to a wall. Beside every bed are a flashlight, a radio, and a hard hat. Now, on Nob Hill, Shepherd and company decide to eat the oysters and the shrimp before fleeing, but to leave the champagne. There was a phone message earlier, from her astrologer. Please call. Shepherd didn't call. Now she is wondering what the astrologer had in mind.

A stairway collapses between the tenth and eleventh floors of an office building in Oakland. Three people are trapped. When they discover that there is no way to shout for help, one of them will dial her daughter in Fairfax County, Virginia. The daughter will dial 911. Fairfax County Police will teletype the Oakland police, who will climb the building, knock down a wall, and make the rescue.

Meanwhile, at sea off Point Reyes, the U.S. Naval Ship Walter S. Diehl is shaking so violently that the officers think they are running aground. Near Monterey, the Moss Landing Marine Laboratory has been destroyed. A sea cliff has fallen in Big Sur—eighty-one miles south of the epicenter. In another minute, clothes in closets will be swinging on their hangers in Reno. Soon thereafter, water will form confused ripples in San Fernando Valley swimming pools. The skyscrapers of Los Angeles will sway.

After the earthquake on the Hayward Fault in 1868, geologists clearly saw that dangers varied with the geologic map, and they wrote in a State Earthquake Investigation Commission Report, "The portion of the city which suffered most was . . . on made ground." In one minute in 1906, made ground in San Francisco sank as much as three feet. Where landfill touched natural terrain, cable-car rails bent down. Maps printed and distributed well before 1989—stippled and cross-hatched where geologists saw the greatest violence to come—singled out not only the Nimitz Freeway in Oakland but also, in San Francisco, the Marina district, the Embarcadero, and the Laocoönic freeways near Second and Stillman. Generally speaking, shaking declines with distance from the hypocenter, but where landfill lies on loose sediment the shaking can amplify, as if it were an explosion set off from afar with a plunger and a wire. If a lot of water is present in the sediment and the fill, they can be changed in an instant into gray quicksand—the effect known as liquefaction. Compared with what happens on bedrock, the damage can be something like a hundredfold, as it was on the lakefill of Mexico City in 1985, even though the hypocenter was far to the west, under the Pacific shore.

In a plane that has just landed at San Francisco International Airport, passengers standing up to remove luggage from the overhead racks have the luggage removed for them by the earthquake. Ceilings fall in the control tower, and windows break. The airport is on landfill, as is Oakland International, across the bay. Sand boils break out all over both airfields. In downtown San Francisco, big cracks appear in the elevated I-280, the Embarcadero Freeway, and U.S. 101, where they rest on bayfill and on filled-in tidal creek and filled-in riparian bog. They do not collapse. Across the bay, but west of the natural shoreline, the Cypress section of the Nimitz Freeway—the double-decked I-880—is vibrating at the same frequency as the landfill mud it sits on. This coincidence produces a shaking amplification near eight hundred per cent. Concrete support columns begin to fail. Reinforcing rods an inch and a half thick spring out of them like wires. The highway is not of recent construction. At the tops of the columns, where they meet the upper deck, the joints have inadequate shear reinforcement. By a large margin, they would not meet present codes. This is well known to state engineers, who have blueprinted the reinforcements, but the work has not been done, for lack of funds.

The under road is northbound, and so is disaster. One after the last, the slabs of the upper roadway are falling. Each weighs six hundred tons. Reinforcing rods connect them, and seem to be helping to pull the highway down. Some drivers on the under road, seeing or sensing what is happening behind them, stop, set their emergency brakes, leave their cars, run toward daylight, and are killed by other cars. Some drivers apparently decide that the very columns that are about to give way are possible locations of safety, like doorjambs. They pull over, hover by the columns, and are crushed. A bank customer-service representative whose 1968 Mustang has just come out of a repair shop feels the jolting roadway and decides that the shop has done a terrible job, that her power steering is about to fail, and that she had better get off this high-speed road as fast as she can. A ramp presents itself. She swerves onto it and off the freeway. She hears a huge sound. In her rearview mirror she sees the upper roadway crash flat upon the lower.

As the immense slabs fall, people in cars below hold up their hands to try to stop them. A man eating peanuts in his white pickup feels what he thinks are two flat tires. A moment later, his pickup is two feet high. Somehow, he survives. In an airport shuttle, everyone dies. A man in another car guns his engine, keeps his foot to the floor, and races the slabs that are successively falling behind him. His wife is yelling, "Get out of here! Get out of here!" Miraculously, he gets out of here. Many race the slabs, but few escape.

Through twenty-two hundred yards the slabs fall. They began falling where the highway leaves natural sediments and goes onto a bed of landfill. They stop where the highway leaves landfill and returns to natural sediments.

Five minutes after five, and San Francisco's Red Cross Volunteer Disaster Services Committee is in the middle of a disaster-preparedness meeting. The Red Cross Building is shivering. The committee has reconvened underneath its table.

In yards and parks in the Marina, sand boils are spitting muds from orifices that resemble the bell rims of bugles. In architectural terminology, the Marina at street level is full of soft stories. A soft story has at least one open wall and is not well supported. Numerous ground floors in the Marina are garages. As buildings collapse upon themselves, the soft stories vanish. In a fourth-floor apartment, a woman in her kitchen has been cooking Rice-A-Roni. She has put on long johns and a sweatshirt and turned on the television to watch the World Series. As the building shakes, she moves with experience into a doorway and grips the jamb. Nevertheless, the vibrations are so intense that she is thrown to the floor. When the shaking stops, she will notice a man's legs, standing upright, outside her fourth-story window, as if he were floating in air. She will think that she is hallucinating. But the three floors below her no longer exist, and the collapsing building has carried her apartment to the sidewalk. Aqueducts are breaking, and water pressure is falling. Flames from broken gas mains will rise two hundred feet. As in 1906, water to fight fires will be scarce. There are numbers of deaths in the Marina, including a man and a woman later found hand in hand. A man feels the ground move under his bicycle. When he returns from his ride, he will find his wife severely injured and his infant son dead. An apartment building at Fillmore and Bay has pitched forward onto the street. Beds inside the building are standing on end.

The Marina in 1906 was a salt lagoon. After the Panama Canal opened, in 1914, San Francisco planned its Panama-Pacific International Exposition for the following year, not only to demonstrate that the city had recovered from the great earthquake to end all earthquakes but also to show itself off as a golden destination for shipping. The site chosen for the Exposition was the lagoon. To fill it up, fine sands were hydraulically pumped into it and mixed with miscellaneous debris, creating the hundred and sixty-five dry acres that flourished under the Exposition and are now the Marina. Nearly a minute has passed since the rock slipped at the hypocenter. In San Francisco, the tremors this time will last fifteen seconds. As the ground violently shakes and the sand boils of the Marina discharge material from the liquefying depths, the things they spit up include tarpaper and bits of redwood—the charred remains of houses from the earthquake of 1906.

An earthquake.

A small flex of mobility in a planetary shell so mobile that nothing on it resembles itself as it was some years before, when nothing on it resembled itself as it was some years before that, when nothing on it . . .

Not long ago, at Mussel Rock, a man named Araullo was fishing. He had a long pole that looked European. He seemed not so much to be casting his lure as sweeping it through the sea. His home was near the top of the cliff. He pointed proudly. The one nearest the view.

He had come down the trail and jumped over water to a wide, flat boulder. The seismic crack that came down the cliff ran into the water and under the boulder. He was fishing the San Andreas Fault, and he was having no luck.

I asked him, "What are you after?"

He said, "Sea perch. I also get salmon and striped bass here. Now I don't know where they are. Someday, they come."

He said that he felt very fortunate to have a house so close to the fish and the ocean, to have been able to afford it. He had bought it six months before. In this particular location, real estate was cheap. He had bought the house for a hundred and seventy thousand. I could barely hear him over the sound of the waves.

"If it going to go down, it going to go down," he shouted, and he flailed the green sea. "You never know what going to happen.

Only God knows. Hey, we got the whole view of the ocean. We got the Mussel Rock. What else we need for? This is life. If it go down, we go down with it."

The cormorants were present, and the pelicans. The big fishing boulder was echeloned with shears. From somewhere near Araullo's house, a hang glider had left the jumpy earth and now hovered safely above us.

Araullo ignored the hang glider and kept on swinging his pole.

"I don't know where they are," he said again. "But someday they come. They always come."

Book 5

Crossing the Craton

CANADA
North Dakota
Minnesota
South Dakota
Wisconsin
Michigan
Iowa
Nebraska
80
Illinois
Indiana
Kansas
Missouri
Kentucky
Tennessee
Oklahoma
Arkansas
Mississippi
Alabama

Crossing the craton—the Stable Interior Craton, core of the continent: Illinois, Iowa, Nebraska, and friends—you don't see a lot of rock. You see in the roadcuts emerald vetch moving under wind like wheat. You see the dark fine loess of airborne hills. You know, of course, that you are riding over subcrop, tens and hundreds and even thousands of feet down, but seldom does it outcrop, and, where it does, it is such an event that it is likely to have been named a state park. In Starved Rock State Park, in La Salle County, Illinois (Ordovician marine sand, exposed in a bluff of the Illinois River), you rub your hand against a massive quartz sandstone so lightly bound that white sand rains down the rockwall and onto your toes. Near Davenport, in Iowa, you stop to photograph a small farm on a drumlin, its barn like a biscuit on the summit of a loaf of bread. Under the hay and the windmills of Iowa is large-groundswell chocolate land—glaciated terrain and nothing flat about it, and a wind that sounds like white water as it moves through isolated stands of trees. A Devonian limestone outcrops at 1100 Dubuque Street in Iowa City and is three stories high. West of Skunk River comes a steady, eight-mile climb to the Des Moines lobe of the Wisconsinan ice sheet—younger country now and without relief. A bubble would center and rest on the horizon. In an overhanging streamcut at Pammel State Park is a limestone younger than the last. "Expect to see rattlesnakes," you are told; but you see what you expect: brachiopods, nautiloids, and crinoids, of Carboniferous time. Ledges State Park, near Ames, is a streamcut lithic sandstone, meaning there's a lot in it besides quartz and feldspar. The whole exposure is so weathered you can't get a decent sample. It is punk rock. There are carbon patches in the rock. The bluffs of Council Bluffs are loess that came in on the wind, and rising toward their summits are the terracettes called catsteps. From a river bar of the braided Platte, in Nebraska, you collect varied pebbles from scattered sources hundreds of miles away. All this notwithstanding, in fifteen hundred miles of the midcontinent there is a great deal less rock to see on the surface than you would see in Wyoming if you opened one eye. In *The Evolution of North America* (Princeton, 1959)—the definitive volume of its time—Philip King, of the United States Geological Survey, encapsulated the Midwest in one memorable sentence: "The rather monotonous geologic features of the Interior Lowlands would seem of less interest than the complex rocks and structures of the Canadian Shield, the Appalachians, or the [western] Cordilleras, nor can we, in this book, devote space to them commensurate with their surface area."

There would be more to tell if you could sense what you can't see. All the ledges and bluffs have been cut in the limestones, shales, and sands that came over the midcontinent in the twelve periods known in geology as Phanerozoic time—the five hundred and forty-four million years since creatures with hard skeletal parts first came into the world and began to leave their fossils in rock. In the American Midwest, the rock of Phanerozoic time rests in thin veneer on basement rock from deeper time. Some of it is a southern extension of the vast Canadian Shield. Geologists long taught that the basement was a platform at the edges of which the continent grew and that the platform had been there forever, down through the eons of Precambrian time. The Precambrian—beginning, as it does, with the beginning of the earth—covers more than four billion years. This is seven-eighths of earth history. Geological textbooks, nevertheless, would typically give the Precambrian one short chapter. First there was the basement, and on that grew the world.

In the three hundred and forty-two pages of *Our Mobile Earth* (Scribner's, 1926), Reginald Aldworth Daly, of Harvard University, composed one page on Precambrian rocks and one page on "Pre-Paleozoic eras." Daly was a giant in the geology of his time. He summarized Precambrian lithology as "the crumpled Basement Complex." Even Reverend H. N. Hutchinson, B.A., F.G.S., did a little better than that in *The Autobiography of the Earth* (Appleton, 1891). Referring to an undifferentiated mass of "fundamental gneiss," Hutchinson devoted one of his sixteen chapters to the Precambrian—fifteen pages in two hundred and eighty-three—calling it "An Archaic Era." Sixty years later, in 1951, A. J. Eardley, of the University of Utah, gave the first eighty-eight per cent of the history of the earth one chapter among the forty-three chapters of his *Structural Geology of North America*. He opened the chapter with this sentence: "The continent of North America is made up in a broad way of a stable interior and surrounding belts of deformed, intruded, and metamorphosed rocks." Of its oldest component, the Canadian Shield, he went on to say, "The time has not yet arrived . . . when the vast region can be broken down into divisions with confidence." Even as the twentieth century began to fade, three Canadian geologists—Colin W. Stearn, Robert L. Carroll, and Thomas H. Clark, of McGill University—reserved twenty pages in five hundred and forty-nine for Precambrian events in their *Geological Evolution of North America* (Wiley, 1979). W. Randall Van Schmus, of the University of Kansas, who did his doctoral dissertation on Canadian rock 2.4 billion years old, has remarked that to some extent geologists still tend to divide the history of the earth into two units, giving attention to the first in inverse proportion to its overwhelming eons. "There's basement, and there's Phanerozoic. There exists a long-standing prejudice that the Precambrian has to be different. The only difference is that it's older. And there's no bugs in it. No fossils that we can use for stratigraphic correlation. Isotopes are our fossils. There's always *something* in the rocks that will give you the answer you are looking for. You need to wait for the development of techniques. We're still dealing with a frontier in continental evolution. The information's there; we just have to figure how to get it out."

The development of techniques in recent years has sent light into deep time, revealing structures that could not have been imagined before. The basement of the continent is no longer the undifferentiated mass it was through most of the history of the science. There have been inventions or advances in metamorphic petrology, in samarium/neodymium geochronology, in argon/argon thermochronology, in uranium/lead dating, in zircon dating, in aeromagnetic mapping, in filtered-gravity mapping, in trace-element geochemistry, and in the isotopic monitoring of the crustal history of rocks and their origins in the mantle. Not to mention—at the two ends of the technological spectrum—the novelty of the ion-probe mass spectrometer and the undiminished relevance of oil wells. Many advances date only from the early nineteen-eighties, brought on by the evolution of computers and the programming to process the data. The collective result has been a new and rapid sketching-in of whole Precambrian scenes.

As you cross Iowa and approach Des Moines, nothing on the surface—not a streamcourse, a fault line, an outcrop, a rise—so much as hints at what is now beneath you. Six hundred feet down is the eastern edge of a great tectonic rift—a rupture of the lithosphere—that reposes there like a sunken boat in the waters of a lake. Filled in during the Precambrian and covered over by sediments of Paleozoic time, the central rift is about thirty miles wide, and trends southwest. If you are on Interstate 80, you angle across it. At Lincoln, Nebraska, you reach the far side. In many places, the walls of the rift are three thousand feet sheer. In the other direction, the buried rift valley runs far to the north under Iowa and under Wisconsin and under Lake Superior, where it forms a triple junction with a rift that trends off through Michigan to the southeast and a third but incomplete rift (a mere crack known as a failed arm) that goes north-northwest into Canada. This great rifting of the "stable" craton—basement of the continent—began eleven hundred and eight million years before the present and ended a thousand and eighty-six million years before the present. Continental in scale, it split North America right up the middle and down one side, threatening to scatter it to who knows what distant corners of the globe. On modern gravity maps and magnetic maps of North America, it is the single most prominent feature that you see. Rifts meeting at a triple junction are a signature of plate tectonics and can be seen all over the modern world—in the far-south Atlantic, where the African, Antarctic, and South American plates conjoin; in the Azores, where the African, Eurasian, and North American plates conjoin; in the Indian Ocean; at the Galápagos Islands; at Cape Mendocino in California. To sense most clearly, though, the Precambrian rift system that lies under the middle of North America, look at a map of Arabia and Africa.

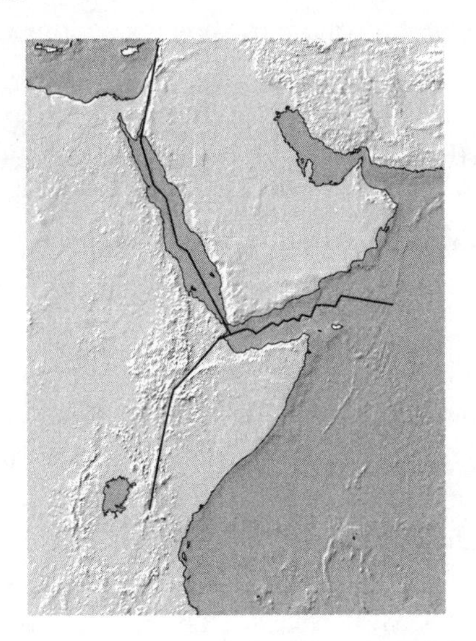

The Red Sea, the Gulf of Aden, and the East African Rift Valley—with its rift-provoked volcanoes (Kilimanjaro) and rift-depression lakes (Tanganyika, Victoria)—meet in a triple junction of plates off the southern tip of Arabia. The splitting is young and has been going on only about twenty million years, but the relic of something similar lies under the Middle West and was not seen to be what it is until the third quarter of the twentieth century. In the context of this discussion, the eleven hundred million years since the North American rifting occurred is not, in its own way, a particularly large number, for its date is much closer to the end than to the beginning of Precambrian time, with fully three-quarters of the earth's history stretching back before it. The fresh insights and technological advances mentioned above have gone far deeper into time and wider in geography than the relatively modern Midcontinent Rift. They have reached back to the beginning of the Proterozoic Eon (twenty-five hundred million years before the present) and beyond that to the early Archean Eon, when the North American craton, once assumed to have been in place forever, evidently did not exist. By some geologists, the first six hundred million years of the earth's history have recently been given status of their own as the Hadean Eon. Why geologists have decided that the earth's beginnings were in Hades is for them to say, but their technologically informed guesswork in describing former scenes does extend, in a general way, into seascapes older than four billion years.

That, as it happens, is the rounded age of the oldest rock ever found on earth (actually 3.96). The oldest rock ever found on earth is found every couple of years, it seems, as still another crustal fragment, radiometrically dated, comes nearer to 4.0. The oldest rock in the United States is in the Minnesota River Valley—about 3.5 billion years. There is rock in West Greenland that is 3.8, in Australia about 3.5. Some rocks in Africa are as old as 3.6. No contemporary continent is anywhere near as old as these Archean dates, but it is interesting that the oldest known rock comes from North America. Discovered and dated by Samuel A. Bowring, who is now at M.I.T., the incumbent oldest rock ever found on earth is east of Great Bear Lake, in the Canadian Northwest Territories, almost exactly on the Arctic Circle. The University of Kansas geochronologist Randy Van Schmus, who has vetted and guided this essay, continues the description: "It's a very strongly deformed foliated gneiss. None of the primary rock structure or texture is preserved. It's in a basement block of [the mountain-building events known in geology as] the Wopmay Orogen. Chemically, it's an evolved rock, in the sense that it was derived from the partial melting of something that predated it. So it's clearly not the oldest. And we know from zircons in sandstones in Australia that there were igneous rocks crystallizing 4.2 billion years ago. So there were older rocks, but they've either been destroyed or not found yet."

The scenes that lie before the oldest rocks and go back to the beginnings of Hadean time rest on isotopic and chemical signatures, cosmological data, and conjecture. From an interstellar gas cloud, evidently, the solar system began to form about 4.56 billion years ago. The first eleven verses of Genesis cover more than four billion of those years—the entire Precambrian and the first hundred and fifty million years of Phanerozoic time. Meanwhile, gravity, a shock wave from a supernova—or something—caused the gas cloud to collapse, becoming incandescent vapor, in which minerals formed dust. According to present theory, planetesimals formed from the dust and were swept up into planets. The collecting and compacting of the earth happened quickly—in a few tens of millions of years—and included cometary material that had water in it. From the beginning, water would have been expelled into the earth's atmosphere or onto its surface. Meteors kept on showering, accreting, increasing the size of the earth. Around 3.9 billion years before the present, meteor impacts were particularly intense. Many of the meteors were large objects hundreds of kilometres across. Stable continental crust could not develop until the earth to some extent cooled and meteor bombardment stopped. It stopped because so much debris had been gathered into planets. The present asteroid belt seems to be a planet that never formed. An object the size of Mars is thought to have collided with the very early earth, sending vaporized material into orbit. It cooled and coalesced as the moon.

On the early Archean earth, was the face of the waters an unfeatured globe-girdling sea? Theoreticians consider that possible, but highly improbable. Most likely, something would have been sticking up, such as a global mob of islands, collectively representing about twenty-five per cent of the surface of the earth. Geophysicists have calculated the production of heat from decaying uranium, potassium, and thorium in the early Archean, and the heat seems to have been three or four times the amount that the earth produces today. For the most part, the modern earth vents heat through geophysical hot spots like Etna, Yellowstone, and Hawaii, through tectonic spreading centers like the Mid-Atlantic Ridge and the East Pacific Rise, or through the volcanism of subduction zones, where one lithospheric plate slides beneath another and to some extent melts, while the resulting magma breaks the surface as a Mt. Rainier, a Mt. St. Helens, an Aconcagua, a Fujiyama. The partial melting of ocean crust would draw off a magma chemically different from the ocean crust, and it would harden as less dense, lighter, continental-style rock, examples of which are andesite and granite. To rid itself of four times as much heat, the early Archean earth must have had many more places where the heat could get out, and at this point the conversation reaches the center of an unresolved question about the look of the Archean world. Some think that lithospheric plates, similar in size to the modern plates, moved a good deal faster, with much more volcanism at the plate boundaries. A majority leans toward a picture of the earth with its eggshell more shattered, with several times as many plates as exist at the moment—smaller ones, of course, and a vastly greater linear aggregate of plate boundaries, all venting heat. The role of hot spots—plumes of heat rising to the surface from deep in the mantle—is a third and considerable factor. They could have been all over the earth in great numbers, accounting for a very large percentage of the vented heat, while early plate motions were less significant. Some as-yet-unknown proportion of these three views composes the picture that geophysical science is trying to see.

"There may have been hundreds of thousands of volcanic islands, or chains of volcanic islands," Van Schmus has remarked. "So instead of large continents you may have seen nothing more than a whole series of island arcs going around the globe—the whole earth like the South Pacific. They gradually amalgamated until they became Japans and later behaved as continents. More than likely there were fairly large continents in the early Archean, but no major or supercontinents. Or we may have had supercontinents and we lost all record of them."

The oldest dated rocks are bits and pieces of Archean crust that cannot be put into a global context because they are so isolated. Are they the scattered fragments of continental rock that was almost wholly recycled into the mantle, or are they remnants of the early forming of continental material on a small scale, a process by which continental crust grew slowly across a very long stretch of time? Did vanished original continents occupy about the same percentage of the surface of the earth that continents occupy now? Or did the growth of continents begin slowly, with granitic scums and gradual accretions? Modern continents are essentially unsinkable, because their suites of rock are light and buoyant and will jam a trench and not go all the way down into it. But they are modern. Original lands were of the early Archean, when global heat and global tectonics were different. Who knows what may have disappeared into the mantle and been recycled? "Modern analogues are good, but we have to avoid falling into the trap of making everything fit them.

They are guidelines." The recent advances of technique and insight may have enabled the science to see deeper into the Precambrian than ever before, but not that deep. The question is still very open.

On the late Archean, which began three thousand million years ago, debate continues, but the picture has more focus. By then, many granitic microcontinents were spread about the globe, and they began to coalesce. Between "2800" and "2700" (as Precambrian geologists are wont to say, meaning from 2.8 to 2.7 billion years ago) came the major growth of continents, or the first major preservation of continental material, depending on whose thesis you espouse. In any case, the numerous new continents were still, by modern dimensions, small. The shield, the core, the basement of North America was nowhere near assembled; but its components were approaching one another and would come together in collisions that built topographies of rugged relief. Spread through the ocean in a paisley way, these modest continents—now called Nain, Rae, Slave, Wyoming, Superior, and Hearne—are known in North American geology as the Archean cratons.

The end of the Archean and the beginning of the Proterozoic Eon came twenty-five hundred million years before the present, about halfway through Precambrian time. This round number represented a tectonic divide. Whatever the tectonics may have been in the Archean, the style of plate tectonics that we see in the modern world can be said to have begun with the new eon, about two and a half billion years ago. In the great historic argument between those geologists who see the earth's processes as most importantly cyclical and repetitive (the concept of uniformitarianism, with its logo: "The present is the key to the past") and those who see the earth's history as a predominantly linear narrative, there has not been a clearer or firmer example of the irreversible and the unrepeatable than the changes that took place approximately twenty-five hundred million years ago.

"It does look like the Archean-Proterozoic transition is a real threshold in the behavior of the earth," Van Schmus remarked one day while I was visiting him at the University of Kansas geological field camp in southern Colorado. "If you want to define plate tectonics strictly on the modern model, then you have to coin another term for the Archean tectonics. People sometimes make the mistake of saying that plate tectonics was not active in the Archean. I would say that modern-style plate tectonics was not, but there was some kind of dynamic earth behavior, whatever you want to call it. You might propose microplate tectonics, or that it was dominated by hotspot activity rather than lateral plate motions. Things do seem to be somewhat different. We have not pieced together enough record to say exactly how different. The Archean-Proterozoic transition is a tectonic transition, a crustal transition, deriving from within the earth, not the surface."

The surface expressed the great transition of twenty-five hundred million years ago in its own way. Although life had begun in the form of anaerobic bacteria early in the Archean Eon, photosynthetic bacteria did not appear until the middle Archean and were not abundant until the start of the Proterozoic. The bacteria emitted oxygen. The atmosphere changed. The oceans changed. The oceans had been rich in dissolved ferrous iron, in large part put into the seas by the extruding lavas of two billion years. Now with the added oxygen the iron became ferric, insoluble, and dense. Precipitating out, it sank to the bottom as ferric sludge, where it joined the lime muds and silica muds and other seafloor sediments to form, worldwide, the banded-iron formations that were destined to become rivets, motorcars, and cannons. This was the iron of the Mesabi Range, the Australian iron of the Hammerslee Basin, the iron of Michigan, Wisconsin, Brazil. More than ninety per cent of the iron ever mined in the world has come from Precambrian banded-iron formations. Their ages date broadly from twenty-five hundred to two thousand million years before the present. The transition that produced them—from a reducing to an oxidizing atmosphere and the associated radical change in the chemistry of the oceans—would be unique. It would never repeat itself. The earth would not go through that experience twice.

Around 2500—the Archean-Proterozoic transition—there was, as well, a distinct change in the over-all chemical composition of continental crust, in that it became distinctly more potassic after the

Archean. Evidently, the earth had cooled enough for this chemical change to occur, and it had also cooled enough for tectonics to assume a more sedate and modern form. By 2300, its surface had cooled enough to support continental ice sheets, perhaps for the first time. Van Schmus continues: "The earth as a whole is producing progressively less heat from radioactive decay, and that, at some point, is going to have a profound effect. In the future, the profound effect is going to be that the plates will stop moving. We'll be a very static Earth, much like Venus. Isolated hot-spot activity will go on for a while, and then die out. And you'll basically have a very sedentary Earth. Sometime, it's going to slow down and stop."

At 2000, or thereabout, an event occurred in the Wyoming continent—the drifting Wyoming craton—of especial significance to the eventual development of North America. Evidently, something peeled away from Wyoming, took off on a transform fault (the two sides slipping horizontally, like the two sides of the Alpine Fault, the Denali Fault, the San Andreas Fault), and that side of the Wyoming continent became a sharp, clean, and almost straight line. Something quite similar happened in the Cretaceous, about nineteen hundred million years later, when India separated from Madagascar and went off on its rapid journey toward collision with Tibet. Together—as Eldridge Moores once noted—the eastern coast of Madagascar and the Malabar Coast of India make the sharp, clean, straight, and matching line where the countries were conjoined.

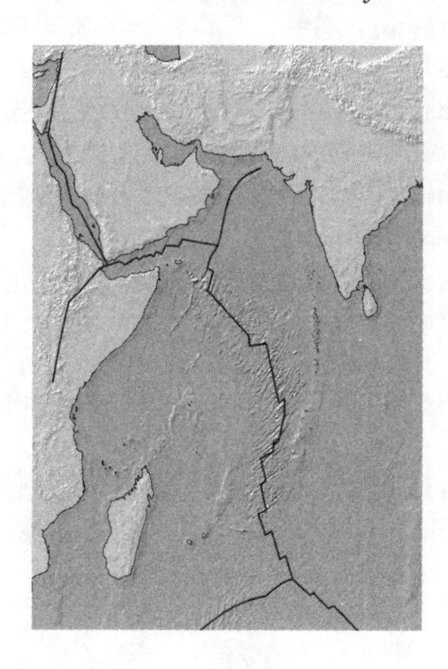

The sheared Precambrian coastline of the Wyoming microcontinent has become known as the Cheyenne Belt. If you could have stood on it 1.9 billion years ago, on rock that is now in the Laramie Range, you would have looked out over blue ocean, not over an epicontinental sea—shallow water lying on submerged land, like Hudson Bay—but abyssal ocean resting on ocean-crustal rock. Beyond, there would not have been a North American midcontinent rift system, because as yet there existed no North America to rift. Soon, though, the Archean cratons began to collide, building mountains where they hit, and sometimes trapping between them volcanic-island arcs—a series of events that took place in the geologic brevity of a hundred million years. About 1850 (1.85 billion years before the present), the Wyoming, Hearne, and Superior microcontinents came together, and the belt of deformation that held them tight is known in the science as the Trans-Hudson Orogen. A closed ocean, it is full of the crunched remains of oceanic islands. In 1900, they might have looked much like Indonesia today, but without its vegetation. By 1830, they would have resembled what Indonesia will look like after it collides with Asia. By 1800, over-all, the Canadian Shield was complete. From 1800 to 1400, most of North America grew on its margins.

The Canadian Shield was named in the nineteenth century by the Austrian geologist Eduard Suess, whose *The Face of the Earth* was first published in English in 1906. He described the extensive and exposed basement rocks of Canada as a "table-land not unlike a flat shield," and formally declared, "It is to the exposed Archean surface that we give the name of the Canadian Shield." Thereafter, the use of the word "shield" to describe the cores of continents became a tradition in geology and also developed into something of a misnomer. Geologists wanted their shields, thousands of miles wide, to resemble shields on warriors' arms—gently convex. "Like a strong plate, this broad expanse of the earth's crust has refused to yield and become folded," wrote Reginald Aldworth Daly, of Harvard, in 1926. "It is shaped roughly like a shield; hence it has been called the Canadian Shield. Like a buckler, it is stiff and strong." Stearn, Carroll, and Clark, of McGill University, wrote in their textbook, in 1979: "The term 'shield' is used as a crude description of its form—a low broad dome."

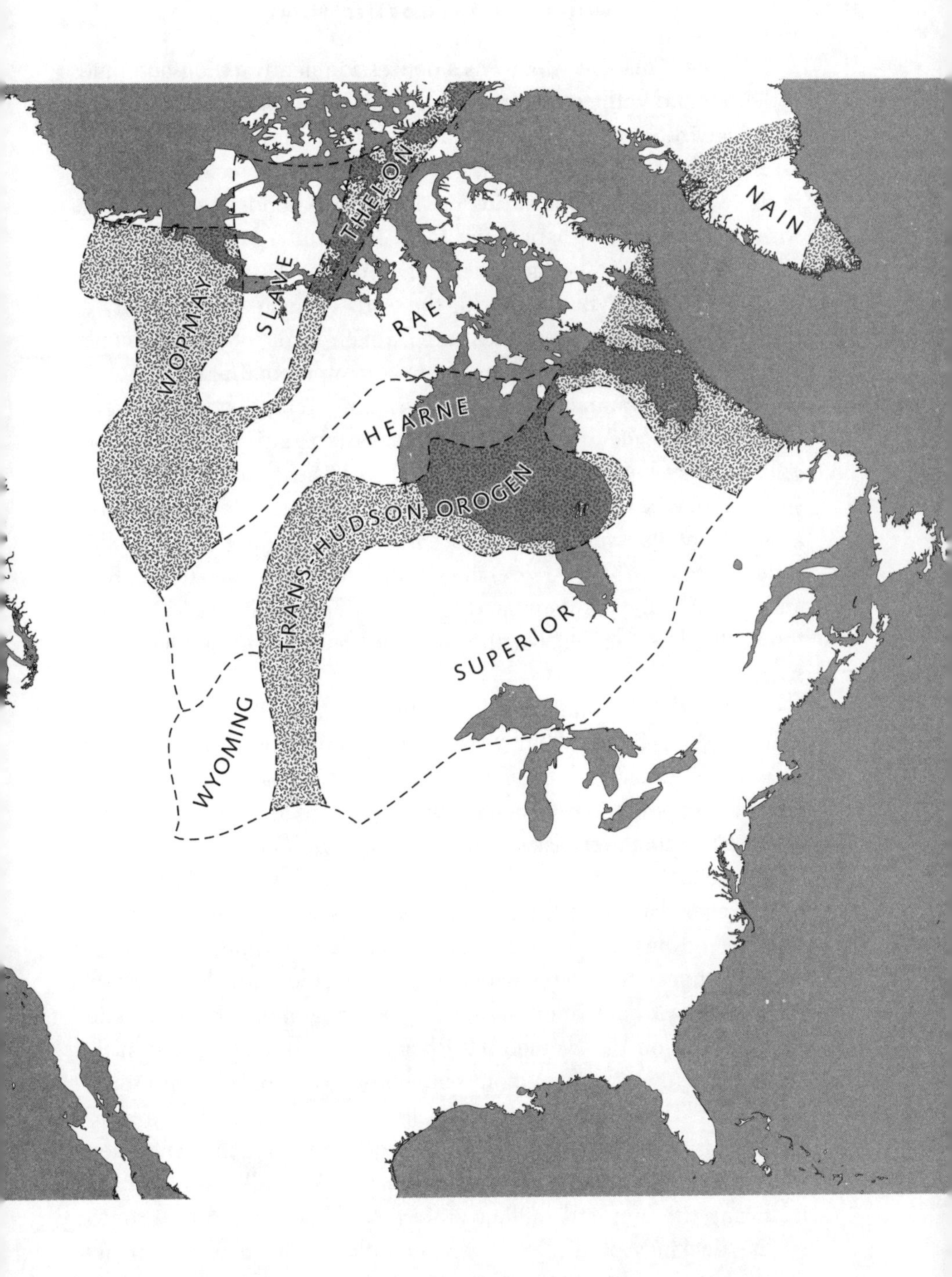

The Archean cratons—Nain, Rae, Slave, Wyoming, Superior, and Hearne—and collisional orogenic belts (Greenland separated from North America in early Cenozoic time)

In the Canadian "dome" is a depression nearly a thousand miles wide, flooded with seawater. And the Canadian Shield is a museum of folded rock, as a result of the collisions through which it was assembled. Stiff and strong it surely is. Like few erosional surfaces in the world, it has been sitting pat for sixteen hundred million years with nothing serious happening to it. Its counterparts are the Brazilian Shield, the Siberian Shield, the Indian, Australian, and Antarctic shields, the Baltic Shield, the Afro-Arabian Shield. Many of these—tectonically separated—are broken parts of other shields. They may have fitted together in ancient supercontinents.

Rugged mountains were built as the Archean cratons assembled, and the mountains slowly downwasted until, by about 1750 million years before the present, they were bevelled to their roots. In some places, rhyolite lavas that came up at that time spread across a relatively flat surface; and ocean water flooding the bevelled shield—transgressing, regressing—deposited blanket sands. Bare and equatorial, the new and growing continent would at that time have been much like the modern planet's Empty Quarter of Arabia.

Beyond the sheared edge of the Wyoming craton, seventeen hundred and fifty million years ago, islands stood in the ocean: volcanic-island arcs—Japans in motion. They came in one after the other. We know them as conterminous country: southeastern Wyoming, Colorado, Nebraska. Moving northerly, a crowd of accreting arcs probably extended as far as Chicago and beyond. Their radiometric ages date from their beginnings as islands, when their liquid substance came up off the mantle to solidify as juvenile crust. Their docking times—the dates when they connected with the Wyoming craton and with one another—are a good deal more difficult to determine, but on the average the islands seem to have existed in the ocean for ten to twenty million years before they collided with something else. The oldest age of the collected island arcs that are now the primary basement of Colorado is 1790—1.79 billion years. There was no Colorado and no Nebraska in 1800. Virtually all of it came in during the hundred million years that followed, and by 1700 or so the dockings of the arcs were complete. The new continental coastline trended from southern New Mexico through Indiana and on toward Labrador.

In the relatively modern Jurassic period, about a hundred and fifty million years ago, the assembling American Far West was much like the assembling Colorado sixteen hundred million years before. The accretionary development of the Far West first became apparent in the early days of plate-tectonic theory. And now, a few decades later, it has become apparent that the central basement of the continent—the platform, the stable craton, the once immemorial core—came together in the same way. The sharp line of the Cheyenne Belt, where southeastern Wyoming and Colorado joined the ancient shield, retains its character today. On the average, it is a couple of miles wide. Also known as the Wyoming Shear Zone, it is a line so fine that in some places you can all but straddle it. Where Interstate 80 crosses the Laramie Range west of Cheyenne, the highway is resting on arc-derived Proterozoic granite, the northern extension of the Colorado province. A few miles to the north is the edge of the Archean craton. The dividing line is sharp, and south of it are no rocks of Archean age. The northern half of the Medicine Bow Mountains—the next range west—is well over a billion years older than the southern half, and the jump takes place across the same narrow line. The line is buried under basin sediments where the interstate crosses it, just west of Laramie.

I met Randy Van Schmus a couple of years ago in what he calls "the post-1800-million-year accretionary complex" and most people call Colorado. The University of Kansas, where he teaches, has a geological field camp in Fremont County, Colorado. In the universities of the midcontinent, not a lot of geology is outside the door. The Precambrian basement is absolutely buried. To see what lies deep beneath the midcontinent, a student needs to be taken to a place like Colorado, where the same suite of rocks has been bent up into the air, where the Precambrian is the core and crown of the great foreland ranges.

On the day I joined him, Van Schmus took his students up to moderately high ground northwest of Cañon City, where their assignment was to map quartzites and other rocks that were inferred to be 1750 million years old and were pendant in a granite that had been precisely dated 1705. The quartzite, which was there first, had hung down in the soft intruding granite like a finger in honey. Quartzite being sandstone altered by heat and pressure, its sedimentary structures were somewhat pentimental and difficult to read. Where was the crossbedding? Which way was up? While the students, looking for "tops," fanned out across alpine meadows and scrambled through junipers and across serrated ridges, Van Schmus had time to sit on a ledge and review the Big Picture. He was a reasonably tall man with an easy and quiet, unexcitable manner, whose boots and jeans and potato-chip hat fitted him into the country. I had sought him out because Eldridge Moores had described him to me as "about as knowledgeable a person as you can find on the midcontinent below the sediments—an age-dater who broadened out," adding that in a recent compendium on North American Precambrian geology Van Schmus was the senior author of most of the papers that had to do with the Midwest. Van Schmus described himself as "a geochemist specializing in geochronology, specializing in the Proterozoic history of the earth." He grew up in Naperville, Illinois (his father was a trust officer in a Chicago bank), and went to Caltech to become a chemist. Geology tends to draw certain people from the less eclectic sciences, and it drew him to the 2.4-billion-year-old shield rock north of Lake Huron, where he did his doctoral research for U.C.L.A. At that time—the early nineteen-sixties, before plate tectonics was established and when many major advances in radiometric dating were some years away—a great deal of Precambrian rock was shown on tectonic and geologic maps as, in his words, "one big green blob." He has helped to fill in details of the blob. His name—his ancestors emigrated from a German town on the Dutch border—is pronounced like deuce and moose. He works in dry rangelands of the Brazilian Shield part of each year, "trying to identify old crustal blocks using isotopes"—the same kind of identification that has brought forth from obscurity the scenery of Archean Wyoming and Proterozoic Colorado. Not to mention Kansas.

As I have attempted to suggest, when Van Schmus and his colleagues talk about dates like 1640 and 1790 and 1850, they say them so familiarly and offhandedly that they might be discussing Oliver Cromwell, the American Revolution, and the publication of *Moby-Dick*. If they speak of, say, something that happened 1.745 billion years ago, they say "1745." "Virtually all of Colorado is 1700 to 1790," Van Schmus said, meaning that the primary crust of Colorado—the collected island arcs—ranges in age from 1700 to 1790 million years. On that first day in Colorado, one or another date like that would roll off Van Schmus's tongue with an ease and a familiarity that now and again caused me to pause, look up, and think, Good Lord, he means sixteen hundred and fifty million years ago. When he said that some evolving process had taken place, for example, from "1750 to 1770"—he would present the dates, as he and his Precambrian colleagues almost always do, in everyday sequence from the smaller to the larger number. Tripping on his own humanity, he was reversing the arrow of time.

"To my way of looking, this is an accretionary complex of arc terrane," he repeated. "Think of the South Pacific—all that stuff waiting to be accreted. Parts of Fiji are forty million years old and are still waiting to dock. The Precambrian of Colorado has not been studied extensively enough. Geologists of the next decade will take it apart and tell the story. More advanced techniques will develop. This is the new frontier. To get to new frontiers you go backward in time. The isotopes are telling us that this is new crust, and about the only way you get new crust is in arc environments. If you want to see what's under Interstate 80 in Nebraska, you're seeing it right here. There's basically no substitute for being able to see the rocks and walk over them. Nebraska is like *this*—a few thousand feet down."

I have also visited him in Lawrence, Kansas, and seen the primary resources that have enabled him and others to see, infer, extrapolate, conjure, discover, and describe some of the long-veiled and innumerable worlds of the Precambrian eons: the university's large archive of drilled Precambrian rock, for example, and the hospital-like rooms of the "Rb/Sr and Sm/Nd Clean Lab," where even a bit of rock powder too fine to be called silt will yield its age and origin to rubidium/strontium and samarium/neodymium dating. Now and again in Colorado, between his field sessions with the students, we travelled around Fremont County so that I could attempt, in the presence of the terrain as it has come down to us, to see its origins through his eyes. The field camp was a collection of cabins far up a gravel road in an amphitheatrical cul-de-sac known to geology as the Cañon City Embayment. Quite near the camp were Jurassic mudstones and sands where a schoolteacher named Ormel Lucas, on his summer vacation in 1876 (the summer of Little Bighorn), discovered numerous large bones of which a memorable roster was to follow: the twenty-four-foot stegosaurus in the Denver Museum of Natural History, the seventy-two-foot haplocanthosaurus in the Cleveland Museum of Natural History, the sixty-nine-foot apatosaurus in the Wagner Free Institute (in Philadelphia), the fifty-nine-foot camarasaurus in the American Museum of Natural History, and the twenty-foot ceratosaurus and the thirty-nine-foot allosaurus and the eighty-eight-foot diplodocus in the Smithsonian. Van Schmus, raising dust, went straight past their beds with an amiable nod. These creatures, on the full scale of time, were so close to the here and now that they might as well have been his students. He ran up toward Cripple Creek on a road ten feet wide that had been chipped into the sheer wall of the embayment. It climbed higher and higher up the wall—four hundred, five hundred feet—and the view over the side was perfect, unobstructed, railless. The rock of the embayment wall was three hundred million years older than the dinosaurs but still 1.3 billion years younger than the rock on which it rested. That would get us back to 1750, when, in the primordial sense, Colorado crust was first assembling. Van Schmus said that a fisherman had found a severely injured motorist inside his automobile at the bottom of this canyon a few days before. "Do you see a car yet? It could have been here."

The western reach of Fremont County—some fifty miles on the Arkansas River between Cañon City and Salida—is particularly memorable for the melodrama of its landscapes and the beauty of its structure. The Arkansas runs fast and white most of the way, maybe a hundred feet wide. The rock around it towers. As the river approaches its debouchment into the flats of eastern Colorado, it makes one last slice that by an order of magnitude is the deepest of all. Like a knife disappearing into a large loaf of bread, it has cut an extremely narrow gorge whose rims are more than a thousand feet above the rapids.

Remains of the original arcs are the rock of the gorge and the general rock of the region, almost everywhere deformed. Quartzites, migmatites, gneisses, and schists, they have been metamorphosed by so many events that their histories are close to inscrutable. Thirty miles up the canyon, however, is an elongate section of primary crust that is the lithic equivalent of virgin forest. Somehow it managed to hide out in the strain shadows of all the tectonic events that followed its own appearance on this scene. Van Schmus said, "Here you are looking at the birth of Colorado, the primary crust. When it arrived, there was basically nothing here."

"When was that?"

"1740."

He went on to say, "The primary basement here evolved over plus-or-minus a hundred million years. Plutons came along more or less in parallel with the volcanism, younging to the south and trending to the southwest." The plutons were the intrusive magmas that, for the most part, cooled as granite. In a closely analogous way, the granites of the Sierra Nevada would come into the country rock there, more than a billion and a half years later.

A coal train went by on the far side of the river. It refocused his vision—flicked his mind forward—to Carboniferous time. "There was very little coal in Africa and South America in the Paleozoic," he said cryptically. They were "stuck near the South Pole" when the big trees, elsewhere, were kings. North America was on the equator then. Hence the coal train.

The train was just passing through, however, and its screeches and rumbles and the contents of its gondolas represented points in time that were not in this conversation. In an instant, the geologist's mind returned to the era of primary midcontinental America—eighteen hundred to fourteen hundred million years before the present, a band in the earth's history about a third of the way from the planet's earliest beginnings to the tick-ticking of the present day. Representing only twenty per cent of the Proterozoic Eon, 1800–1400 was nonetheless the central frame of his professional absorption. If one part of Precambrian time was his specialty, this was it. His own lifetime—beginning in the calendar year 1938 and expecting at least four score and ten—was a submicroscopic speck at the end of a widening shaft of information and thought that could reach to and bracket those four hundred million years. The difference between one human lifetime and four hundred million years would seem to be a difference between time incomprehensible and time infinitesimal, but what brings them together is that the smaller unit—bridging in the mind the intervening eons—can imagine and virtually see the larger one.

A quarter-horse jockey learns to think of a twenty-second race as if it were occurring across twenty minutes—in distinct parts, spaced in his consciousness. Each nuance of the ride comes to him slowly as he builds his race. If you can do the opposite with deep time, living in it and thinking in it until the large numbers settle into place, you can sense how swiftly the initial earth packed itself together, how swiftly continents have assembled and come apart, how far and rapidly continents travel, how quickly mountains rise and how quickly they distintegrate and disappear. No matter how impressive or extensive the data might be, Van Schmus said, it is well to bear in mind, as you imagine something as comprehensive as the original building of the center of North America, that the developed picture is authenticated by the best current hypothesis and nothing more. "You are never going to write the definitive answer in geology; it's the nature of the field."

A suspension bridge—"The World's Highest"—crosses the profound gorge, mainly for tourists on foot, and if a car happens along while you are walking in mid-span the suspended deck undulates under your feet as if it were a raft in the rapids a thousand feet below. There is a wind so stiff that you lean hard against it. It feels as if it is about to blow you over the side. For anyone with a fear of height, cars do not add to the terror, because the terror is already so complete that there is no room for more. The whorls and convolutions in the gorge walls—metavolcanic and metasedimentary rocks, migmatites and gneisses thoroughly cooked at least twice—are all but undecipherable under any conditions and glazed under these. Then a young woman happens by, wheeling a baby carriage and clucking fondly at the contents. She should go into geology.

In the parkland beside the gorge, Van Schmus remarked that the extensive alteration of the primary crust we were standing on probably resulted from the high-grade metamorphism that would have occurred while arcs collided. He said it was possible that this was part of the first terrane to attach to the Wyoming craton. "The discrete terranes have not been identified," he went on. "But the principle is valid. We may agree that we are looking at a collage of accreted terranes analogous to what went on in California in the Mesozoic, but we're not at the stage where we can identify terrane boundaries. We don't know the age of the gorge. I strongly suggest that it's no older than 1800. The problem with the gorge rock is that it has been overprinted with metamorphism so many times. Sorting all this out and figuring what it was originally is the present frontier, and will lead to an understanding of what things were: back-arc basins like the Sea of Japan, accretionary wedges like the Coast Ranges of California, inter-arc rift basins, fore-arc basins, fore-deeps." As he spoke, I was scribbling notes and he was looking down into a pegmatite quarry. When we looked up, we saw Pikes Peak to the north-northeast, above the trees.

Near Salida, where we saw gabbros, pillow basalts, and other subaqueous volcanics, he said, "With confidence, you would associate them with arc rocks. Although we're still dealing with a frontier in continental evolution, the isotopes tell us very clearly that this stuff had a very short crustal residence time." By that he meant that when magma arose from within the earth to cool as lithospheric crust, the crust had not travelled long or far before it went into a subduction zone to melt anew. The timetable of decay of radioactive elements within the initial magma would mark the event when the magma first emerged. "The clock starts ticking when the stuff comes out of the mantle into the crust," he explained. "The arc, in turn, may get melted and turn into granite, or whatever, but the clock never stops ticking. The story is there in the rocks. The problem for us is to figure out how to read them."

Isotopes that have become particularly useful in monitoring the crustal history of rocks and their origins in the mantle belong to the elements samarium and neodymium. Because samarium radioactively decays and becomes neodymium at a fixed rate, the two serve as a chronometer, like rubidium and strontium, like uranium and lead. "In the late seventies, early eighties, samarium/neodymium as an analytical technique was perfected," Van Schmus said. "With advances in instrumentation, the technique became a sort of everyman's tool. What it tells us basically is when rocks come out of the mantle. It doesn't provide ages, necessarily, but it does provide an isotopic tracer that allows us to track the history of continental crust, and particularly to determine how long the material in a piece of continental crust has been separated from the mantle. We use other methods to get the crystallization age of rocks, and samarium/neodymium analyses to tell how long those particular rocks have been part of the continental system."

Samarium/neodymium ages are accurate but they are not very precise. This distinction—between accuracy and precision—has a difference, and is less of a split hair than it may at first appear to be. For example, if you say that George Washington submitted his last expense account when he was forty-seven, plus or minus twenty years, you are accurate but you are not at all precise. If you say that Santa Claus came down the chimney at 12:26:09 A.M. on Halloween you are precise but almost surely inaccurate. "You are accurate if you are within the stated limits of uncertainty," Van Schmus said, defining the goal of geochronology. "You want to be both accurate and precise. You want your limits narrow, and you want the window to include the correct answer." He added, "Precise means good lab technique. You can be very precise and very inaccurate. The idea is to narrow the window of accuracy and be, at the same time, precise." In Precambrian geochronology, a window of two million years is an extremely narrow one. A date like 1746—plus or minus two—is very precise.

The rock component most widely employed in the quest for precision in deepest time is zirconium orthosilicate, the mineral zircon. As a result of advances in laboratory technique, zircons have become, in Van Schmus's phrase, "the workhorse of the Precambrian," because they yield with greater accuracy and considerable precision the primary crystallization ages of the rocks they come from. Zircons for centuries have had an independent status as gems. They are pyramids or prisms, and have an adamantine lustre and considerable variety of color. Starlite is a blue zircon from Thailand.

The sometimes smoky or colorless or pale yellow zircon of Sri Lanka is a gemstone called jargon. The word "zircon" is derived from the Old French *jargon*, which had the same ultimate source as "gargle."

Most zircons are not of a size to flash bright color from jewelry, or even to be easily seen. They are typically less than a tenth of a millimetre in their longest dimension. They are in sandstones, schists, gneisses. They are in nearly the whole family of solidified magmas of which granite is the most familiar. They are in rhyolite —the rock result of granitic magma erupting on the surface of the earth. And of course they are in all the sedimentary rocks and the beaches and the river placers that derive from the granites and the other sources. Unfortunately, they are rarely in basaltic rocks— the essence of, among other things, island arcs and ocean crust.

Van Schmus and his associates in their "clean lab" at the University of Kansas grind rock to powder, concentrate the heavy minerals, and pour them into a flask of bromoform, a colorless liquid consisting of carbon, hydrogen, and bromine, closely analogous to chloroform. The quartz and other relatively light materials float on the bromoform. The heavies—sphene, pyrite, zircons, magnetite, apatite, hornblende, garnet—fall to the bottom. The heavies go into a Frantz separator, where a very strongly focused magnetic field pulls out the magnetite, hornblende, and garnet. Remaining are sphene, apatite, pyrite, and zircons, all of which are dropped into methylene iodide. In methylene iodide, apatite floats. It is removed, and now we are down to zircons, pyrite, and sphene. An acid dissolves the pyrite, ignoring the zircons and the sphene. A Frantz separator, its power adjusted, takes the sphene away from the zircons. Thus isolated and concentrated, a tiny pile of zircons looked to me like heavy, glinting dust. Under a microscope, they were elongate tetragonal pyramids, of a light clear amber. They resembled capsules of Vitamin E.

As magma cools—as, for example, a great intruded batholith slowly and progressively hardens and differing magmatic juices sequentially develop into various minerals and rocks—the attraction of zirconium ions for silicon and oxygen is what makes zircons. "They bond very strongly," Van Schmus said. "Other things sneak in— uranium, hafnium, thorium. Lead begins forming from the uranium and thorium by radioactive decay. To make zircons, there has to be a lot of silicon around, so zircons are commonly associated with quartz-bearing igneous rocks."

Although they are semi-microscopic, zircons are nonetheless polished in the lab to grind away their outer parts—where disruption might have occurred in various ways—and get down to pristine samples. This improves the accuracy of the age analysis that follows. In a process that takes a few days, the zircons are dissolved by hydrofluoric acid in teflon bombs. The bombs are not unlike pressure cookers. Ion-exchange chromatography, a chemical extraction process, takes the uranium and lead out of the solution. Analysis of the abundance and composition of the lead results in the date when the zircon first crystallized. This method—developed by Thomas E. Krogh at the Carnegie Institution in Washington—has dramatically simplified a task that used to be very difficult. A full uranium/lead analysis can be done with zircons aggregately weighing five-millionths to ten-millionths of a gram.

Among the several characteristics that have enlofted zircons to the state of the art of Precambrian dating is the fact that once they form they do not easily recrystallize. In huge tectonic events across time, they can go through medium-grade and even high-grade metamorphism and survive. In the clean lab, three or four weeks may be needed to get an age from one rock. There's no reasonable alternative, Van Schmus remarked. "Nowadays you don't feel that you have a totally accurate date unless you've done it with zircons."

Cycles of mountain building that were long thought to have taken place over a period of three hundred million years have recently been narrowed down to a few tens of millions of years through zircon dating. The approximate age of the oldest known rock on earth—3.96 billion years—was determined from zircons within the rock. Because zircons are rare in basaltic rocks, they seldom can be used directly for the dating of remnants of Precambrian ocean crust. Near Salida, however, those gabbros, pillow basalts, and other ocean-crustal rocks from the beginnings of Colorado were interlayered with rhyolite tuff that would have erupted on an island surface. The rhyolite contained zircons. A date obtained from the zircons by Van Schmus's colleague M. E. Bickford had inferentially provided the dates of origin of the gabbros and basalts: 1728—plus or minus six million years.

"We cannot make any correlations in the Precambrian based just on what the rocks look like," Van Schmus reiterated. "We have no index fossils. So the only thing we have to connect various disjointed pieces of Precambrian crust are radiometric dating techniques. Isotopes have given us a great mapping tool. Beginning in the fifties and early sixties, two principal techniques were potassium/argon and rubidium/strontium, which had limitations in terms of both accuracy and precision. They were fairly easy to carry out, and a great deal of early information on the Precambrian, particularly of North America, was done with those two dating techniques. A smaller but significant amount of work was done on uranium/lead dating in zircons—the technique was very difficult to work with and there were only a few practitioners in North America then. The methodological breakthroughs of the seventies made the uranium/lead dating technique very convenient. Since then, most of our precise age information has come from uranium/lead dating of zircons in igneous rocks. Argon/argon thermochronology has matured substantially in the last decade, and it's being used extensively for the study of young rocks and for the study of metamorphic rocks, looking particularly at the last phase of metamorphism in tectonic environments. That is not something that concerns us very much with the Precambrian basement."

Van Schmus said that he and several others were going to date xenoliths in volcanic necks in the Four Corners area, where Colorado, Utah, Arizona, and New Mexico all touch. That is where Shiprock is, and Shiprock (like Devil's Tower, in Wyoming) is a volcanic neck. Also called a volcanic chimney, a volcanic neck is a conduit through which magma rises toward eruption on the surface of the earth. After everything freezes and epochs pass and erosion tears down the volcano and surrounding land, the volcanic neck may be left standing high because the frozen magma it is made of is so much tougher and more durable than the rock that once lay around it. Differential erosion. A xenolith, in this instance, is a rock from around the margins of the neck that fell into the magma when it was soft. Like a bit of chocolate in a cookie, it is a foreign body with its own age and its own history, distinct from the stuff it went into. The xenoliths in volcanic necks come from all depths of the crust and upper mantle, and the Four Corners project is meant to yield a crustal profile that is the equivalent of a drilled hole thirty miles deep. It is a project that summarizes in itself the new revelation of Precambrian scenes—a revelation made feasible by techniques and technology that were not available ten years ago. Advances in metamorphic petrology—specifically, in understanding the temperatures and pressures at which minerals form—will enable the team to determine the depths from which the xenoliths derive. Advances in trace-element geochemistry will allow them to associate the xenoliths with their original geologic environments. Trace-element indicators will suggest, for example, whether the environment was oceanic or continental. Samarium/neodymium dating will yield crustal histories. Refinements in uranium/lead analyses will allow information to be gathered from units as small as a hundred-thousandth of a gram of zircon.

In what would become Kansas, Nebraska, Colorado, and thereabout, the last of the island-arc dockings occurred around 1700—seventeen hundred million years ago. This final collision further cooked and deformed the rock now visible in the walls of Fremont County's Arkansas River gorge. The coast of the continent, in 1700, ran from southwestern Texas through Oklahoma, Missouri, Illinois, and Michigan. South of Denver and Cañon City, it was somewhere in southern New Mexico. By 1650, the continental margin had evidently developed into something much like the modern west coast of South America, which is paralleled most of the way by the Peru-Chile Trench, the boundary between the South American and the Nazca tectonic plates. The Peru-Chile Trench is as much as twenty-six thousand feet deep, a number that suggests the grandeur of the collision that is taking place as South America, moving west, rides over the Nazca Plate, which consists of nearly ten million square miles of ocean crust. Below the trench, the subduction zone dips to the east and under South America. Down there, the descending crust significantly melts, and the resulting magma has risen through the rock of the edge of South America to emerge as Volcán San Pedro, Volcán Llullaillaco, Volcán El Potro, del Toro, Domuyo, Mercedario, Aconcagua—the Andes. The vertical difference between Aconcagua and the bottom of the trench not far away is forty-nine thousand two hundred and eighty-eight feet. That was the style (the altitudes are unknown) of the mountains of Kansas 1.65 billion years ago, and of the North American coast, running through southern Oklahoma—an Andean margin, with ocean crust subducting and melting beneath it, making Mercedarios and Aconcaguas.

Of the various forms of geophysical data—seismic reflections, gravity anomalies, and so forth—the most useful in the illumination of the Precambrian have been measurements of varying magnetic fields. Data are collected mainly by air, and their effect is to strip off the Phanerozoic cover and show you the Precambrian as if nothing else were there. By 1980, magneticists had reached the point where they felt they could correctly identify rock types from the rocks' magnetic signatures. Strong magnetic fields showed up on their maps in varied intensities of red, and weak ones in shades of blue and green. The granite family was essentially blue and green, but granites containing magnetite were red. The Midcontinent Rift showed up almost luridly, because of the strong signal of its iron-rich basaltic rock. Where drill holes existed, identifications were compared with and supported by drill cores. In 1982, the publication of Isidore Zietz's Composite Magnetic Anomaly Map of the United States represented what Van Schmus describes as a "major breakthrough" in Precambrian geology. An accomplishment of, among other things, advances in computer programming, it assessed, related, and blended data that had previously been limited to regions and states. To be sure, it was not perfect. Data assembled by one state's geological survey could be thin and sketchy, while the state next door had been gone over with a fine magnetic comb. The results of such contrasts were apparent geologic schisms along certain state lines, known to sarcastic Precambrian geologists as boundary faults.

Van Schmus would like to see the boundary faults disappear as a result of a dream federal project flying magnetometers over the entire country between the Appalachians and the Rockies with one-kilometre spacing. "In the past decade, we've made a tremendous first-order advance in the understanding of the basement," he said.

"To take it to the next level of refinement is to take it to more orders of magnitude in effort and cost. The price of one fighter-bomber would advance our knowledge of the basement by an order of magnitude. The full survey would require three thousand trips across the country the short way, at ten dollars a mile. That would come to fifty million dollars, maybe a hundred million before you're through. What would you learn? You don't know before you see it, but you would understand the structure of the continent."

Across the Composite Magnetic Anomaly Map of the United States from one side to the other of the Proterozoic continent runs a series of red bull's-eyes that have an average date of fourteen hundred and fifty million years before the present. These are the so-called 1450 plutons, an enigmatic event in the behavior of the earth, unprecedented, unrepeated, and unexplained. Something partially melted the whole accretionary belt and the continent was stitched with twenty-five hundred miles of granite plutons. They were granites that tended to be rich in magnetite, hence the red signature. Alternatively known as the 1450 batholiths, they included the Sherman granite of the Laramie Range, the Silver Plume granites of Colorado, the St. Francois Mountains of southern Missouri, and, in Wisconsin, the Wolf River batholith. They cooked and altered yet again the rock that now flanks the deep gorge at Cañon City. The basement of Illinois—very much of a piece with the 1800-to-1650 accretionary complex that lay under the states to the west—became covered with rhyolite lavas of the 1450 events.

Plutons and batholiths are almost by definition a stage and consequence of great orogenies—the building of mountains. Leon Silver, a geologist at Caltech, has called the 1450 plutons the "Anorogenic Perforation of North America." That is the title of a scientific paper by Silver and others calling attention to the signal mystery of these huge bodies of igneous rock: they entered the earth unaccompanied by mountains. The common wisdom about the 1450 phenomenon has been that it happened in the pressure release of an extensional regime. That is, a supercontinent was coming apart, stretching, thinning, breaking elsewhere to form oceans, and while the event pulled, extended, stretched, and thinned crust that is now North American it brought the melting heat of the mantle closer to the surface, producing anorogenically not only the 1450 plutons but also vast terranes of granite and rhyolite that filled in and covered the upper crust along the eastern and southern margin of the continent. It is even possible that a large piece of North America was torn away in the continental stretching and is now somewhere in the world, its origin in North America as yet unrecognized.

In a very different scenario for the 1450 plutons and related events, some theorists hypothesize a large stable continent in which heat flow is inhibited because there is no active volcanism or rifting. Temperatures under the continent gradually build up with heat from the mantle, and the lower crust undergoes partial melting. As a result, the plutons form and rise. The one explanation seems as logical as the other, and the cause of the 1450s is not understood. The train of plutons is not the track of a geophysical hot spot, like the volcanic chains of a Tristan da Cunha, a Réunion, or an Hawaii. The 1450s did not intrude the old shield. Some geophysicists have proposed that while the lower crust of the whole continent probably became plenty hot, only the post-1800 accretionary part of North America —the midcontinent south of the shield—was fertile enough to give rise to granites, because everything granitoid had long since been distilled out of the old Archean cratons. Geologists have also mentioned the possibility that the "new" crust—the accreted crust—was enriched in uranium and thorium and potassium, and therefore radioactively melted itself. Van Schmus does not have a favorite guess. He will only shake his head and say, "There is nowhere else in the world where a string of plutons of essentially the same age goes across four thousand kilometres."

Gravity anomalies are another scope into very deep time. Whereas magnetic anomalies are measured with airplanes, gravity anomalies are measured by people in cars. They stop at every section corner with a gravity meter and take a reading. When missiles mattered a good deal more than they do now, gravity readings did, too, because data on varying gravity fields were essential in the calculations of ballistic trajectories. Even so, collection of gravity data has been nationally and globally spotty. Every square mile of a planet is a lot of section corners. "They haven't really developed a good airborne gravity meter yet," Van Schmus remarked. "When they do, it will be nice, because you can fly remote areas like the Brazilian Shield and get a good gravity map." Magnetic maps include more detail, but they do not show what is in the crust in a complete and three-dimensional way, as gravity maps do. In Van Schmus's words, "The magnetic field is sampling the shallow crust, the upper few thousand feet. The gravity field is averaging, basically, the whole lithosphere. When you look at a gravity map you are seeing deeper features of the crust." A gravity meter measures the densities of the rocks below it. A gravity low will be a response to granites or a sedimentary basin—light continental rock. A gravity high is an indication of the densest material: ocean crust right out of the mantle, or even the mantle itself. The Midcontinent Rift, with its basaltic rocks analogous to an ocean spreading center, registers on the meters as a strong gravity high. For interpreting basement geology in the United States, gravity maps are most useful between the Appalachians and the Rockies, because the confusions that are caused by rugged topography do not interfere.

In many ways, the geophysical data from gravity meters and magnetometers are only as good as the nearest tangible rock, the control point that gives credence to the interpretation of numbers. If Precambrian Kansas or Precambrian Nebraska happens to be your target, the nearest rock may be two or three thousand feet below the surface, and you will depend on the drill cores of wells. They anchor the insights. In combination with magnetic maps, the petrology, chemistry, and radiometric dating done on rock fragments from wells has provided the most powerful current method of looking great distances back into time, with gravity maps as a supplement.

Van Schmus has said of the magnetic signature, "It only works if you've got some rock to test it with." When oil companies go particularly deep, academic geologists have a way of appearing and asking for chips. Since oil does not form in Precambrian rock (nothing there to make it of), most oil drilling stops when the Precambrian level is reached. In a few places, though, where faults were thought to have caused oil and gas to go down into the Precambrian, drill holes have followed. Texaco went twelve thousand feet into Kansas—far into the dry sediments of the center of the Midcontinent Rift. In the heart of Iowa and on the flank of the rift, Amoco drilled seventeen thousand feet on the strength of a show of hydrocarbons in similar rock near Lake Superior—producing no oil but substantial amounts of data. Elsewhere, drilling rigs have gone several hundred feet into the Precambrian because the operators did not know where they were, or, thinking they were hitting a Precambrian wedge thrust over younger rock, tried to go through it. When rig operators hit the Precambrian, university geologists sometimes pay them to continue—a smiling face between industry and academia known as piggyback drilling. For a hundred and fifty dollars an hour, or so, the drillers keep going until the bit wears out. For five hundred to a thousand dollars, geologists go away with several tubs full of rock chips. Companies sometimes drill for an extra hour "just to be nice."

Nebraska wells have confirmed that the Nebraska basement is much the same as Colorado's—accreted island arcs, and so forth. Van Schmus remarked, "However, there are no good control points in Iowa. It is either an eastern extension of Nebraska or a southwestern extension of Wisconsin. The early Proterozoic history of Iowa remains a great puzzle. The Iowa basement is our biggest gap in knowledge. If I had fifty million bucks for scientific drill holes, I'd spend it there." An explanation for the gap in knowledge is contained in an expression oft heard in geology: "You could just about drink all the oil that ever came out of Iowa."

The most comprehensive rock archive of the basement of the Midwest is sort of where it belongs—in the basement of the Geology Department of the University of Kansas. Full of well cores and cuttings from every midwestern state, the archive is about thirty years old. Among the direct results of its existence are maps of the Precambrian midcontinent based less on inference than on fact. Randomly picking up and examining a core there from Buffalo County, Nebraska, Van Schmus said it was a tonalitic gneiss, age 1790. Its biotite and hornblende held its zircons. "The basement of Nebraska is more interesting than the basement of Kansas," he remarked. "Nebraska has more juvenile, primitive arcs." For Precambrian purposes, the four best-drilled states in the archive are Oklahoma, Missouri, Kansas, and Nebraska. "In Texas, they get so much oil before they get to the basement they quit," he said.

The German Republic, in an attempt to be the leader in the field of scientific deep drilling, made a two-hundred-million-dollar hole in Bavaria thirty-two thousand feet deep. On the Kola Peninsula, between the White Sea and the Arctic Ocean, Russia drilled a hole thirty-five thousand feet deep. Beyond the hit-and-miss, oil-driven, scrounging mode, scientific drilling in the United States has so far been modest.

Deep in the basement archive seemed as good a place as any to ask Van Schmus to say in his own words what anyone would hope to learn by drilling seven miles through Precambrian worlds.

"First, of course, the Precambrian is nearly ninety per cent of earth history," he said. "You've got to understand it if you're going to understand the moment when Phanerozoic geology begins. Second, Precambrian shield areas—particularly Archean shield areas—are the hosts of major economic resources: gold, copper, iron, nickel, lead. Diamond pipes are relatively young but they are found only in Precambrian cratons. It has something to do with building up the necessary pressure to form the kimberlites from the mantle"—kimberlite, the species of volcanic neck that is the matrix rock of diamonds. "The Precambrian preserves certain aspects of the earth's dynamic systems that we can see only in old rocks, because they have been eroded to great depths," he continued. "We can look at the roots of mountain belts in the Precambrian, whereas in the Phanerozoic all we can see, really, are the middle or upper portions of mountain belts. We don't know what the roots of the Appalachians look like. We've got the folded core of the Appalachians but not the deep collisional roots. We have to look at Precambrian foldbelts to try to understand what goes on in younger mountains at depth. An important thing to recognize now is that with dating techniques we have a way of establishing stratigraphic order in the Precambrian. With mineralogical and geochemical techniques, we have a way of understanding the so-called protolith—or preexisting rock types. We can begin to decipher geologic histories that relate directly to the original formational environments. There are large areas around the world where the Precambrian rocks are preserved in an almost pristine state. They have preserved with them, in local areas, fossil assemblages going back to three and a half billion years or more. They are the only record we have for tracking the evolution of life."

In Geology 101, at U-Name-It University, Professor Lucius P. Aenigmatite, scorner of continental drift, used to teach that the Precambrian eons left no fossils. That was virtually the definition of the Precambrian. The arrival of fossils—that is, the abrupt and explosive and unprecedented development of creatures with hard parts—marked the beginning of Phanerozoic time. "There is a large soft-bodied fauna known as Ediacara that closely precedes the Cambrian," Van Schmus continued. "What we see before that are just complex bacterial forms, algae, small single-celled organisms. Without that record, and without understanding the environments from which these things evolved, we cannot really understand evolution. There are many so-called chemical fossils in the record, too. Isotopic composition of carbon, for example. It can be monitored through time. It is a reflection of biologic activity on the earth. There's a number of things in the Precambrian we can look at that sort of get us up to the traditional base-of-the-Cambrian starting point for modern geology."

Precambrian landscapes had a barrenness beset by weather without vegetal control. The rock summits of high mountains would have looked like the summits of present time, the bare slopes piled in fans of deep scree, like the ranges of Antarctica. Texture rested in topography, color in rock, braided rivers running over the rock. The cycle through which rock is torn apart, ground up, set down, stratified, and made into fresh new rock was unimpeded by so much as a root or a stem, and therefore cycled more rapidly. Only gravity—within its angle of repose—held boulders and gravels to inclined ground. Silts and sands washed down quickly to lakes and seas. Unadorned, unembellished, severely simple, a picture of the Precambrian would present to us the incongruity of desert landscape invaded by white rivers drenched in rain.

If you could have travelled westward from the site of Chicago eleven hundred million years before the present, you would have traversed, along the modern route of Interstate 80, most of what geologists now see in the Precambrian basement. In Illinois, you would have been among the bevelled rhyolites and buried granites of the once high and Andean continental margin. In eastern Iowa, you would have crossed a plutonic belt, also bevelled by erosion, dating from the anorogenic perforation of North America.

If you could have come the other way, also on I-80, eleven hundred million years ago, you would have passed through the sites of Rock Springs and Rawlins on the worn-down shield rock of the Archean Wyoming craton. Near Laramie—after running along and then jumping across the old Cheyenne Belt, edge of the continent in 1800—you would have moved onto the compacted island-arc complex whose dockings filled in Colorado and Nebraska, and much of the rest of the United States. You would not have encountered there the pink granite (today's front-range granite) that came into the arc complex around 1450 as one of the mysterious plutons. It was still buried too deep. On post-1800 metavolcanic rock, you would have continued across Nebraska through North Platte and Kearney to Lincoln.

These eleven-hundred-million-year-old North American time lines have now all but met in the middle—at the future sites of Lincoln, Omaha, Des Moines. In 1108, when the rifting began, the three sites were a good deal closer together. Now, around 1100, they were still moving apart and would continue to move apart for fourteen million years. Something under the core of North America was tearing North America apart, threatening its continuing existence as an integral continent. It seems likely that the cause of the Midcontinent Rift was a thermal plume from deep in the mantle, a geophysical hot spot doming the crust and then cracking it. Flood basalts filled the rifting valley. At night above the lava fountains the whole sky was red. At about 1100, the triple junction broke open under Lake Superior, connecting this southwestern arm and the arm that ran through Michigan. If the rifting had gone on long enough, the country between the two active arms—including at least half of what is now called the Midwest—would have departed from North America to end up who knows where and in how many pieces. In the middle of North America, a great bay would have developed, with a shoreline of a thousand miles.

Between Lincoln and Omaha, I-80 runs directly over the center of the rift. It gradually slides toward the rift's eastern flank at Des Moines. On I-80, the whole western half of Iowa is over the rift itself or its flanking basins. To follow this cartographically, you would need the Composite Magnetic Anomaly Map or an isostatic residual gravity map of the United States. On the magnetic and gravity maps the Midcontinent Rift is the most prominent feature you see. A quarter-century ago, it was as unknown in scientific mapping as it still is in road mapping. It was referred to as "the midcontinent gravity anomaly" or "the midcontinent gravity high"—descriptions merely, without implication or sense of cause. If the rifting had continued even for a couple of hundred million years, as the Mid-Atlantic rifting has done, Lincoln and Des Moines would be as far apart as Jersey City and Casablanca, whose sites were once as close as Lincoln and Des Moines. Yet that did not happen. The midcontinent rift system did not in the end play a major role in the evolution of the continent, because the rifting stopped—or was stopped—more or less abruptly. The rift system's oldest rocks date from 1108, the youngest from 1086, so the rifting lasted twenty-two million years—not much by comparison with an Atlantic Ocean, but (to date) about three times the rifting of the Gulf of California and longer than the rifting of the Red Sea.

Something seems to have snuffed out the young hot spot, leaving the midcontinent intact. Where crustal blocks had dropped in the middle of the rift as it widened, they now were subjected to a compressional force so great that the middle of the rift rose up to a position higher than the sides. In the language of geology, grabens were squeezed upward and became horsts. It was as if the Red Sea were to stop widening, while its floor came up to stand higher than the shores. The compressional force that stopped the rift in Proterozoic North America is believed to have been the Grenville Orogeny. This name has been given to a continent-to-continent collision, completed by about 1050, that brought the West African Craton and the Amazonian Craton against the eastern and southern margins of North America to create the supercontinent Rodinia, hundreds of millions of years before Pangaea, the most recent of supercontinents. In Grenville time, the African and South American cratons were neither configured nor juxtaposed as they would be in the next eon. They were not put together as they are now. They resembled their modern forms about as little as North America did. From Texas to Labrador, the Grenville Orogeny built the beginnings of eastern America.

As a kind of exclamation point at the end of these events, an isolated plume under Colorado seems related to the Grenville Orogeny. Dating from the same time as the continent-to-continent collision and the stopping of the rift, a batholith intruded Colorado, its presence otherwise inexplicable. Nothing else was happening, or was about to happen, in or near Colorado at that time. Then suddenly appeared this maverick granite—the granite of Pikes Peak—distinct in age and texture from all other Rocky Mountain granites. In Van Schmus's words, "It cooled off, and that was it. The Pikes Peak batholith just sits out there all by its lonesome." After that—to the end of the Precambrian—the midcontinent was quiet for half a billion years.

List of Maps

Index

List of Maps

Index

Index

Geologic Time Scale

THE CENOZOIC ERA

65 MILLION YEARS

MILLIONS OF YEARS BEFORE THE PRESENT	*System* / *Period*	*Series* / *Epoch*	*Stage* / *Age* EUROPE	*Stage* / *Age* NORTH AMERICA
	QUATERNARY	HOLOCENE		
0.01		PLEISTOCENE	TYRRHENIAN	WISCONSINAN
			MILAZZIAN	SANGAMONIAN
				ILLINOIAN
			SICILIAN	YARMOUTHIAN
			EMILIAN	KANSAN
			CALABRIAN	AFTONIAN
				NEBRASKAN
				BLANCAN
1.64	TERTIARY	PLIOCENE	PIACENZIAN	
			ZANCLEAN	HEMPHILLIAN
5		MIOCENE	MESSINIAN	
			TORTONIAN	CLARENDONIAN
			SERRAVALLIAN	BARSTOVIAN
			LANGHIAN	
			BURDIGALIAN	HEMINGFORDIAN
			AQUITANIAN	
23		OLIGOCENE	CHATTIAN	ARIKAREEAN
				WHITNEYAN
				ORELLAN
			RUPELIAN	CHADRONIAN
35		EOCENE	BARTONIAN	DUCHESNEAN
				UINTAN
			LUTETIAN	BRIDGERIAN
			YPRESIAN	WASATCHIAN
56		PALEOCENE	THANETIAN	CLARKFORKIAN
			MONTIAN	TIFFANIAN
				TORREJONIAN
			DANIAN	DRAGONIAN
				PUERCAN
65				

THE MESOZOIC ERA

185 MILLION YEARS

MILLIONS OF YEARS BEFORE THE PRESENT	*System* / *Period*	*Stage* / *Age*: EUROPE			NORTH AMERICA
65	CRETACEOUS		MAASTRICHTIAN		GULFIAN
			CAMPANIAN		
			SANTONIAN		
			CONIACIAN		
			TURONIAN		
			CENOMANIAN		
100			ALBIAN		COMANCHEAN
			APTIAN	GARGASIAN	
				BEDOULIAN	
			BARREMIAN		
		NEOCOMIAN	HAUTERIVIAN		
			VALANGINIAN		
			BERRIASIAN		
145	JURASSIC		TITHONIAN		
			KIMMERIDGIAN		
			OXFORDIAN		
			CALLOVIAN		
			BATHONIAN		
			BAJOCIAN		
			AALENIAN		
			LIASSIC		
208	TRIASSIC		RHAETIAN		
			NORIAN		
			CARNIAN		
			LADINIAN		
			ANISIAN		
			SCYTHIAN		
250					

THE PALEOZOIC ERA

ABOUT 300 MILLION YEARS

Millions of Years Before the Present	System / Period	Stage: Europe	Age: North America
250	Permian	Tatarian	Ochoan
		Kazanian	Guadalupian
		Kungurian	Leonardian
		Artinskian	
		Sakmarian	Wolfcampian
290	Pennsylvanian (Carboniferous)	Stephanian	Virgilian
			Missourian
		Westphalian	Desmoinesian
			Atokan
			Morrowan
323	Mississippian (Carboniferous)	Namurian	Chesterian
		Visean	Meramecian
			Osagean
		Tournaisian	Kinderhookian
360	Devonian	Famennian	Chautauquan
		Frasnian	Senecan
		Givetian	Erian
		Couvinian	
		Emsian	Onandagan
		Siegenian	Oriskanyan
		Gedinnian	Helderbergian
408	Silurian	Ludlovian	Cayugan
		Wenlockian	Niagaran
		Llandoverian	Medinan
439	Ordovician	Ashgillian	Cincinnatian
		Caradocian	Trentonian
		Llandeilian	Blackriveran
		Llanvirnian	Chazyan
		Arenigian	Canadian
		Tremadocian	
490	Cambrian	Dolgellian	Croixian
		Festiniogian	
		Maentwrogian	
		Menevian	Albertan
		Solvan	
		Caerfaian	Waucoban
544			

PRECAMBRIAN TIME

ABOUT 4000 MILLION YEARS

MILLIONS OF YEARS BEFORE THE PRESENT	Eon	ERAS	Canadian Time Scale	
544	THE PROTEROZOIC EON	NEOPROTEROZOIC	HADRYNIAN	
1000		MESOPROTEROZOIC	HELIKIAN	NEO-HELIKIAN
1600		PALEOPROTEROZOIC		PALEO-HELIKIAN
1750			APHEBIAN	
2000				
2500	THE ARCHEAN EON			
4560